W0257222

Tagung III/1980 am 10./11. 10.1980 in Konstanz. 303 Seiten, DM 54,–/ÖS 394,–/SFr. 49,–

Band 7: **Nehmer, Implementierungssprachen für nichtsequentielle Programmsysteme**
Tagung I/1981 am 20. 2. 1981 in Kaiserslautern. 208 Seiten, DM 38,–/ÖS 277,–/SFr. 34,–

Band 8: **Schlier, Personal Computing**
Tagung II/1981 am 12. 10. 1981 in Freiburg i. Br. 195 Seiten, DM 40,–/ÖS 292,–/SFr. 36,–

Band 10: **Kulisch/Ullrich, Wissenschaftliches Rechnen und Programmiersprachen**
Fachseminar am 2./3. 4. 1982 in Karlsruhe. 231 Seiten, DM 52,–/ÖS 380,–/SFr. 47,–

Band 11: **Langmaack/Schlender/Schmidt, Implementierung PASCAL-artiger
Programmiersprachen**
Tagung II/1982 am 12. 7. 1982 in Kiel. 221 Seiten, DM 46,–/ÖS 336,–/SFr. 41,–

Band 13: **Schneider, Proceedings of the International Computing Symposium 1983
on Application Systems Development**
March 22 – 24, 1983 Nürnberg. 528 Seiten, DM 90,–/ÖS 657,–/SFr. 81,–

Band 18: **Morgenbrod/Sammer, Programmierumgebungen und Compiler**
Tagung I/1984 vom 2. bis 4. 4. 1984 in München. 293 Seiten, DM 56,–/ÖS 409,–/SFr. 50,–

Band 20: **Gorny/Kilian, Computer-Software und Sachmängelhaftung**
Workshop am 29./30. 11. 1984 in Hannover. 208 Seiten, DM 48,–/ÖS 350,–/SFr. 43,–

Band 21: **Kölsch/Schmidt/Schweiggert, Wirtschaftsgut Software**
Tagung I/1985 am 26./27. 3. 1985 in Ulm. 318 Seiten, DM 58,–/ÖS 423,–/SFr. 52,–

Band 22: **Molzberger/Zemanek, Software-Entwicklung:
Kreativer Prozeß oder formales Problem?**
Seminar am 20. 3. 1985 in Neubiberg. 176 Seiten, DM 42,–/ÖS 307,–/SFr. 38,–

Band 23: **Klopcic/Marty/Rothauser, Arbeitsplatzrechner in der Unternehmung**
Tagung II/1985 am 12./13. 9. 1985 in Zürich. 355 Seiten, DM 66,–/ÖS 482,–/SFr. 59,–

Band 24: **Bullinger, Software-Ergonomie '85 Mensch-Computer-Interaktion**
Tagung III/1985 am 24./25. 9. 1985 in Stuttgart. 482 Seiten, DM 78,–/ÖS 569,– SFr. 70,–

Band 25: **Wedekind/Kratzer, Büroautomation '85**
Tagung IV/1985 vom 2. bis 4. 10. 1985 in Erlangen. 280 Seiten, DM 56,–/ÖS 409,–/SFr. 50,–

Band 27: **Remmele/Sommer, Arbeitsplätze morgen**
Tagung II/1986 vom 11. bis 14. 3. 1986 in Marburg. 431 Seiten, DM 78,–/ÖS 569,–/SFr. 70,–

Band 28: **Balzert/Heyer/Lutze, Expertensysteme '87**
Tagung I/1987 am 7./8. 4. 1987 in Nürnberg. 493 Seiten, DM 82,–/ÖS 599,–/SFr. 74,–

Band 29: **Schönpflug/Wittstock, Software-Ergonomie '87**
Tagung II/1987 vom 27. bis 29. 4. 1987 in Berlin. 512 Seiten, DM 82,–/ÖS 599,–/SFr. 74,–

Band 30: **Winkler, Proceedings of the International Workshop on Software Version
and Configuration Control**
January 27 – 29, 1988 Grassau. 478 Seiten, DM 78,–/ÖS 569,–/SFr. 70,–

Band 31: **Dillmann/Swiderski, WIMPEL '88**
Tagung I/1988 vom 28. bis 30. 6. 1988 in München. 479 Seiten, DM 78,–/ÖS 569,–/SFr. 70,–

Band 32: **Maaß/Oberquelle, Software-Ergonomie '89**
Fachtagung vom 29. bis 31. 3. 1989 in Hamburg. 509 Seiten, DM 88,–/ÖS 642,–/SFr. 79,–

Fortsetzung 3. Umschlagseite

Berichte des German Chapter
of the ACM   49

R. Liskowsky / B.M. Velichkovsky /
W. Wünschmann (Hrsg.)
Software Ergonomie '97

Berichte des German Chapter of the ACM

Im Auftrag des German Chapter
of the ACM herausgegeben durch den Vorstand

Chairman
Wolf-Rüdiger Gawron, BMW AG, Petuelring 130, 80788 München

Vice Chairman
Prof. Dr. Günter Riedewald, Universität Rostock, Einsteinstraße 21,
18052 Rostock

Treasurer
Eckhard Jaus, CSC Ploenzke Consulting GmbH, Zettachring 2,
70567 Stuttgart

Secretary
Roland Dürre, Interface Connection GmbH, Leipziger Straße 16,
82008 Unterhaching

Band 49

Die Reihe dient der schnellen und weiten Verbreitung neuer, für die Praxis relevanter Entwicklungen in der Informatik. Hierbei sollen alle Gebiete der Informatik sowie ihre Anwendungen angemessen berücksichtigt werden.

Bevorzugt werden in dieser Reihe die Tagungsberichte der vom German Chapter allein oder gemeinsam mit anderen Gesellschaften veranstalteten Tagungen veröffentlicht. Darüber hinaus sollen wichtige Forschungs- und Übersichtsberichte in dieser Reihe aufgenommen werden.

Aktualität und Qualität sind entscheidend für die Veröffentlichung. Die Herausgeber nehmen Manuskripte in deutscher und englischer Sprache entgegen.

Software-Ergonomie '97

**Usability Engineering:
Integration von Mensch-Computer-Interaktion
und Software-Entwicklung**

Herausgegeben von

Prof. Dr. habil. Rüdiger Liskowsky
Prof. Dr. habil. Boris M. Velichkovsky
Prof. Dr. habil. Wolfgang Wünschmann
Technische Universität Dresden

B. G. Teubner Stuttgart 1997

Die Deutsche Bibliothek – CIP-Einheitsaufnahme

Software-Ergonomie '97 : Usability Engineering: Integration
von Mensch – Computer – Interaktion und Software-Entwicklung;
[gemeinsame Fachtagung des German Chapter of the ACM,
der Gesellschaft für Informatik (GI) und der Technischen
Universität Dresden vom 3. bis 6. März 1997 in Dresden] /
hrsg. von Rüdiger Litkowsky ; Boris M. Velichkowsky ;
Wolfgang Wünschmann. –
Stuttgart : Teubner, 1997
  (Berichte des German Chapter of the ACM ; Bd. 49)
   ISBN-13: 978-3-519-02690-7   e-ISBN-13: 978-3-322-86782-7
   DOI: 10.1007/978-3-322-86782-7
NE: Liskowsky; Rüdiger [Hrsg.] ; Association for Computing Machinery /
  German Chapter : Berichte des German ...

Gesamtherstellung: Präzis-Druck GmbH, Karlsruhe

Programmkomitee

Udo Arend, SAP AG Walldorf
Astrid Beck, GUI Design Stuttgart
Heinz-Dieter Böcker, GMD Darmstadt
Peter Gorny, Universität Oldenburg
Winfried Hacker, TU Dresden
Thorsten Hasbargen, Siemens AG München
Heidi Heilmann, Universität Stuttgart
Andreas Heinecke, hi-soft Hamburg
Christian Kasten, DLR Bonn
Reinhard Keil-Slawik, GH-Paderborn
Ulrich Klotz, IG Metall Frankfurt a.M.
Helmut Krueger, ETH-Zürich
Rüdiger Liskowsky (Vorsitz), TU Dresden
Uwe Malinowski, Siemens AG München
Klaus Meißner, TU Dresden
Horst Oberquelle, Universität Hamburg
Matthias Rauterberg, ETH-Zürich
Harald Reiterer, Universität Wien
Eric Schoop, TU Dresden
Christoph Thomas, GMD, Sankt Augustin
Claus Unger, Fernuniversität Hagen
Boris Velichkovsky, TU Dresden
Axel Viereck, Hochschule Bremen
Hartmut Wandke, Humboldt-Universität Berlin
Wolfgang Wünschmann, TU Dresden
Volker Wulf, Universität Bonn
Jürgen Ziegler, FhG-IAO Stuttgart

Sponsoren

Die Fachtagung fand freundliche Unterstützung durch das
Sächsische Staatsministerium für Wissenschaft und Kunst
und die S A P Aktiengesellschaft, Walldorf

Vorwort der Herausgeber

Erstmals nach der Wiedervereinigung Deutschlands findet die nunmehr 8.Fachtagung "Software-Ergonomie" in einem neuen deutschen Bundesland, in Dresden statt. Damit wurden die langjährigen Traditionen des Mitveranstalters dieser Tagung, der Technischen Universität Dresden, auf dem Gebiet der psychologischen Arbeitsanalyse und der Bewertung von Software an der Benutzungsschnittstelle gewürdigt. Auf der anderen Seite sollte die Wahl des Tagungsortes dazu anregen, verstärkt osteuropäische Fachkollegen in die Tagungsdiskussionen einzubeziehen.

Die Erkenntnisse und Fragenstellungen zu computergestützten Tätigkeiten des Menschen wachsen ständig, was sich unter anderem durch die zunehmende Anzahl wissenschaftlicher Veranstaltungen und Publikationen ausdrückt. Diesen Trend begleitet logischerweise ein sich ausweitendes Spektrum beteiligter Fachdisziplinen. Bei computergestützter Tätigkeit in Beruf und Alltag gilt es, philosophische Aspekte ebenso einzubeziehen wie psychologische, soziologische, technische, ökonomische, medizinische und andere mehr. Bei all dieser Vielfalt kann weiterhin als gesichert angesehen werden, daß in dem Spektrum beteiligter Fachdisziplinen zur Beantwortung praktischer Aufgaben bei der Gestaltung computergestützter menschlicher Tätigkeit interdisziplinäre Herangehensweisen im allgemeinen und das Zusammenwirken von Informatikern, Psychologen und Arbeitswissenschaftlern im besonderen ein dringendes Gebot darstellen.

Ein wichtiger Grund dafür liegt sicherlich in der wachsenden Komplexität der aktuellen weltweiten gesellschaftlichen Entwicklungsprozesse insgesamt. In diesem Sinn besteht kein Zweifel an der Notwendigkeit, die Reihe von Fachtagungen zur Software-Ergonomie langfristig fortzuführen. Es besteht aber die zunehmende Herausforderung, sie mit aktuellen Schwerpunktthemen entsprechend den neuesten Entwicklungstrends an der Benutzungsschnittstelle fortzuschreiben.

Bedingt durch die enorm gewachsene und weiter zunehmende Menge von Produkten, die Funktionsgruppen automatischer Datenverarbeitung enthalten, rücken Fragen nach deren Qualitätsplanung und Qualitätsbeurteilung zunehmend in den Vordergrund wettbewerblicher Marktaktivitäten. Usability (Gebrauchstauglichkeit) von Produkten als ein Bewertungs- und Gestaltungsziel einerseits, Engineering als eine auf Innovation und Wiederverwendbarkeit orientierte wissenschaftliche Vorgehensweise der Produktentwicklung andererseits, sind als gedanklicher Hintergrund des Leitspruches der 8. Fachtagung "Software-Ergonomie '97" zu verstehen.

Integration von Mensch-Computer-Interaktion und Software-Entwicklung steht thesenhaft für eine notwendige ganzheitliche Orientierung sowohl bei der Entwicklung wissenschaftlicher Grundlagen computergestützter Tätigkeiten als auch bei der Sicherung hoher Praxiswirksamkeit rechnerenthaltender Hilfsmittel des Menschen im weitesten Sinne. Aus dem Thema der Tagung läßt sich somit der Appell ableiten, die Erweiterung technischer Möglichkeiten automatischer Datenverarbeitung für die Steigerung von Lebensqualität zu nutzen. Dies kann um so besser gelingen, je klarer die Grundsätze angegeben werden können, nach denen der Mensch seine Zufriedenheit im Umgang mit technischen Hilfsmitteln entwickelt und je deutlicher dabei Effekte gesellschaftlicher Integration unterschiedlicher Nutzergruppen entstehen.

Um in der Tagungsvorbereitung möglichst viele Interessenten ansprechen zu können, wurden die elektronischen Medien des Internet genutzt, über die regelmäßig die neuesten Informationen und Richtlinien für die Einreichung von Beiträgen bereitgestellt wurden. In Erweiterung der bisher zur Tagungsvorbereitung eingesetzten Technologien war es möglich, daß sich Teilnehmer direkt über das Internet anmelden konnten, woran sich die gesamte elektronische Tagungsorganisation lückenlos anschloß. An dieser organisatorischen Integration hat die Firma Multimedia-Systeme Dresden durch Bereitstellung des WWW-Systems „International Conference Organizer" mitgeholfen. Es ist vorgesehen, die Internet-Informationsquelle:

http://www.inf.tu-dresden.de/TU/Informatik/SE97/

mit Querverbindungen zu den Autoren und den Mitgliedern des Programmkomitees auch über den Zeitpunkt der Tagung hinaus zu pflegen. Interessierte können so Zugriff zu von Vortragsfolien bzw. den von den Autoren übergebenen Präsentationen ihrer Beiträge (Poster, Systemvorführungen) erhalten.

Das Programm der 8. Fachtagung spiegelt den Stand der Forschung sowie exemplarisch den der industriellen Anwendung und öffentlichen Praxis im deutschen Sprachraum, wenn nicht gar im europäischen Gebiet wider. Diese territoriale Begrenzung schafft günstigere Möglichkeiten der fachlichen Diskussion, als sie nordamerikanische Großveranstaltungen, wie z. B. die CHI-Konferenzen, bieten. Es ist leichter in einem kleineren Kreis Partner zu finden, mit denen man schnell in einen intensiven Erfahrungsaustausch treten kann.

Die eingeladenen Beiträge setzen das Motto der Tagung um. Ausgehend vom gegenwärtigen Stand des Usability Engineering (Bob Glass, Sun Microsystems) über Probleme des Anwendungsentwurfs zu neuen Innovationen in interaktiven Systemen (William Newman, Xerox Parc), der Rückwirkung von Qualität auf die Bildung von Designern (Pelle Ehn, Universität Stockholm) bis zur praktischen Umsetzung in einem gebrauchstauglichen „elektronischen Haus" (Roland Schoeffler, Bosch-Siemens) berichten.

Von den insgesamt 47 eingereichten Arbeiten zum wissenschaftlichen Vortragsprogramm wurden auf der Basis von jeweils vier Gutachten 26 Vorträge für die Tagung ausgewählt. Sie umspannen schwerpunktmäßig Probleme der Dialogentwicklung, bei der objektorientierte Beschreibungsmittel, Probleme der Wiederverwendung von Bausteinen und werkzeuggestützte Vorgehensweisen dominieren. Andere Aspekte der Mensch-Computer-Interaktion, wie Schnittstellen für Menschen mit besonderen Bedürfnissen, die Gestaltung von Gruppenarbeitssystemen, die Adaptivität sowie multimediale Oberflächen und Probleme des interaktiven Fernsehens gehören ebenfalls dazu. In recht breitem Umfang wurden Beiträge zur softwareergonomischen Evaluation und zum Usability- Testing eingereicht. Dieser Umstand folgt sicher aus der Tatsache, daß die EU-Bildschirmrichtlinie 90/270/EWG im Jahr 1997 Gesetzeskraft erhält und demzufolge viele Beiträge Konsequenzen aus ihrer Umsetzung aufzeigen.

Neben eingeladenen und eingereichten Beiträgen wird die 8. Fachtagung traditionsgemäß durch Tutorien, Workshops, Poster und Demonstrationsveranstaltungen ergänzt. Damit wird dem Wunsch vieler Teilnehmer nach weiteren Formen aktiver Diskussion und Beteiligung entsprochen. Die Tutorien setzen die "Schulen der Software-Ergonomie" fort, indem sie aktuelle Themen für den Einsteiger bis zum Praktiker bieten. Die Workshops fügen sich nahtlos in das Tagungsmotto ein. In dem vorliegenden Tagungsband sind neben den Vorträgen weitere Beitragsarten in Kurzdarstellungen enthalten. Mit der Aufnahme von Kurzfassungen zu Posterbeiträgen soll deren weitere Diskussion über den Zeitraum der Tagung hinaus angeregt werden.

Die Vorbereitung und Durchführung der Tagung ist als Ergebnis vieler aufeinander abgestimmter Einzelaktivitäten anzusehen. An erster Stelle sei in diesem Zusammenhang den Autoren gedankt, die dem Aufruf zur Tagung durch das Einreichen von Beiträgen gefolgt waren. Hervorzuheben ist gleichfalls das Engagement der Mitglieder des Programmkomitees, ohne deren Mitwirkung die aktuelle Orientierung sowie die Bewertung und Auswahl der Tagungsbeiträge nicht denkbar gewesen wäre.

Von der frühen konzeptionellen Phase bis zur Planung des praktischen Ablaufes der Tagung ist Herr Peter Gorny dem Organisationskomitee und den Herausgebern des Tagungsbandes ein stets einsatzbereiter und kompetenter Berater gewesen. Herr Hans-Dieter Böcker vermittelte uns auf kollegiale Art Erfahrungen aus der vorangegangenen Fachtagung.

Die Technische Universität Dresden und darin insbesondere die Fakultät Informatik hat in unkomplizierter Art und Weise die technischen Rahmenbedingungen der Tagung gesichert. Einzelheiten koordinierten Herr Siegfried Lattig und Herr Hans-Günther Dierigen. Stellvertretend für viele Helferinnen und Helfer, die den Tagungsablauf unterstützt haben, sei Frau Monika Schubert für ihre korrespondierende und koordinierende Tätigkeit im Tagungssekretariat sowie Frau Constanze Liebers für die Vorbereitung des Rahmenprogramms sehr herzlich gedankt. Frau Rosemarie Matthes hat stets auf die finanziellen Rahmenbedingungen geachtet. Frau Kerstin Baldauf, Frau Elvira Bartzsch, Frau Brita Heinze, Frau Kerstin Knochenhauer und Herr Falk Liesegang unterstützten mit hoher Einsatzbereitschaft die Erstellung aller notwendigen Tagungsdokumente.

Besonderer Dank gilt auch dem Teubner Verlag, insbesondere Herrn Dr. Spuhler, für die unkomplizierte Zusammenarbeit bei der Herausgabe des Tagungsbandes.

Dresden, Dezember 1996   Rüdiger Liskowsky
                                          Boris Velichkovsky
                                          Wolfgang Wünschmann

Inhaltsverzeichnis

12

Workshops (Kurzfassungen)

Poster (Kurzfassungen)

Swept Away in a Sea of Evolution:
New Challenges and Opportunities for Usability Professionals

Dr. Bob Glass
Futurist & Director of Strategic Technology
Sun Microsystems, Inc.

In this paper I will explore the future and examine the role of usability professionals. I will then address some practical considerations and some real life examples of usability engineering. Finally I will look at opportunities for us in the coming decade.

THE FUTURE . . .

As we approach the millennium, the importance of Usability Engineering to Society will grow. Opportunities and challenges will create a demand for Usability Professionals never before seen . The pace for product cycles is quicker, the tolerance for bad design is less. Products are evolving in ways we had not imagined. Who would have guessed that the basic telephone and telephone industry would evolve into ubiquitous integrated services, the first really big effort to combine a variety of technologies into a useful, portable device?

Breakthroughs in memory density, display technology and speed of computers will turn them into "communicators", not number crunchers-(i.e."computers"). The integration of electronic devices will finally yield the ultimate computer experience, that is when the concept of computer as we know it will disappear into the background. They will be integrated and intelligent. A quality experience, one in which products work as expected, correctly fresh out of the box. They may even be joyful to use. As the future unfolds the demand for quality will increase. People will demand experiences that are simple and elegant, in a word: wonderful!

Competition will be fierce. For every bad experience, customers will immediately move to select other choices. Usability will contribute significantly to the winners. Shoddy workmanship will not be tolerated. Software driving technology during this phase of the millennium will become reliable and useful and operate as expected. Productivity and quality of life will improve thanks in large part to wide scale use of Human Factors, including usability engineering.

A glimpse into the near future...smart rooms, smart clothes

Scientists and engineers are finally beginning to talk about "Smart Rooms" and "Liquid Computers" in a regular fashion. Up until recently, computers have been essentially blind and deaf. Humans and our pets are able to recognize us and make sense of what we are thinking. Yet computers were not able to easily and cheaply do the same task.

For the past five years Alex Pentland and his students and associates at MIT have been experimenting with computers that recognize faces, expressions and gestures. Imaging living in a house that always knows where the children are whether they are getting into trouble. Imagine working in an office space that is aware of you, that can interact with you or protect you if you are in a special meeting.

Perhaps the room will automatically handle interruptions such as the phone or visitors and take messages. Or if you are alone in your smart room and are talking, perhaps it will sense that you are alone (not on the phone and not with a visitors) and will try to understand what you are saying. In Sun Microsystems future vision film STARFIRE we tried to address some of the experience associated with. Imagine the opportunities for Usability Engineers over the next decade!

What is the future like?

When Ive talked to people about their ideas of the future, these words have profound implications for technology:

Simpler
Smaller
Smarter
Stronger
Faster
Lighter
Cheaper
Bolder
Happier

Some of the words in the list are a surprise. I added happier because I believe that products of the future should celebrate life! They should be a joy to use. I predict that Joy of Use will become an important factor in product development and product success. As technology becomes more efficient, products will become commodities. That is, they will be cheap enough that it will be hard to tell one product from another. Products seeking to differentiate themselves will draw on the skills of the usability community. New metrics and techniques to measure joy of use will need to be developed.

More Bandwidth and innovation means true sensory integration!

Over the next 10 years or so, as bandwidth becomes more available and cheaper, and compression and other techniques improve, we will begin to see the evolution of products like the present day computer (it computes bits using numbers). In the future the word computer will disappear from our language. The computer will evolve into an integrated personal aura and communications device- which I shall call an Aura Device (AD). It will become a personal envelope that surrounds each of us and helps us in all situations. It will enable fast, and easy ability to communicate with others. Our Aura Devices will be like an invisible envelope, extending in all directions around the person. (radar detectors in our cars are an example of such a device)

If we have global video services available there will still always be the problem of time zones....if you are communicating with someone 12 times zones away, there will continue to be the question of exactly who gets up in the middle of the night to have the meeting. Perhaps advances in asynchronous communication will help solve the problem. Video mail will replace e-mail.

Eventually, perhaps in 15-20 years we will evolve out of the dark ages of computers. Today we can use visual media and auditory media with our computers. But what about the other senses? What about smell, taste, touch? What about liquid computers which will enable us to feel new sensations? (we are liquid computers!). As bandwidth and compression techniques develop, we will finally begin to get true sensory experiences which will include smell, taste, pressure as well as other aspects of people meetings which we cannot yet quantify. Perhaps rates of endorphins will be able to be varied, thus giving the participant a very realistic experience. For example in the future if I am watching a sporting event, like a running marathon, I should be able to smell the sweat and feel the endorphins of the winner pumping though my own body [Hence the excitement!].

When my vision of the future happens, the world will be an augmented and joyful place! Imagine the usability opportunities! New ways of measuring our satisfaction will have to developed. Usability engineering will help the products succeed or fail on the global market!

Today, in 1997 we are beginning to see the evolution of smart card technology. We are entering the age of sensors. Devices that measure and integrate will become common. Smart cards largely introduced this year, are typically financial transaction devices with information about the person stored on them. Smart cards of the future will become our biometric trust. The will have information, like our DNA on them. Networks will have an Integrated Sensorium-not just sight and sound, but true sensory experiences including smelling, tasting, touching being transmitted to us. Eventually they will record and replay our total experiences. Linked to our personal Aura Devices (ADs) they may even include our dreams!

Computers will become secondary to the appearance of how we live our lives, because they will blend into our environment. No longer getting between us and the thing we are trying to do! They will be integrated and knowledgeable. Operating behind the scenes they will be able to augment and improve our lives. Imagine wearing clothes that are intelligent. Perhaps your hat will have a built in recognizer, and everytime you see someone, their face is recognized and their name and other relevant information is presented directly to your brain (of course at first there will be tiny displays !). You need never appear to forget again!

The environment forming your room will communicate with your Aura Device, and will recognize when you are alone, versus in an important meeting. Perhaps it will perceive that the meeting is so important that it will take all calls automatically based on its own decision. We show a scene alluding to that idea in our movie, Starfire.

THE PRESENT . . .

Now I'd like to bring us back from the future and look a bit at the present and the state of being a usability professional in today's times. One of the problems we have had in the field of usability engineering is justifying our existence. Although we know we are important to successful product development, it has been hard to show how we contribute to the bottom line of our companies. Concrete data that can be presented to the financial justification side of our organizations has been scarce. In the future it will become obvious.

Lets look at history. Only 12 years ago (1985) we were in the Iron Age of computing. The experience was primarily of text based searching. Low speeds were used. Data bases like DIALOG (~1974) had a limited amount of scientific titles and abstracts. In the medical community, MEDLINE and MEDLARS were a decade old by 1985.

Looking back three years to 1994, we find that we were in the Bronze Age of computing. It was the first time you probably ever heard the word web or home page. We saw the development of on line shopping malls (with text and photos). We learned a horrible new interface: (http://www.*xxx.com/*   ) Whats your URL? first entered human languages as mankinds newest line. :-)

By last year, 1996, we had entered the Silver Age of computing. This period was characterized by multiple media, higher speeds (28.8, 56Kb etc.) and finally a critical mass audience of perhaps 12,000,000 ! For the first time we saw original content being developed. New channels to serve new markets.. Network initiated, purchased or produced shows. In 1996 it was common to have a personal web site.

Over the next decade a dramatic increase in the number of personal home pages is likely to evolve. There will be lots of opportunities for usability engineering!

Within another two years (1999) we will enter what may be the Golden Age of computing. We will begin to see lots of data streaming. Full motion video. The next thing: cable , telephony , data communications and
Hollywood blended together. We will see a proliferation of advertising. In fact a lot of the support for the net will be due to advertising. We will see more consumers using PCs in the prime time than TVs. Of course this vision of the future implies that the problems of speed and performance on the network has been solved.

One of my colleagues, Dr. Jakob Nielsen has a list of Usability Slogans which we have put to good use in our engineering community:

      Your best guess is not good enough
      The user is always right
      The user is not always right
      Users are not designers
      Designers are not typical users

Less is more
Details matter
Help doesnt

Time doesn't permit me to expand each item, but you get the idea.

THE OPPORTUNITIES OF USABILITY ENGINEERING

*There is a theory that the amount of intelligence
in the Universe is constant...
Unfortunately, the population is growing!*
-author unknown

As we move into the next millenium we note a number of changes required for successful products. Customers are already changing what is acceptable in a product. My short list is:

1) low tolerance for poor usability
2) immediate gratification
3) support fast learning
4) must be engaging
5) No Fear
6) no error messages
7) supports both passive and active use
8) ENJOYMENT ! - Joy of Use
  ...which means
  Simplicity & Elegance
  A Quality Experience

Incorporating these concepts into products will help make them win in the coming decade.

Over the course of my career I've had the opportunity to be involved with the development of several of the major operating systems. My team's work on usability has led to the development of some excellent products.

Designing System 7 for the Macintosh- a real life example:

Sometimes in the design of interfaces to products we have to draw on other fields. In the design of various operating systems that Ive helped develop, weve had to draw on the principles of magic to improve the overall experience. During the development of Apples Macintosh System 7 (the first time the interface was ever changed) we were doing usability testing on some aspect of the interface. We had designed the hard drive icon to open in a list (View Menu by name) view. I wanted a way of each folder to show the contents without opening another window to see the contents of the folder, so we put a radio button beside each folder. The person clicks the radio button and the folder folds open.

As we reviewed the video tapes from the usability lab we noticed that every person always made a comment like gee --- this is slow when they clicked on the radio button (a subsequent investigation noted that CPU speed did not appear to make any noticeable difference.) . We were not testing the speed of the system, but were looking for data on how people used the trash can.

I went to the team that owned that part of the Operating System (OS) Code and asked them to speed up the rate at which the folder unfolded. After about three days of evaluation, they came back and told me that if they did it, it would break all applications since it would violate Inside Macintosh our technical specifications .   Clearly the perception of the user lost out if nothing was done. However not one to be defeated, I went back to my team and asked them to brainstorm with me on how to improve the quality of the experience. I pointed out that perceived performance was as important as performance, and there must be some way to solve the problem.

During the course of our discussions we examined several other fields for fresh ideas, and decided that our user interface problem had a lot in common with Stage Magic. So as we talked about what do magicians do?. We noticed that one thing they did often, was to divert your attention from one place to another. So we decided to draw on their techniques. Since part of the problem was that a radio button has a sense of immediacy associated with it (you touch it and it changes), we needed a delaying tactic. We invented the triangle that animates downward when you touch it (the user interface item currently in the interface). We never sped up the performance, we merely gave the user something to occupy their time. We never heard any complaints about the performance again!

The lessons learned from this example, include the fact that sometimes you have to draw on other fields to improve your design.

Principles of Magic and their application to Human Interface Design.

One of my colleagues, Bruce Tognazzini, human interface designer and magician, wrote a paper about similarities of magic and human interface design. He discusses ideas which all of us should consider in our work. See Principles, Techniques, and Ethics of Stage Magic and Their Application to Human Interface Design, proceedings of InterCHI, 1993 (Amersterdam, The Netherlands, April 24-29, 1993). ACM, New York , 1993, pp. 355-362.

Virtual Realities

Both human interface designers and magicians create virtual realities. Magicians bring their act alive on stage. It is really an alternate reality. They appear to defy laws of physics and nature right before your eyes. In the same way, human interface designers bring their act alive on the display. Human Interface Design is really *the user illusion* .
Examine some of the principles that Magicians use, and you will see the similarities:
> *Consistency*
>> You make people believe if you are consistent

Unity
> Tying things together - we are part of a system

Use of Real World Metaphors
> Magician's tools should be disguised to look like objects
> in the real world. User Interface designers follow the same principle

Perform user testing
> Both for magic and interface design

Showmanship
> Presentation is part of perception

Smoothness
> Fit and finish or silky smoothness makes a finished product.

Get to the point
> - be brief. Elegance and Simplicity have become a mainstay.

WHAT ARE THE CHALLENGES FOR THE FUTURE?

New Devices require new designs. Lets look at the near future and guess at some things that will be evolving.

Ubiquitous computing finally!
> Computers everywhere! Even more than now! And they communicate with each other too!

Intelligent Offices and environments.
> They adjust your environment according to your needs. Personal Videoconferencing will become ubiquitous. Well communicate with everyone with a visual call! or video mail!

Integration of personal & work life
> Systems will develop that allow us to be in contact with our personal life and work life at the same time. Cellular phones are the first example of technology being used this way.

We will begin to see different displays arrive on the screen including Brain Displays-Displays imaging directly to your brain. We will also see flexible, lightweight flat panel displays everywhere. We may also begin to see "Globe displays"- The ability to utilize holographic spherical projections for many activities will make our lives safer. Consider the use of three dimensional displays for air traffic control systems. We can visualize problems well in advance of a real problem.

Finally we will see the creation of devices that are usable, in part because of their increased intelligence and in part because of our use of usability engineering . People of the future no longer will tolerate shoddy workmanship, or imperfect products, thus giving the Usability professional a lifetime of employment.

What can we expect over the next several years? Looking at the toy industry it, we will begin to see different input devices. Some of these devices will be controllable by our own nervous system. NASA has worked with eye gaze control for more than a decade. We are finally seeing a lot of breakthroughs happening. Perhaps in the not too distant future we will be able think about something and actually interact with it. The work of Usability Professionals is never done!

In the not too distant future we will see the emergence of brain-interface devices. Devices where we interact with them by "thinking". For and example information see: "Controlling Computers with Neural Signals"- (Scientific American, October 1996)

Neural Interface design will become increasingly important. It will help people with physical disabilities take charge of their environment. Eventually it will help all of us. We are TABs- Temporarily able-bodied. There is always the chance that something could happen to make some of our sensory processes stop (ie. blindness). As we grow older the need for reading glasses increases. It is important to consider usability access for all individuals, because the older you get, the more likely you are to require them.

If we begin to look beyond the next decade we will see devices that are truly integrated with our lives. They will have intelligence and natural interfaces. Finally they will also include our other sensory processes.

Trends and Issues

Over the next few years we will begin to see more bandwidth and compression schemes develop. As bandwidth increases, the integration of products and natural (and augmented) user interfaces will need to be developed.

Already we are becoming overwhelmed with the vast amount of data and information available. We will see the emergence of a valuation system for data, probably by our Active Agent who will rate each piece of material for integrity. As we sleep at night, our active Agents will roam the universe (OK ...initially just the internet :-)) in search of trends and events of importance and interest to us. Of course each of these agents will be followed by a financial agent who will tally up the nanomoney transactions (i.e.. nano-cent or nano mark) and take them from our accounts and deposit them in the appropriate places.

We will begin to see the world drive to a financially driven market, one in which you will also be paid (in nano-money) to read your e-mail or deal with your asynchronous video mail. In fact one prediction that Im sure of , is that in 10 years from now you pay money not to be on net. A whole new business model will evolve in *access transactions*. Your agent, or room will guard you from unwanted intrusions because you paid a fee for the night not to be disturbed by outside forces. Already Sun Microsystems has released Java commerce, a way to transfer small fractions of money around the net. A corollary model is one in which you raise the bar of access to yourself. For example, you may tell your guard agent that you will only accept e-mail if $1000 US Dollars are deposited in your account. Or you will only accept that phone call from your boss if you get a raise! We will become a financially driven world over the next 20 years.

Of course this scenario is only for the technologically wired. Vast segments of the population will rebel against this type of life style. They will demand all of the good things that technology can deliver, but none of the straight jacket things.

Usability engineering for the Web

There will be tremendous opportunities for usability engineering related to the web and home page design. Shortly there will be millions of people attempting to design user interfaces. The creation of a successful web page includes a tremendous amount of usability engineering. In fact, roughly the same amount of effort of developing an application, can easily go into the development of a web site.

User interface design and all its methodology will become more important. Even the designers will evolve. Today, video designers (as in advertisers of shows that we watch on TV) are designing un-interactive sites. Their goal is to keep your finger off of the remote control button that changes the channel- so you watch their compelling advertisement. An Interactive Visual Designer has different skills. Here their goal is to keep you interacting with the system. Your finger (or other sensory modality) is always on the button!

International Usability Engineering

Dont overlook internationalization aspects of usability, especially as it relates to the Web. A good source of information is: International User Interfaces, 1996, Elisa M. del Galdo and Jakob Nielsen, editors, New York: John Wiley & Sons, Inc.

Usability Engineering at Sun Microsystems

In late 1991, Sun recruited me from Apple Computer (where I was Manager of the Human Interface Group - the ones making all of the design decisions on the Macintosh) to come and build a Human Factors Department. Included was enough commitment for me to shape the future of the company and its products. I obtained funding to develop several state of the art usability centers at a variety of Sun Microsystems sites in the United States.

Sun is committed to usability engineering all of our products. As a whole, the computer industry has seen benefits of human interface design and usability engineering in:
   Increased Productivity
   Increased customer satisfaction
   Increased Sales and Revenues
   Decreased Training and Support Costs
   Reduced Development Time and Costs

USABILITY REQUIREMENTS ARE CHANGING

Minimal Instructions
- more reliance on software intelligence. Devices that will set themselves up, or talk to you to help set them up.

No tolerance for obvious bugs
- we have become far too accepting of bad products from the computer world, especially in software design. You wouldnt use your telephone, if every third time you dialed you actually got the right number! Consumers of the future will be much more demanding. The way the computer industry runs will not be acceptable. Products will be expected to work the first time correctly and as the consumer expected.

Commodities and Joy of Use
- Intelligent Products become commodities. They are everywhere and we dont think of them. You will wake up in the morning and buy coffee and electronic devices without thinking about them. You will tend to buy devices that you like, so joy of use will become an even more important issue in the future.

Users demographics are changing

They are much younger and wiser with respect to electronic computer related devices. After all they always grew up with computers and electronic complex devices.

They are competent with vast amounts of information and their demands are different. They have no tolerance for errors . Its also best to realize that HELP doesnt!

We are also seeing the beginning of an older population using devices. They have the time and are beginning to explore. Good visual support material and simple instructions are essential. Simplicity and elegance will be a characteristic of the winning products.

Conclusion

The future will increase the opportunity for Usability Engineers. Fierce competition will drive out less usable products. The winners will not only be usable, but they will be a joy to use!

Autor

Dr. Bob Glass
Futurist & Director of Strategic Technology
Sun Microsystems, Inc.
E-Mail: bob.glass@sun.com

What Application Designers Know: Some Thoughts on Innovation in Interactive Systems

William Newman

Rank Xerox Research Centre, Cambridge UK

Abstract

This paper takes as a starting point Walter Vincenti's seminal book, "What Engineers Know and How They Know It," and explores the implications for interactive system design of some of his observations on engineering knowledge. In particular, it discusses the obstacles designers face in accessing relevant knowledge, and their consequent difficulties in engaging in design. The paper suggests that a crucial step in building knowledge about interactive system design knowledge is the establishment of *critical parameters* concerning the human activity that the application supports. It concludes with a discussion of the implications for innovative system design.

1   Introduction

Those who design and build software systems for direct use by people and organisations play an increasingly important role in society. These are, almost universally, *interactive systems* that are operated directly by their users via the system's user interface. Their designers must try to respond to users' demands for improved services and productivity by developing new applications. They must be prepared to follow up, when their designs are successful, with more powerful and reliable systems that keep pace with users' expanding needs. There is constant pressure on the designers of interactive systems to do more and to do it better.

In this paper I am concerned, not so much with whether software engineers are doing a good enough job of designing interactive systems, but how in the first place they can acquire the knowledge they need to do their job. In effect, I take as my starting point a paraphrase of a remark of Walter Vincenti's [27]: *what software engineers do depends on what they know*. My interest is to understand what software designers need to know, where this knowledge is to be found, whether there are significant barriers to the growth of knowledge about interactive systems and, if so, what effects these barriers might have.

To address this topic, I have chosen to try to apply general theories about engineering design knowledge to the special case of software design. There are dangers in adopting this approach, for it is not unanimously accepted that software design can be treated as an engineering exercise [29]. However, I try in the course of this paper to provide evidence that my approach is a sound one, and that it is valid to talk about interactive system design as "software engineering."

I therefore begin by outlining a basic taxonomy of engineering design knowledge, and then provide some evidence that it applies to software design. When applied to the design of interactive systems it yields interesting results, especially when contrasted with the design of *embedded software*, i.e., the non-interactive systems and components that support interactive software. I conclude that it is particularly hard for designers of interactive systems to acquire the knowledge they need. I have some final comments on how this affects innovation in interactive applications.

2   A taxonomy of engineering design knowledge

As I have pointed out, the question of how growth in engineering knowledge occurs is in effect two questions: what kinds of knowledge do engineers depend upon, and what causes this knowledge to accumulate? My first task is therefore to answer the first sub-question, by enumerating some major categories of engineering design knowledge, and elaborating on the categories in discussing how they contribute to design. I then compare the taxonomy with the findings from studies of software designers' knowledge.

2.1   A taxonomy

As a first cut at the categorization of engineering knowledge, I offer the following five categories, drawn from a number of previous analyses of engineering design [24, 23, 27]:

1   Domain knowledge relating primarily to designed artefacts and their constituent technologies
2   Domain knowledge relating primarily to environments in which artefacts are used
3   Knowledge of representations, particularly those applied to describing aspects of the domain
4   Known techniques for analysing the design or for simulating the proposed artefact's behaviour
5   Knowledge of the critical parameters against which the artefact's performance is measured.

The list is not exhaustive. It deliberately omits two categories of knowledge that play roles in software design, namely knowledge of procedures and processes, and general knowhow and rules of thumb. I do not regard this omission as crucial to the arguments I will lay out here.

2.2   Fitting the taxonomy to descriptions of engineering design

Accounts of engineering design refer frequently to the five categories of knowledge listed above. Henry Petroski [21], for example, describes structural engineering design thus (my own category numbers in brackets): "As each hypothetical arrangement of parts [1] is sketched either literally or figuratively on the calculation pad or computer screen [3], the candidate structure must be checked by analysis [4]. The analysis consists of a series of questions about the behavior of the parts under the imagined conditions of use after construction [2]." In a similar vein, Rogers identifies the engineer's need "to determine the size and shape of a piece of equipment [1] to perform a specified duty [5], or to predict the performance of the design [4] when it is called upon to function under other operating conditions [2]" [23].

A particularly thorough treatment of design can be found in Herbert Simon's book *The Sciences of the Artificial*, in which he identifies roles for all of the taxonomy's components [24]. He places particular emphasis on the distinction between the "inner" environment of design—the substance and organization of the artifact itself [1]—and the "outer" environment in which the artefact operates [2]. He goes on to discuss designers' frequent difficulty in predicting how their designs will behave [4], the influence of choice of representation on design problem solving [3] and so-called "figures of merit" that permit comparison between designs [5].

2.3   How the taxonomy fits software design

Do these accounts of engineering design apply to the design of software? Or is software design fundamentally different in its knowledge requirements? Answers to these questions can be found in the studies of programmers and software designers by Adelson and Soloway [1], Visser [28], Guindon [12] and others. To a large extent the studies confirm that software designers do rely on the same five categories of knowledge and, in this respect, go about their work like other engineering designers.

For example, reliance on knowledge of existing designs can be seen in virtually every study of software design. Designers tend to start with a schema or model of the solution, which they construct from past experience or from a supplied solution. If the problem is familiar to them, the schema is likely to be well-formed and easily specified [1]. If the problem is unfamiliar, software designers are still likely to reuse an existing tested design, even though its adaptation to the new problem may be timeconsuming [12].

Software designers also rely on their knowledge of the environment of use. Like the engineers described by Petroski and Rogers, they apply their knowledge to constructing scenarios of use, which are then used in simulating the performance of the design. Guindon, for example, describes how designers of a lift control system created hypothetical configurations of the lifts for the purpose of testing designs [12]. Scenario-based simulation can be observed helping the designer to understand the behaviour of the system as a whole. Not surprisingly, designers who lack familiarity with the environment of use have difficulties constructing scenarios and are less able to simulate their designs' performance [1].

Knowledge of notations and other representations can also be seen contributing to software design. Experienced designers will often search among the notations familiar to them in order to choose one suitable for the problem [12]. If they are less experienced, they may go to the extreme of adopting a familiar notation to prepare the design, subsequently translating the finished design to the required notation [28].

2.4   Where the taxonomy falls short

There is plenty of evidence, therefore, of software designers' reliance on domain knowledge and on familiarity with representations; but what about techniques for analysis and simulation, and what about critical parameters? These kinds of knowledge are also in evidence, but not in great quantities.

The designers studied by Adelson and others showed no inclination to apply analytical techniques to the evaluation of their designs. Their simulation methods can only be regarded as primitive. They relied on simple inspections and scenario walkthroughs, whose purpose was just to check the functioning of the designs, and not to predict their performance. The almost total lack of attention to performance analysis is surprising, because programmers are known to take pride in squeezing speed improvements out of programs.

Software designers also showed little awareness of critical performance parameters. No "figures of merit" are mentioned in the studies; instead designers appear to adopt their own preferred evaluation criteria, such as reliability or simplicity. They use these to guide the choice of solution, but never apply any overall performance targets [12].

Again, the reason may lie in the designers' lack of experience with similar problems. Although presented with a precise problem definition, they mostly tended to continue elaborating on the stated requirements and constraints during design, illustrating software designers' well-known proclivity for "requirements drift." By continually adjusting the functional requirements for the system, they lost track of critical performance parameters as potential design targets.

3 Software designers' acquisition of knowledge

On the basis that it is valid to apply the knowledge taxonomy to software design, I will now look at the means available to software designers to acquire the knowledge they need. This is an exercise that has been carried out before, e.g., by Denning [7] and Curtis *et al.* [6]. However, these studies have not distinguished between the design of embedded software and interactive systems. Do designers in these two areas face the same problems in accessing information?

In answering these questions I have had to switch to a more subjective approach, because there is so little published information on how software designers acquire knowledge. I hope nevertheless to convince readers that the two problems of information access, concerning embedded software and interactive systems respectively, are significantly different problems, because of fundamental differences in the nature of the information.

3.1 Knowledge in support of embedded software design

What do the designers of embedded software need to know? And how can they come about this knowledge? To answer these questions I will look in turn at the taxonomy's five main categories of knowledge.

Domain knowledge about embedded software artefacts can be acquired from published accounts. Textbooks lay out a range of solutions, traditionally as coded algorithms but increasingly as object descriptions [9]. There are journals and conferences devoted to specific types of software systems, such as operating systems, distributed systems, image processing, computer graphics, databases, and so forth. Articles and papers describe new algorithms and compare them with past solutions. By reading this literature the software designer can gain familiarity with a range of designs, from which he or she can develop a schema for a new design problem. The fact that designs are presented as algorithms or patterns assists this process.

The same literature provides examples of the environments in which embedded software artefacts are used, because these environments are software systems too. They may not always be described to the same level of detail as are components, but they support the generation of scenarios for use in testing the design of the embedded component. A superficial knowledge of a CAD system will, for example, enable the designer of a graphics package to simulate the latter's performance in generating a display.

The representations needed in embedded software design are essentially those for the specification of software. They include programming languages, data representations, state-transition notations and the various other graphical notations offered by the software engineering methodologies [14]. Software designers get to know these representations early in their careers.

Techniques for the analysis of embedded software designs form a less explicit part of the designer's knowledge. There are, as I pointed out earlier, techniques for inspecting code and conducting walk-through analyses [2]. More rigorous methods of algorithmic analysis are also available, although empirical evidence suggests that these are used comparatively rarely. From these analyses the designer can gain a rough idea of the software system's performance. However, an alternative to analysis is always available, in the form of prototyping and testing. Through a combination of these, the designer is generally able to evaluate the design to an adequate level.

Finally, the critical parameters against which embedded software is tested appear to be widely understood. A number of universal parameters are applied to most software designs: speed of execution, reliability, use of resources. Speed, often the primary consideration, may be measured in terms of performance of standard benchmarks, or in terms of dimensions of the problem domain; thus speed may be proportional to n^2 or $\log(n)$. The existence of critical parameters that are tacitly agreed among designers, and localised to the embedded component in question, enables designers to set and achieve specific performance targets.

3.2 Knowledge in support of interactive system design

For the designer of an interactive system, acquisition of relevant knowledge is a very different matter. Again, the differences can be appreciated by looking at each type of knowledge in turn.

Like embedded systems, interactive systems are described in the literature. Furthermore, and unlike most embedded systems, they can sometimes be experienced first-hand, for example in the form of cash machines and on-line library catalogues; and some of them can even be purchased in computer stores for personal use. In a sense, therefore, knowledge about interactive systems is particularly accessible to designers. However, the availability of this knowledge is distinctly uneven: if the application cannot be purchased or accessed publicly, information about it is likely to be very hard to obtain. The "organisational" systems used by banks, police services, hospitals, government offices and military personnel are kept under wraps for the most part. Very few descriptions can be found in the literature, and then they are usually very superficial. Even the most celebrated organisational applications, such as the SAGE air defence system and the Sabre airline reservation system, have been documented to only a most perfunctory degree [8, 5]. As a result, designers' knowledge is steeply skewed towards those designs that are readily accessible.

The environments in which interactive systems are used are environments of human activity, altogether different from the software-systems environments in which embedded components are used. Learning about these environments is, again, relatively easy when access to the environments is itself easy. However, the difficulty here for the designer is to know what are suitable scenarios against which to test the design. It is tempting for the designer to say, "This is how *I* would use the system," but this carries a risk of applying a scenario that is weighted in the system's favour. Discovering patterns of behaviour in application environments is a specialised task, and few software designers have the skill or the time to carry it out. As a result there is a constant danger that interactive software will be designed on a basis of relatively superficial knowledge of the environment of use.

One of the barriers to disseminating domain information about interactive systems is the relative lack of suitable representations. The designs of interactive systems are difficult to describe because they involve many linked software components and have complex, dynamic user interfaces. The notations for describing such systems are less well developed, in comparison with the notations for algorithms and software objects. So even when details of interactive systems are published, weaknesses in the descriptive representations may prevent designers from finding out what they need to know. For example, when details of the Xerox

Star system were made public in a number of articles [13, 25, 26], the descriptions appear to have been inadequate for those who tried to replicate what Xerox had built [30].

Designers of interactive systems are poorly equipped with analytical tools. The techniques available to the embedded software designer do not apply here, because the problem is to analyse and predict the behaviour of an interactive system in the hand of its users. This is a complex domain of analysis, and there is a tendency for analysis to be timeconsuming and difficult to learn. Some simple methods have been developed, including Keystroke-Level Analysis and Cognitive Walkthroughs [3, 22, 17]. However, they do not appear to have achieved widespread use.

Perhaps the most pervasive problem for the designers of interactive systems is the lack of known critical parameters, analogous to the execution-speed parameters that govern the design of most embedded software components. In the case of specialised applications, such as cash machines, critical parameters do tend to be known within the organisations that develop them, but are unavailable to other designers. A well-documented example of this is the study, known as "Project Ernestine," of workstations for toll-and assistance operators, or TAOs [11]. In this application the time taken by the TAO to complete each call is a critical parameter in the workstation's design; however, the times are different for each type of call, and information about times and call-types is considered confidential by telephone companies [10]. Each application has its own critical parameters; in many cases, the research has not yet been done to establish what they are.

In summary, therefore, the design of interactive systems is significantly less well provided with essential information, and levels of knowledge in designers are bound to be lower as a result. Particular problems are patchiness in knowledge of existing designs, superficial knowledge about application environments, and an almost complete dearth of knowledge about the critical parameters that apply to individual applications.

3.3 Is this engineering?

With so many barriers in the way of acquiring knowledge, we might find it hard to view interactive system design as an engineering activity. After all, if engineering design is a knowledge-dependent activity, and the knowledge is unavailable, can engineering design be said to take place?

The answer lies, I believe, in the evidence that application designers try, in almost every case, to access the information they need to design and build systems in an organised way; and this organised use of available knowledge is the ultimate stamp of the engineer [23]. Even if they abandon the search for information in some areas, e.g., for information regarding existing solutions to the problem at hand, they will still attempt to search methodically for it in others. They will not start from scratch unless there is no alternative. I would claim, therefore, that we are discussing a design practice that exhibits the main features of an engineering activity. As such, we can compare it with other engineering practices and use them as a basis for understanding how interactive system design might develop into a stronger discipline.

3.4 The search for critical parameters

I have mentioned the problem of identifying critical parameters for interactive systems, and I will wrap up this section by expanding on this problem and suggesting how it might be attacked.

The essence of a critical parameter is that it provides a basis for quantifying a requirement. Vincenti quotes an example in his study of the development of requirements for aircraft flying qualities [27]. By means of extensive testing, engineers were able to isolate the parameter *stick force per* g *acceleration* as a basic determinant of maneuverability. Once this parameter had been identified, and accepted by the aircraft industry, it became a standard means of specifying flying qualities. Vincenti quotes figures of six pounds per *g* on fighter-type aircraft and up to 50 pounds per *g* on bombers and transports. He adds, "Aeronautical engineers today express amazement that any meneuverability criterion besides stick force per *g* ever existed." One of the hallmarks of engineering is, I claim, this tendency for critical parameters to become tacitly and universally accepted [18].

As I have said, critical parameters are different for each application. The parameter governing the design of TAO workstations—call-completion time—cannot be assumed to apply to calls to directory-assistance or ambulance-service operators [15]. The reason for this variation lies in the dependence of application-design parameters on the *supported activity*. If the activity changes, critical parameters may no longer apply.

In comparison with other fields of engineering, very little research has been done into establishing critical parameters for software systems, of any category. As regards embedded systems, the tendency has been to assume that the standard parameters—speed, reliability, use of resources, etc.—apply in all cases. In the design of interactive systems, the main concern has been to provide adequate functionality (where "adequate" can mean "more than the competitor's"). A secondary concern has been usability; but the establishment of an interactive system's usability often includes deciding what to measure, and different parameters—not necessarily critical to the application—may be selected on an arbitrary basis for each usability evaluation.

The difficulty of determining critical parameters for interactive systems should not be underestimated. The concept that human activity is sufficiently repeatable to allow measurement of recurring parameters may seem controversial; some might argue that such parameters cannot be found, or cannot be used as a basis for design. Certainly my own experience has been that deliberate attempts to find them often fail, and that the parameters sometimes emerge in the course of looking for something else [19]. The strongest arguments for their existence are that they have been detected in some applications, e.g., in TAO activities [11], and that people's ability to plan their activities suggests an ability to make their own estimates of how long the activities will take [20]. If we can tap into this tacit knowledge, we may be able to discover the critical parameters we need to know.

Difficult though critical parameters are to determine, there is a strong argument for trying to do so, because critical parameters lie at the root of the acquisition of software engineering design knowledge. They provide a basis for enhancing designs, for without them the decision as to whether enhancement has been achieved becomes arbitrary. They also provide a means of selecting scenarios for testing designs: if the scenario is couched in terms of critical parameters, e.g., it describes a task whose performance-time is critical, then it will provide a valid basis for testing. Critical parameters also help us to understand what kinds of analyses we need to carry out, and therefore what tools need to be developed. I would suggest, therefore, that if more attention is paid to the identification of these parameters, faster progress may be achieved towards providing interactive system designers with the knowledge they need to do their job. I will conclude with some remarks on this point.

4 Thoughts on innovation in interactive systems

To work in the computer field is to experience unceasing innovation on a massive scale. This is true whether one works in hardware, embedded software, interactive systems or some other related area. In a sense, innovations must proceed in tandem in all areas because advances in one area depend on advances in another.

What does it mean, then, if one area—interactive systems—experiences particular difficulties in acquiring essential knowledge? How can this area keep pace with the rest? I have worked in the interactive systems area for some years, and have meanwhile kept an eye on developments in other areas, and my sense is that innovations in interactive systems do *not* keep pace with the rest, in the true sense. Although novel ideas for interactive systems are generated at the same rate as in other areas, or even faster [16], the *innovation* process by which these ideas are brought into general use often falters.

Innovation is indeed a process, not an instantaneous "eureka"-type event. Ideas for new designs do not work perfectly first time, unless they are very minor enhancements to existing designs. The ideas for interactive systems that flood the computer business tend to be relatively radical. They offer totally new ways to do things, e.g., to access library books, to hold meetings, to carry out surgical operations. Ideas of this kind tend to bring with them a host of side-effects and performance problems, which need to be worked out over time. The goal of this innovative process is, as Constant has put it, to reduce the original "radical technology" to a generally accepted and established "normal technology" [4]. During the innovative process, designers and users alike are motivated by the prospect of gaining an advantage over the previous technology. Thus reading books on-line should be advantageous in comparison with reading them on paper, holding a videoconference should offer advantages over holding it face-to-face, remote surgery should be preferable, at least some of the time, to traditional surgery.

As I have pointed out, knowing whether we have improved on an existing design involves knowing the critical parameters for that design problem; and in the design of interactive applications the critical parameters are generally unknown. There is also a need to know about existing designs, and about techniques for analysing and predicting their performance. The lack of these forms of knowledge has two undesirable effects.

First, *radical ideas may be pursued when existing solutions could easily be improved.* A new idea for an interactive system has an attraction all of its own; because it is new, it generates interest in its further development. But it may not offer any real advantages over existing systems. There may be simple ways of enhancing an existing design. We may not know enough about the existing design to tell. Unless we know the critical parameters against which to assess the two approaches, and are familiar with the other design, we cannot make an informed choice here.

Second, *the innovative process may degrade rather than improve performance.* Innovation involves a long series of design changes. Since the critical parameters are unknown, there is no way to tell whether the design is being changed for the better. Even if they are known, analytical techniques may be inadequate, as in the case of Project Ernestine [10]. What can designers do in these circumstances? One common recourse is to add features, because these are seen as "improvements" in the competitive sense. Interactive systems thus gradually become more feature-rich and resource-demanding; but do they offer improved support to their users?

There is a solution to these problems, in the form of research to build the kinds of knowledge that I have suggested is so hard for interactive system designers to acquire. Foremost among these is to establish the critical parameters of interactive applications. In tandem, work is needed to develop analytical models for similating and predicting the outcomes of design. Efforts need to be made to document existing applications. As I have said, these are hard areas of research; but they have the potential to offer considerable benefits. I believe they could enable the process of innovation in interactive systems to proceed a lot better than it currently does.

Acknowledgements

I am grateful to the many people with whom I have had helpful discussions and correspondence on the topics covered in this paper. They include Walter Vincenti, Wayne Gray, Bonnie John, Mary Shaw, Stuart Card, Richard Harper, Marge Eldridge, Bob Anderson, Hervé Gallaire, Thomas Green, Ruven Brooks, Patrick Waterson, Susan Gasson, Robert Rist, Willemien Visser and Terrie Shaft.

References

[1]   B. Adelson and E. Soloway: The Role of Domain Experience in Software Design. In: IEEE Trans. on Software Engineering, Vol SE-11, no. 11 (November 1985), pp. 1351-1360.

[2]   J. L. Bentley: Writing Efficient Programs . Englewood Cliffs NJ: Prentice-Hall, 1985.

[3]   S. K. Card, T. P. Moran and A. Newell: The Psychology of Human Computer Interaction. Hillsdale, NJ: Lawrence Erlbaum Associates, 1983.

[4]   E. W. Constant II: The Origins of the Turbojet Revolution. Johns Hopkins Univ. Press, Baltimore, MD, 1980.

[5]   D. G. Copeland, R. O. Mason and J. L. McKenney: Sabre: The Development of Information-Based Competence and Execution of Information-Based Competition. In: IEEE Annals of the History of Computing, Vol. 17 no 3 (1995), pp 30-57.

[6]   B. Curtis, H. Krasner and N. Iscoe: A Field Study of the Software Design Process for Large Systems. In: CACM Vol 31, no. 11 (1988), pp. 1268-1287.

[7]   P. J. Denning: Computing, Applications and Computational Science. In: CACM Vol. 34, no. 1 (October 1991) 129-131.

[8]   R. R. Everett, C. A. Zraket and H. D. Bennington: SAGE: A Data Processing System for Air Defence. In: Proc. Eastern Computer Conf. (1957), pp. 144-157.

[9]   E. Gamma, R. Helm, R. Johnson and J. Vlissides: Design Patterns. Reading MA: Addison-Wesley, 1995.

[10] W. D. Gray, B. E. John and M. E. Atwood: The Précis of Project Ernestine or, An Overview of a Validation of GOMS. In: Proceedings of CHI '92 Human Factors in Computing Systems (May 3-7, 1992, Monterey, CA) ACM/SIGCHI, N.Y., pp. 307-312.

[11] W. D. Gray, B. E. John and M. E. Atwood: Project Ernestine: Validating a GOMS Analysis for Predicting and Explaining Real-World Task Performance. In: Human Computer Interaction Vol 8 (1993), pp 237-309.

[12] R. Guindon: Knowledge exploited by experts during software system design. In: Intnl. J. of Man-Machine Studies, Vol. 33, no. 3 (1990), pp. 279-304.

[13] D. E. Lipkie, S. R. Evans, J. K. Newlin and R. L. Weissman: Star graphics: An object-oriented implementation. In: Computer Graphics (SIGGRAPH '82 Proceedings) Vol 16. no 3 (1982). pp. 115-124.

[14] J. Martin: Recommended diagramming standards for analysts and programmers. Englewood Cliffs, NJ: Prentice-Hall, 1987.

[15] M. J. Muller, R. Carr, C. Ashworth, B. Diekmann, C. Wharton, C. Eickstaedt and J. Clonts: Telephone Operators as Knowledge Workers: Consultants Who Meet Customer Needs. In: Proceedings of CHI '96 Human Factors in Computing Systems (April 13-18, 1995, Vancouver BC) ACM/SIGCHI, N.Y., pp. 130-137.

[16] W. M. Newman: A Preliminary Analysis of the Products of HCI Research, Based on Pro Forma Abstracts. In: Proceedings of CHI '94 Human Factors in Computing Systems (April 24-28, 1994, Boston, MA) ACM/SIGCHI, N.Y., pp. 278-284.

[17] W. M. Newman and Lamming M. G. (1995) Interactive System Design. Wokingham: Addison-Wesley.

[18] W. M. Newman: The Place of Interactive Computing in Tomorrow's Computer Science. In: Wand I. and Milner R., eds.: Computing Tomorrow. Cambridge: Cambridge Univ. Press, 1996.

[19] W. M. Newman: Models of Work Practice: Can they Support the Analysis of System Designs? Conference Companion, CHI '96 Human Factors in Computing Systems (April 13-18, 1996, Vancouver BC) ACM/SIGCHI, N.Y., pp. 216.

[20] W. M. Newman, M. A. Eldridge and R. H. R. Harper: Modelling Last-minute Authoring: Does Technology Add Value or Encourage Tinkering? In: Conference Companion, CHI '96 Human Factors in Computing Systems (April 13-18, 1996, Vancouver BC) ACM/SIGCHI, N.Y., pp. 221-222.

[21] H. Petroski: To Engineer is Human: The Role of Failure in Successful Design. New York: St Martin's Press, 1985.

[22] P. G. Polson and C. H. Lewis: Theory-based design for easily learned interfaces. In: Human Computer Interaction., Vol 5 (1990), pp 191-220.

[23] G. F. C. Rogers: The Nature of Engineering: a Philosophy of Technology. London: Macmillan, 1983.

[24] H. A. Simon: The Sciences of the Artificial, Second edition. Cambridge MA: MIT Press, 1981.

[25] D. C. Smith, C. Irby, R. M. Kimball, E. Harslem and H. L. Morgan: The Star User Interface: An Overview. In: Proc. AFIPS National Comp. Conf., Vol. 51 (June 1982), pp. 515-528.

[26] D. C. Smith, C. Irby, R. M. Kimball, W. Verplank and E. Harslem: Designing the Star User Interface. In: BYTE, vol. 7 no. 4 (April 1982).

[27] W. G. Vincenti: What Engineers Know and How They Know It: Analytical Studies from Aeronautical History. Baltimore: Johns Hopkins Univ. Press, 1991.

[28] W. Visser: More or less following a plan during design: opportunistic deviations in specification. In: Intnl. J. of Man-Machine Studies, Vol. 33, no. 3 (1990), pp. 247-304.

[29] T. Winograd, ed.: Bringing Design to Software. Reading MA: Addison-Wesley, 1996.

[30] G. H. Woodmansee: The Visi On Experience—From Concept to Marketplace. In: Human Computer Interaction—INTERACT '84 (Shackel B., ed.), pp. 871-875. Amsterdam: North Holland, 1984.

Address of author

Dr William Newman
Rank Xerox Research Centre
61 Regent Street, Cambridge UK CB2 1AB
Email: newman@cambridge.rxrc.xerox.com

Usability Engineering am Beispiel des Home Electronic System von Siemens und Bosch

Roland Schoeffel

Ergonomie-Berater, Siemens AG, München

Zusammenfassung

Die Einbringung der Ergonomie bei der Softwareentwicklung wird anhand der Entwicklungsphasen des Home Electronic Systems von Siemens und Bosch beispielhaft beschrieben. Die vorgestellten softwareergonomischen Aktivitäten schließen Machbarkeitsstudie, Benchmarking, Review, Akzeptanztests, Usability Test, Requirement Engineering, interdisziplinäre Detailentwicklung und Style Guide ein. Die methodische Problematik der softwareergonomischen Aktivitäten in jeder Phase wird diskutiert und Wünsche an eine zukünftige Entwicklung der Softwareergonomie werden vorgetragen. Insgesamt wird gezeigt, wie die systematische Einbringung der Ergonomie zu einer hohen Produktakzeptanz und zu meßbarer Benutzerfreundlichkeit führt.

1 Einleitung

Dieser Artikel soll zeigen, wie softwareergonomische Aspekte bei einem Produkt der Bosch Siemens Hausgeräte GmbH, einer Tochtergesellschaft von Siemens und Bosch, über die verschiedenen Phasen der Produktentwicklung eingebracht wurden. Einerseits wird dargestellt, welchen Stand die Softwareergonomie bei diesem Projekt hatte, andererseits werden Wünsche an die Weiterentwicklung der Softwareergonomie als Wissenschaft vorgebracht.

Die softwareergonomische Beratung erfolgte dabei durch die Siemens AG. Die Siemens AG hat seit 1936 eine Designabteilung und seit 1955 Mitarbeiter, die sich ausschließlich mit Fragen der Ergonomie befassen. Heute arbeiten verschiedene Abteilungen an unterschiedlichen Teilgebieten der Ergonomie. Die Abteilung für Corporate Design hat 40 Mitarbeiter und entwirft die User Interfaces der Siemens Produkte. In der Zentralabteilung Technik gibt es eine Fachabteilung für Produktergonomie mit 10 Mitarbeitern, die ein Usability Lab betreiben und die Customer Sessions durchführen. Es gibt eine Fachabteilung für die arbeitswissenschaftliche Arbeitsplatzgestaltung mit 10 Mitarbeitern, die u.a. mit CAD-Modellen von Personen neue Arbeitsplätze simulieren, und es gibt eine Abteilung mit 40 Mitarbeitern, die neuartige Mensch-Maschine-Schnittstellen wie z.B. Spracherkennung und Gestik erforschen.

Die Ergonomen sind dabei von ihrer universitären Vorbildung her meist Ingenieure, aber zunehmend auch Psychologen, Biologen, Soziologen und Informatiker. Leider aber gibt es kaum welche, die mehrere relevante Gebiete studiert haben. Wenn sich ein Psychologe Änderungen einer Oberflächengestaltung wünscht, dann sagen ihm die Programmierer manchmal, daß das mit dem eingesetzten Tool nicht geht und das der Style nur ganz bestimmte Gestaltungen erlaubt. Die Nichtprogrammierer unter den Ergonomen werden so leicht durch programmiertechnischen Details ausgehebelt. Ohne Kenntnis dessen, was geht und was nicht geht, haben sie einen schweren Stand.

Es ist deswegen zu begrüßen, daß sich heute auch Informatiker für das Thema Softwareergonomie interessieren. Die haben es nun jedoch in bezug auf die andere Seite der Schnittstelle ähnlich schwer, und das sollte nicht unterschätzt werden. Die Informatiker müssen sich nun erst alles aneignen, was über den Menschen zu wissen ergonomisch wichtig ist - und der

Mensch ist keine einfache Maschine, vor allem Bau und Funktion des menschlichen Zentralprozessors sind außerordentlich komplex.

Um eine gute Schnittstelle zum Menschen zu schaffen, muß Wissen aus der Gestalt- und Wahrnehmungspsychologie, der Denkpsychologie und der Sinnesphysiologie einbezogen werden. Die Begutachtung von Bildschirmoberflächen zeigt, daß ohne Beachtung der Psychologie und Physiologie des Menschen erhebliche Fehler passieren, die Oberflächen schier unbedienbar machen können: Da werden z.B. Farbkontraste nicht beachtet oder die Zeiten für das Sehen, für Aufmerksamkeits- und für Entscheidungsprozesse, bleiben unberücksichtigt. Es ist deswegen begrüßenswert, daß auf dieser Tagung auch Psychologen beteiligt sind.

Rundum erscheint es immer notwendiger, für die Ergonomie und für die Softwareergonomie auch in Deutschland Spezialisten auszubilden, mit allem, was es dafür braucht, so wie es das Ausland tut. So kann man in den USA einen Doktorgrad in Human Factors erwerben, und auch in England, Frankreich, den Niederlanden, in Japan und in Südafrika gibt es Spezialausbildung zu in etwa einem "Diplom-Ergonomen". Hier in Deutschland gibt es dergleichen nicht, und für den Standort Deutschland bedeutet es einen Nachteil, diese Profession nicht bieten und einsetzen zu können.

Der Standort Deutschland braucht speziell ausgebildete Softwareergonomen.

2   Das Home Electronic System von Siemens und Bosch

Das Home Electronic System von Siemens und Bosch beruht auf einem vernetzten Haus, in dem alle Hausgeräte miteinander kommunizieren können. Man spricht in diesem Zusammenhang auch von "Smart Homes" und von "intelligentem Wohnen". Im Zentrum des vernetzten Hauses befindet sich ein Rechner, der die eingehenden Informationen integriert und über den alle angeschlossenen Geräte kontrolliert und geschaltet werden können. Über eine Telefonleitung können dabei auch Außenverbindungen geschaffen werden. Dieser zentrale Rechner ist der sog. "Home Assistant" und wird auch schon mal als "elektronischer Butler" bezeichnet.

Die Vernetzung erfolgt über den sog. "Instabus" der European Installation Bus Association. Etwa 100 Firmen sind Mitglied dieser Vereinigung. Bei diesem Bus sind die Stromleitungen eines Hauses um einen Informationsstrang ergänzt. Die ans Stromnetz angeschlossenen Geräte sind dann auch gleichzeitig am Informationsnetz angeschlossen. Jedes Gerät trägt dann einen Adresscode und ist nun von überall im Netz her schaltbar, insbesondere natürlich vom zentralen Rechner aus.

Der Nutzen für den Verbraucher ist vielfältig. Um das Licht in der Garage auszuschalten muß der Hausbewohner nun nicht mehr unbedingt in die Garage gehen, das kann er nun auch von der Küche aus tun. Beim Verlassen des Hauses kann er auf einem Display sehen, wo überall im Haus das Licht noch brennt, welche Türen und Fenster noch offen und welche Geräte noch eingeschaltet sind. Mit einer Zentralverriegelungsfunktion, ähnlich der eines Kraftfahrzeuges, kann er nun mit einem Tastendruck das ganze Haus in den gewünschten Zustand für die Abwesenheit bringen, eingeschlossen dem Schalten der Heizung auf Frostschutz, dem Einschalten der Alarmanlage und des Anrufbeantworters. Unterwegs kann über das Telefon der Zustand einzelner Geräte abgefragt werden und vor der Rückkehr aus dem Urlaub kann das Haus vorbereitet werden, der Nutzer kann die Heizung aus der Ferne hochfahren. Sollte während der Abwesenheit ein Einbruchsensor ansprechen, so kann der Home Assistant den Besitzer über ein Handy informieren, und wenn das nicht klappt, selbständig einen Nachbarn oder einen

Schutzdienst anrufen. Zu Hause kann der Verbraucher den Home Assistant dazu nutzen, mit einem Tastendruck ganzheitliche Zustände über das Haus zu schalten, wie Nacht- oder Partybetrieb. Man kann am Display des Home Assistant aber auch Bestellungen aufgeben, von einer CD Rezepte lesen und auch gleich direkt aus dem Rezept den Backofen ansteuern, kann am Bildschirm des Home Assistent ein Fernsehprogramm verfolgen oder eine CD hören.

Von besonderem allgemeinen Interesse sind all die Funktionen des Home Assistant, die Wege einsparen und die die Kontrolle verbessern. Der Home Assistant kann so nämlich Senioren helfen, einen Haushalt bis ins hohe Alter selbständig zu führen.

Aus ergonomischer Sicht vereint der Home Assistant alle Bedienoberflächen eines Hauses in sich, weil er es erlaubt, alle angeschlossene Geräte in einem Haus zu kontrollieren. Darüber hinaus bietet er eine Vielzahl von Extrafunktionen, so daß insgesamt in den Oberflächenmasken leicht über 1000 Befehlstasten realisiert sein können. Der Home Assistant beflügelt so nicht nur die Möglichkeiten der Hauskontrolle, er stellt auch das komplexeste Gerät des Hauses dar, und das war von Anfang an eine ergonomische Herausforderung, denn leicht bedienbar mußte er trotzdem sein.

3   Softwareergonomie beim Home Assistant

Die softwareergonomischen Arbeiten am Home Assistant umfaßten verschiedene Schritte, die Tabelle 1 im Überblick zeigt. Die Reihenfolge der Schritte kann durchaus als Leitfaden für den Entstehungsgang vieler anderer softwareergonomischer Projekte gelten, doch es gibt auch Besonderheiten.

Am Anfang der Entwicklung eines neuen Produktes sollte eine softwareergonomische Machbarkeitsanalyse stehen. Nicht nur die technische Machbarkeit, sondern auch die psychophysiologische Machbarkeit sollte untersucht werden. Dies wird immer wichtiger, weil die Komplexität der MM-Schnittstellen weiter zunimmt. Es ist nicht immer selbstverständlich, daß Menschen Schnittstellen auch bedienen können. Wir haben deswegen in unserer Welt auf Bahnhöfen Fahrkartenautomaten, mit denen viele Passagiere nicht klarkommen, und die schlechte Bedienbarkeit der Videorecorder ist sprichwörtlich geworden, woran auch Show View nichts geändert hat. Geräte können auch völlig unbedienbar sein und dann ein Produkt zum Scheitern bringen und zu einem Verlustgeschäft machen. Eine psychophysiologische Machbarkeitsanalyse gehört deswegen an den Anfang jedes komplexeren Schnittstellenentwurfs.

Beim Benchmarking werden schon vorhandene ähnliche und verwandte Bedienschnittstellen analysiert und gute und negative Details werden registriert. Wenn die Machbarkeitsanalyse negativ ausfällt, fehlen meist auch Produkte zum Benchmarking; gibt es Produkte, kann eine Machbarkeitsanalyse wegfallen und man kann sie über das Benchmarking erledigen. Das Benchmarking kann aber auch bei positiver Machbarkeitsanalyse Einzelaspekte untersuchen.

Sobald ein erster Prototyp geschaffen ist, sollte er einem Review durch einen Experten für Softwareergonomie unterzogen werden. Man kann statt dessen auch einen Usability Test durchführen. Die Einbeziehung des Experten ist aber meist das ökonomischere Verfahren und resultiert in nützlicheren Verbesserungsvorschlägen. Wird statt dessen ein iteratives Rapid Prototyping durchgeführt, so können prospektive Nutzer schon bei der Entwicklung einbezogen werden. Das geht allerdings nur, wo die Nutzer schon eine Vorstellung von dem Produkt haben. Bei ganz neuen Produkten geben naive Nutzer im allgemeinen keine nützlichen Hinweise.

Ist der Prototyp voll funktionsfähig, kann geprüft werden, ob die Nutzer aus der Zielgruppe mit der Bedienoberfläche zurecht kommen und ob sie das Produkt akzeptieren würden. Dies kann über Messestudien bei Ausstellungen oder über einen Usability Test erfolgen. Usability Tests haben dabei den Vorteil, die Bedienbarkeit der ganzen Auslegung systematisch transparent zu machen, während die Besucher bei Messestudien sich durchweg nur an der Oberfläche eines Produktes aufhalten.

Mit dem Requirement Engineering sollen die Anforderungen an ein Produkt aus Kundensicht, aus technischer und aus ergonomischer Sicht gefunden und festgelegt werden. Das Requirement Engineering gehört deswegen ganz an den Anfang. Es ist allerdings eine aufwendige Aktivität und deswegen in der Startphase eines Projekte nicht immer finanzierbar. Wird es jedoch erst in einer späten Phase durchgeführt, kann es erhebliche Veränderungen an einem Produkt bewirken, die dann sehr teuer werden können.Die interdisziplinäre Detailentwicklung stellt eine Produktdefinition am runden Tisch dar, bei der Techniker und Programmier, Designer und Ergonomen, Marketing- und Vertriebsspezialisten zusammen die Auslegung bestimmen. Früher wurde dieser Prozeß sequentiell durchgeführt. Der interdisziplinäre Entwicklungsprozeß stellt hohe Anforderungen, spart aber Zeit und Kosten und führt letztlich zu besseren Produkten.

ENTWICKLUNGSPHASE		ZEIT	SOFTWAREERGONOMISCHE AKTIVITÄT
Konzept-Phase	1.	Ende 92	Machbarkeitsanalyse
	2.	Feb 93	Benchmarking
Prototyp-Phase	3.	Dez 94	Review der Bedienoberfläche
	4.	Feb 95	Akzeptanztests
	5.	Jun 95	Usability Test
	6.	Dez 95	Requirement Engineering
Definitions-Phase	7.	96	Interdisziplinäre Detailentwicklung
	8.	97	Styleguide

Tab. 1: Softwareergonomische Aktivitäten bei der Entwicklung des Home Assistant.

Die letzte softwareergonomische Aktivität einer Produktentwicklung ist der Styleguide. In einem Styleguide werden die besonderen Gestaltungsmerkmale, aber auch die softwareergonomischen wichtigen Details, vereinheitlicht. Ein Styleguide gehört ans Ende einer Entwicklung, wenn weitere Entwicklungen geplant sind und auch andere Entwicklungsteams Anwendungen für eine Softwareumgebung schreiben sollen.

Abgesehen vom späten Requirement Engineering kann also der Entwicklungsgang beim Home Electronic System als beispielhaft für andere Entwicklungen gelten.

Abschließend sei an dieser Stelle auf eine besondere Problematik aufmerksam gemacht. Größere Softwareprojekte für den öffentlichen Dienst werden typischerweise nach bestimmten Vorgehensmodellen entwickelt, so nach dem sog. V-Modell [1]. Das softwareergonomische Problem beim V-Modell besteht darin, daß es der Softwareergonomie keinen Platz einräumt und so die Entwicklung "ergonomiefreier" Oberflächen festschreibt. Im V-Modell wird bis hin zum Spezialisten für die Dokumentation alles genau beschrieben, aber der Softwareergonomie wird keine Zeile eingeräumt. Wenn eine Entwicklung nach diesem Modell vereinbart wird, dann ist es seitens der Industrie sehr schwer, Softwareergonomie in ein Projekt einzubringen, denn es gibt keine Finanzierung für Aktivitäten, die nicht über das V-Modell vereinbart sind. Im Sinne einer allgemeinen Brauchbarkeit der Produkte wäre eine Überarbeitung des V-Modells oder eine Erarbeitung von Vorgehensmodellen mit softwareergonomischen Aktivitäten eine vordringliche Aufgabe.

Die vorhandenen Verfahrensmodelle müssen um die Aspekte der Softwareergonomie ergänzt werden.

3.1 Machbarkeitsanalyse: Konzept für eine Dialogschnittstelle des Home Assistant

Schon ganz zu Anfang der Entwicklung des Home Assistant wurde die Siemens-Fachabteilung für Produktergonomie einbezogen. Angesichts der Komplexität der Software ergab sich schon vor dem eigentlichen Projektbeginn die Frage, ob man einfachen Nutzern die Komplexität eines Home Assistant zumuten könnte. Das nächste Problem war die Eingabetechnologie. Man dachte daran, den Home Assistant in der Küche anzubringen, aber die Brauchbarkeit einer Tastatur und die Akzeptanz einer Maus in der Küche erschienen zweifelhaft.

Die Einbeziehung der Softwareergonomie erfolgte 1992, und Touchscreens schienen damals eine mögliche Lösung für den Anbringungsort zu sein. Touchscreens kamen damals gerade auf den Markt und man mußte sie in den Städten noch suchen. Es gab keine softwareergonomischen Erfahrungen und nur wenige erste Artikel zum Thema. Die Beobachtung von Personen beim Hantieren mit diesen Touchscreens waren nicht sonderlich vertrauenerweckend. Das Drücken der Tasten funktionierte nicht immer und die Leute verloren schon bei geringen Menütiefen die Übersicht und Orientierung.

Bei alledem waren die ersten Fragen der Machbarkeitsanalyse trotzdem zunächst technisch orientiert. Wie waren die Auflösung der Touchflächen beschaffen und wie korrelierten diese mit der sichtbaren Softwareoberfläche? Welche Flexibilitäten und Einschränkungen bestanden? Man mußte damals die Hersteller der Touchscreens selbst aufsuchen, um über die Details der Touchscreens Erkenntnisse zu gewinnen. Die vier damals gängigen Arten von Touchscreens wurden damals ermittelt, verglichen, und ihre Vor- und Nachteile wurden aufgelistet [9].

Die Machbarkeitsanalyse führte dann zu Studien, wie die Bedienoberflächen von Touchscreens aussehen müßten. Es wurden Untersuchungen dazu durchgeführt, wie groß die Auflagefläche einer Fingerkuppe auf dem Bildschirm ist und wie groß die Sensorflächen mindestens sein müssen. Diese Erkenntnisse sind später in die ISO 9241 eingeflossen. Die besonderen Probleme der Parallaxe und der Berührzeiten wurden ermittelt. Die Möglichkeit verschiedener Dialogarten wurden spezifiziert. Am Ende der Untersuchungen wurde seitens der Ergonomie ein Freizeichen für den Dialog per Touchscreen gegeben. Touchscreens waren zwar neu, aber sie schienen durchaus als seriöse Dialogschnittstellen realisierbar. Es wurde eine interne Richtlinie für die Gestaltung von Touchscreenoberflächen erstellt [7].

Die softwareergonomischen Machbarkeitsanalysen zu Beginn eines Projektes sind eine Aktivität von nicht zu unterschätzender Bedeutung. Liegt es schließlich an der Bedienbarkeit eines Produktes, daß sich dieses nicht vermarkten läßt, so trägt die Ergonomie die Verantwortung für die Kosten einer fehlgeschlagenen Produktentwicklung. Und tatsächlich gibt es Softwareprodukte, die sich wegen unzureichender Bedienbarkeit nicht lange am Markt gehalten haben.

Die Machbarkeitsanalysen haben häufig ihren Gegenstand im Hardwarebereich, aber auch softwareergonomische Fragen treten auf. Die Fragen zum Hardwarebereich werden häufig von den Einsatzbedingungen einer Applikation diktiert, so wenn Bildschirmanzeigen und Eingabemittel unter schwierigen Beleuchtungs-, Vibrations- oder Klimabedingungen eingesetzt werden müssen. Neben der Bewertung und Auswahl von Alternativen kommt es hier darauf an, den technischen Möglichkeiten die menschlichen Fähigkeiten und Belastungsgrenzen gegenüberzustellen. Welche Auflicht- und Kontrastbedingungen sind noch zumutbar und ab wann werden Streßerscheinungen wie wahrscheinlich?

Softwareergonomen brauchen gründliche Kenntnisse der menschlichen Physiologie.

Bei den softwareergonomischen Machbarkeitsanalysen geht es darum, ob Menschen von ihren Fähigkeiten her bestimmte Dialoge ausführen könnten und wie die Dialogstrukturen aussehen müßten, um eine effiziente und streßfreie Mensch-Computer-Kommunikation zu ermöglichen. Hier finden sich bei den heutigen Applikationen noch sehr viele Fehler. Die bekannten und zum Teil in der ergonomischen Normung auch veröffentlichten Daten werden nicht immer berücksichtigt. Fehleinschätzungen der menschlichen Leistungsfähigkeit können Produkte völlig unbrauchbar machen.

Die Krönung bei der Gestaltung der Mensch-Rechner-Kommunikation ist es sicher, die Aufgabenverteilung zu gestalten. Dies ist die Stelle, an der durch Mittel der künstlichen Intelligenz die Beziehung auch quasimenschlich gestaltet werden kann. Schon gibt es ja telefonische Auskunftssysteme für Zugfahrpläne. Den KI-Interessierten wird hier das Programm ELIZA von Weizenbaum [12] ein Begriff sein, das einen ersten Versuch darstellte, einen quasimenschlichen Dialog rechnergestützt zu realisieren. Sollte der Home Assistant also vielleicht wie ein elektronischer Butler mit künstlicher Sprache versehen werden und vielleicht mit einer synthetischen Stimme sprechen? Sollte sich vielleicht auf dem Display ein Gesicht zeigen, daß durch Morphingtechniken Stimmungen ausdrücken kann? Für den Home Assistant ergab sich hier auch die Frage, inwiefern das System aktiv in das Geschehen eingreifen sollte. Sollte der Home Assistant die Bewohner aktiv beobachten? Eine George-Orwell-Vision sollte sicher nicht realisiert werden. Dann aber wiederum erschien es durchaus wünschenswert, daß das System bei einem Feuer im Haus selbständig aktiv wird.

Von der Konzeption her sind auf der Seite der Sensorik Videokameras, Mikrofone, Verschlußsensoren, Rauchsensoren, Strommeßgeräte und viele andere Sensoren prinzipiell anschließbar, so daß der Home Assistant mit mehr Sinnen als ein Mensch ausgestattet werden kann. Auch in bezug auf das, was der Home Assistant betätigen kann, gibt es prinzipiell kaum Grenzen. Schwierig wird es jedoch bei der Erkennung der Wünsche des Nutzers [8].

Die Überlegungen zum Home Assistant gingen dahin, dieses System ein "Gerät" sein zu lassen und daraus keinen "elektronischen Butler" zu machen. Der Home Assistant erhielt deswegen auch eine technische Bedienoberfläche, und es zeigt sich kein künstliches Gesicht. Das System sollte aber nicht ganz passiv sein, sondern im Hintergrund bereitstehen und definierte Funktionen wahrnehmen. Es sollte ein stiller aber hocheffizienter elektronischer Partner sein. In diesem Sinne ist das Gerät nun so ausgelegt, daß nur bestimmte detektierte Gefahren wie Feuer, Einbruch, Notfall zu Durchbruchmeldungen führen, das Gerät aber ansonsten in einem wachsamen Ruhezustand verbleibt. Für den Notfall ist das System so ausgelegt, daß es erst versucht, die Bewohner im Haus zu alarmieren, indem es über das Display optische und akustische Meldungen ausgibt und auf laufenden Fernsehern Meldungen generiert, bevor es dann schließlich externe Nummern anzurufen beginnt. Der Nutzer kann aber jederzeit sehr mächtige Funktionen anstoßen wie Zentralverriegelung und Szenen, die eine Vielzahl von Geräten gleichzeitig schalten. Er kann das Gerät auch bei bestimmten Bedingungen Szenen anstoßen lassen und kann so selbst Verhaltensregeln an das System weitergeben. Nötigenfalls kann der Home Assistant einen Arzt oder die Feuerwehr informieren. Das System bietet hier sogar den Vorteil, nicht unter Streß zu vergessen, bei einer Notfallmeldung die Adresse mit Name, Straße und Hausnummer genau anzugeben.

3.2 Benchmarking von Dialogvarianten: 4-Ebenen-Modell der Kommunikation

Nach der Entscheidung für die Machbarkeit aus technischer und ergonomischer Sicht begann der Entwurf konkreter Bedienoberflächen. Angesichts der großen Zahl der Bedienfunktionen und der manchmal schwierigen Bedienung einzelner Geräte im Haus ergab sich die Frage, ob man die Bedienoberflächen der Hausgeräte 1:1 auf dem Bildschirm abbilden sollte oder ob man die Bedienkonzepte auch verbessern dürfte oder sogar verbessern sollte. Aus softwareergonomischer Sicht sprach die Forderung nach Konsistenz für die Übernahme 1:1, andererseits aber sind wohl Tasten auf Touchscreens 1:1 übernehmbar, nicht aber Drehknöpfe. Diese Problematik führte zu der allgemeineren Frage, wie gut die Ergonomie der heutigen Bedienblenden von Hausgeräten allgemein ist. Es wurde deswegen zunächst ein Benchmarking der Dialogvarianten von Hausgeräten in Angriff genommen. Es resultierte ein Modell der Mensch-Maschine-Kommunikation für Hausgeräte, das aber auch softwareergonomisch interessant ist, wenn man in Betracht zieht, daß sich die Softwareergonomie bei den aufkommenden Virtual Realities auch den Fragen der 3-dimensionalen Gestaltung stellen müssen wird.

Das Benchmarking zu den Dialogkonzepten bei Hausgeräten fand auf der Domotechnica 1993 in Köln statt. Insbesondere Herde und Mikrowellen, Geschirrspüler, Kühlschränke, Waschmaschinen und Trockner wurden einbezogen. Interessanterweise fanden wir auf dieser Messe auch schon einen Einbauherd mit einem Display, der mit einer Zahlentastatur zu bedienen war. Die Beobachtung der Besucher vor diesem Display ließ allerdings manchen Zweifel an der Eignung von Displays für Hausgeräte aufkommen.

<table>
<tr><td rowspan="5" style="writing-mode:vertical-lr;">Ebenen der Mensch-Maschine-Kommunikation</td><td></td><td>PSYCHOPHYSIOLOGISCHE FAKTOREN</td><td>GESTALTUNGSDETAILS</td><td>TYPISCHE FEHLER</td></tr>
<tr><td>Ebene der Aufgabenverteilung Mensch-Maschine</td><td>• Stärken und Schwächen des Menschen
• Wünsche des Menschen
• Organisatorische Anforderungen</td><td>• Automatisierungs-konzept
• Optionskonzept</td><td>• Technikorientierte Gestaltung
• Überflüssige Optionen</td></tr>
<tr><td>Kognitiv-mentale Regulationsebene</td><td>• Gedächtnis
• Wissen und Kenntnisse
• Intellektes Vermögen und Denkschemata</td><td>• Bedienkonzept</td><td>• Fälschliche Vorausset-zung von bestimmten Vorkenntnissen und Fähigkeiten</td></tr>
<tr><td>Perzeptive Ebene</td><td>• Visuelle Fähigkeiten
• Auditive Fähigkeiten
• Taktile Fähigkeiten
• Unterscheidungs-vermögen
• Erkennungsvermögen</td><td>• Kodierung, Farbe u. Form von Schriften, Piktogrammen und Anzeigen
• Lautstärke und Modula-tion von Signaltönen
• Taktile Rückmeldung</td><td>• Unverständliche Informationskodierung
• Optische Irreführung
• Nichtbeachtung von Sehräumen und Randbedingungen</td></tr>
<tr><td>Manipulative Ebene</td><td>• Anthropometrische Gestalt des Menschen
• Greifräume
• Bewegungsvermögen
• Kräfte</td><td>• Auswahl, Formgebung, Positionierung und Anordnung von Stellteilen</td><td>• Kompatibilitätsfehler
• Schwer zugängliche Schaltelemente
• Unangemessene Ansprechcharakteristik</td></tr>
</table>

Tab. 2: 4-Ebenen-Modell der Mensch-Maschine-Kommunikation.

Aus der Beurteilung der Geräte und der Beobachtung von Besuchern bei der Handhabung der Geräte entwickelte sich ein Bewertungsmodell über 4 Ebenen. Tabelle 2 zeigt dieses Modell und differenziert die Besonderheiten. Das 4-Ebenen-Modell hat inhaltlich Ähnlichkeit mit dem GOMS-Modell [2], weist jedoch keine zeitliche Sequenzierung auf sondern eine hierachische Staffelung. Mit dieser wird einer Besonderheit der Bedienkonzepten von Hausgeräten Rechnung getragen, nämlich der hierachischen Abhängigkeit der Ebenen.

Wir fanden, daß als oberste Ebene der Gestaltung die Aufgabenverteilung zwischen Mensch und Gerät zu sehen wäre und daß hier über die optionalen Funktionen eines Gerätes nachgedacht werden müßte. Die Funktionen müßten am Bedarf der Zielgruppe orientiert sein, und die Aufgabenverteilung müßte die Stärken und Schwächen des Menschen im Vergleich zum Gerät berücksichtigen. Die Festlegung der Funktionen sollte sich dabei an den Wünschen des Menschen orientieren und nicht an technischen Erfordernissen. Am besten erschienen Geräte, die alle nötigen und technischen Einstellungen automatisch vollziehen und die Einstellungen orientiert am Geschmack des Gebrauchers ermöglichen. Spiegelreflexkameras sind hier ein gutes Beispiel gelungener Automatisierung: Die Entfernungseinstellung und Belichtung erfolgt automatisch besser und schneller als ein Mensch das könnte, aber der Fotograf kann nach seinem Geschmack einstellen, ob er lieber die Tiefenschärfe oder den Nahbereich betont hätte. Der Fotograf wird so von einem Kämpfer mit der Kameratechnik zu einem Komponisten von Bildern.

Je nachdem, welche Funktionen und Optionen resultieren, bleiben mehr oder weniger viele Einzelfunktionen übrig, für die auf der kognitiv-mentalen Regulationsebene ein Bedienkonzept zu entwickeln ist. Vollautomatische Geräte würden nur eine Ein/Aus-Funktion nötig machen.

Derzeit bleiben aber in der Regel mehr oder weniger viele Einzelfunktionen übrig. Diese Einzelfunktionen sind über das Bedienkonzept in einen schlüssigen, selbsterklärungsfähigen und benutzerfreundlichen Ablauf zu bringen. Und hier kann man ganz unterschiedliche Wege einschlagen. Man kann Funktionen separat Drehknebeln zuordnen, sie auf einem einzigen Drehknebel integrieren, alles mit Tasten machen oder per Bildschirm ein Menüsteuerung vorsehen. Diese unterschiedlichen Bedienkonzepte erweisen sich als sehr unterschiedlich benutzerfreundlich. Auf dieser Ebene sind insbesondere die kognitive Leistungsfähigkeit der Zielgruppe zu berücksichtigen, ihre Vorerwartungen, die Vorkenntnisse, die Gedächtnisspanne, die Denkweise und die Fehlermöglichkeiten. Hier ist insbesondere der Psychologe als Berater gefragt.

Das auf der kognitiv-mentalen Regulationsebene entwickelte Bedienkonzept bedingt die Ebene der Perzeption und die Ebene der Manipulation. Je nach Bedienkonzept sind hier ganz unterschiedliche Gestaltungen vorzunehmen. Auf der Ebene der Perzeption geht es darum, das Bedienkonzept als ganzes und alle Einzelfunktionen selbsterklärend darzustellen. Die Beschriftungen müssen groß genug sein und die verwendeten Symbole müssen verständlich sein. Bei einigen Geräten auf dem Markt ist nicht einmal erkennbar, wo man sie einschalten kann. Auf der Manipulationsebene kommt es darauf an, das richtige Stellteil auszuwählen und eine angemessene Auslegung vorzunehmen. Die Kräfte von Stellteilen dürfen nicht zu hoch sein und die Ansprechcharakteristik muß den Erfordernissen entsprechen. Die perzeptiven Fähigkeiten des Menschen, die Kräfte und die Bewegungsmöglichkeiten seiner Hände müssen einbezogen werden.

Nun kann man noch die sog. "Barrierefreiheit" einbeziehen. Diese stellt eine Ausweitung aller Ebenen um die Belange der Älteren dar, und wir beziehen diesen Gesichtspunkt seit neuem in unsere Betrachtungen ein. Der Anteil der Älteren an der Bevölkerung wird sich in der Zukunft nämlich dramatisch vergrößern, und so muß ihren Bedürfnissen verstärkt Rechnung getragen werden. Die Lebenserwartung wird noch steigen, nur 20% der Älteren werden pflegebürftig sein, und die übrigen werden am Leben intensiv teilnehmen und werden auch Rechner benutzen können. Während nun aber für die arbeitende Bevölkerung ergonomische Werte ausreichend vorliegen, fehlen ergonomische Werte überall noch für die Älteren und werden erst in einigen Jahren ausreichend vorhanden sein. Wir unterscheiden den Gesichtspunkt der Barrierefreiheit deswegen derzeit noch bewußt, weil die Umsetzung derzeit ungleich viel mehr Aufwand bedeutet. Benutzerfreundliche Geräte entlang der vier Ebenen werden wir bald haben, aber auf barrierefreie Geräte werden wir noch einige Zeit warten müssen.

Die Funktionsabnahme beim Älterwerden ist erheblich und muß berücksichtigt werden. Bei Älteren verändert sich z.B. der Nahpunkt des Sehens, also die Entfernung, die etwas für das Ablesen mindestens haben muß, von 50 cm bei 50-jährigen auf 1 m bei 60-jährigen. Derzeit sind noch alle Bildschirmarbeitsplätze auf 50 cm Sehentfernung ausgelegt, und hier wird sich sehr viel ändern müssen, viele Normen und Standards werden umgeschrieben werden müssen. Was dabei genau Ältere sind, bestimmt sich für den einzelnen weniger durch sein Lebensalter als durch seinen individuellen Stand der Funktionsabnahme.

Zurück zum Beispiel Home Assistant. Anhand des dargestellten Bewertungsmodells wurden Bildschirmseiten entworfen, bei denen für jede Funktion im einzelnen entschieden wurde, ob sie 1:1 konsistent übernommen werden sollte oder ob eine alternative Gestaltung benutzerfreundlicher wäre.

3.3 Review der Bedienoberfläche des Home Assistant: Erstes Design und Expert Ratings

Hat man einen ersten Papierentwurf der Bedienoberfläche, so können softwareergonomisch schon die gröbsten Fehler bereinigt werden. Inkonsistenzen, zu kleine Buchstabengrößen, unzureichende Farbkontraste, unverständliche Beschriftungen und Symbole sind schon erkennbar. Die Dialogführung ist auf Papiervorlagen natürlich schwerer nachvollziehbar, aber bei vollständigen Entwürfen kann der Dialograum durchaus beurteilt werden. Es fehlen jedoch die wichtigen technisch bedingten Reaktions- und Rückmeldezeiten. Die Informationsverarbeitungszeiten, die ein Nutzer braucht, um sich auf einer Seite zu orientieren, sind demgegenüber schon erkennbar, wenn man Nutzern Papierentwürfe vorlegt. Es hat sich so insgesamt als sinnvoll erwiesen, schon in der Entwurfsphase papierene Vorlagen softwareergonomisch durchzuprüfen.

Moderne *Prototyping Tools* haben es zwar sehr leicht gemacht, Ideen schnell umzusetzen, beim Einsatz dieser Tools ist jedoch zu bedenken, daß sie binden und zu bestimmten Lösungen verleiten, während Papierentwürfe ein völlig freies Gestalten der Oberfläche erlauben. Die Toolbindung kann im Einzelfall durchaus zu softwareergonomisch unguten Resultaten führen. So zwingen die Prototyping Tools für Motif sehr kleine Buttons auf, die für schnelles Bedienen ungeeignet sind. Softwareergonomisch verlangen schnelle Bedienbewegungen nach größere Schaltflächen, wie Fitts' Law [3] verdeutlicht. Bei der Gestaltung von Fluglotsenarbeitsplätzen z.B. sind solche Details softwareergonomisch durchaus wichtig.

Softwareergonomisch optimierte Prototyping Tools sind aber kaum zu finden. Tatsächlich orientieren sich die heutigen Tools primär an Zielbetriebssystemen und an Büroerfordernissen. Es fehlen Tools, die softwareergonomsch gute Lösungen für zeitkritische Anwendungen zu erstellen helfen. Dazu kommt, das die Bedienbarkeit der Tools selbst häufig zu wünschen übrig läßt.

Es fehlen gut bedienbare Prototyping Tools für zeitkritische Anwendungen, die softwarergonomisch gute Lösungen leicht machen.

Sollen Applikationen allerdings einem bestimmten Style, wie z.B. Windows, folgen, können Prototyping Tools sehr nützlich sein. Während nämlich Papiervorlagen mit großer Exaktheit erst aufwendig entwickelt werden müssen, kann ein Prototyping Tool heute schon der schnellere Weg zu einer brauchbaren Darstellung sein, und Doppelarbeit kann eingespart werden.

Manuelle Papierlayouts bieten aber weiterhin bei der Einbeziehung von Kundenwünschen Vorteile. Bei der kundenorientierten Entwicklung von Software ist es zweckmäßig, in *Customer Sessions* die Vorstellungen über die Inhalte und Bedienelemente von Bildschirmseiten auf einen Nenner zu bringen. Bei solchen Customer Sessions moderiert ein Softwareergonom eine Besprechung mit Entwicklern und Kunden. Es werden Bildschirmseiten auf Papier oder Folie entworfen und im Detail durchgesprochen. Wenngleich das Prototyping heute schon sehr schnell ist, so ist es erfahrungsgemäß doch noch nicht schnell genug, um dem schnellen Gedankenwechsel der Teilnehmer unmittelbar folgen zu können. Das Entwerfen und Verändern von manuellen Zeichnungen ist hier unerreicht. Oft sind es auch Datendarstellungen und -aufbereitungen besonderer Art, auf die es ankommt, und die zu programmieren es länger braucht. Zeichnungen und Skizzen lassen sich hier schnell adaptieren. Ein gewisses graphisches Talent ist nützlich und auch das Einbeziehen eines Graphikers empfiehlt sich an dieser Stelle.

Sehr wichtig ist die interdisziplinäre Zusammensetzung der Gruppe aus Kunden, Software-entwickler, Grafiker und Softwareergonom. Nur so kann allen Anforderungen sofort und unmittelbar entsprochen werden und es sind Entwürfe erzielbar, die allen Anforderungen gerecht werden. Der Kunde bringt dabei seine Wünsche ein, der Softwareentwickler beurteilt die Programmierbarkeit und die Kosten, der Grafiker setzt erste Ideen visuell um, und der Softwa-reergonom geht auf die aus seiner Sicht vorhandenen Notwendigkeiten ein, um die Benutzer-freundlichkeit sicherzustellen. In der Praxis erhalten so abgestimmte Entwürfe eine hohe Wer-tigkeit. Jeder Teilnehmer solch einer Customer Session kann am Ende eine Kopie der abge-stimmten Layouts mitnehmen. Diese Rohentwürfe können dann in ersten Oberflächen umge-setzt werden, wobei dann schon die Zielsprache für die Oberfläche verwendet werden kann. So abgestimmte Entwürfe sind dann meist im Zuge der weiteren Entwicklung nur noch geringfü-gig zu ändern, die Kundenkritik bleibt meist ganz aus und es zeigt sich eher Freude über die Umsetzung.

Beim Home Assistant wurde die erste Oberfläche nach einer Spezifikation durch Entwickler von einem Grafikbüro entwickelt. Kunden und Softwareergonomen waren damals noch nicht gleich mit einbezogen. Erst vor dem ersten Messeauftritt des Home Assistant wurde dann ein softwareergonomische Review durchgeführt.

Das Review wurde anhand von sowohl Papierlayouts als auch anhand von Programmsegmen-ten durchgeführt [10]. Die Papierlayouts waren nötig, weil noch nicht alle Programmsegmente fertig programmiert waren. Das Review wurde tätigkeitsanalytisch durchgeführt und an den vorgegebenen Zielgruppen orientiert. Für den Home Assistant waren die Zielgruppen sowohl DINKS (Double Income No Kids) als auch ältere Personen, und junge Familien. Bei dieser Vorgehensweise kommt es darauf an, die Bedürfnisse und psychischen Fähigkeiten der Nutzer auf das Vorgehen bei der Aufgabendurchführung zu projizieren und wahrscheinliche Probleme bei der Erstbegegnung aufzudecken. Es ist dabei von Vorteil, wenn der Experte das System selbst noch nicht zu gut kennt. Er muß aber trotzdem darauf achten, seine Interessen nicht mit denen der prospektiven Nutzer zu verwechseln und darf seine Fähigkeiten auch nicht mit de-nen der Zielgruppen gleichsetzen. Sowas ist natürlich insbesondere schwierig, wenn der Ergo-nom einen Expertenarbeitsplatz zu beurteilen hat. Hier muß er versuchen, die relevanten menschlichen Grenzwerte der Informationsverarbeitung heranzuziehen. Dabei kann er sich an Nervenleitgeschwindigkeiten, an entscheidungstheoretischen Grundregeln oder an informati-onstheoretischen Modellen orientieren.

Beim tätigkeitsanalytischen Vorgehen werden Aufgaben angesetzt wie sie bei der Nutzung der Software auftreten. Diese Aufgaben werden dann theoretisch oder praktisch ausgeführt, und der Experte projiziert Bedürfnisse und Nutzerfähigkeiten auf jeden Ausführungsschritt. Es ist dabei durchaus nützlich, jeden Schritt im Sinne eines Kommunikationsmodells, wie dem GOMS-Modell [2] oder dem Modell von Herzceg [3], zu sehen. Es ist bei der Prüfung dann zu hinterfragen, wie der Nutzer mit einer Aufgabe im Sinn (*"Goal" bei GOMS*) vorgehen wird (*welche GOMS-Operationen er ausführen wird*). Auch die Mittel (*"Methods" bei GOMS*) und die Einzelbewegungen (*"Selection Rules" bei GOMS*) können dann hinterfragt werden. Dieses Vorgehen deckt bei einem genügend erfahrenem Softwareergonomen 80% bis 90% der vor-handenen Probleme auf [5].

Das tätigkeitsanalytische Review entspricht der Vorgehensweise bei der Prüfung und Bewertung der Benutzerfreundlichkeit von Kommunikationsendgeräten und -systemen nach der VDE-ITG-Richtlinie [13]. Nach dieser Richtlinie werden für ein Gerät von drei Ergonomieexperten Aufgaben formuliert, über die ein Gerät dann geprüft wird. Bei beiden Vorgehensweisen können und sollten dabei auch die Kriterien der ISO 9241 einbezogen werden. Eine Checklist-Vorgehensweise nach der ISO 9241 ist demgegenüber in der Industrie derzeit aber nur probeweise praktiziert worden. Diese Vorgehensweise erwies sich als relativ zeitaufwendig. Beide Richtlinien haben industriell prinzipiell Bedeutung, es gibt aber bislang nur wenige, exemplarisch geprüfte Applikationen.

Die tätigkeitsanalytische Vorgehensweise kann dabei Ergebnisse produzieren, die durchaus im Widerspruch zu Kriterien der ISO 9241 stehen. So fanden wir wiederholt, daß die Forderung nach Konsistenz oft nur ästhetischen Wert hatte und die Benutzbarkeit bei inkonsistenten Dialogen nicht litt.

3.4 Akzeptanztests: Beobachtungen zur Benutzerfreundlichkeit

Bei einem System, das so neuartig wie das Home Electronic System ist, stellt sich natürlich industriell die Frage, ob bei den Kunden genügend Interesse an solch einem Produkt besteht. Zwar gibt es ähnliche Produktvorhaben in den USA und in Japan, und auch die Europäische Gemeinschaft ist an dem Thema Smart Homes sehr interessiert, das alles heißt aber noch nicht, daß ein Home Assistant Käufer finden würde. So wurde das System auf der Domotechnica 95 öffentlich vorgestellt. Diese Vorstellung bot auch softwareergonomisch eine Gelegenheit, Besucher beim Umgang mit dem Produkt zu beobachten und sie zur Benutzerfreundlichkeit zu befragen.

Wichtig ist, das Thema Akzeptanz deutlich vom Thema Benutzerfreundlichkeit zu unterscheiden. Bei der Akzeptanz eines Produktes geht es um die Akzeptanz der Funktionen und Optionen sowie des äußeren Designs und des dafür zu zahlenden Preises. Marktforschungsinstitute sind spezialisiert darauf, vergleichende Befragungen durchzuführen und den Herstellern zu sagen, welches Produkt eine Chance hätte oder was an einem Produkt verändert werden müßte. Sie bedienen sich dabei repräsentativer Befragungen, oft in verschiedenen Landesregionen, befragen prospektive Kunden zu Hause oder in Testlabors in Städten, und sie analysieren die Ergebnisse mit speziellen statistischen Verfahren.

Die Benutzerfreundlichkeit eines Produktes stellt einen separat zu bewertenden Faktor dar, wenngleich nicht ganz unabhängig davon. Die Benutzerfreundlichkeit eines Produktes beeinflußt natürlich die Akzeptanz. Untersuchungen im Bereich der Consumer Electronics belegen sogar, daß für den heutigen Verbraucher die Benutzbarkeit eines Produktes gleich nach dem Preis kommt. Es gibt allerdings viele Produkte, deren hohe Akzeptanz zu breitem Verkauf führte, gleichwohl ihre Benutzerfreundlichkeit nicht sonderlich gut war. Die Videorecorder mit ihrer legendär schwierigen Bedienung sind hierfür ein Beispiel.

Die Benutzerfreundlichkeit eines Produktes ist mit Akzeptanzfragen nicht sonderlich gut bewertbar. Zu unterschiedlich sind die Ausgangspositionen, Interessen und Nutzungstiefen der Gebraucher. Zwar gehen einige Testhäuser so vor, daß sie ihren Versuchspersonen ein Produkt zur Betrachtung überlassen und sie hinterher um ein Gesamturteil bitten, doch solche Urteile sind zwangsläufig sehr grob und nur eingeschränkt nutzbar.

So war es auch mit den Ergebnissen der systematischen Befragung zur Benutzerfreundlichkeit des Home Assistant. Die 100 Befragten kreuzten auf einer Karte mit 20 Kategorien an, daß ihnen besonders die hohe Benutzerfreundlichkeit des Produktes gefiel. Die parallel laufenden Beobachtungen zeigten derweilen aber, daß es sehr wohl an einigen Stellen Bedienschwierigkeiten gab.

Auch wenn noch keine voll funktionierenden Modelle gezeigt werden können, sind erste Aussagen zur Bedienbarkeit möglich. Soweit Akzeptanzuntersuchungen durchgeführt werden, ist es sicherlich zweckmäßig auch Fragen der Softwareergonomie mit aufzunehmen. Die Ergebnisse aus Akzeptanzuntersuchungen für die Benutzerfreundlichkeit müssen jedoch äußerst vorsichtig interpretiert werden.

3.5 Usability Testing

Nachdem die Benutzerfreundlichkeit des Home Assistant gelobt wurde, die eigenen Beobachtungen aber Schwachstellen zeigten, wurde das System einem systematischen Usability Test unter Laborbedingungen unterzogen.

Usability Tests sind nötig, um ein qualifiziertes Urteil zur Bedienbarkeit eines Produktes zu erhalten. Diese Tests werden sowohl bei Hardware- als auch bei Softwareprodukten eingesetzt. Tatsächlich sind solche Tests seit den Anfängen der Ergonomie vor 50 Jahren bekannt. Bei der Bundeswehr hießen sie "Gebrauchstauglichkeitsuntersuchungen" und "Truppenversuche". Erst seit neuem werden diese Tests als Usability Tests nun auch zur Bewertung von kommerziellen Produkten eingesetzt.

Bei einem Usability-Test von Software wird eine Versuchspersonen an einem Rechner bei der Durchführung von Aufgaben systematisch beobachtet. Der Test ermöglicht sowohl ein heuristisch-qualitatives Auffinden der groben Fehler einer Bedienoberfläche, als auch quantitative Angaben zur Bedienbarkeit, zur Bediengeschwindigkeit und zur emotionalen Benutzerfreundlichkeit.

In einem frühen Entwicklungsstadium genügen wenige Versuchspersonen, um die groben Fehler einer Auslegung zu finden. Genauso gut, schneller, besser und preiswerter findet diese Fehler allerdings ein erfahrener Softwareergonom schon bei der Durchsicht papierener Konzepte, so daß hier prinzipiell schon Programmierkosten einsparbar sind [12]. Erfahrene Softwareergonomen stehen aber nicht überall zur Verfügung, und so bleibt vielerorts nur der Weg der heuristischen Usability Tests. Soweit es nur um heuristisch-qualitative Fehlersuche geht und um die Frage, ob eine Versuchsperson eine Aufgabe überhaupt ausführen kann, wirkt sich die Anwesenheit eines geschulten Beobachters nicht negativ aus, und ein Usability Test braucht kein Labor und kann durchaus auch in normaler Umgebung durch einen daneben sitzenden Beobachter ausgeführt werden.

Industriell sind bei Usability Tests insbesondere quantitative Angaben, die sog. Bedienbarkeitsquoten, interessant.

Die Bedienbarkeitsquoten geben an, wieviel Prozent der Nutzer eine Bedienfunktion bedienen können, ohne daß eine Gebrauchsanleitung zur Verfügung steht. Die meisten Nutzer ziehen heute keine Gebrauchsanleitungen mehr heran, und so läßt sich über die Bedienbarkeitsquoten frühzeitig feststellen, wieviele Nutzer wahrscheinlich die Hot Lines belasten oder den Service anrufen werden, und wahrscheinliche Folgekosten können kalkuliert werden.

Zur Messung feiner Unterschiede der Bediengeschwindigkeiten alternativer Oberflächenauslegungen ist dagegen eine strenge Kontrolle der Rahmenbedingungen notwendig, d.h. ein spezielles Labor wird notwendig, in dem die Versuchsperson Standardbedingungen vorfinden, und die Ergebnisse von Versuchsabläufen müssen technisch aufgezeichnet werden. Diese Aufzeichnung kann je nach Fragestellung mit Keyboard Tracern oder per Video erfolgen. Keyboard Tracer sind genauer, Videoaufzeichnungen sind aufwendiger in der Auswertung, machen aber eine umfassendere Auswertung möglich und erleichtern die nachträgliche Situationsbewertung.

In den neuesten Usability Labs in Japan will man noch einen Schritt weitergehen und auch die bei der Benutzung einer Bedienoberfläche auftretenden Empfindungen mit messen. So will man herausfinden, welche Stellen eines MM-Dialoges stressen und welche Freude bereiten. Inwiefern solche Messungen gelingen werden, ist noch offen.

3.6 Requirement Engineering: RFA-Analysen und Focus Groups

Ein Requirement Engineering dient der Festlegung der Anforderungen an ein Produkt. Diese Phase wird in der deutschen Softwareentwicklung auch als *Anforderungsanalyse der Anwenderschnittstelle* bezeichnet. Das Requirement Engineering für das Home Electronic System wurde seitens der Ergonomie sehr früh angeboten, als das Projekt erst in der Konzeptphase war. Das war technisch sicherlich der richtige Moment für dieses Angebot, gilt es doch möglichst früh die für das Gesamtaussehen entscheidenden Dinge zusammenzustellen, da spätere Veränderungen immer teurer kommen. Doch in der frühen Konzeptphase arbeiteten lediglich zwei Mitarbeiter an dem Projekt, und ein voll ausgereiftes softwareergonomisches Requirement Engineering hätte das damalige Budget gesprengt. So reifte das Produkt heran, und nach Review, Akzeptanzbefragung und Usability Test ergab sich die Situation, daß man zwar ein softwareergonomisch gutes Produkt hatte, daß aber keiner wußte, ob die gut bedienbaren Funktionen auch die sind, die ein Nutzer braucht. Erst als die Grundfunktionen und Hauptmenüs schon standen, als nach 3 Jahren nun etwa 15 Mitarbeiter an dem Projekt arbeiteten, wurde untersucht, welche Funktionen der Nutzer braucht und wie sie angeordnet sein sollten. Auf der Einstiegsseite war Platz für 9 Hauptfunktionen oder Funktionsgruppen, welche sollten das sein?

Ist es nun schon nicht leicht, für ein bestehendes Produkt solche Requirements zu klären, so ist das um so schwerer bei einem Produkt, das noch keiner kennt und das noch nicht auf dem Markt ist. Wir sind der Frage mit einer fiktiven RFA- [6] und Häufigkeitsanalyse nachgegangen, die wir dann über Focus Groups validiert und quantifiziert haben. Zunächst also überlegten Ergonomie-Experten anhand fiktiver Rollen, Funktionen, Aufgaben und anhand der Räume eines angenommenen Hauses, welche Tätigkeiten dort ablaufen, die durch ein Home Electronic System unterstützt werden könnten. Dann wurden diesen Unterstützungsfunktionen Nutzungshäufigkeiten zugeordnet. So fanden wir, daß an der Haustür eine Zentralverriegelung eine nützliche Funktion sein dürfte und daß in der Diele ein Statusboard nützlich wäre, welches die für das Verlassen eines Hauses wichtigen Funktionen anzeigt: ob die Heizung richtig geschaltet ist, Türen oder Fenster offen sind, Verbraucher oder Lichter an sind und der Anrufbeantworter auf Ein geschaltet ist. So wurden alle Räume analysiert, und es resultierten als nützlich anzusehende Funktionen [11].

(Sicherheit)	(Licht und Strom)	(Tel/Fax)
(Wetter)	(Geräte)	(Radio/Wecker)
(Heizung)	(TV/Video)	(HES-System)

Tab. 3: Die neun Hauptmenüfunktionen des Home Assistant als Ergebnis der RFA-Analyse.

Im nächsten Schritt wurden drei Gruppen von jeweils 10 Teilnehmern eingeladen, die repräsentativ für die Zielgruppen des Produktes waren. Sie wurden anhand eines Modells mit dem Home Electronic System vertraut gemacht. Dann wurde ihnen Gelegenheit gegeben, sich selbst Funktionen für das System zu wünschen. Das machten die Gruppen mit unterschiedlicher Kreativität. Schließlich wurden auch die in der Tätigkeitsanalyse gefundenen Funktionen mit ihnen durchgesprochen. Zwischen Tätigkeitsanalyse und den Ideen der Focus Groups gab es zwar einige Übereinstimmungen, problematisch an den Ideen der Teilnehmer war aber, daß vielen die technische Realisierbarkeit fehlte (Z.B. *Bäder und Küchen, die sich selbst reinigen*). Interessant war, daß die Teilnehmer den Funktionen aus der RFA-Analyse schließlich Vorzug vor den selbst gewünschten gaben. Die Ergebnisse führten schließlich zu einer Neuordnung der Menüinhalte, so daß schließlich neben der Bedienbarkeit auch eine hohe Nützlichkeit des Systems gegeben erschien.

Ob wir zu den richtigen Funktionen vorgestoßen sind, muß jetzt die Praxis zeigen. Die ersten prototypischen Systeme sind in Häuser eingebaut und wir warten gespannt auf die Aussagen und Bewertungen durch die Bewohner.

3.7 Interdisziplinäre Produktentwicklung

Die einzelnen Funktionen und Bildschirmseiten mußten nun durchentwickelt werden. Dies wurde interdisziplinär getan. Die Seitenvorschläge kamen zwar von einem Kernteam von Entwicklern, es erfolgten dann aber gemeinsame Treffen und eine Abstimmung der Oberflächen zwischen Entwicklung, Ergonomie, Design und Marketing.

Bei anderen Softwareentwicklungen waren in solche interdisziplinären Teams auch Kunden einbezogen. Dies macht jedoch bei völlig neuartigen Systemen, wie dem Home Assistant, keinen Sinn.

Die interdisziplinäre Produktentwicklung stellt hohe Ansprüche, ist aber eine sehr effiziente Vorgehensweise. Während früher die Entwicklung die Disziplinen nacheinander erreichte und viele Iterationsschritte nötig waren, erfolgt über die interdisziplinäre Produktentwicklung heute sofort eine Definition von Oberflächen, mit der alle einverstanden sind.

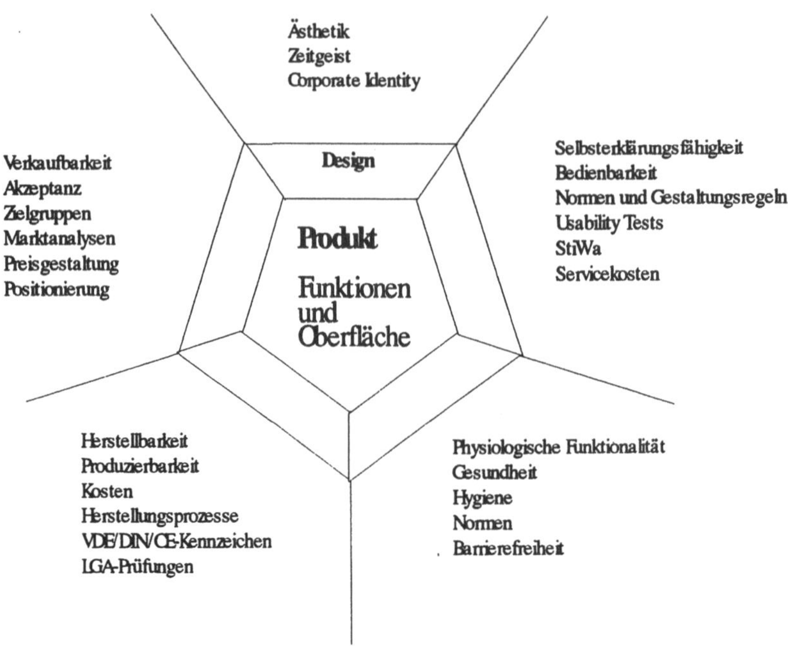

Abb. 1: Verantwortungsbereiche bei der Interdisziplinäre Produktentwicklung

Die Abbildung zeigt, wofür die einzelnen Disziplinen Verantwortung zu übernehmen haben. Das Schema gilt für die Bedienoberflächen von Geräten im Haushalt, durchaus also auch für Mikrowellen und für Waschmaschinen.

Die Ergonomie verantwortet die Selbsterklärungsfähigkeit und die Bedienbarkeit, die Einhaltung von Normen, und Gestaltungsregeln. Das Produkt soll schließlich einen möglichen Test durch externe Institute oder Zeitschriften bestehen, und es soll auch keine Servicekosten aufwerfen. Der verantwortliche Ergonom muß sich in diesen Besprechungen einbringen, aber auch zurückhalten und nur auf Probleme aufmerksam machen, indem er Anforderungen formuliert. Für die eigentliche Oberflächengestaltung soll nämlich der Designer möglichst freie Hand haben. Natürlich müssen die Techniker und Softwareentwickler beachten, daß die Herstellbarkeit gewährleistet bleibt, und bei allem was Kosten verursacht ist es dann Sache des Marketings, die Auswirkungen auf die Verkaufbarkeit zu beurteilen.

Interessant ist im Haushaltsbereich die Einbeziehung der Ökotrophologie. Diese Spezialdisziplin bringt zum Beispiel bei Waschmaschinen ihr Wissen darüber ein, bei welchen Temperaturen bestimmte Enzyme zu wirken beginnen und hilft so, der Technik optimale Zielsetzungen zu verschaffen, besser als jeder Verbraucher es könnte.

3.8 Style Guide

Das Home Electronic System hat inzwischen eine Reife erlangt, daß andere Firmen Interesse bekundet haben, Software für den Home Assistant zu schreiben. Das System soll auch nicht proprietär sondern offen für jeden Produkthersteller sein. Nun wirft diese Offenheit natürlich das softwareergonomische Problem auf, daß jetzt Hersteller mit Software in das System kommen können, die nicht dem vorhandenen softwareergonomischen Niveau entspricht, so daß womöglich Anwender in Fremdapplikationen hängen bleiben.

Es wird deswegen derzeit ein Style Guide geschrieben, der hier Sicherheit bieten soll, indem eine minimale Einheitlichkeit der wichtigsten Dialogfunktionen festgeschrieben wird. Dieser Styleguide soll gleichzeitig ein Leitfaden und eine Hilfestellung für die Entwicklung von neuen Applikationen sein und wird Standardlösungen anbieten.

Die Aufgabe der Softwareergonomie bei der Entwicklung von Styleguides ist es zu beachten, daß auch die ergonomisch relevanten Details aufgenommen werden und nicht nur äußere Merkmale.

Im Styleguide müssen eindeutige und prüfbare Regeln aufgestellt werden, die die Benutzerfreundlichkeit möglichst sicherstellen. Ein gewisses Maß an Benutzerfreundlichkeit kann so mit einem Styleguide festgeschrieben werden. Erfahrungsgemäß bleibt aber immer noch genügend Spielraum, um auch dann noch Oberflächen unbenutzbar zu machen. Programmierer zeigen in dieser Beziehung erstaunliche Kreativität.

Literaturverzeichnis

[1] A.-P. Bröhl & W. Dröschel: Das V-Modell. Oldenbourg, München, 1993

[2] S.K. Card, T.P. Morell & A. Newell: The Psychology of Human Computer Interaction. Lawrence Erlbaum Associates, Hillsdale NJ, 1983

[3] M. Herczeg: Software-Ergonomie. Addison Wesley, Bonn, 1994

[4] MacKenzie, I.S. und Buxton, B.,A.: Tool for the Rapid Evaluation of Input Devices using Fitts' Law Models. ACM SIGCHI Bulletin, Band 25(3), 1993, 58-63

[5] J. Nielsen: Usability Engineering. Academic Press, London, 1993

[6] H. Oberquelle: Sprachkonzepte für benutzergerechte Systeme. Springer, Berlin, 1987

[7] R. Schoeffel: Gestaltungsregeln für berührungssensitive Oberflächen. Siemens Projektbericht, Erlangen, 1994

[8] R. Schoeffel: Sensitive screen HCIs in recent Siemens telecommunication products. Proceedings of the HCI, 1995

[9] R. Schoeffel & U. Pawlak: Touchscreens für Hausgeräte. Siemens Projektbericht, Erlangen, 1993

[10] R. Schoeffel & T. Hasbargen: BSHG Home Assistant. Siemens Projektbericht, München, 1994

[11] R. Schoeffel & S. Ortlieb: Requirement Engineering für den BSHG Home Assistant. Siemens Projektbericht, München, 1996

[12] J. Weizenbaum: ELIZA - A program for the study of natural language communication between man and machine. CACM, 1966, 9, 36-45

[13] VDE-ITG-Richtlinie: Prüfung und Bewertung der Benutzerfreundlichkeit von Kommunikationsendgeräten. VDE-Verlag, Berlin, 1996

Adresse des Autors

Dipl.-Psych. Dr. rer. nat. Roland Schoeffel
Ergonomie-Berater
Siemens Corporate Design
St.-Martin-Str. 76, 81541 München
100671.3403@compuserve.com

Ergonomie-Reviews und Usability-Testing als Beratungs- und Qualifizierungsinstrumente

Peter Ansorge und Uwe Haupt

Universität Bremen, Technologie-Zentrum Informatik,
Institut für Software-Ergonomie und Informationsmanagement[1]

Zusammenfassung

Das Projekt „Software-Ergonomie-Transfer" (SET) berät auf regionaler Ebene in einem Verbundvorhaben kleine und mittlere Softwarehäuser zur Fragen der Software-Ergonomie. Die dabei in allen Phasen der Software-Entwicklung eingesetzten punktuellen Reviews und Usability-Tests fungieren keineswegs nur als Prüfverfahren, sondern zusätzlich als Qualifizierungsinstrumente für die Software-Entwickler.

Für die Phasen der Anforderungsdefinition, des Entwurfs und der Implementierung wurden Reviews mit einer Kombination von Walkthrough und checklistenbasierter Analyse erfolgreich erprobt. Sie können Usability-Tests, in denen weitere, in der Regel schwerwiegendere ergonomische Defizite zutage treten, vorbereiten und effizienter gestalten.

1 Ausgangslage: Veränderung des Qualitätsbewußtseins

Im Rahmen allgemeiner Qualitätssicherungsverfahren (Stichwort ISO 9000) werden Bewertungsverfahren auch von kleinen und mittleren Softwarehäusern zunehmend als unumgänglich angesehen und akzeptiert. Allerdings stehen hierbei Tests und Erprobungen auf der *funktionalen* Ebene immer noch Vordergrund, während Ergonomie-Überprüfungen weitestgehend vernachlässigt werden. Mit steigender, aber immer ähnlicher werdender Funktionalität von Programmen wird die Benutzbarkeit der Software als wichtiges Qualitätsmerkmal erkannt und so zu einem die Kaufentscheidung wesentlich beeinflussenden Faktor. Die Forderungen der EU-Bildschirmrichtlinie, der Arbeitsschutzgesetze, der Bildschirmarbeitsverordnung und der hierdurch referenzierten Normen werden – zwar noch zögerlich, aber in wachsendem Maße – von Anwendern und Benutzern gegenüber Softwareherstellern geltend gemacht. Ergonomische Konformitätsnachweise, z. B. als Zertifizierungen, werden am Abschluß eines Software-Entwicklungsprozesses gefordert und von Herstellern nicht nur passiv als notwendiges Übel hingenommen, sondern aktiv als Vermarktungshilfe akzeptiert und benötigt.

Unabhängig von der Motivation werden Produktbewertungen als fester Bestandteil einer Software-Entwicklung im Produktionsplan als Meilenstein berücksichtigt, auf den hingearbeitet wird. Anders verhält es sich mit ergonomischen Schulungs- und Qualifizierungsvorhaben für Software-Entwickler, die zwar nach Überblicksveranstaltungen als durchaus notwendig angesehen werden, deren Durchführung aber nicht selten dem allgemeinen Termindruck der Softwarebranche anheim fällt.

---

[1] Diese Arbeit entstand im Rahmen des Projektes *SOFTWARE-ERGONOMIE-TRANSFER (SET)*, gefördert vom Landesprogramm Arbeit und Technik des Landes Bremen.

Eine Kombination von Reviews und Qualifizierungsmaßnahmen, von Akzeptiertem und Notwendigem, erweist sich nicht nur für kleine und mittlere Softwarehäuser als effektives und effizientes Verfahren zur Steigerung der ergonomischen Produktqualität. Hierbei sind allerdings einige Besonderheiten zu beachten, die einerseits durch die Beteiligten, andererseits durch die Verfahren bedingt sind.

2 Software-Ergonomie-Transfer

Im Forschungsprojekt Software-Ergonomie-Transfer (SET) der Universität Bremen, Informatik, werden Möglichkeiten des Transfers software-ergonomischer und arbeitswissenschaftlicher Erkenntnisse untersucht und erprobt. Adressaten der Transfermaßnahmen sind Benutzer, Anwender und insbesondere die Hersteller von Software. Innerhalb der dreijährigen Projektlaufzeit sollen die Grundlagen für eine dauerhaft Beratungseinrichtung für Software-Ergonomie-Transfer gelegt werden.

2.1 Akteure

Die Kooperation von Softwareherstellern und Software-Ergonomie-Spezialisten birgt eine Reihe von Problemen und setzt daher eine präzise Kenntnis der jeweiligen Bedingungen bei den einzelnen Akteuren voraus [vgl. auch 9]. Dies gilt insbesondere dann, wenn Software-Ergonomie-Berater als Externe den Software-Entwicklern gegenübertreten, wie etwa bei kleinen und mittleren Unternehmen, die sich noch keine eigenen Ergonomen leisten (können), und wenn durch die Kontroll- und Kritikfunktion eines Reviews Befürchtungen und Abwehrhaltungen auf seiten der Software-Entwickler induziert werden könnten.

Die kleinen und mittelgroßen Unternehmen der Softwarebranche wickeln typischerweise kleine und mittelgroße Projekte ab, sehr häufig als Branchenlösungen. Langjährige und persönliche Kontakte zur Anwenderorganisation sind als Stärken zu nennen. Wesentliche Kennzeichen sind:

- informelle Kooperationsverfahren zwischen den beteiligten Entwicklern und teilweise auch in der Kundenbeziehung,
- geringer Einsatz analytischer Verfahren,
- Dominanz funktionaler Sichten,
- geringe ergonomische Kenntnisse der Entwickler und geringe Fähigkeiten zur Operationalisierung ergonomischer Ansprüche,
- ständige Zeitknappheit,
- praktisch nur schwach ausgeprägte Phasenabgrenzung,
- weitgehenden Verzicht auf Dokumentation.

Externe Berater neigen dazu, vor Vielschichtigkeit und Facettenreichtum der Software-Ergonomie die akuten Nöte und Bedürfnisse der Entwickler aus den Augen zu verlieren. Sie lassen sich charakterisieren durch:

- mangelnde Vertrautheit mit der aktuellen Entwicklungsaufgabe,
- nicht per se vorhandenem Wissen um die zu lösenden Arbeitsaufgaben,
- keine Kenntnisse über Anwender und Benutzer,
- geringe Kenntnisse über den Wissensstand der Entwickler,

- breites Wissen zu allgemeinen ergonomischen Fragen, Vorschriften und Regeln,
- Erfahrung durch unternehmensübergreifende Betrachtungen.

2.2 Maßnahmen

Maßnahmen von SET zum Transfer software-ergonomischer und arbeitswissenschaftlicher Erkenntnisse sind u. a.:

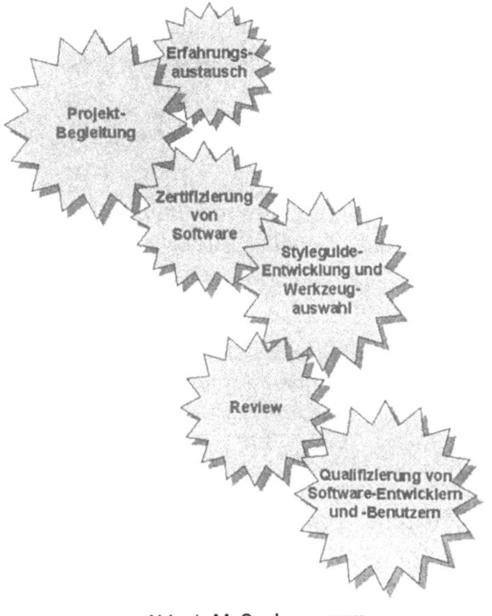

Abb. 1: Maßnahmen zum
Software-Ergonomie-Transfer

- Review bereits erstellter Softwareprodukte in Bezug auf Formularlayout, Dialoggestaltung und arbeitsorganisatorische Erfordernisse

- Vermittlung von Review-Strategien für den späteren selbständigen Review von Softwareprodukten durch die Hersteller

- Projektbegleitende Beratung zur software-ergonomischen Erstellung und Gestaltung eines neuen Softwareproduktes

- Entwicklung von software-ergonomischen unternehmensspezifischen Styleguides und Beratung und Auswahl zum Einsatz von Entwicklungswerkzeugen

- Qualifizierung von Mitarbeitern in firmeninternen Schulungen oder öffentlichen Seminaren

- Seminare zur Verbesserung der innerbetrieblichen Kommunikationsstrukturen

Bei den zentralen Transfermaßnahmen Qualifizierung und Review treten im Kontext mit kleinen und mittelständischen Softwareherstellern folgende Probleme auf:

2.2.1 Probleme klassischer Qualifizierung

Betriebliche Weiterbildung ist in kleinen und mittleren Unternehmen im Vergleich zu Großunternehmen sowohl bezüglich Quantität als auch hinsichtlich einer systematischen Planung nur schwach ausgeprägt. Es dominieren ad-hoc-Qualifizierungsmaßnahmen am Anfang von Entwicklungsaufträgen, in denen es um das Erlernen neuer Werkzeuge bzw. Sprachen geht. Selbstinitiiertes Lernen durch Lektüre von Handbüchern oder Ausprobieren wird häufig gar nicht als solches erkannt und spielt trotz der immens hohen – allerdings im allgemeinen Personalbudget versteckten – Kosten eine überproportional große Rolle. „Klassische Kurse" werden wegen des überschaubaren zeitlichen, finanziellen und personellen Aufwands nur zögernd wahrgenommen. Hinzu kommen die relativ geringen Wirkungen derartiger Qualifizierungsmaßnahmen, die durch Streuverluste aufgrund nicht angepaßten Lernstoffes sowie durch die Überbetonung reiner Kenntnisvermittlung gegenüber der Aneignung von Fähigkeiten verursacht wird. Durch die zeitliche Entkopplung von der Umsetzungsphase sinkt der Nutzen weiter.

Die ohnehin problematischen, aber in kleinen und mittleren Unternehmen vorherrschenden Formen der Weiterbildung sind für software-ergonomische Themen in besonderem Maße ungeeignet: Software-Ergonomie erfordert grundsätzlich andere Denk- und Herangehensweisen, die sich von den gewohnten, technisch dominierten Denkweisen der Software-Entwickler deutlich unterscheiden. Anschlußlernen ist somit nicht möglich. Damit wird die Einarbeitung ohne fremde Unterstützung besonders aufwendig. Hinzu kommt, daß auch heute noch selbst die Grundzüge der Software-Ergonomie nur an wenigen deutschsprachigen Hochschulen vermittelt werden.

2.2.2 Defizite von Reviews

Produktreviews, Audits und Bewertungen werden ex-post durchgeführt, um die Qualität von Produkten anhand von Vorschriften und Vorgaben *festzustellen* und sie insbesondere gegenüber Kunden *nachzuweisen*. Adressaten eines Reviews sind in der Regel also nicht die Software-Entwickler.

Voraussetzungen für Softwareprüfungen sind insbesondere das ablauffähige Programm, die Dokumentationen zum Programm, Prüfvorschriften, Anforderungen, Spezifikationen, Gesetze, Normen, Styleguides und der Prüfauftrag.

Abschließende Beurteilungen kommen für eine Einwirkung auf die Produktqualität zu spät. Verbesserungen lassen sich erst in Nachfolgeprodukten / -versionen realisieren oder durch kostenintensive Nacharbeit – was bei groben ergonomischen Fehlern mit einem Neuanfang gleichzusetzen ist.

Die Durchführung von fundierten Ergonomie-Überprüfungen stellt sich durch das Fehlen oder den mangelnden Bekanntheitsgrad geeigneter Ergonomie-Prüf- und Beratungseinrichtungen de facto als ein Problem dar. Methoden der Selbstkritik und -beurteilung sind zwar interessante Ansätze, dennoch kann auf unbeteiligte Prüfer nicht verzichtet werden. In kleinen und mittleren Softwarehäusern fehlt aber häufig selbst die Kapazität für hausinterne Bewertungen durch Unbeteiligte, die nicht der Gefahr der „Betriebsblindheit" unterliegen.

3 Entwicklungsbegleitende Reviews

Die „Verbesserung der ergonomischen Qualität" von Programmen setzt die Qualifizierung der Entwickler voraus [vgl. z. B. 14]. Um effizient wirksam werden zu können, muß fachliche Beratung an den konkreten Problemen in der Entwicklung aufsetzen; die Probleme lassen sich am besten an den entstehenden Programmen durch ergonomische Inspektionen ermitteln.

3.1 Informationsgewinnung für Prüfer – Qualifikation für Entwickler

DIN 66 285 [5] schreibt vor, ohne spezielle Unterlagen dürfe nicht geprüft werden, aber diese strenge Forderung, die ihre Berechtigung bei Testat-erteilenden Beurteilungen wie Zertifizierungen hat, ist wenig praxisgerecht, wenn es um Begutachtungen im Zusammenhang mit Beratungen und Schulungen geht. Realiter wird aber in fast allen kleinen und mittleren Entwicklungsvorhaben auf die Erstellung spezieller Ergonomie-Anforderungen und -Spezifikationen („ErgoSpec") bisher verzichtet.

Prüfer sind dennoch auf diese Angaben angewiesen. Die Erhebung dieser Informationen bei den Software-Entwicklern hat einen eigenständigen Wert als Qualifizierungsinstrument: Sie macht nicht nur Berater/Prüfer mit dem entstehenden Programm vertraut und versetzt sie in die Lage, eine Minimal-/Ersatzspezifikation anzufertigen, sondern verdeutlicht den Entwicklern, welche Punkte klärungsbedürftig, mindestens aber mißverständlich sind. Hier wirkt allein die Übermittlung der bisher möglicherweise nur implizit unterstellten Annahmen erkenntnisfördernd.

Neben den spezifischen Dokumenten müssen selbstverständlich allgemeingültige wissenschaftliche Erkenntnisse, Gesetze und Normen, ggf. plattformspezifische Regelungen („Betriebssystem-Styleguides") herangezogen werden. Hier können aus der DIN EN ISO 9241 [6] abgeleitete und operationalisierte Kriteriensysteme Verwendung finden.

3.2 Formulierung einer Prüfaufgabe – Vertrauensbildung

In der Regel steht bei kleinen und mittleren Entwicklungsvorhaben zunächst die Unterstützung *einer* speziellen Arbeitsaufgabe im Vordergrund oder dominiert die Anwendung auch bei einer Weiterentwicklung und Erweiterung noch stark, so daß Externen der Zugang nicht durch übergroße Komplexität erschwert wird. Hieraus kristallisiert sich bei beharrlichem Nachfragen mindestens eine während des Entwicklungsprozesses als typisch angesehene Programmnutzung heraus. Leitfragen zur Ermittlung können dabei z. B. Fragen nach der geschätzten Nutzungshäufigkeit oder einer Beeinträchtigung bei Funktionsverlust sein. Als zweckmäßig zur Abschätzung der unterstellten Relevanz haben sich auch Erhebungen über die Menge zugrundeliegender / verwendeter (Papier)Formulare oder sonstiger Datenquellen bzw. über die Menge der Ausgaben herausgestellt.

Aus den erhaltenen Hinweisen über den Zweck und Anleitungen zur Programmnutzung (meist müssen während der Entwicklung noch besondere Vorkehrungen zur Nutzung auf Fremdrechnern vorgenommen werden) und einer Grobsichtung des Programms läßt sich ein Eindruck über den Funktionsumfang und die geplante Nutzung gewinnen. Danach besteht die Notwendigkeit, die Prüfung zu strukturieren, da sich eine Totaluntersuchung meist aus Kapazitätsmangel verbietet. Hierzu formuliert der Prüfer in Zusammenarbeit mit Anwendern und Benutzern eine *Prüfaufgabe*, mit der das Programm untersucht werden soll. Hierdurch wird eine Beliebigkeit ausgeschlossen; eine Annäherung an die spätere Nutzungssituation wird ermöglicht, und die Entwickler erhalten Vertrauen, daß nicht willkürlich überall nach Fehlern gesucht wird. Dabei wird festgelegt, welche Arbeitsaufgabe mit welchen Testdaten erledigt werden soll und welche Schritte als notwendig angesehen werden. Vorzugsweise handelt es sich um *einen* vertikalen Durchlauf durch das Programm. Die Vorgaben werden den Entwicklern vorab vorgelegt und mit ihnen abgestimmt. Nach deren Zustimmung – die hier besonders wichtig ist, um nicht möglicherweise in noch nicht implementierte Module zu gelangen – wird der Prüfplan festgelegt.

3.3 Prüfungsdurchführung

Es hat sich bewährt, die Prüfaufgabe zunächst einmal ohne jegliche Anmerkungen oder Aufzeichnungen vom Prüfer durcharbeiten zu lassen – auch Prüfer müssen sich mit dem System vertraut machen. Im zweiten Durchgang erfolgt die eigentliche Prüfung anhand der gestellten Aufgabe. Hierzu werden die Arbeitsschritte minutiös dokumentiert – sinnvollerweise durch technische Unterstützung der Verlaufsprotokollierung. Hier kommen Verfahren zum Einsatz, wie sie üblicherweise bei Protokollierungen des Nutzungsverhaltens in Usability-Tests eingesetzt werden, z. B. statische Dokumentation durch „Screenshots", dynamische Aufzeichnung der Präsentationen durch „Screengrabber", Aufzeichnung der Interaktionen mit Registrierung der Eingaben und Festhalten einer „Mausspur" (vgl. Abb. 2), Audioanmerkungen, Videoaufzeichnung. Anders als bei der Beobachtung späterer Benutzer dienen diese Instrumente lediglich der eigenen Auswertung durch den Prüfer selbst bzw. der Illustration seines Berichts.

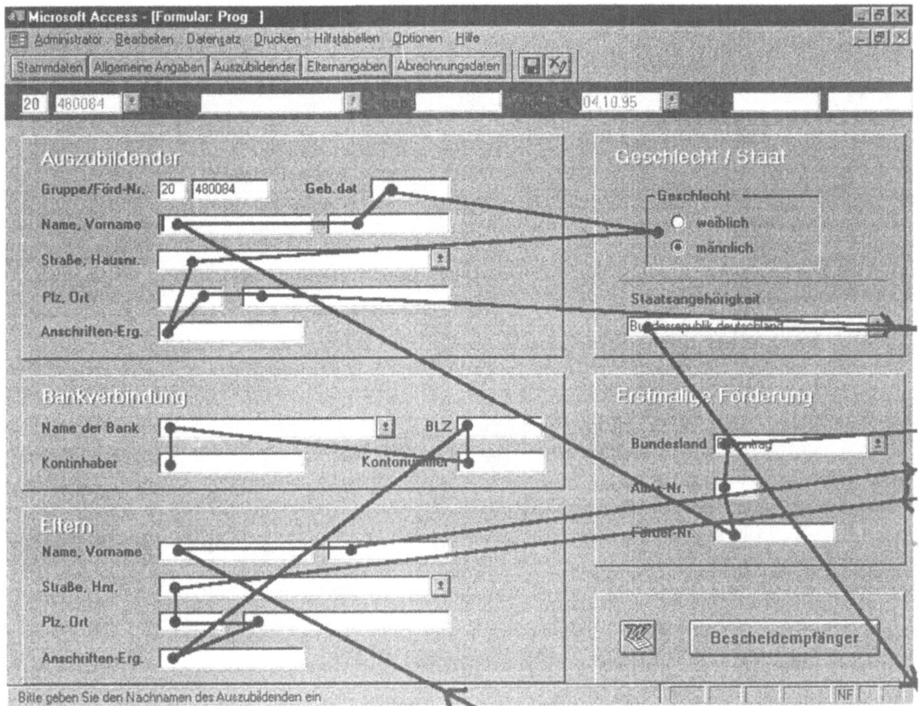

Abb. 2: Ergebnis eines Reviews
Sachbearbeiter hatten die Aufgabe, Daten von einem (bundeseinheitlichen) Papierformular in den Rechner einzugeben. Die Feldanordnung der Vorlage und des Bildschirmformulars unterschieden sich deutlich. Die Eingabe der Daten in der Reihenfolge des Papierformulars erforderte einen Navigation über den Bildschirm entlang des oben markierten, komplizierten Weges.

Die Bewertung erfolgt anhand eines Kriteriensystems. Der Prüfer dokumentiert jeden durchgeführten Schritt in einem Prüfbericht. Insbesondere sind Formular-/Maskenübergänge, Feldpositionierungen, Eingaben und sonstige Interaktionen festzuhalten. Als zweckmäßig hat sich gezeigt, die Kriterienliste („Checkliste") zunächst zu *jedem* neuen Bildschirmformular einmal zu durchlaufen und sämtliche Prüfkriterien zu beantworten, wobei festgestellte Verstöße begründet zu protokollieren sind. Zur Unterstützung dieser Tätigkeit hat sich der Einsatz von Datenbanksystemen bewährt, bei dem einerseits die Kriterien durch Beispiele konkretisiert werden, andererseits alle bisher jemals zu diesem Kriterium gemachten Angaben (auch bei anderen zu prüfenden Programmen) abgerufen werden können. Hierbei werden interpersonale Bewertungsspielräume relativiert, außerdem entsteht mit zunehmender Prüfungsanzahl entsprechende Bewertungssicherheit.

Bei abschießenden Bewertungen, deren Ergebnisse sich vorzugsweise an Dritte wenden, sind Aussagen wie „Kriterium erfüllt" bzw. „nicht erfüllt" vollkommen ausreichend. Zur Qualifikation werden aber weitere Elemente benötigt: So ist sehr exakt zu benennen, welche Kriterien aus welchen Gründen nicht erfüllt werden. Hierbei ist auf die entsprechenden Stellen in den der Prüfung zugrundeliegenden Dokumenten wie Normen und Styleguides zu verweisen – diese Dokumente sollen ja später von vornherein beachtet werden, sind also Qualifizierungsgegenstand.

3.4 Konstruktive Kritik

Beurteilungsfähigkeit setzt besondere Qualifikationen der Prüfer voraus. Einem erfahrenen Prüfer fällt nicht nur aufgrund von Checklisten ein Kriterienverstoß auf, sondern auch durch Kenntnis von anderen Programmen und durch eigene Nutzungserfahrungen. Es liegen also auch bei Prüfern, die mit der konkreten Arbeitsaufgabe weniger vertraut sind als die späteren Anwendungsspezialisten, bestimmte Erwartungshaltungen mindestens über weniger aufgabenabhängige Teile vor.

27. Es ist nicht ersichtlich, in welchen Feldern Eingaben unbedingt erforderlich und welche optional sind.

> *Mußfelder sind von Kannfeldern deutlich zu unterscheiden. Dies kann durch Verwendung unterschiedlicher Hintergrundfarben erfolgen, z.B. Mußfelder gelb (Selbstbeschreibungsfähigkeit).*

28. Die Feldbezeichnungen sind nicht mit Schnellschaltelementen (Mnemonics) versehen. Die Felder müssen deshalb in der vom Programm vorgesehenen Reihenfolge eingegeben oder mit der Maus angesteuert werden. Bei reiner Tastaturbedienung kann eine Weiterschaltung nur mit ⇆ erfolgen.

> *Die Feldbezeichnungen sind möglichst mit eindeutigen Mnemonics zu versehen, z.B. „Name, Vorname", „Kontoinhaber" (Steuerbarkeit; Erwartungskonformität; Windows-Styleguide).*

Abb. 3: Ausriß aus einem Prüfbericht

Hilfreich für die Korrektur von Fehlern ist also die Benennung der üblicherweise zu erwartenden Elemente der Benutzungsschnittstelle. Verfügen die Prüfer darüber hinaus auch über eine Gestaltungsfähigkeit, so können sie nicht nur analytisch Hinweise über den Grund von Fehlern („Warum?") und auf das „Was?", sondern auch konstruktiv über die Mittel („Wie?") geben (vgl. Abb. 3). Werden Angaben zur ergonomischen Verbesserungen von Programmen angegeben, so wird mindestens verdeutlicht, daß es nicht nur den *einen* von den Entwicklern verfolgten Weg gibt. Wird die Alternative auch noch durch ablauffähige Prinziplösungen mindestens als machbar belegt, so wird die Zurückweisung, ergonomische Forderungen seien unrealistisch, von vornherein vermieden.

3.5 Präsentation und Vermittlung

Die Ergebnisse eines entwicklungsbegleitenden Reviews werden in einem kompakten Bericht, bestehend aus der Dokumentation des Walkthroughs anhand der Prüfaufgabe, einer checklistenbasierten Bewertung und einer exemplarischen Einzellösung eines zentralen Problems des geprüften Programmes zusammengefaßt. Die Präsentation dieser Dokumentation stellt einen zentralen Vermittlungsschritt innerhalb des Qualifizierungsvorhabens dar. In einer halbtägigen Präsentation des Reviews vor den Software-Entwicklern läßt sich sowohl an der Praxis der Entwickler orientiertes ergonomisches Fachwissen vermitteln als auch gleichzeitig ein konkreter Beitrag zum laufenden Entwicklungsprojekt leisten.

Reviewpräsentationen weisen ein erhebliches Konfliktpotential auf, zumal Schwächen der Software deutlich zutage treten, so daß Software-Entwickler zunächst eine vernichtende, unproduktive Kritik ihrer Arbeit befürchten. Die Diskussion kann nicht allein durch den Duktus der schriftlichen Dokumentation und der Präsentation von der Ebene einer Schuldzu- und -zurückweisung ferngehalten werden; vielmehr gelingt dies durch Konzentration auf konstruktive Elemente. Wider Erwarten beeinträchtigte die Beteiligung der zuständigen Geschäftsführer und auch der Pilotanwender nicht die produktive Auseinandersetzung.

Kritikbereitschaft wird durch die Darstellung von konkreten Lösungen und die Erläuterung möglicher Ursachen hergestellt: Häufig gelingt es, Folgefehler defizitärer Entwicklungswerkzeuge oder unzureichende Vorgaben der Anwender zu identifizieren oder anhand von Hintergrundmaterial aufzuzeigen, welcher Aufwand zur Herstellung ergonomisch geeigneterer Lösungen bei anderen Entwicklungsvorhaben notwendig war. Besonders einprägsam sind Beispiele durch Veränderungen des Originalprogramms, allerdings erfordert dies vom Reviewer nicht nur ein hohes Maß an Beurteilungs-, sondern auch an Gestaltungskompetenz.

3.6 Reaktionen auf Reviews

In den unterschiedlichen Ebenen der Softwarehäuser wurden gravierende Konsequenzen aus den von uns durchgeführten Reviews gezogen. Die monierten Defizite wurden von Entwicklern durchgängig ernst genommen und in der Regel unverzüglich beseitigt. Lediglich nach sehr spät im Entwicklungsverfahren ansetzenden Erstprüfungen konnten in wenigen Punkten Korrekturen nicht mehr durchgeführt werden, so daß Lerneffekte erst in Folgeprojekten wirksam wurden.

Auf der Ebene der Entscheidungsträger wurden – auch aufgrund des positiven Echos bei den Entwicklern – häufig deutlich weitreichendere Konsequenzen gezogen, die bis hin zur vollständigen Neuentwicklung einzelner Systeme reichten. Teilweise folgte auf den Review eine grundlegende Umstellung der Entwicklungsmethode z. B. auf objektorientierte Techniken, wobei unternehmensspezifische Klassenbibliotheken in Kooperation mit den Reviewern erstellt wurden. Auch die Ergonomie der eigenen Werkzeuge wurde in Frage gestellt. Die Bereitschaft zur Beteiligung an systematischen Qualifikationsveranstaltungen wurde erhöht.

Innerhalb der Softwarehäuser traten durchweg positive Effekte auf: Das Verhältnis zwischen kundenorientierten Mitarbeitern – meist mit langjährigen Erfahrungen mit Vorversionen – und den Methodenspezialisten – häufig mit noch geringer Projekterfahrung – verbesserte sich nachhaltig. Auf bereits geplante Umorganisationen durch strenge Aufteilung in Vertrieb und Programmierung wurde zugunsten eines integrierten Entwicklungsteams verzichtet.

Gerade die Diskussion ergonomischer Fragen anhand der „eigenen" Produkte befähigte die Führungskräfte, die Programmverbesserungen auch bei Kundenkontakten aufzugreifen und in Abgrenzung zu Mitbewerbern verkaufswirksam einzusetzen.

3.7 Reichweite des entwicklungsbegleitenden Reviews als Qualifizierungs- und Bewertungsinstrument

Kriterien wie Fehlertoleranz und Steuerbarkeit, eingeschränkt auch Individualisierbarkeit und Lernförderlichkeit, sowie Fragen der Anordnung und Codierung lassen sich weitgehend durch Experten abprüfen. Bei der Beurteilung der Aufgabenangemessenheit, Selbstbeschreibungsfähigkeit und der Erwartungskonformität können Experten allerdings nicht mehr auf der Basis von allgemeingültigen ergonomischen Erkenntnissen auf Benutzerverhalten und -belastung rückschließen, so daß diese Aspekte unmittelbar mit Benutzern zusammen erhoben werden müssen.

Die punktuellen Reviews und deren Präsentationen ersetzen auch in der hier skizzierten Form keine software-ergonomischen Grundlagenschulungen, sondern bereiten sie vor. Mit vertretbarem zeitlichen Aufwand werden produktbezogene und aufgabenangemessene Empfehlungen gegeben, wodurch die Glaubwürdigkeit unterstrichen und die Akzeptanz der vertretenen Ansätze deutlich erhöht wird. Die Aufnahmebereitschaft für theoretische Grundlagen wird geweckt.

Viele Defizite können auch externe Experten nicht erkennen, insbesondere wenn keine oder nur geringe Angaben über Benutzer, Benutzungssituationen und organisatorisches, technisches und qualifikatorisches Umfeld vorliegen. Durch ihre umfangreiche Erfahrung im Umgang mit Rechnern verkennen Experten ebenso wie Entwickler schnell Probleme von Rechnernovizen und können fast nie beurteilen, wie Arbeitsgänge realiter ablaufen, ob nicht durch nichtformalisierbare Usancen falsche Dialogstrukturen und Werkzeugzuordnungen vorgenommen werden. Fragen danach werden durch einen Review provoziert, wobei von den Entwicklern erkannt wird, daß es einen „one-best-way" nicht gibt und zur Vermeidung von Umwegentwicklungen eine frühzeitige Benutzerbeteiligung unverzichtbar ist. Benutzerbeteiligung darf sich aber nicht auf Ermittlung von Vorgaben beschränken, sondern muß zur Validierung auch Usability-Tests einbeziehen.

Eine benutzerzentrierte und aufgabenangemessene Problemlösung hat weitreichende Konsequenzen für die strategische Ausrichtung von Softwarehäusern: Aus den Reviews wird deutlich, daß der Entwicklungsprozeß neu strukturiert werden muß, daß wesentliche qualitäts- und damit wertsteigernde Aktivitäten nicht mehr in der Programmierphase, sondern in vorgelagerten Analysephasen zu erbringen sind. Da die Budgets in der Regel festgelegt sind bzw. nicht beliebig erweitert werden können, müssen Maßnahmen zur Kostenumverteilung innerhalb der Projekte erfolgen, z. B. durch Einsatz effizienterer Entwicklungswerkzeuge. Der dadurch geschaffene finanzielle Spielraum für Analysearbeiten positioniert Softwarehäuser in neuen, dem Wettbewerb weniger stark ausgesetzten Kompetenzfeldern.

Verstanden sich Softwarehäuser bis dahin nicht selten als reine „Software-Ersteller", so wurden ihre Aktivitäten, provoziert durch die Auseinandersetzung mit „Aufgabenangemessenheit" und den konkreten Anwendungssituationen, zwangsläufig erweitert. Beratungsdienstleistungen werden wesentlich stärker mit den Kunden integriert als dies bei einfachen Zulieferungen der Fall ist. Hierdurch können weitergehende ergonomische Forderungen berücksichtigt werden.

4 Usability-Test – Einbeziehen der Benutzer

Eine Grunderkenntnis der Software-Ergonomie ist, daß sie ohne Beteiligung der Benutzerinnen und Benutzer kaum zufriedenstellend umgesetzt werden kann. Ein Transfer-Konzept darf sich daher nicht auf die Unterstützung bei der Lösung eines „ergonomischen Programmierproblems" durch Expertenberatung beschränken. Es muß vielmehr die arbeitswissenschaftliche Begleitung von der Aufgabenanalyse über die software-ergonomische Spezifikation bis zur Realisierung und Einführung umfassen. Dabei wird deutlich, daß auch die Kunden eines Softwarehauses und die dort arbeitenden späteren Benutzer der Software in den Transferprozeß einbezogen werden müssen.

Eine wesentliche Rolle spielt hierbei die Diskussion der Prüfaufgabe mit den Entwicklern, bei der allen Beteiligten deutlich wird, daß eine Perspektivenübernahme – also das Hineinversetzen in die Benutzer – notwendig, aber nicht vollends möglich ist. Daher sind Usability-Tests, also die direkte Konfrontation der Benutzer mit Programmen (oder Vorläufern) unverzichtbar.

4.1 Stand der Forschung

Mit expertengestützten Prüfmethoden wie heuristischer Evaluation [19] oder „Cognitive Walkthrough" [18] kann eine Vielzahl ergonomischer Gestaltungsfehler aufgedeckt werden. So erwies sich in Untersuchungen von JEFFRIES, MILLER, WHARTON und UYEDA [17] die expertengestützte, heuristische Evaluation als diejenige Methode, mit der die größte Anzahl ergonomischer Gestaltungsfehler entdeckt werden konnte. Die unmittelbare Einbeziehung der Benutzer in die Softwareüberprüfung (Usability-Testing) fördert demgegenüber in der Regel zwar eine geringere Anzahl von ergonomischen Gestaltungsfehlern zutage, doch sind die beim Usability-Test entdeckten Fehler schwerwiegender und „globaler". Sie treten häufiger auf als die von den Experten entdeckten Fehler. Die Hälfte der bei Usability-Tests gefundenen Fehler blieb bei Anwendung expertenbasierter Prüfmethoden unentdeckt [vgl. 16]. Daraus ergibt sich die Notwendigkeit zur Kombination expertengestützter Verfahren und Usability-Tests.

Ausschließlich expertenorientierte Verfahren wie z. B. EVADIS II [20] und die ausschließlich auf Benutzerbefragungen basierenden Verfahren wie QUIS [24], SUMI [21], ISONORM 9241-10 [22] und das Verfahren von RAVDEN und JOHNSON [23] werden um kombinierte Verfahren ergänzt, die expertenbasierte Evaluationen mit Benutzerbefragungen und Usability-Tests im Labor verbinden (z. B. die Verfahren von CLEGG et. al [2] und von ENGLISCH [10] sowie der ERGOguide von DZIDA, WIETHOFF und ARNOLD [8]).

Im Rahmen von Arbeits- und Aufgabenanalysen sollen zwar die aus der Forderung nach Aufgabenangemessenheit und Erwartungskonformität resultierenden Anforderungen ex-ante erhoben werden, jedoch erweist sich dieser Ansatz für sich allein als unzureichend [vgl. 11]. Er kann demnach lediglich am Anfang eines Systementwicklungsprozesses stehen, der in seinem weiteren Verlauf auch benutzerorientierte Evaluationsschritte beinhaltet.

Die in der ISO 9000, Teil 3, dem Leitfaden für die Anwendung von ISO 9001 auf die Entwicklung, Lieferung und Wartung von Software, niedergelegten Regelungen beziehen „usability tests" als Testempfehlung ausdrücklich mit ein [15, S. 19]; allerdings fehlen in der Norm hierzu noch Maßnahmen zur Umsetzung. Hier besteht ein akuter Handlungsbedarf bezüglich eines wissenschaftlich fundierten Methodentransfers.

Das Kriterium Benutzbarkeit wird in den derzeit gängigen, ausschließlich produktbezogenen Qualitätssicherungsmethoden für die Software-Entwicklung nicht ausreichend berücksichtigt. Beispielsweise beschränkt DUNN [7, S. 179f] Benutzbarkeitstests auf die Auslieferung von Alpha- und Betaversionen. Prüfungen nach DIN 66 285 umfassen nur Teilaspekte der DIN 66 234, Teil 8 (resp. DIN EN ISO 9241, Teil 10), und in den Verfahren nach DGQ-NTG [3] sowie nach ASAM, DRENKHARDT und MAIER [1] nimmt der Teil zur Benutzbarkeit nur einen sehr geringen Raum ein [vgl. 13, S. 49-52].

4.2 Usability-Tests – vorbereitet durch Experten-Reviews

Im Vergleich zu entwicklungsbegleitenden Reviews sind Usability-Tests hinsichtlich des zeitlichen, finanziellen und personellen Ressourcenbedarfs sehr aufwendig. Dieser Aufwand kann bei systematischer Vorbereitung insbesondere durch einen entwicklungsbegleitenden Review deutlich gesenkt bzw. die Ergebnisqualität der Usability-Tests erheblich gesteigert werden.

In allen Phasen der Systementwicklung, in denen Benutzer mit frühen Versionen des Endproduktes oder mit Prototypen konfrontiert werden, läßt sich zuvor mit Hilfe eines entwicklungsbegleitenden Reviews mit relativ geringem Aufwand ein Eindruck über die ergonomische Qualität der Software gewinnen. Die hierdurch möglich gewordene Behebung offensichtlicher Ergonomie-Defizite noch *vor* einem Usability-Test macht den Blick frei auf die wesentlichen, nur mit dieser Methode erkennbaren Fehler. Die konstruktiven Hinweise im Review erlauben meist eine umgehende Behebung der Bugs. Bestandene Reviews dienen somit einer (auch formalen) „Freigabe zum Usability-Test". Ohne die Vorschaltung besteht die Gefahr, sich während des aufwendigen Usability-Tests in der Erkennung von Fehlern bei Anordnung und inkonsistenten Bezeichnungen o. ä. zu verlieren.

Der Review kann darüber hinaus Hinweise auf offene Fragen enthalten und damit einen Beitrag für die Planung der Usability-Tests leisten. Die in den Usability-Tests zu bearbeitende Aufgabe sollte beispielsweise bereits als Review-Prüfaufgabe formuliert sein. Auf diese Weise erhält auch der Reviewer als Nebenergebnis eine Rückmeldung über die Validität seiner Kritik und der vorgeschlagenen exemplarischen Musterlösungen.

4.3 Usability-Tests in sämtlichen Entwicklungsphasen

Usability-Tests lassen sich sowohl als Feldversuche am Arbeitsplatz als auch im Labor unter kontrollierten Bedingungen durchführen. In Feldversuchen muß bereits ein lauffähiges, produktiv einsetzbares System vorliegen. Typischerweise ist dies erst kurz vor oder am Ende der Entwicklung der Fall, wenn ausgereifte Beta-Versionen bzw. schon das Endprodukt zur Verfügung stehen. Zu diesem Zeitpunkt ist der Gestaltungskorridor bereits soweit verengt, daß nur noch Geringfügigkeiten im Endprodukt verbessert werden können.

Um die spezifische Kompetenz der Benutzer bereits in möglichst frühe Entwicklungsphasen in den Entwicklungsprozeß einfließen zu lassen, sollte in Usability-Tests nicht nur eine Konfrontation mit dem Endprodukt bzw. Prototypen durchgeführt werden, sondern auch mit Alt- und Alternativsystemen.

Am Anfang eines Entwicklungsprozesses liegt häufig mit einem abzulösenden *Altsystem* eine bisher zu gering genutzte Informationsquelle vor. Ein Feld-Usability-Test gibt hier wertvolle Aufschlüsse über die zu lösenden Aufgaben und vor allem über das Arbeitsumfeld (z. B. Unterbrechungen, Umgebungsbedingungen). Eine so erstellte Analyse der Arbeitsaufgaben und der Arbeitsumfeldes liefert durch ihren konkreten Bezug zu Programmen einen von Entwicklern besser umsetzbaren Beitrag zu Anforderungsdefinitionen und Spezifikationen. Diese Dokumente werden ihrerseits dann in den folgenden Reviews zur Formulierung von Prüfaufgaben herangezogen.

Während die Qualifizierungsfunktion entwicklungsbegleitender Reviews im wesentlichen auf die Software-Entwickler gerichtet ist, kann mit Usability-Tests ein Beitrag zur Qualifizierung aller Beteiligten (Benutzer und Entwickler) geleistet werden. Gerade am Anfang beteiligungsorientierter Projekte ist die Herstellung von Beurteilungs- und Beteiligungskompetenz wichtig. Durch Usability-Tests, hier in der Ausprägung einer *vergleichenden Betrachtung* unterschiedlicher, auf dem Markt für ähnliche oder gleichartige Zwecke angebotenen Programme, können Benutzer wie Entwickler zu wesentlich aussagekräftigeren Festlegungen gelangen.

Eine auch als informelle Version geeignete Variante sind Ad-hoc-Usability-Tests, beispielsweise durch Einladung von Benutzern zu Entwicklerbesprechungen und dem gemeinsamen Durchgehen von Skizzen und Prototypen. Sie erfordern geringen Aufwand und führen in der Regel zu Ergebnissen, wie sie in streng formalisierten Verfahren nicht zu gewinnen sind.

4.4 Usability-Test-Methoden

Usability-Tests – verstanden als Konfrontation von Benutzern mit Programmen oder Vorläufern hierzu – kann mit sehr unterschiedlichen, durchaus dem jeweiligen Entwicklungsstand angepaßten Methoden durchgeführt werden. Die folgende Liste erhebt nicht den Anspruch der Vollständigkeit, bietet aber Orientierung schon bei ganz frühen Phasen bis hin zur Einführung.

- Rapid Prototyping („low fidelity"-Prototypen)
  Entwickler und Benutzer skizzieren auf Papier (nicht auf Rechnern, das begünstigt die Entwickler) Bildschirmformulare für typische Arbeitsaufgaben.

- Prototypenentwicklung
  Entwickler führen den Benutzern Bildschirmentwürfe vor. Benutzer kommentieren die Entwürfe direkt. Die Ausführungen werden aufgezeichnet oder protokolliert.

- „Papier und Bleistift"-Simulationen
  Entwickler legen Benutzern Sätze von ausgedruckten Bildschirmformularen vor. Benutzer arbeiten an einer typischen Sachaufgabe, tragen Ergebnisse in Felder ein und skizzieren durch Linien die typischerweise mit der Maus zurückzulegenden Strecken. Viele „Zacken" in den Linienzügen lassen auf eine nichtangepaßte Feldaufteilung schließen, häufige Übergänge zu anderen Formularen zeigen Probleme des Maskenzuschnittes auf.

- Lautes Denken
  Bei der Erprobung von Prototypen werden alle Aktionen mit dem Programm laut vorgetragen und aufgezeichnet / protokolliert. Die direkte Kommentierung beeinflußt stark den Prüfablauf und sollte auf Einzelaspekte, wie z. B. Wahrnehmbarkeit von Interaktionselementen, beschränkt werden.

- Benchmark-Test (bezogen auf die Benutzung)
  Bei Benchmark-Tests wird eine Programmentwicklung dadurch überprüft, daß mehrere Benutzer unter reproduzierbaren Umfeldbedingungen typische Arbeitsaufgaben erledigen. Mit objektivierbaren Methoden wie Zeitmessungen, Interaktionsverfolgung und Fehleranalysen lassen sich Hinweise auf Ergonomiedefizite gewinnen. Sinnvoll ist hierbei die Überprüfung unterschiedlicher Programmvariationen („Welches Modul zeigt die größten Fehlerhäufigkeiten?").

- Videokonfrontation (im Labor oder der tatsächlichen Arbeitsumgebung)
  Benutzer führen am Rechner eine definierte Prüfaufgabe mit dem zu bewertenden Softwareprodukt durch. Sämtliche Interaktionen werden aufgezeichnet, im Anschluß den Benutzern vorgeführt und von ihnen kommentiert. Angaben wie „Hier hatte ich vor, ... zu tun.", „Von Programm ... bin ich gewohnt, ..." oder „Das konnte ich nicht erkennen." geben Gestaltungshinweise zu allen Ebenen der Benutzungsschnittstelle.

5 Resümee

Der hier vorgestellte Methodenmix, die inzwischen mehrfach von uns praktisch erprobten Reviews und die Usability-Tests, versprechen sich derart zu ergänzen, daß trotz Reduzierung des Gesamtaufwands für Prüfungen und Qualifizierungen eine präzise, schnelle und frühzeitige Fehlererkennung und als Lernerfolg eine nachhaltige Fehlervermeidung zu qualitativen Verbesserungen bei der Softwareherstellung beiträgt.

Der Methodenverbund von Experten-Reviews und Usability-Tests führt zu folgenden Synergien:

- Durch vorgeschaltete Experten-Reviews wird sichergestellt, daß Benutzer sich intensiv mit Fragen zur Aufgabenangemessenheit befassen können und Usability-Tests damit nicht auf „Bugsuche" reduziert werden.

- Die Ergebnisse der Usablity-Tests stehen für eine zweiten Review-Zyklus zur Verfügung und erlauben den Experten eine deutlich bessere Perspektivenübernahme.

Beide Prüfverfahren leisten ihren spezifischen Beitrag zur Qualifizierung der Beteiligten:

- Entwickler erlernen bei der Auseinandersetzung mit dem Prüfungen ihrer konkreten Projekte effizient die Grundzüge der Software-Ergonomie und erkennen den Bedarf für eine tiefergehende Auseinandersetzung mit diesem Gebiet.

- Anwender und Benutzer entwickeln im Zuge von Usabilitiy-Tests eine gesteigerte Beurteilungskompetenz.

Der Einsatz des Methodenmixes erfordert von den Experten sowohl software-ergonomische Beurteilungs- und Gestaltungskompetenz (Review) als auch die Fähigkeit zur Durchführung der Usability-Tests, insbesondere zur Moderation des Kommunikationsprozesses zwischen den sich meist unstrukturiert äußernden Benutzern und den Entwicklern, die diesen Äußerungen in der Regel skeptisch gegenüberstehen.

Während einige Prüfmethoden von Entwicklern selbständig durchgeführt oder initiiert werden können, so bedarf es für die anvancierteren Usability-Tests entsprechender Infrastruktur, die von kleinen und mittelständischen Softwareherstellern allein kaum vorgehalten werden kann. Gemeinschaftliche Einrichtungen, z. B. von Universitäten und Softwarehäusern, stellen hier einen Beitrag zur Problemlösung dar.

6 Literatur

[1] Asam, R.; Drenkardt, N.; Maier, H.-H.: Qualitätsprüfung in Softwareprodukten. Berlin, 1986

[2] Clegg, C.W.; Warr, P.; Green ,T.; Monk, A.; Kamp, G.; Allison, G.; Lansdale, M.: People and Computers – How to Evaluate Your Companys New Technology. Chichester, 1988

[3] DGQ-NTG Deutsche Gesellschaft für Qualität e.V. (DGQ) und Nachrichtentechnische Gesellschaft im VDE (NTG): Software-Qualitätssicherung. Berlin, 1986

[4] DIN 66 234: Bildschirmarbeitsplätze, Grundsätze ergonomischer Dialoggestaltung. 1988

[5] DIN 66 285: Anwendungssoftware – Gütebedingungen und Prüfbestimmungen. August 1990

[6] DIN EN ISO 9241: Ergonomische Anforderungen für Bürotätigkeiten mit Bildschirmgeräten; Teil 10: Grundsätze der Dialoggestaltung. 1995 (noch als Entwurf)

[7] Dunn, R.H.: Software Qualität. München u.a., 1993

[8] Dzida, W.; Wiethoff, M.; Arnold , A.G.: ERGOguide. The Quality Assessment Guide to Ergonomic Software. St. Augustin, 1993

[9] Ehrlich, K.; Rohn, J.A.: Cost Justification of Usability Engineering: A Vendors Perspective. In: Bias, R.G.; Mayhew, D.J.: Cost Justifying Usability. Boston u.a., 1994. 71-110

[10] Englisch; J.: Systematische Entwicklung von Evaluationsverfahren zur Beurteilung der Benutzungsfreundlichkeit von CAD-Systemen. Forschungsbericht aus dem Institut für Arbeitswissenschaft und Betriebsorganisation der Universität Karlsruhe Band 4 (1992)

[11] Hamborg, K.-C.; Schweppenhäuser, A.: Zur Bedeutung psychologischer Arbeits- und Aufgabenanalyse für die Softwaregestaltung. In: Rödiger, K.-H. (Hrsg.): Software-Ergonomie '93. Stuttgart, 1993. 227-235

[12] Hampe-Neteler, W.: Software-ergonomische Bewertung zwischen Arbeitsgestaltung und Softwareentwicklung (Diss.), Bremen, 1994

[13] Hampe-Neteler, W.; Rödiger, K.-H.: Software-Ergonomie. Verfahren der Evaluierung und Standards zur Entwicklung von Benutzungsoberflächen. Bericht 2/1992, Fachbereich Mathematik/Informatik, Universität Bremen, 1992

[14] Hüttner, J.; Wandke, H.: What do system designers know about software-ergonomic and how to improve their knowledge. In: H Luczak; A.Cakir; G. Cakir: Work With Display Units. New York u.a., 1992. 304-308

[15] ISO 9000, Teil 3: Leitfaden für die Anwendung von ISO 9001 auf die Entwicklung, Lieferung und Wartung von Software. Juni 1992

[16] Jeffries, R.; Desurvire, H.: Usability Testing vs. Heuristic Evaluation: Was there a contest? In: SIGCHI Bulletin Vol 24, Nr.4. October 1992

[17] Jeffries, R.; Miller, J.R.; Wharton, C.; Uyeda, K.M.: User interface evaluation in the real world: A comparison of four techniques. In: Proceedings of CHI'91. New York, 1991. 119-124

[18] Lewis, C.; Polson, P.; Wharton, C.; Rieman, J.: Testing a Walkthrough Methodology for Theory-Based Design of Walk-Up-an-Use Interface. In: Proceedings of CHI'91. 1991. 235-242

[19] Nielsen, J.; Molich, R.: Heuristic Evaluations of User Interfaces. In: Proceedings of CHI'90. 1990. 249-256.

[20] Oppermann, R; Murchner, B.; Reiterer, H.; Koch, M.: Software-ergonomische Evaluation. Der Leitfaden EVADIS II. Berlin u.a., 1992

[21] Porteus, M; Kirakowski, J.; Corbett, M.: SUMI-Handboook. Human Factor research Group. Cork, 1993

[22] Prümper, J.; Anft, M.: Die Evaluation von Software auf Grundlage des Entwurfs zur internationalen Ergonomie-Norm ISO 9241 Teil 10 als Beitrag zur partizipativen Systemgestaltung - ein Fallbeispiel. In: K.-H. Rödiger (Hrsg.): Software-Ergonomie '93. Stuttgart, 1993. 145-156

[23] Ravden, S.; Johnson, G.: Evaluating Usability of Human Computer Interfaces. Chichester, 1989

[24] Wallace, D.F.; Norman, T.J.: Approaches to Interface-Design. In: Interacting with Computers No. 5., 1993. 259-278 s. a. http://www.lap.umd.edu/QUISFolder/quisHome.html

Adressen der Autoren

Dipl.-Inform. Peter Ansorge
Universität Bremen
Fachbereich Mathematik / Informatik
Technologie-Zentrum Informatik
Institut für Software-Ergonomie
und Informationsmanagement
Bibliothekstraße 1
28359 Bremen

E-Mail: ansorge@informatik.uni-bremen.de
http://selab24.informatik.uni-bremen.de/ansorge

Dipl. oec. Dipl.-Inform. Uwe Haupt
Universität Bremen
Fachbereich Mathematik / Informatik
Technologie-Zentrum Informatik
Institut für Software-Ergonomie
und Informationsmanagement
Bibliothekstraße 1
28359 Bremen

E-Mail: haupt@informatik.uni-bremen.de
http://selab24.informatik.uni-bremen.de/haupt

Konfiguration und Präsentation komplexer Anwendungen

Udo Arend und Edmund Eberleh

SAP AG, Usability Group

Zusammenfassung

In diesem Beitrag wird ein Ansatz zur modellbasierten Konfiguration eines Endbenutzersystems vorgestellt, der bestimmte Nachteile von adaptierbaren Systemen, adaptiven Systemen oder auch intelligenten Agenten vermeidet. Aufgrund von Metadaten zur betriebswirtschaftlichen Struktur einer Organisation als auch aufgrund von Aufgabenbeschreibungen können Systemeinstellungen explizit vorgenommen werden. Der Endbenutzer wählt diejenige Sicht auf den Anwendungsumfang der Software, der für ihn relevant ist, und erhält dadurch eine einfach zu bedienende Benutzungsoberfläche. Die in diesem Beitrag beschriebene neuartige Präsentationskomponente, der sogenannte „Session Manager", bietet neben dem einfachen Einstieg in Anwendungen noch eine intuitive, weitergehende Individualisierungsmöglichkeit. Die Vorteile des gewählten Ansatzes, die Software an die Arbeitsabläufe innerhalb von Organisationen anzupassen, werden dargestellt.

1 Problemstellung

Die Zahl der verfügbaren Anwendungsprogramme und Utilities nimmt ständig zu, ebenso wie die Funktionen der Anwendungsprogramme und Betriebssysteme selbst. Gleichzeitig steigt die Zahl der Benutzer von weltweit angebotener Software in den dreistelligen Millionenbereich. Die Softwarehersteller stehen vor folgenden Problemen: der Benutzer soll „seine" Programme schnell innerhalb der großen Menge anderer ebenfalls vorhandener finden, er soll innerhalb dieser Programme „seine" Funktionen schnell auslösen können, und er soll Aussehen und Organisation der Programme und Funktionen an seine ästhetischen und kognitven Strukturen anpassen können. Der software-ergonomischen Lösung dieser Probleme kommt beim Neuentwurf eines Produktes mittlerweile hohe Bedeutung zu [18]. Es lassen sich zur Zeit mindestens zwei gegensätzliche Lösungsstrategien hierfür identifizieren: Einstellung durch direkten Eingriff seitens des Benutzers und Einstellung durch „intelligente" Systemagenten. Wir argumentieren, daß beide Strategien aus einem arbeitswissenschaftlichen Blickwinkel betrachtet wesentliche Aspekte software-ergonomischer Systemgestaltung vernachlässigen. Nach einer Diskussion dieser Ansätze werden wir am Beispiel des SAP R/3-Systems einen dritten Weg zur Konfiguration und Präsentation komplexer Anwendungssysteme aufzeigen, der auf Aufgabenbeschreibungen basiert.

Diese Arbeit knüpft bzgl. des Präsentationsaspekts an Überlegungen von Eberleh & Meinke [4] an. Die nachfolgend dargestellte Arbeit verändert und konkretisiert die dort beschriebenen Überlegungen allerdings in wesentlichen Teilen und verbindet sie mit einem Konfigurationsverfahren, welches bisher noch nicht thematisiert wurde. Im folgenden wird daher nur eine Übersicht über bestehende Ansätze und Probleme bei der Konfiguration komplexer Anwendungssysteme gegeben. Für eine Übersicht über die Präsentationsprobleme und Lösungen unterschiedlicher Betriebssysteme sei auf Eberleh & Meinke [4] verwiesen.

2 Bestehende Ansätze zur Konfiguration komplexer Anwendungssysteme

2.1 Put the User in Control !?

Aus der psychologischen Forschung, z.B. [17], ist seit längerem bekannt, daß subjektiv wahrgenommene fehlende Kontrolle über die Umwelt irritierend, belastend und motivationsmindernd wirkt. Diese Erkenntnis liegt letztlich den software-ergonomischen Forderungen nach Steuerbarkeit, Transparenz und Erwartungskonformität von Softwaresystemen zugrunde [5, 19] und hat im Prinzip der „Direkten Manipulation" und im „WYSIWYG-Prinzip" Einfluß auf die Gestaltung heutiger Softwaresysteme genommen [1]. Die Verletzung dieser Prinzipien oder „usability heuristics" trägt nach einer Analyse von Nielsen [14] wesentlich zur Erklärung von Benutzbarkeitsproblemen bei. Die erklärungsstärksten ersten vier Faktoren hierfür lauten: (1) Visibility of system status, (2) Match between system and real word, (3) User control and freedom, (4) Consistency and standards.

Nach diesen Prinzipien gestaltete Systeme bieten dem Benutzer grafisch eine möglichst vertraute Abbildung seiner bekannten Umwelt und deren Objekte an und erlauben ihm eine möglichst direkte und sofort sichtbare Änderung dieser Abbildung. Programmsymbole können in diesem Paradigma direkt an den gewünschten Ort verschoben werden und auf diese Weise sichtbarer und zugreifbarer gemacht werden oder versteckt werden, weil weniger wichtig oder häufig genutzt. Da die direkte Manipulation in vielen Fällen an Grenzen stößt, wird das Sichtbar- bzw. Unsichtbarmachen von Programmen oder Funktionen oft zusätzlich oder ausschließlich indirekt mittels Menüs und Funktionstasten angeboten. Neben dieser Wahl der Anordnung und Sichtbarseins von Arbeitsobjekten kann der Benutzer in der Regeln noch die grafische Repräsentation vieler Objekte beeinflussen. Zentrales Merkmal eines derartigen Systems ist, daß das System solange unverändert bleibt, bis der Benutzer mittels der genannten Möglichkeiten explizit die Systemkonfiguration ändert, es somit ein adaptierbares System ist [15]. Dieses Prinzip der Adaptierbarkeit führt aus mehreren Gründen zu Problemen, wie am Beispiel von Windows 95 deutlich wird (Abb. 1):

1. Es werden redundante Möglichkeiten für die gleiche Aufgabe angeboten: Programme können gleichzeitig im Startmenü, in der Office-Shortcutleiste, in der Task-Bar, im Benutzermenü des Startmenüs, auf dem Desktop und im Ordnerfenster abgelegt werden, ermöglicht durch Objektkopien. Es ist fraglich, ob Benutzer mit diesem Konzept der Objektreferenzen tatsächlich eine Kontrolle der Umwelt bekommen.

2. Viele Einstellungsprobleme werden auf den Benutzer verlagert, in der Annahme, daß der Benutzer selbst am besten weiß, was für ihn optimal geeignet ist. Dieses ist unserer Meinung nach bei der Art und Fülle heute angebotener Einstellungsmöglichkeiten zu hinterfragen. In den Prüfgrundsätzen der Unfallverhütungsvorschrift [7] geht man folgerichtig nur noch davon aus, daß ein System in *einer* Einstellung der Vorschrift genügen muß, um das Prüfsiegel zu bekommen. Was dann vor Ort der Benutzer am System verändert, entzieht sich völlig der Kontrolle des Herstellers, so daß ein ergonomie-geprüftes System in der Praxis völlig zum Ungünstigen hin verstellt werden kann.

3. Die Einstellungsdialoge erhöhen zumindest initial den Aufwand im Bereich des Interaktionsproblems. Inwieweit längerfristig dadurch tatsächlich die beabsichtigte Optimierung im Arbeiten mit dem Softwaresystem eintritt, ist gerade unter Berücksichtigung von Punkt 2 fraglich.

4. Die Einstellungen müssen von jedem Benutzer individuell neu vorgenommen werden. In manchen Fällen werden zwar Voreinstellungen oder Paletten mitgeliefert. Diese bieten eine gewisse Erleichterung bei der Einzelauswahl, geben aber trotzdem kaum einen Hinweis auf die Angemessenheit der Wahl.

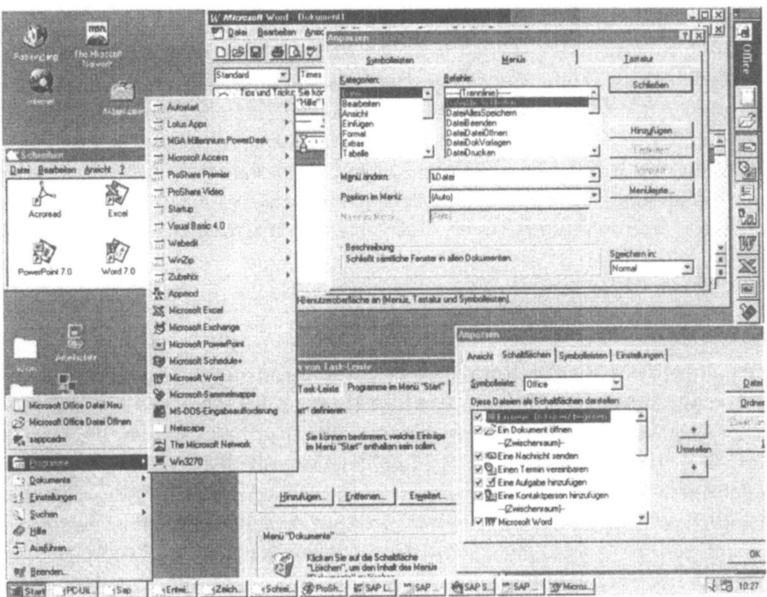

Abb. 1: Konfigurationsmöglichkeiten von Anwendungen in Windows 95

Insgesamt stehen den positiven Möglichkeiten einer adaptierbaren Benutzungsschnittstelle somit vielfältige Kosten gegenüber, die ein exaktes Bewerten dieses Gestaltungsprinzips auch bei empirischen Untersuchungen erschweren und bisher unklar lassen [11].

2.2 Don´t bother the User with all Tasks !?

Eine weitere, bereits seit längerem bestehende arbeitspsychologische Forderung besteht darin, die Aufgaben- und Funktionsverteilung zwischen Mensch und Rechner ausdrücklich im Gestaltungsprozeß eines Mensch-Rechner-Systems mit zu berücksichtigen [8, 10]. Ein Teil dieser Forderung ist in das Prinzip der Aufgabenangemessenheit eingeflossen, welches u.a. aussagt, daß der Benutzer bei seiner Aufgabenerledigung durch das System adäquat unterstützt wird, indem es ihm bestimmte Routinetätigkeiten oder für die Sachaufgabe nicht notwendige Tätigkeiten abnimmt [5]. Um dieses zu erreichen, ist ein Modell über den Aufgabenbereich und -ablauf und den Benutzer erforderlich. Derartige Modelle können zur Erreichung von Aufgabenangemessenheit in zwei Richtungen eingesetzt werden: (1) zur Verringerung des Interaktionsproblems, indem der Computer die Präsentation und Interaktion selbständig an den Benutzer optimal anpaßt (adaptives System), und (2) zur Unterstützung bei der Sachaufgabe, indem der Computer dem Benutzer schwierige oder langweilige Tätigkeiten selbständig abnimmt (intelligente Agenten). In beiden Fällen verliert der Benutzer mehr oder weniger die Kontrolle über das Verhalten des Computersystems. Es liegt somit ein Konflikt bezüglich verschiedener software-ergonomischer Gestaltungskriterien vor. Ein Vorschlag zur Lösung des Konfliktes zwischen Flexibilität und Transparenz besteht im Anbieten einer „Schalttafel", wo sämtliche Modifikationsmöglichkeiten mit dem aktuellen Zustand ersichtlich sind und auf Wunsch geändert werden können [6]. Eine ausschließliche, automatische Anpassung und Einstellung der Benutzungsoberfläche durch das System selbst wird denn auch als kritisch eingeschätzt [16], ist jedoch trotzdem Gegenstand aktueller Arbeiten zur Softwareergonomie und kontextsensitven Hilfesystemen [2]. Kontrovers verhält es sich ebenfalls bezüglich der Arbeiten zu intelligenten Agenten. In einer Reihe von Studien werden Agenten als Lösung zur Bewältigung der Informationsflut verwendet [12] und werden auch zur systemgesteuerten Auswahl der für den Benutzer sinnvollen Funktionalität eingesetzt [3]. Der Benutzer gibt somit als Preis einer effektiven Aufgabenerledigung einen wesentlichen Teil seiner Autonomie an das System ab, was von anderen Autoren als äußerst kritisch gesehen wird [9].

2.3 The User wants to get his Job done !

Im folgenden möchten wir einen dritten Weg zur Unterstützung des Benutzers beim Einstellen seines Systems aufzeigen. Diese Alternative basiert auf der Annahme, daß ein Benutzer in einer Arbeitssituation (nur dieser Anwendungskontext liegt den weiteren Überlegungen zugrunde) in der Regel nicht die Zeit und Lust hat, sich mit den vielfältigen Individualisierungsmöglichkeiten seines Rechners intensiv zu beschäftigen, da diese Tätigkeiten nichts mit seiner eigentlichen Aufgabe zu tun haben. Idealerweise sollte der Benutzer eine Software vorfinden, welche für seine persönliche Arbeitstätigkeit bereits so gut wie möglich angepaßt ist, so daß er nur noch geringfügige und aktiv kontrollierte Anpassungen vornehmen muß. Die Vorkonfiguration des Systems sollte nicht in kaum nachvollziehbarer Weise dynamisch durch einen „intelligenten" Agenten geschehen, sondern auf der Grundlage einer expliziten und damit diskutierbaren Aufgabenbeschreibung des Benutzers. Die „Intelligenz" eines Softwaresystems sollte unserer Meinung nach ausschließlich dazu dienen, die von den Benutzern als sinnvoll definierte Organisations- und Aufgabenstruktur in ein leicht benutzbares Softwarewerkzeug zur Unterstützung dieser Inhalte umzusetzen. Wie eine derartige Lösung konkret aussehen kann, stellen wir nachfolgend am Beispiel des SAP R/3-Systems dar.

3 Kurze Charakterisierung des R/3-Systems und Probleme

Das R/3-System der SAP AG ist ein Softwarepaket für integrierte betriebswirtschaftliche Standardanwendungen. Es ist modular aufgebaut und deckt den gesamten Bereich betrieblicher Funktionen ab. Während das R/3-System nach außen als eine abgeschlossene Einheit im Desktop des jeweiligen Front-End-Geräts erscheint, stellt es nach innen ein sehr komplexes Paket vieler miteinander vernetzter Anwendungen und Daten dar. In der Vergangenheit (bis Release 3.0) wählte ein Benutzer seine Funktionen folgendermaßen aus: wenn er den Namen (Transaktionscode) der Anwendung kennt, konnte er ihn direkt in einem entsprechenden Feld eingeben (Gedächtnisproblem!). Der normale Weg war jedoch, die Anwendung durch Klikken mit der Maus aus einem Menübaum herauszusuchen (vgl. Abb. 2) wie beim Startmenü von Windows 95.

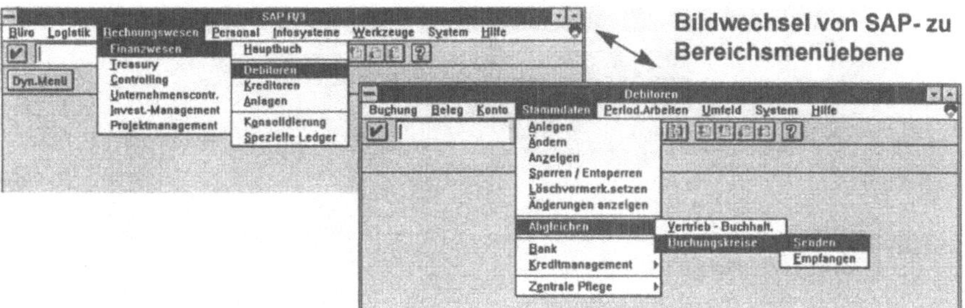

Abb. 2: Herkömmlicher Menüeinstieg in R/3-Anwendungen

Bei einer maximalen Zahl von ca. zehntausend (!) Anwendungs- und Berichtsvarianten war dies für den Endbenutzer nicht immer eine einfache Aufgabe, selbst wenn die Menüebenen bereits aufgabenangemessen strukturiert waren [4]. Ein durchschnittlicher Benutzer verwendet nur 10 bis 50 Anwendungen aus der Gesamtmenge, die allerdings in der Regel sehr häufig (Sachbearbeiter). Allerdings gibt es noch eine zweite Gruppe von Benutzern, die sehr viel mehr Anwendungen nutzen, diese jedoch eher unregelmäßig und seltener (Führungskräfte) [13]. Mit dem nachfolgend dargestellten Ansatz wollten wir drei Probleme lösen:

1. wie können Benutzer die von ihnen genutzten Funktionen aus einem umfangreichen Gesamtsystem auswählen,
2. wie kann das System sie dabei unterstützen, so daß nicht jeder Benutzer diese Auswahl alleine und immer wieder neu treffen muß,
3. wie können Benutzer eine einmal getroffene Auswahl individuell und schnell anpassen

4 Ein Ansatz zur modellbasierten Konfiguration

Das Spektrum der Anwender einer komplexen betriebswirtschaftlichen Software reicht vom mittelständischen Betrieb bis zum multinationalen Konzern. Ebenso sind alle Branchen vertreten, die das R/3-System einsetzen, seien es z.B. traditionelle Maschinenbauunternehmen, die Chemiebranche, die Autoindustrie, Hitech-Unternehmen, Energieversorger, Versandhandel, Medienkonzerne oder auch öffentliche Unternehmen. Entsprechend unterschiedlich sind die Nutzungsbedürfnisse der Anwender. Daraus ergibt sich die Notwendigkeit, die Standardsoftware auf die jeweiligen Bedürfnisse des Kunden einzurichten. Um dies zu ermöglichen, können die Einstelltransaktionen des Systems topdown aufgrund von einer sogenannten „Anwendungskomponentenhierarchie" ausgewählt werden. Dieser Komponentenbaum bildet eine generalisierte betriebswirtschaftliche Organisationstruktur ab (s. Abb. 3). Die Menge der angezeigten Anwendungen ändert sich jedoch nicht.

Die Idee für die im folgenden beschriebene neuartige Präsentationskomponente des R/3-Systems, den sogenannten „Session Manager", ist es, diese Metadatenstruktur der Komponentenhierarchie für die Auswahl der betriebswirtschaftlichen Anwendungen auszunutzen. Die angestrebte Lösung soll die Komplexität der bisherigen Auswahl von Anwendungen für den Endbenutzer (vgl. Kapitel 3) dadurch reduzieren, daß die Gesamtmenge der R/3-Anwendungen zweifach reduziert wird:

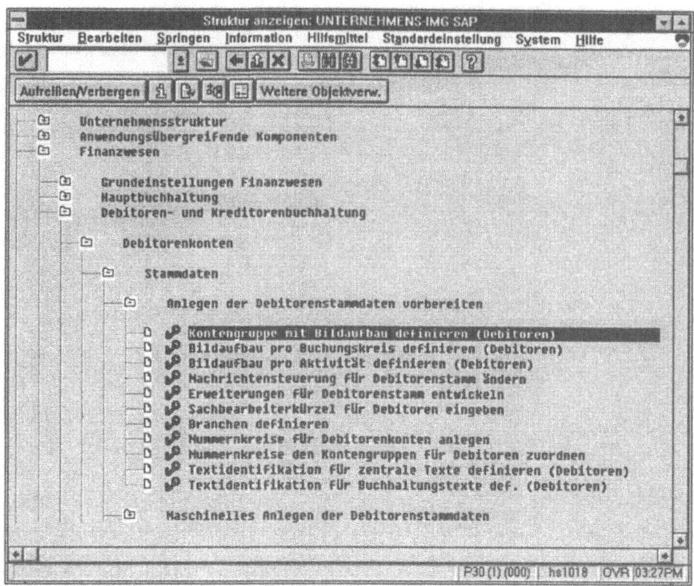

Abb. 3: Ein kleiner Ausschnitt der Anwendungskomponentenhierarchie

1. Ziel: Bereitstellung eines *Unternehmensmenüs* durch eine modellbasierte *Reduktion* der Gesamtzahl von Anwendungen im R/3-System auf diejenigen, die der Kunde tatsächlich nutzen möchte, und durch *Konstruktion* eines sichtbaren Menübaumes, der die Anwendungen tätigkeitsorientiert darstellt. Ansatz: Nutzung der Darstellungsstruktur des „alten" R/3-Menübaumes (vgl. Abb. 2), an dem alle Anwendungen hängen, und zusätzliche Zuordnung aller R/3-Anwendungen zu den Anwendungskomponenten der Komponentenhierarchie. Bei der Systemeinführung werden diejenigen Zweige aus der Komponentenhierarchie ausgewählt, die aktiv werden sollen. Dadurch bestimmt sich die Menge der Anwendungen, die das „unternehmensspezifische Menü" bilden. Aus der reduzierten Menge der Anwendungen wird der sichtbare Baum konstruiert.

2. Ziel: Bereitsstellung eines *Benutzermenüs* durch eine modellbasierte *Reduktion* der im Unternehmen eingesetzten Anwendungen, auf diejenigen, die ein bestimmter Endbenutzer ausführen darf und soll und ebenfalls durch *Konstruktion* eines sichtbaren Menübaumes wie unter 1. beschrieben. Ansatz: Organisationen lassen sich dadurch charakterisieren, daß Mitarbeiter auf ihnen zugeordneten Stellen bestimmte Aufgaben ausführen. Die Idee ist, im R/3-System *Stellen* mit zugehörigen *Aufgaben* zu definieren, d.h. einer Stelle R/3-Anwendungen zuzuordnen und diese einem Stelleninhaber mitzugeben. Aus der Menge der Aufgaben einer Stelle (=R/3-Anwendungen) läßt sich dann ein benutzerspezifisches Menü generieren.

Abbildung 4 verdeutlicht den Ansatz: R/3-Anwendungen werden einmal der Komponentenhierarchie zugeordnet und zum anderen vordefinierten Stellen. Aufgrund der Auswahl bestimmter Bereiche aus der Komponentenhierarchie können die benötigten Anwendungen bestimmt werden, und mit Hilfe der Struktur des alten Menübaumes kann das Unternehmensmenü konstruiert werden. Mit Hilfe einer Stelle werden die Anwendungen pro Benutzer bestimmt, und daraus wird aufgrund der Menübaumstruktur das Benutzermenü konstruiert.

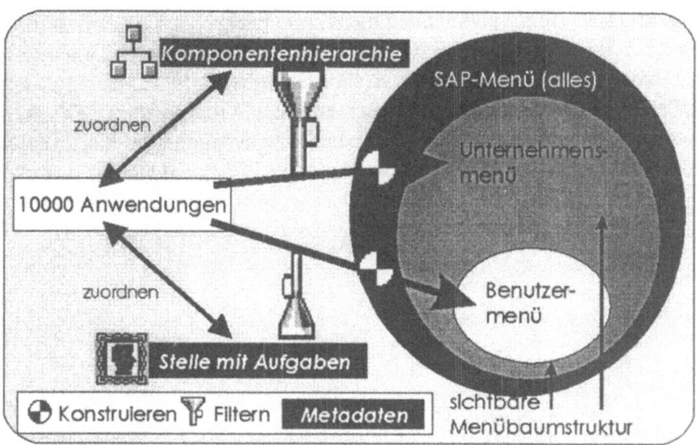

Abb. 4: Struktur zur modellbasierten Konfiguration

Das folgende Kapitel erläutert die von uns gewählte Lösung im Detail.

5 Die Konstruktionskomponenten des Session Managers

5.1 Die sichtbare Menübaumstruktur und die zugrundeliegende Struktur

Die Komponentenhierarchie bildet eine betriebswirtschaftliche Gliederung eines Unternehmens. Der Endbenutzer dagegen muß bestimmte Aufgaben an seinem Arbeitsplatz bearbeiten, die eine zeitliche aber auch inhaltliche Gliederung besitzen. Welche Aufgaben der Endbenutzer bearbeiten darf, hängt zusätzlich von seiner Stellung im Unternehmen ab. Die Struktur des eigentlichen Menübaums mit allen Anwendungen, den der Endbenutzer sieht, soll aber seinen Arbeitsbereich möglichst adäquat abbilden. Diese Information liegt im System bereits in Form der „alten" Menübäume vor. Die Menübäume wurden so definiert, daß sie möglichst ganzheitlich bestimmte Arbeitsbereiche (Bereichsmenü) abbilden [4], d.h. die Menübäume eignen sich zur Darstellung des „unternehmensspezifischen" als auch „benutzerspezifischen" Menüs. Die Verbindung zwischen den sichtbaren Menübäumen und der Komponentenhierarchie wurde von uns durch Zuordnung aller Anwendungen zur Komponentenhierarchie geschaffen. Damit sind alle Daten abgelegt, um aus einer vom Kunden reduzierten Komponentenhierarchie ein sichtbares Unternehmensmenü zu erzeugen.

5.2 Stellen, Aufgaben und Tätigkeitsklassen

Für die Erzeugung des Benutzermenüs sind weitere Daten im System abzulegen. Dabei muß der Prozeß der Definition von Stellen mit Aufgaben sinnvoll unterstützt werden. Eine einfache Auswahl von Anwendungen aus der Gesamtmenge von 10000 Anwendungen würde bei dem Stellendefinierer voraussetzen, daß er genau weiß, welche Anwendungen für welche Stelle benötigt werden. Man kann nicht davon ausgehen, daß dieses Wissen vorhanden ist. Was allerdings vorhanden ist, ist ein Wissen über die Struktur der Komponentenhierarchie und über die hierarchische Position der zu definierenden Stelle innerhalb der Organisation, z.B. Führungskraft. Mit Hilfe von Anwendungskomponenten der Komponentenhierarchie läßt sich eine Stelle bzgl. ihrer organisatorischen Einbettung beschreiben, z.B. Debitorenbuchhaltung. Durch Auswahl der hierarchischen Position innerhalb der Organisation dagegen kann das Spektrum der Aufgaben beschrieben werden. Mit diesen beiden Dimensionen lassen sich die Aufgaben einer Stelle genau beschreiben. Zur Festlegung der hierarchischen Position ist eine Zuordnung der Anwendungen zu sogenannten „*Tätigkeitsklassen*" erforderlich.

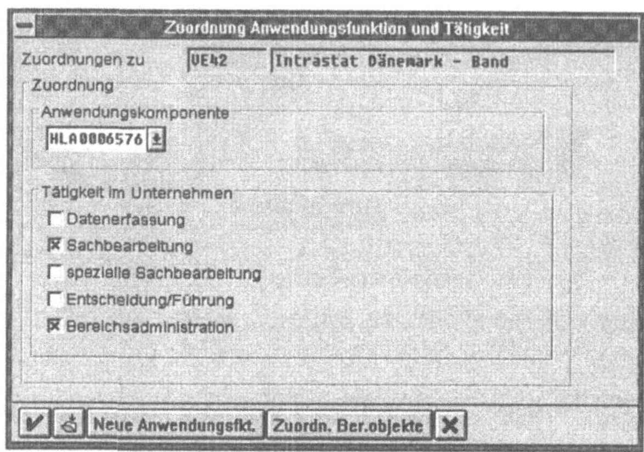

Abb. 5: Definition von Tätigkeitsklassen

Fünf Tätigkeitsklassen reichen nach unserem Verständnis dazu aus (vgl. Abb. 5). Alle Anwendungen wurden gezielt diesen Tätigkeitsklassen zugeordnet. Beide Dimensionen sind zueinander orthogonal. Mit dem gewählten Modell ist z.B. die Situation abbildbar, daß ein Sachbearbeiter im Bereich Controlling anspruchsvollere Anwendungen ausführen kann als ein Sachbearbeiter im Rechnungswesen.

5.3 Filterung und Konstruktion

Durch die Zuordnung aller Anwendungen zu Anwendungskomponenten und Tätigkeitsklassen können sowohl das Unternehmensmenü und aufgrund einer Stellenbeschreibung das Benutzermenü automatisch erzeugt werden. Dazu wird ein zweistufiger Prozeß benutzt: ein *Analyseprozeß* und ein nachfolgender *Konstruktionsprozeß*.

Der *Analyseprozeß für das Unternehmensmenü*: der Kunde wählt aus, welche Zweige der Komponentenhierarchie von ihm gewünscht sind. Anhand der Tabelle, die die Zuordnungen Anwendungen zu Anwendungskomponente enthält, werden diejenigen Anwendungen vermerkt, die den ausgewählten Anwendungskomponenten zugeordnet wurden.

Der *Analyseprozeß für das Benutzermenü*: in einer Stellenbeschreibung werden Tupel aus Anwendungskomponenten und Tätigkeitsklassen festgelegt. Aufgrund dieser Tupel werden anhand der Zuordnungstabelle die Anwendungen vermerkt, die den entsprechenden Tupeln entsprechen.

Der *Konstruktionsprozeß*: aufgrund der übriggebliebenen Anwendungen werden alle die Zweige aus dem zur Präsentation anstehenden Gesamtmenübaum eliminiert, die an ihren

Blättern keine Anwendungen mehr enthalten. Es entsteht ein Menübaum, der nur noch die gewünschten Anwendungen enthält.

Der entstandene Menübaum, sei es das Unternehmensmenü oder das Benutzermenü, wird von der Präsentationskomponente automatisch angefordert und auf dem PC gepuffert.

6 Die Präsentationskomponente des Session Managers

Der Session Manager präsentiert sich dem Endbenutzer als eigenes paralleles Fenster. Wenn man sich am R/3-System anmelden möchte, startet man den Session Manager und gibt Benutzernamen, Kennwort, Mandant und Sprache an und wählt eine Systemkennung aus.

Nach erfolgter Anmeldung kommt jeweils pro Systemanmeldung ein eigenes Register hoch, welches die eigentlichen Komponenten zur zentralen Steuerung der Anwendungen enthält (vgl. Abb. 6).

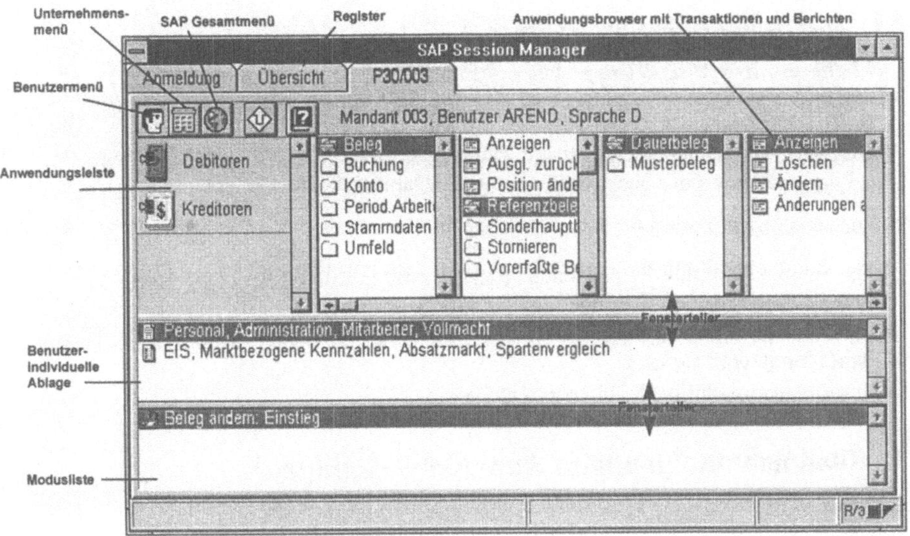

Abb. 6: Aufbau der Systemregister des Session Managers

Die oberste Reihe wird durch die Systemregister gebildet. Hier kann der Benutzer zwischen Anmeldebild, Übersichtsbild und den einzelnen Systemen wechseln. Darunter erscheint eine Kontrolleiste mit den Funktionen zur Auswahl einer bestimmten Sicht auf die R/3-Anwendungen (Benutzermenü, Unternehmensmenü und SAP-Gesamtmenü), der Abmeldenfunktion und der Funktion zum Aufruf der Hilfe zum Session Manager. Die Anwendungsleiste enthält die Einstiegsbereiche in das R/3-System. Im Falle des SAP-Menüs sind dies die Bereiche „Finanzwesen", „Personal", „Informationssysteme-Katalog", „Logistik", „Büroanwendungen" und „Werkzeuge" (Workbench, Administration). Der Anwendungsbrowser neben der Anwendungsleiste enthält dann die jeweiligen Äste der Anwendungen zu dem gewählten Anwendungsbereich.

Ein Beispiel möge die Funktionsweise verdeutlichen: wählt man das Gebäude-Icon aus der Kontrolleiste aus, wird das aufgrund des Einführungsprozesses erzeugte Unternehmensmenü geladen und angezeigt. Es erscheinen in der Liste links und im Browser nur noch die Äste, die aufgrund der Auswahl von Komponenten übriggeblieben sind. Der Benutzer klickt in der Anwendungsleiste einen Bereich an. Von links nach rechts wählt er sich bis zur gewünschten Anwendung und startet diese durch Doppelklick mit der Maus oder durch Ziehen in den Bereich der Moduslist. Die browserartige Darstellung eines Baums hat gegenüber der baumartigen Darstellung im Microsoft Explorer den Vorteil, daß der jeweilige Kontext zu einer getroffenen Wahl auf der nächsthöheren Hierarchieebene des Baumes erhalten bleibt.

Der mittlere Bereich des Session Managers beinhaltet die „benutzerspezifische Ablage". Hierein kann der Benutzer mittels Drag und Drop häufig genutzte Anwendungen hineinziehen. Die Anwendungen werden genauso gestartet wie im Browser. Der untere Bereich, die „Moduslist", zeigt die aktiven Anwendungen. Mit Hilfe der Fensterteiler lassen sich die Bereiche Browser, Ablage und Moduslist in der Größe den Wünschen des Benutzers anpassen.

6.1   Das benutzerspezifische Menü

Das Benutzermenü wird geladen, wenn man auf das Icon mit dem „Kopf" klickt. Links erscheinen in der Anwendungsleiste als Einstieg nun die aufgrund der Stellendefinition erzeugten Bereichsmenüs (vgl. Abb. 6). Wählt man eines aus, wie in der Abbildung „Debitoren", erscheinen rechts im Anwendungsbrowser die zugehörigen Hierarchiestufen. Man klickt sich von links nach rechts durch und wählt eine Anwendung aus.

Der entscheidende Vorteil des Benutzermenüs ist:

- der Baum enthält nur die Anwendungen, die dem jeweiligen Benutzer zugeordnet wurden und für die er auch eine Berechtigung hat
- die Zahl der Hierarchiestufen ist klein, da bereits auf Bereichsmenüebene eingestiegen wird (Breite vor Tiefe)
- das Benutzermenü ist PC-unabhängig.

7   Resümee und geplante Weiterentwicklungen

Grundlage zur Entwicklung des Session Managers waren zwei Forderungen:

- die Konfiguration eines komplexen Systems soll transparent und topdown durch den Kunden erfolgen und zwar auf Grundlage des Arbeitsablaufs bei der Einrichtung [10]
- der Konfigurationsprozeß soll durch die Nutzung von Metadaten unterstützt werden.

Die erste Forderung wurde umgesetzt, indem zum einen eine Sicht der Organisation geschaffen wurde (betriebliche Komponentenhierarchie und Stellenbeschreibungen) und zum anderen davon losgelöst eine Sicht des Benutzers auf das System (Benutzermenü und Unternehmensmenü im Session Manager). Die zweite Forderung wurde umgesetzt, indem alle Anwendungen durch Experten (SAP-Berater mit guten Praxiskenntnissen) klassifiziert wurden, d.h. Anwendungskomponenten und Tätigkeitsklassen zugeordnet wurden.

Der Nutzen ergibt sich für alle Beteiligte: das Projektteam, daß ein R/3-System einführt, kann transparent alle Komponenten abschalten, die nicht eingesetzt werden. Die Bereichsadministratoren können mit einem Minimum an Aufwand Stellen für ihre Benutzer definieren und haben dabei gleichzeitig eine Kontrolle darüber, was der Benutzer soll und darf. Der Endbenutzer selbst enthält genau die Anwendungen in einem stark reduzierten Menübaum, die für ihn relevant sind. Mit einer einfach strukturierten Benutzungsoberfläche erhält er nicht nur die Kontrolle über gestartete Prozesse, sondern kann noch weitere Individualisierungen durch einfaches Drag und Drop vornehmen. Damit ist eine sinnvolle Kombination von direkter Manipulation und modellbasierter Steuerung gefunden worden. Die Resonanz von Beratern und Nutzern ist entsprechend positiv.

Aufwand entsteht bei dem gewählten Ansatz an einer Stelle - darauf soll abschließend hingewiesen werden: die erzeugten Metadaten müssen ständig aktuell gehalten werden (wenn z.B. neue Anwendungen ergänzt werden) und die Klassifikation einer Anwendung muß valide sein. Um hierbei auf der sicheren Seite zu sein, wird bei dem zur Zeit ausgelieferten Session Manager auf die Auswertung der Zuordnung zu Tätigkeitsklassen verzichtet, um von den Kunden Daten über den Prozeß der Stellendefinition zu sammeln. Wenn Erfahrungen vorliegen, kann eine entsprechende Erweiterung vorgenommen werden.

Naheliegend ist auch, daß vom Session Manager nicht nur SAP-Anwendungen aufgerufen werden können, sondern auch beliebige andere Desktop-Applikationen. Dem Benutzer stünde dann ein vollständig integrierter Desktop zur Verfügung. Entsprechende Anforderungen von SAP-Kunden liegen bereits vor.

Literaturverzeichnis

[1] Apple Computer. Macintosh Human Interface Guidelines. Addison-Wesley, Reading, MA, 1992.

[2] D. Benyon, T. Kühme, U. Malinowski, S. Piyawadee: Computer-aided adaptation of user interfaces. An INTERCHI '93 Workshop Report, SIGCHI Bulletin, Vol. 26, Jan. 1994, p. 25.

[3] A.E. Blandford: An agent-theoretic approach to computer participation in dialogue. Int. J. Man-Machine Studies, 39, 1993, pp. 965-998.

[4] E. Eberleh, F. Meinke: OASE: Eine Arbeitsplatzumgebung für komplexe Anwendungssysteme. In: H.D. Böcker (Hg.): Software-Ergonomie '95. Teubner, Stuttgart, 1995, 125-141.

[5] DIN 66234, Teil 8: Bildschirmarbeitsplätze - Grundsätze ergonomischer Dialoggestaltung, Berlin, Beuth, 1988.

[6] T. Greutmann, D. Ackermann: Zielkonflikte bei Software-Gestaltungskriterien. In: S. Maaß, H. Oberquelle (Hg.): Software-Ergonomie '89. Teubner, Stuttgart, 1989, 144-152.

[7] Grundsätze für die Prüfung der Arbeitssicherheit von Elektronischen Datenverarbeitungs- und Bildschimgeräten, Entwurf 04/96. Fachausschuß Verwaltung der Berufsgenossenschaften, Hamburg, 1996.

[8] W. Hacker: Arbeits- und organisationspsychologische Grundlagen der Software-Ergonomie. In: E. Eberleh et al. (Hg.): Einführung in die Software-Ergonomie. DeGruyter, Berlin, 1994, 53-94.

[9] G. Hellbardt: Die Ethik von Agenten. In: Informatik-Spektrum, 19, 1996, 87-90.

[10] ISO 13407. Human centred design processes for interactive systems. Committee Draft International Standard, 1996-02.

[11] C. Karger, R. Oppermann: Empirische Nutzungsuntersuchungen adaptierbarer Schnittstelleneigenschaften. In: D. Ackermann & E. Ulich (Hg.): Software-Ergonomie '91. Teubner, Stuttgart, 1991, 273-280.

[12] P. Maes, A. Wexelblatt: Interface Agents. Workshop at CHI '96 in Vancouver, BC, Canada, April 13-18, 1996.

[13] C. Neugebauer: Arbeitsorganisation & mentale Repräsentation betriebswirtschaftlicher Software. - Wissensstrukturen als Grundlage benutzerorientierten Interface-Designs. Unveröffentlichte Diplomarbeit, RWTH Aachen, 1991.

[14] J. Nielsen: Enhancing the explanatory power of usability heuristics. In: Proceedings of the CHI, 1994 (Boston, MA, USA, April 24-28, 1996) ACM, New York, 1994, pp. 152-158.

[15] R. Oppermann: Individualisierung von Benutzungsschnittstellen. In: E. Eberleh et al. (Hg.): Einführung in die Software-Ergonomie. DeGruyter, Berlin, 1994, 235-269.

[16] R. Oppermann, C. Thomas: Auf dem Weg zur benutzerspezifischen Systemanpassung. In: W. Coy, P. Gorny, I. Koop, C. Skarpelis (Hg.): Menschengerechte Software als Wettbewerbsfaktor. Teubner, Stuttgart, 1993, 442-462.

[17] M.E.P. Seligman: Erlernte Hilflosigkeit. München, Urban & Schwarzenberg, 1983.

[18] K. Sullivan: The Windows 95 Interface: A Case Study in Usability Engineering. In: Proceedings of the CHI '96 (Vancouver, BC, Canada, April 13-18, 1996) ACM, New York, 1996, pp. 473-480.

[19] E. Ulich: Von der Benutzungsoberfläche zur Arbeitsgestaltung. In: K.H. Rödiger (Hg.): Software-Ergonomie '93. Teubner, Stuttgart, 1993, 19-30.

Adressen der Autoren

Dr. Udo Arend
Usability Group
SAP AG
Neurottstr. 16
69190 Walldorf/Baden
Email: udo.arend@sap-ag.de

Dr. Edmund Eberleh
Usability Group
SAP AG
Neurottstr. 16
69190 Walldorf/Baden
Email: edmund.eberleh@sap-ag.de

Einsatz elektronischer Lehr- und Lernumgebungen in der Software-Ergonomie-Ausbildung

Andreas Brennecke und Reinhard Keil-Slawik

Heinz Nixdorf Institut, Universität-GH Paderborn

Zusammenfassung

In der Software-Ergonomie stellt die notwendige praktische Ausbildung ein besonderes Problem dar. Betreuungsintensive Praktika mit kleinen Gruppen Studierender sind oft nicht durchführbar, weil die Personalkapazitäten nicht ausreichen.

In diesem Beitrag wird über den Einsatz einer Multimediainfrastruktur zur effektiven Durchführung einer Software-Ergonomie-Veranstaltung berichtet. Die hierbei gemachten Erfahrungen ermutigen zum Weitermachen, zeigen aber auch, an welchen Stellen noch Verbesserungen im System sowie Veränderungen im organisatorischen Ablauf und didaktischen Aufbau notwendig sind.

1 Einleitung

Software-Ergonomie wird in der universitären Ausbildung immer noch stiefmütterlich behandelt. Obwohl es bereits ein ausgearbeitetes Software-Ergonomie-Curriculum gibt (siehe Maaß et al. [11]) ist dieses in den wenigsten Fällen in bestehende Studien- und Prüfungsordnungen integriert. So hängt die Ausbildung auf dem Gebiet der Software-Ergonomie sehr stark von dem Engagement einzelner Lehrpersonen ab und kann von diesen oft auch nicht im notwendigen Umfang durchgeführt werden.

Die Konsequenz ist, daß viele Informatikstudierende die Universitäten verlassen, ohne das geringste Wissen über die Prinzipien software-ergonomischer Gestaltung zu haben. Insofern ist es nicht verwunderlich, daß in der Praxis oft noch nicht einmal ein schon in Normen verankertes Wissen, z. B. über die Ausrichtung von Bildschirmelementen DIN 66234 Teil 3 (Anforderungen an Visuelle Anzeigen), beachtet wird (wie beispielsweise ein Anfang 1996 von der DGB Technologieberatung Berlin erstelltes Software-Ergonomie-Gutachten (Brennecke, Holl [5]) zeigt). Um in der Praxis solch unzureichend gestaltete Systeme zu vermeiden, muß wenigstens ein minimales Maß an Software-Ergonomie-Ausbildung stattfinden, auch wenn es bisher kaum Lehrstühle zur Software-Ergonomie gibt.

Eine praxisbezogene Ausbildung stellt aufgrund der hohen Studierendenzahlen ein Problem dar. Deshalb werden beispielsweise im Rahmen der Software-Ergonomie-Ausbildung in Bremen neben einer Vorlesung auch ein Software-Ergonomie-Praktikum und eine Software-Ergonomie-Werkstatt durchgeführt (siehe Ansorge, Friedrich, Streibl [1]). Solche Veranstaltungen erfordern zwar einen recht hohen Betreuungsaufwand, werden aber ausdrücklich in den Empfehlungen des Fachausschusses 2.3 und des Fachbereichs 2 der Gesellschaft für Informatik für die Software-Ergonomie-Ausbildung gefordert: „Im Fachgebiet Software-Ergonomie ist die Vermittlung praktischer Fähigkeiten ein wesentliches Ausbildungsziel." (Maaß et al. [11]).

Dennoch ist eine praktische Ausbildung in Spezialveranstaltungen nicht überall etabliert. Deshalb ist es wichtig, auch im Bereich traditioneller Lehrveranstaltungen (Vorlesungen und Übungen) praxisrelevant vorzugehen. In der Informatikausbildung an der Universität-Gesamthochschule Paderborn bietet unsere Fachgruppe „Informatik und Gesellschaft" Lehrveranstaltungen zur Software-Ergonomie an. Darin werden Themen behandelt, die von gesetzlichen Rahmenbedingungen und Gestaltungskriterien bis hin zu Arbeitsbewertungsverfahren reichen. Die Vermittlung von Gestaltungswissen durch praktisches Handeln stellt ein besonderes Problem dar, weil praktische Aufgaben in den normalen Vorlesungs- und Übungsbetrieb nur unzureichend integriert werden können.

Zur Unterstützung des Lehrbetriebs haben wir in unserer Fachgruppe eine technische Infrastruktur basierend auf einem Hypermediasystem aufgebaut. Damit werden sowohl Unterlagen bereitgestellt, als auch die Durchführung praktischer Übungen unterstützt. Dabei ergeben sich spezielle Anforderungen an die Gestaltung und den Einsatz eines solchen Systems. Neben software-ergonomischen Kriterien zur Oberflächengestaltung sind aber darüber hinaus andere Aspekte wie die Verfügbarkeit der Materialien oder die Vermeidung von Medienbrüchen zu berücksichtigen.

Im Vordergrund sollen in diesem Beitrag keine theoretischen Überlegungen zum Einsatz von Hypermediasystemen stehen, sondern es werden die in der Praxis gemachten Erfahrungen geschildert. Es wird über die Umsetzung und die aufgetretenen technischen, inhaltlichen und organisatorischen Probleme berichtet und aufgezeigt, welche Konsequenzen sich hieraus für zukünftige Veranstaltungen ergeben.

2 Anforderungen an das Hypermediasystem

Ziel der Fachgruppe Informatik und Gesellschaft ist es, Wirkungsforschung und Systemgestaltung so miteinander zu verknüpfen, daß die informatikspezifischen Konsequenzen sichtbar werden. Hierbei interessiert uns besonders, welche Wirkungen durch veränderten Technikeinsatz beeinflußbar sind. Ein Forschungsschwerpunkt ist es, den Einsatz und die Wirkungen von Technik in Lehr- und Lernprozessen zu untersuchen, um die tatsächlichen Probleme und Potentiale bestimmen zu können (siehe Engbring, Keil-Slawik, Selke [7]). So werden in der Arbeitsgruppe seit 1994 Hypermediasysteme zur Unterstützung von Lehrveranstaltungen eingesetzt. Mittlerweile wurden 4 große Veranstaltungen (Vorlesungen mit Übungen) sowie diverse Seminare und Projektgruppen mit Hypermediasystemen durchgeführt und evaluiert. Dabei wurde ein besonderes Augenmerk auf die folgenden Fragestellungen gelegt:

- Welche software-ergonomischen Gestaltungsanforderungen ergeben sich speziell für den Einsatz von Hypermediasystemen?

- Läßt sich der Hypermediaeinsatz unter alltagspraktischen Bedingungen durchführen?

- Welche Verbesserungen und Veränderungen müssen an den technischen Mitteln vorgenommen werden?

- Wie muß sich die Organisation und Durchführung der Lehrveranstaltungen verändern?

Im Wintersemester 1995/96 wurde erstmals eine Veranstaltung bis auf die Vorlesung vollständig mit elektronischen Medien durchgeführt. Hierzu war eine multimediale Arbeitsumgebung, sowohl für die Kleingruppenarbeit der Studierenden als auch für die Durchführung der Übungen, geschaffen worden.

Zum Nacharbeiten wurden den Studierenden Materialien (Folien der Vorlesung, wissenschaftliche Texte zur Vertiefung des Stoffes, sowie behandelte Normen, Gesetze und Richtlinien) zur Verfügung gestellt. Weiterhin erfolgte nicht nur die Verteilung der zu bearbeitenden Aufgaben über das System, sondern die Abgabe, Bewertung und Präsentation der Aufgaben in den Übungen fand ebenfalls über das System statt. Daraus ergaben sich besondere Anforderungen für das eingesetzte System:

- Eine längerfristige Pflege und Aktualisierung muß auch bei großen Datenbeständen effektiv durchgeführt werden können.
- Um den effizienten Umgang mit den großen Datenbeständen zu ermöglichen, sind über die Hypertextmöglichkeiten hinaus zusätzliche Navigationshilfen erforderlich. Auch sollte eine Suchfunktion im System vorhanden sein.
- Die Lernenden sollen mit den Unterlagen aktiv arbeiten, also die vorhandenen Materialien erweitern und eigene Dokumente in einen persönlichen Bereich einfügen können. Weiterhin sollen Verweise einfach und interaktiv angelegt werden können und insbesondere Verweise zwischen allen verwendeten Dokumenttypen (Texte, Bilder, Postscript-Dokumente, ...) ermöglicht werden.
- Um Datenschutz- und Copyrightbestimmungen berücksichtigen zu können, muß ein Mechanismus zur Verwaltung von Zugriffsrechten vorhanden sein.
- Das System soll ein dezentrales, plattformunabhängiges Arbeiten an den verschiedenen Workstations und PC's der Universität ermöglichen.

Diese Forderungen konnten durch den Einsatz des Hypermediasystems HyperWave, einem speziellen WWW-Server, der zusätzliche Funktionen bereitstellt, weitgehend berücksichtigt werden (HyperWave war bisher unter dem Namen Hyper-G bekannt, siehe Maurer [12]).

3 Einsatz von HyperWave in Lehrveranstaltungen

Die Veranstaltung „Grundlagen der Systemgestaltung", eine vierstündige Vorlesung mit zweistündigem Übungsanteil, wurde von ca. 100 Studierenden besucht. Da zu der Zeit sowohl die technische Infrastruktur nicht vorhanden war, als auch das System HyperWave noch nicht optimal als Präsentationswerkzeug geeignet ist, fand die Vorlesung in einem gewöhnlichen Hörsaal mit normalen Präsentationsmitteln wie Tafel und Overhead-Folien statt.

Alles Weitere wurde durch den Einsatz einer technischen Infrastruktur basierend auf dem System HyperWave unterstützt. Damit konnte eine plattformunabhängige Bearbeitung innerhalb eines Systems ermöglicht werden, ohne daß ein ständiger Medienwechsel zwischen Papier und Monitor (Aufgaben vom Rechner ausdrucken, Erstellen von Lösungen am Rechner, Ausdruck der Lösungen, Korrektur, Punkteverwaltung im Rechner, Ausdruck der Punktelisten für Aushänge, ...) stattfand. Trotz des Einsatzes von Hypermedia sollte jedoch kein erhöhter Personaleinsatz stattfinden, um die neuen Möglichkeiten und Probleme unter alltagspraktischen Bedingungen zu testen.

Während der Integration der Unterlagen in das System und des kontinuierlichen Einsatzes in der Lehrveranstaltung wurden einige Defizite von HyperWave deutlich. Als vorteilhaft erwies sich jedoch die Offenheit des Systems. So konnten Verbesserungen und Erweiterungen vorgenommen werden, einerseits durch den Kontakt zu den Entwicklern von HyperWave, andererseits durch eigene Entwicklungen beispielsweise im Rahmen von Diplomarbeiten.

Bereitstellung von Materialien

Zur Nachbearbeitung und Vertiefung der Vorlesung wurden Materialien in HyperWave zur Verfügung gestellt. Diese bestanden zum einen aus für die Software-Ergonomie relevanten

Normen, Gesetzen und Richtlinien (DIN/EN 9241, Betriebsverfassungsgesetz, Bundesdaten-schutzgesetz, EU-Bildschirmrichtlinie 90/270 EWG, ...). Des weiteren wurden wissenschaft-liche Artikel zu den aktuellen Themen in die Unterlagen eingebunden. Ein Problem hierbei stellte das Copyright für Normen und wissenschaftliche Artikel dar, das sich mit den Zu-griffsschutzmechanismen von HyperWave aber relativ einfach lösen ließ.

Eine besondere Qualität wurde hierbei erreicht, indem die Studierenden den selektiven Um-gang mit komplexen Regelwerken üben mußten, die sich in Form von Kopiervorlagen wohl nur auszugsweise hätten verteilen lassen. Da die Veranstaltung nur einen Teil des Software-Ergonomie-Curriculums abdeckt, wurden auch mehr Textvorlagen als aktuell benötigt im System abgelegt. Zusammen mit dem Aufbau einer umfangreichen Beispielsammlung ergibt sich so für interessierte Studierende die Möglichkeit zur Vertiefung. Diese Beispiele wurden teilweise von den Studierenden selbst während des Semesters gesucht oder erzeugt und in das System eingebunden und können nun für weitere Veranstaltungen genutzt werden.

Weiterhin wurden die in der Vorlesung verwendeten Folien und die Übungsaufgaben durch Verweise direkt mit den entsprechenden wissenschaftlichen Artikeln verknüpft, was ein ein-faches Auffinden des vertiefenden Stoffes ermöglichte.

Schwierigkeiten bereitete das Einbinden der in der Vorlesung verwendeten Powerpoint-Folien, da kein geeignetes Austauschformat zur Verfügung stand. Die ersten Folien mußten somit einzeln manuell konvertiert und eingebunden werden. Um diesen Aufwand zu minimieren, wurde im folgenden ein Zusatzwerkzeug entwickelt, das nun eine einfache Integration von Powerpoint-Folien erlaubt (Bollmeyer [4]). Der größte Aufwand bestand jedoch nach wie vor in der Integration von Materialien, die nicht in elektronischer Form vorlagen und erst di-gitalisiert werden mußten.

Die Folien in das System einzubinden brachte jedoch zusätzlich zur besseren Verfügbarkeit einige Vorteile mit sich:

- Farbfolien, die gerade im Abschnitt Farbgestaltung zur Veranschaulichung notwendig wa-ren, konnten so kostengünstig, ohne das Anfertigen teurer Farbkopien, verteilt werden.
- Gestaltungskriterien konnten durch Verweisstrukturen mit positiven und negativen Fall-beispielen verknüpft werden. Zusätzlich hatten die Studierenden die Möglichkeit, vorge-gebene sowie auch eigene Beispiele mit den Kriterien zu verknüpfen.
- Folien konnten mehrfach in der Datenbank von HyperWave erscheinen und so nach ver-schiedenen Gesichtspunkten geordnet oder sortiert im System abgelegt werden.

Bearbeitung und Beurteilung von Übungsaufgaben

Neben den Unterlagen wurden auch Übungsaufgaben, die von den Studierenden wöchentlich oder 14-tägig zu bearbeiten waren, über das System verteilt. Die Studierenden konnten von jedem vernetzten Universitätsrechner auf die Unterlagen zugreifen, wobei jedoch Farbmonitore noch nicht in ausreichender Zahl zur Verfügung stehen. Weiterhin bestand die Möglichkeit die Unterlagen als lokale Datenbank auf einem Heim-PC zu installieren oder über Modem auf das System zuzugreifen. Hierbei gab es aber des öfteren noch technische Schwierigkeiten.

Auch die Abgabe der Übungsaufgaben erfolgte durch direktes Einfügen der bearbeiteten Auf-gaben in das System. Die Abgaben konnten dabei Texte, Bilder, Verweisstrukturen oder auch ausführbare Programme sein. Dabei macht es wenig Sinn, ein Hypermediasystem für die Ab-gabe konventioneller Aufgaben, beispielsweise längerer Texte zu benutzen. Hier war es das

Ziel, die dem neuen Medium innewohnenden Qualitäten zu erschließen und diese durch die Verwendung neuer Aufgabentypen zu berücksichtigen.

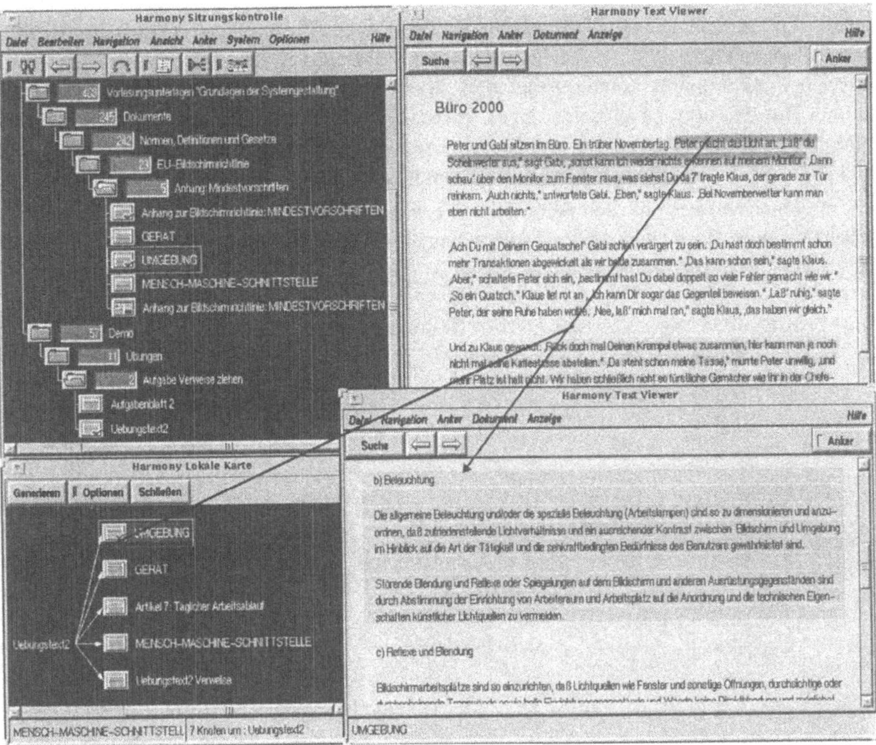

Abb. 1: Das Hypermediasystem HyperWave mit dem Beispiel einer Übungsabgabe. Der schwarze Pfeil deutet einen Verweis zwischen zwei Texten an. Der Verweis findet sich auch in der „Lokalen Karte" wieder, wo alle Verweise des Dokuments angezeigt werden.

Als eine Aufgabe mußte ein vorgegebener Text mit der Beschreibung einer Bürosituation auf problematische Situationen bezüglich des Betriebsverfassungsgesetzes und der EU-Bildschirmrichtlinie hin untersucht werden. Hierzu sollten Verweise zu den entsprechenden Gesetzesabschnitten angelegt und diese kommentiert werden. In Abbildung 1 ist eine entsprechende Abgabe zu sehen. Von dem Text „Büro 2000" (oben rechts) gehen Verweise in andere Dokumente, durch den schwarzen Pfeil angedeutet, in die Mindestvorschriften im Anhang der EU-Bildschirmrichtlinie „b) Beleuchtung" (im Textfenster unten rechts). Links oben ist die „Sitzungskontrolle" zu sehen, die immer die Position des aktuellen Dokuments in den Unterlagen anzeigt.

Die Lokale Karte (unten links) bietet eine sehr effektive Möglichkeit, die Abgaben der Studierenden schnell zu überprüfen. Es läßt sich direkt anzeigen, welche Dokumente in der Abgabe referenziert und ob weitere relevante Texte nicht berücksichtigt wurden.

Ein wesentlicher Teil der Übungsaufgaben bestand im eigenen Entwurf von Benutzungsoberflächen und Systemen. Hierbei sollten die in der Vorlesung vorgestellten Gestaltungskriterien zum Entwurf interaktiver Systeme umgesetzt werden. Die Entwürfe wurden von den Studierenden auf verschiedenen Plattformen mit beliebigen Programmen (Zeichenprogrammen, Oberflächengeneratoren, ...) erstellt und dann als Bitmaps in das System eingebunden. Eine teilweise Funktionalität konnte dabei über Verweisstrukturen realisiert werden, indem bestimmte Buttons oder Menüzeilen durch Verweise mit weiteren Bildschirmabzügen verknüpft wurden. Für die Plattformen PC und Silicon Graphics ließen sich auch ausführbare Programme einbinden und in den Übungen über das System präsentieren.

Die Abgaben wurden von den Mitarbeitern bewertet und falls notwendig mit Kommentaren versehen. Diese Rückmeldung im System war durch Zugriffsrechte eingeschränkt und daher nur für die jeweiligen Studierenden einsehbar.

Übungsdurchführung im elektronischen Seminarraum

Abb. 2: PC-Netzwerk im elektronischen Seminarraum. Mit dem „pädagogischen Netzwerk" (Schalttafel vor dem Monitor unten rechts) können Rechner und Monitore beliebig untereinander verbunden werden.

Eine ausführliche Besprechung und Diskussion der Abgaben fand wöchentlich im elektronischen Seminarraum statt. Dieser ist durch eine kreisförmige Anordnung von sechs Monitoren (siehe Abbildung 2) bestens für Präsentations- und Diskussionszwecke geeignet. Anfangs konnte immer nur wahlweise eine Workstation oder ein PC an den „Monitorring" angeschlossen werden, um während der Übungen Zugriff auf die Unterlagen und die abgegebenen Lösungen der Studierenden zu haben. Mittlerweile ist der elektronische Seminarraum um sechs

PC's und ein „pädagogisches Netzwerk" ergänzt worden, was sowohl lokales Arbeiten als auch Präsentationen von jedem Rechner aus erlaubt.

4 Bewertung

Um Aufschluß über die Nutzung des Systems und die Probleme seitens der Studierenden zu bekommen, wurden zusätzlich zu den Besprechungen in den Übungen zu Beginn und am Ende der Veranstaltung Fragebögen verteilt. Weiterhin werden die Erfahrungen der Mitarbeiter beschrieben, die neben der inhaltlichen Gestaltung der Übungsaufgaben und der Durchführung der Übungen auch die Systembetreuung und das Einbinden der Unterlagen in das System erledigten.

Schwierig in der Bewertung ist es, daß ein unmittelbarer Vergleich (mit und ohne Einsatz des Hypermediasystems und elektronischen Seminarraums) nicht möglich ist. Auch läßt sich nicht auseinanderdividieren, welche Rolle der Technikeinsatz, der Inhalt der Veranstaltung oder die veränderte Durchführung spielen.

Akzeptanz bei den Studierenden

Studierende der Informatik arbeiten zum großen Teil bereits mit Hypertextsystemen, vor allem dem WWW (siehe Tabelle 1). Deshalb lassen sich Hypermediasysteme ohne große Probleme und mit wenig Schulungsaufwand einsetzen. Die Ergebnisse sind somit nur teilweise auf andere Fachbereiche übertragbar, obwohl auch dort zunehmend das WWW eingesetzt wird.

	keine Hypertexterfahrung	Erfahrung mit WWW	Erfahrung mit HyperWave
SS1994	56,1%	30,4%	noch nicht verfügbar
WS1995/96	20,3%	71,6%	16,2%
SS1996	9,4%	78,0%	21,9%

Tab. 1: Erfahrung mit Hypertextsystemen der Teilnehmenden der Lehrveranstaltungen jeweils zu Beginn des Semesters (Differenzen ergeben sich aus der Nutzung mehrerer und anderer Systeme (Notes, Hilfe-Systeme, Gopher, ...)).

Insgesamt wurde sowohl die Bereitstellung der Materialien, als auch die Abgabe der Aufgaben im System von den Studierenden am Ende der Veranstaltung besser bewertet als zu Beginn (siehe Tabelle 2 und 3).

	sehr gut	gut	eher gut	eher schlecht	schlecht	ganz schlecht
zu Beginn	35,1%	33,8%	18,9%	4,1%	5,4%	2,7%
am Ende	35,6%	44,1%	13,6%	5,1%	1,7%	0%

Tab. 2: Antworten auf die Frage: „Wie findest Du die Idee, Unterlagen für Lehrveranstaltungen auf dem Rechnernetz bereitzustellen?" zu Beginn und am Ende der Veranstaltung.

	sehr gut	gut	eher gut	eher schlecht	schlecht	ganz schlecht	keine Angabe
zu Beginn	16,2%	33,1%	29,7%	10,8%	8,1%	1,4%	2,7%
am Ende	35,6%	44,1%	13,6%	5,1%	1,7%	0%	0%

Tab. 3: Antworten auf die Frage: „Wie findest Du die Idee, Übungen in elektronischer Form abzugeben?" zu Beginn und am Ende der Veranstaltung.

74,6% der Studierenden beantworteten die Frage: „Sollte das Hypermediasystem HyperWave in folgenden Veranstaltungen wieder so eingesetzt werden?" mit ja, nur 6,8% sprachen sich gegen einen weiteren Einsatz aus, der Rest machte keine Angaben. Im Gegensatz zu einer Veranstaltung, die in der Fachgruppe Informatik und Gesellschaft im Sommersemester 1994 durchgeführt und ausgewertet worden war und in der die Studierenden das System positiv bewerteten, ohne es intensiv genutzt zu haben (siehe Brennecke, Keil-Slawik [6]), mußten die Studierenden das System diesmal zumindest zur Abgabe und in den Übungsstunden einsetzen.

Das System wurde diesmal nicht nur zum Bereitstellen von Texten verwandt, die bisher nicht ergonomisch angemessen am Rechner lesbar sind. Die besondere Qualität zeichnete sich durch die bessere Aufbereitung der Aufgaben und Materialien für das System und den zusätzlichen Einsatz als Präsentationsmedium in den Übungen aus.

Aufwand für die Betreuenden

Wie schon erwähnt fand der Systemeinsatz ohne zusätzlichen Personaleinsatz statt. Für die beiden beteiligten Mitarbeiter erhöhte sich der Aufwand für die Veranstaltung durch die Verwaltung von Benutzerrechten, sowie das Digitalisieren und Einbinden der Unterlagen. Weiterhin wurden von jedem Mitarbeiter drei anstatt der vorgesehenen zwei Übungen angeboten, da der elektronische Seminarraum nur jeweils 16 Studierenden Platz bietet.

Dagegen reduzierte sich die Arbeit, die normalerweise durch das Verteilen von Kopiervorlagen (laufendes Vervollständigen und Ordnen von Unterlagen) anfällt. Weiterhin wurden die Aufgaben so gestellt, daß die Auswertung durch das System unterstützt wurde und einfacher von den Mitarbeitern zu erledigen war.

Die Durchführung der Veranstaltung ist insgesamt als sehr effektiv zu bezeichnen, insbesondere wenn man davon ausgeht, daß die Veranstaltung wiederholt wird, die Unterlagen dann mehrfach genutzt werden, und sich eine Beispielsammlung anlegen und in weiteren Veranstaltungen nutzen läßt. Aber auch der Mehraufwand bei der erstmaligen Durchführung hielt sich in Grenzen.

Als besonderer Vorteil hat sich herausgestellt, sämtliche Materialien der Veranstaltung (Regelwerke, Folien der Vorlesung, Abgaben der Studierenden, gute Beispiele aus anderen Gruppen usw.) während der Übungen im elektronischen Seminarraum verfügbar zu haben und allen Teilnehmenden präsentieren zu können. So konnte in den Übungen ständig Bezug auf die Vorlesung und bisherige Übungen genommen werden.

Das System HyperWave selbst hat nach software-ergonomischen Gesichtspunkten sicher auch noch einige Defizite, es ist aber verglichen mit anderen WWW-Systemen eher als fortschrittlich zu bezeichnen und zeichnet sich besonders durch zusätzliche Navigationshilfen aus. Da das System offen ist und Erweiterungen zuläßt, werden in unserer Arbeitsgruppe auch Zusatzwerkzeuge entwickelt, implementiert und im Einsatz getestet.

Um diesen Aufwand für die Evaluation und die Weiterentwicklung des Systems möglichst gering zu halten, ist es wichtig mit anderen Gruppen zusammenzuarbeiten. Mittlerweile setzen an der Universität Paderborn mehrere Arbeitsgruppen HyperWave ein und es findet eine Kooperation mit externen Gruppen statt. So wird beispielsweise in Verbindung mit einer Forschungsgruppe in Auckland eine zusätzliche Navigationshilfe entwickelt. Diese „Semantischen Karten" sollen einen inhaltsbezogenen Überblick zu einzelnen Themenbereichen ermöglichen (siehe Klemme [10]).

Insgesamt wurde die Durchführung der Veranstaltung von allen Beteiligten als positiv angesehen, so daß in unserer Arbeitsgruppe weitere Veranstaltungen wieder durch den Einsatz von HyperWave unterstützt werden sollen.

6. Fazit und Ausblick

Die Gestaltung der Hypermedia-Unterlagen und der Einsatz eines Hypermediasystems für die Durchführung von Übungen zur Software-Ergonomie fand anders als in den meisten Hypermedia-Projekten sozusagen „online" während der Veranstaltung statt. Ziel war es, den Hypermediaeinsatz nicht auf Forschungsprojekte zu beschränken, sondern ihn in den Alltag der Lehre zu integrieren.

Die Ergebnisse, die hierbei gemacht wurden, sind sehr erfreulich. Neben der positiven Bewertung durch die Studierenden, wurden auch von den Lehrenden viele Vorteile, wenn auch noch bei erhöhtem Aufwand, gesehen. Dieser wird sich jedoch bei Mehrfachnutzung der Unterlagen und Aufgabensammlungen reduzieren. Besonders fiel auf, daß der Einsatz von Hypermedia/Multimedia nicht unabhängig von der Durchführung der Lehrveranstaltung gesehen werden kann. Ein einfaches Zusatzangebot für die Studierenden würde praktisch nicht genutzt werden, wie es beispielsweise am System VENE festgestellt wurde (Beran et al. [2]). Die didaktische und inhaltliche Aufbereitung von Aufgaben und Materialien müssen also mit dem Einsatz von Technik und der organisatorischen Umsetzung einhergehen.

Aus diesem Grund wird auch die Umsetzung neuer Lehrformen erprobt. So werden gemeinsam mit der Forschungsgruppe „Erwägungskultur", von der Universität Paderborn, Erwägungsseminare (Blanck [3]) im elektronischen Seminarraum durchgeführt.

Der von unserer Arbeitsgruppe eingerichtete elektronische Seminarraum wird auch von weiteren Gruppen genutzt. Mittlerweile hat sich an der Universität Paderborn das Innovationsforum Multimedia gebildet, in dem Arbeitsgruppen aus verschiedenen Fachbereichen über den Einsatz von Multimedia und die notwendigen organisatorischen Veränderungen nachdenken (siehe Keil-Slawik [9]). In Planung ist auch die fachgruppenübergreifende Ausstattung eines elektronischen Hörsaals für größere Veranstaltungen (HNI [8]), für dessen effektive Nutzung ebenfalls neue Konzepte erforderlich sind.

Allerdings ist es bisher nur schwer möglich, Vorlesungen in anderer Form durchzuführen, da einerseits die erforderliche Infrastruktur noch nicht vorhanden ist und zum anderen Prüfungs- und Studienordnungen bestimmte Veranstaltungsformen vorschreiben. Für eine andere Art der Durchführung von Übungen wurde jedoch der erste Schritt getan.

Der Einsatz einer elektronischen Arbeitsumgebung zeigte speziell bei der Software-Ergonomie-Ausbildung einige wesentliche Vorteile gegenüber bisherigen Veranstaltungen:

- Dynamische Prozesse und Farbgestaltung konnten durch den Systemeinsatz erfahrbar und in den Übungen diskutierbar gemacht werden.
- Der selektive Umgang mit großen Texten und Regelwerken wurde ohne umfangreiche Kopiensammlungen und mit zusätzlicher Funktionalität möglich.
- Der Aufbau und die kontinuierliche Aktualisierung einer umfangreichen Beispielsammlung wurde in effektiver Weise möglich.

Um solche Qualitäten in der Software-Ergonomie-Ausbildung berücksichtigen zu können, scheint ein wie in diesem Beitrag beschriebener Ansatz unerläßlich zu sein und ist, wie unsere Erfahrungen zeigten, auch unter praktischen Bedingungen umsetzbar.

Literaturverzeichnis

[1] Ansorge, P., Friedrich, J. und Streibl, R. E.: Gestaltungsfähigkeit als Prinzip der Software-Ergonomie-Lehre in Bremen. Ergonomie & Informatik Nr. 22, Juli 1994,S. 26-32

[2] Beran, H., et al.: VENE – ein System zur Benutzerführung, Suche und Evaluation in einer medizinischen Lehrprogramm-Bibliothek. In: E. Schoop et al. (Hg.): Hypermedia in der Aus- und Weiterbildung. Dresdner Symposium zum computergestützten Lernen. Universitätsverlag Konstanz, 1995, S. 129-135

[3] Blanck, B.: Erwägung und Didaktik. Arbeitspapier 1996-4 der Forschungsgruppe Erwägungskultur, Paderborn, 1996

[4] Bollmeyer, J.: Anforderungen und Kriterien zur Integration multimedialer Standardsoftware in eine hypermediale Lehr- und Lernumgebung. Diplomarbeit, Universität-GH Paderborn, Juni 1996

[5] Brennecke, A. und Holl, F.: Gutachten für den Hauptpersonalrat des Landes Berlin zur Software-ergonomischen Qualität des Software-Systems ProFISKAL. Durchgeführt von der DGB Technologieberatung Berlin, Januar 1996

[6] Brennecke, A. und Keil-Slawik, R.: Alltagspraxis der Hypermediagestaltung: Erfahrungen beim Einsatz des World Wide Web und Mosaic in der Lehre. In: H.-D. Böcker (Hg.): Software-Ergonomie 95. Teubner, Stuttgart, 1995, S. 129-135

[7] Engbring, D., Keil-Slawik, R. und Selke, H.: Neue Qualitäten in der Hochschulausbildung: Lehren und Lernen mit interaktiven Medien. Technischer Bericht Nr. 45, Heinz Nixdorf Institut, Universität-GH Paderborn, Dezember 1995

[8] HNI: KONTAKT: Ein neuer Forschungsschwerpunkt am HNI. HNI Nachrichten – Mitteilungen aus dem Heinz Nixdorf Institut, Universität-GH Paderborn, Ausgabe 5, April 1996, S. 1-2

[9] Keil-Slawik, R.: Innovationsforum Multimedia: Einsatz von Multimedia verbessert Qualität der Lehre. Paderborner Universitäts-Zeitschrift, Nr. 2, 1996, S. 12-13

[10] Klemme, M.: Semantic Spaces: A new Access Paradigm to Hypermedia Systems. Technical Report, University of Auckland Department of Computer Science, No. 134, August 1996

[11] Maaß, S. et al.: Software-Ergonomie-Ausbildung in Informatik-Studiengängen bundesdeutscher Universitäten. Informatik Spektrum 16, Heft 4, August 1993, S. 191-205

[12] Maurer, H. (Hg.): Hyper-G now Hyperwave: The Next Generation Web Solution. Addison-Wesley 1996

Adressen der Autoren

Andreas Brennecke,
Heinz Nixdorf Institut
Universität-GH Paderborn
Fürstenallee 11
33102 Paderborn
Email: anbr@uni-paderborn.de
http://www.uni-paderborn.de/cs/anbr.html
Prof. Dr.-Ing. Reinhard Keil-Slawik
Heinz Nixdorf Institut
Universität-GH Paderborn
Fürstenallee 11
33102 Paderborn
Email: rks@uni-paderborn.de
http://www.uni-paderborn.de/cs/rks.html

Metaphern für interaktives Fernsehen
Eine Fallstudie mit Endbenutzern

Michael Burmester und Franz Koller

Fraunhofer-Institut für Arbeitswirtschaft und Organisation, Stuttgart

Schlüsselworte

Benutzungsoberfläche, interaktives Fernsehen (ITV), Benutzeranforderungen, Informationsdarstellung, Benutzertest, Metaphern

Zusammenfassung

Die in dem vorliegenden Beitrag beschriebene Studie hat zum Ziel, Metaphern für Benutzungsoberflächen für Dienste des interaktiven Fernsehens zu untersuchen. Die Metapher der Benutzungsoberfläche sollte leicht zu verwenden sein sowie die Navigation und die Informationsabfrage innerhalb der zahlreichen, angebotenen Multimedia-Dienstleistungen unterstützen. Letztendlich müssen sie von der Benutzerzielgruppe akzeptiert werden.

1 Benutzungsoberflächen für interaktives Fernsehen:
Eine Frage - viele Antworten

Beim jetzigen Stand der Technik können noch keine exakten Aussagen über die Anforderungen und Wünsche zukünftiger Benutzer des interaktiven Fernsehens gemacht werden. Die fehlenden Vorerfahrungen erschweren Designentscheidungen besonders deshalb, weil interaktives Fernsehen seinen Platz zwischen dem herkömmlichen Fernsehen, dem PC und dem Internet finden muß. Darüber hinaus ist die Akzeptanz potentieller ITV-Nutzer in keiner Weise gewährleistet.

Aus diesem Grund wurde das ACTS-Vorhaben MUSIST (AC010, Multimedia User Interface for Online Services and TV) ins Leben gerufen, um intelligente und komfortable Benutzungsoberflächen für interaktive, multimediale Fernsehdienste zu untersuchen, zu entwickeln und zu testen. Diese Benutzungsoberflächen sollen einfach zu nutzen sein und sollen eine hohe Benutzerakzeptanz hinsichtlich multimedialer Dienste und interaktivem Fernsehen erreichen. Die Navigation und die Informationsdarstellung für die angebotenen Multimedia-Dienste sollen sowohl für TV als auch PC geeignet sein. Deshalb liegt der Schwerpunkt des Vorhabens MUSIST im Bereich Benutzungsoberflächen darauf, die Bedürfnisse und Vorlieben von Benutzern für die Interaktion mit multimedialen Diensten in der häuslichen Umgebung zu untersuchen und zu modellieren, um dann innovative Dialoge und Navigationskonzepte für komplexe Benutzeraufgaben zu überprüfen, zu simulieren und zu implementieren.

Zu Beginn des Projektes, Ende 1995/Anfang 1996, wurde eine Untersuchung vom Fraunhofer-Institut IAO durchgeführt, die zeigte, daß die meisten Feld- und Pilotversuche, die im Bereich multimedialer, interaktiver Dienste und interaktivem Fernsehen existieren, noch in den Kinderschuhen stecken. Die Untersuchung basiert auf Informationen, die aus technischen Zeitungen und Zeitschriften, persönlichen Interviews mit verschiedenen Dienstleistern, Informationen aus dem Internet und aus Fragebögen stammen. Sie stellt eine Zusammenfassung aktueller Entwicklungen im Bereich des interaktiven Fernsehens vor und bietet einen Überblick über existierende Ansätze für interaktive Dienste und interaktives Fernsehen. Aufgrund der Tatsache, daß interaktives Fernsehen eine junge Technologie ist, war es schwierig, geeignete Angaben bezüglich der geplanten Oberflächenkonzepte zu erhalten. Im Gegensatz dazu waren allgemeine Stellungnahmen bezüglich des Netzwerkes und der geplanten Server relativ leicht erhältlich (selbst wenn die technischen Eckdaten ständigen Veränderungen unterliegen).

Basierend auf den Untersuchungsergebnissen wurden Benutzungsoberflächenkonzepte abgeleitet und in einer Fallstudie, die im folgenden beschrieben wird, untersucht.

2 Die Fallstudie

Die Untersuchung hatte zum Ziel, detaillierte Daten über Vorstellungen, Anforderungen und Konzepte zu sammeln, die zukünftige ITV-Nutzer hinsichtlich Benutzeroberflächen, Navigation und Design haben. Von besonderem Interesse war herauszufinden, welche Zugangsmechanismen zu den interaktiven Diensten sowie zu den Informationen in den unterschiedlichen Diensten angemessen sind. Folgende typische Dienste waren in der Fallstudie vertreten: Teleshopping, elektronische Programmnavigator für Pay TV und Informationsdienste.

Zielgruppen für interaktives Fernsehen sind vor allem die privaten Kunden, die elektronischen Dienste von ihrem Heim aus nutzen wollen. Interaktives Fernsehen soll auch Personen ansprechen, die nicht über extensive Computer- und Internet-Erfahrung verfügen. Aus diesem Grunde wurden Personen für die Untersuchung ausgewählt, die keine besonderen Vorkenntnisse im Bereich Multimedia und Internetnutzung haben.

Bei der Durchführung war es besonders interessant, direkte Antworten auf die folgenden Fragen zu bekommen:

- Wie stellt sich der Benutzer geeignete Benutzungsoberflächen für ITV vor;
- Welche Vorstellungen hat der Benutzer bezüglich effektiver Navigation und interaktiven Mechanismen;
- Werden die in der Untersuchung vorzustellenden Eingangsmetaphern akzeptiert und
- welche Verbesserungen bezüglich individueller Prototypen schlägt der Benutzer vor.

Aufgrund der Tatsache, daß sieben von neun der im Interview gezeigten Konzepte statisch waren, wurden hohe Anforderungen an die Fähigkeit der Benutzer gestellt, kreativ zu denken. Der Benutzer wurde ermutigt, eigenen Ideen und Vorstellungen zu entwickeln.

In einem Teil der Studie wurde auf bereits existierende Benutzungsoberflächenkonzepte zurückgegriffen. Zusätzlich dazu wurden neue Konzepte entwickelt und in die Studie aufgenommen. Im folgenden solle die unterschiedlichen Konzepte vorgestellt werden.

3 Die getesteten Konzepte

3.1 Das Apple-Kaufhaus

Das Apple-Kaufhaus ist eine CD-ROM die von der Firma Apple unter Verwendung von Quick-Time VR hergestellt wurde.

Mittels einer Kaufhaus-Metapher wurden in dieser CD alle angebotenen Produkte in einer realistischen Umgebung dargestellt. Started der Benutzer das Programm, betritt er das Kaufhaus, die Tür geht auf und der Kaufhausbummel kann beginnen. Die verschiedenen Produkte sind entweder an der Theke, in Regalen oder Ständen ausgelegt. Mit dem Bewegen der Maus ändert sich die Perspektive, die verschiedenen Bereiche des Kaufhauses können näher betrachtet werden. Sogar eine Drehung um die eigene Achse ist möglich. Außerdem kann man in den Gängen zwischen den Regalen laufen, d. h., die Waren können so schnell wie gewünscht und in beliebiger Reihenfolge näher betrachtet werden.

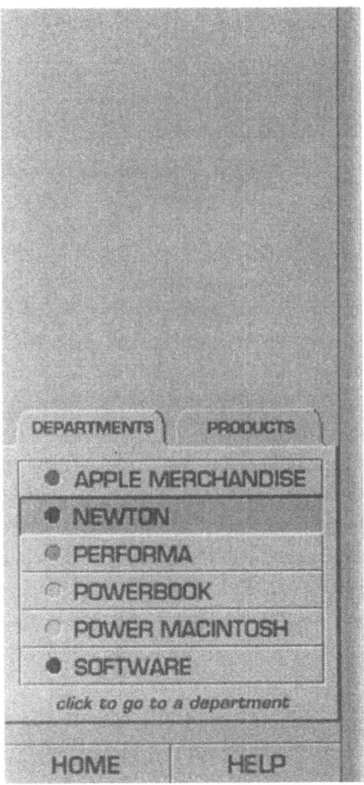

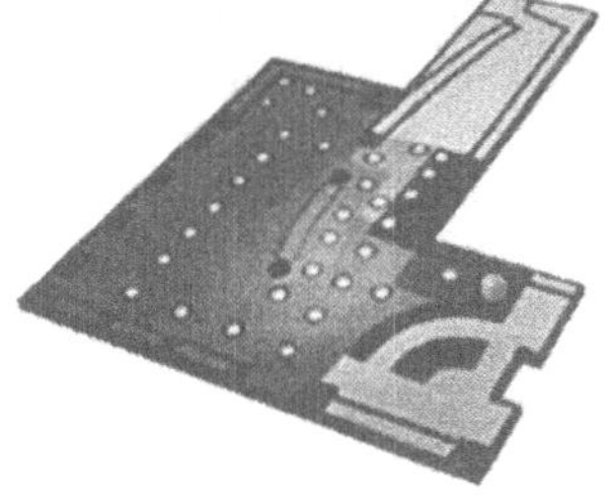

Abb. 1: Das Apple-Kaufhaus (Quelle: Apple Computer Company Store CD, Apple Computer Inc., Cupertino, USA) bietet drei verschiedene Navigationsmöglichkeiten. Die erste Navigationsstrategie ist die Auswahl verschiedener Produkte auf der linken Seite. Die zweite Möglichkeit ist die Navigation durch das Kaufhaus im rechten oberen Fenster. Der dritte Navigationsweg führt zu direktem Zugang spezieller Räume durch die direkte Auswahl auf dem Lageplan.

Das Programm bietet drei Alternativen an, um auf die Waren zuzugreifen (Abb. 1):

- durch das virtuelle Kaufhaus,
- durch einen Lageplan, der unterhalb des Kaufhauses dargestellt ist und der dem Benutzer helfen soll, sich im Kaufhaus zurechtzufinden,
- durch eine alphabetische Liste, die einen direkten Zugriff auf die angebotenen Waren bietet.

Wählt der Benutzer einen Artikel aus der Liste oder dem Regal aus und klickt ihn an, so wird statt des Kaufhauses der Artikel mit der entsprechenden Produktinformation gezeigt. Mit Quick-Time VR kann der Artikel in alle Richtungen gedreht und somit von verschiedenen Seiten betrachtet werden.

Das Apple-Kaufhaus ist ein gutes Beispiel für eine realistische Kaufhausmetapher, die dem Benutzer Zugriff auf verschiedene Produkte ermöglicht. Diese Art von Metapher kann ebenso auf interaktives Fernsehen angewendet werden. Der einzige Unterschied besteht darin, daß das Apple-Kaufhaus über die Maus gesteuert wird, während beim interaktiven Fernsehen die Bedienung eines Zeigegerätes nicht unbedingtvorgesehen ist.

3.2 Der elektronische Programmguide von PREMIERE (EPG)

Die zweite interaktive Alternative war ein elektronischen Programmguide (EPG), der für den Pay-TV-Sender PREMIERE entwickelt wurde und der Teil einer Dienstleistung der Programme des Senders sein wird.

Das Hauptmenü benutzt die obere Hälfte des Bildschirms, um den „Toptip des Tages" mit Bild und Kurztext anzukündigen. Die untere Hälfte des Bildschirms besteht aus acht rechteckigen Feldern für die acht Menüpunkte, die die Informationsbereiche des PREMIERE EPG abdecken (Abb. 2). Die Hauptstruktur und die Navigationsmechanismen sind innerhalb des gesamten Systems einheitlich.Neben anderen Diensten bietet das System Inhalte, die mit denen eines Fernsehmagazins identisch sind, nämlich alle Fernsehsender und Kanäle, detailliertere Informationen (information-on-demand) über bestimmte Programme, die Möglichkeit, zu einem bestimmten Programm notizen zu machen oder den Videorekorder automatisch zu programmieren, wenn ein bestimmtes Programm ausgewählt wird. Zusätzlich ist das System mit einer „Kindersicherung", dem PIN-Code, ausgestattet. Der Benutzer kann durch verschiedene Programme zappen, ohne dabei den vorher ausgewählten Kanal ändern zu müssen. Ein On-Screen-Display erscheint vor dem aktuelle Fernsehbild und zeigt Textinformationen über andere Programme. Im Gegensatz zum Apple-Kaufhaus navigiert der Benutzer hier mittels der Pfeiltasten seiner Fernbedienung, indem er einzelne Felder auswählt und mittels der OK-Taste auswählt.

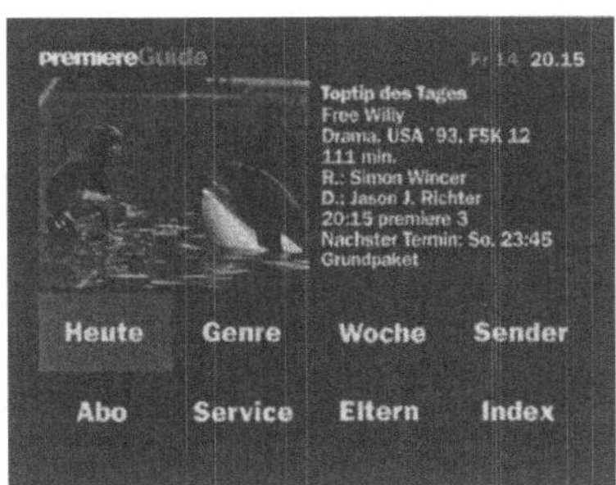

Abb. 2: Das Hauptmenü des PREMIERE EPG (Quelle: PREMIERE Medien GmbH)

Die bevorzugten Farben sind schwarz und grau, gemäß dem PREMIERE-CI [4]. Farbakzente werden nur sehr sparsam gesetzt, lediglich Bilder, die verschiedene Sendungen illustrieren, sind mehrfarbig.

Der PREMIERE EPG wird über eine Fernbedienung betätigt, auf der vier Richtungstasten (Pfeiltasten), eine Bestätigungstaste (OK), eine Exit-Taste für den Zugriff auf Fernsehprogramme und eine Menütaste, um das Hauptmenü aufzurufen, vorhanden sind.

3.3 Der Baum

Diese Alternative zeigt einen realistisch gezeichneten Baum mit vielen Ästen inmitten einer Winterlandschaft (Abb. 3). An den Ästen hängen große Äpfel, auf denen Symbole zu sehen sind, die für die dahinter verborgenen Menüeinträge stehen. Die Äpfel ändern ihre Form und Farbe zu Bällen, auf denen eine bestimmte Rubrik steht, sobald sie ausgewählt werden.

Abb. 3: Die Baummetapher für die Auswahl von Diensten und Informationsbereichen. Fernsehen ist momentan ausgewählt.

Die Früchte können durch die Betätigung der Navigationstasten auf der Fernbedienung ausgewählt werden. Durch die Betätigung der Bestätigungstaste (OK) wird die Auswahl aktiviert. Das Konzept kann ebenso über Zeigegeräte wie einem Track Pad oder einen Fernzeiger bedient werden.

3.4 Das Buch

Benutzer sind an das Layout und die Struktur eines Buches gewöhnt. Es sollte deshalb für den Benutzer einfach sein, das bereits vorhandene Wissen wie mit einem Buch umzugehen ist, auf den Umgang mit einem virtuellen Buch anzuwenden.

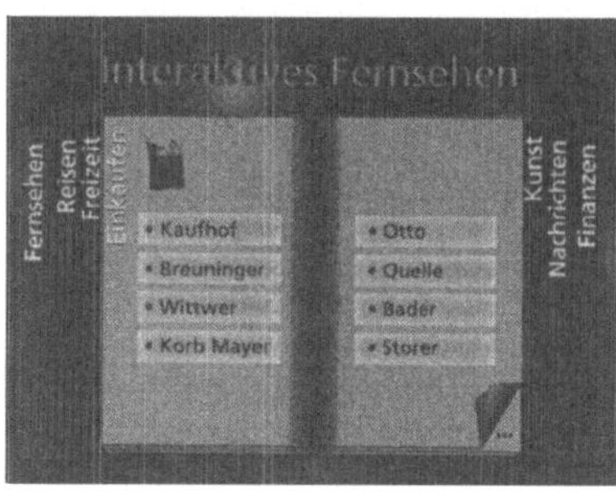

Abb. 4: Die Buchmetapher für die Auswahl von Diensten und Informationsbereichen

Die geteste Alternative zeigt ein geöffnetes Buch auf dem Bildschirm, das in seinen Kapiteln und auf seinen Seiten verschiedene Dienste und Informationsbereiche beinhaltet (Abb. 4). Ein vertikal beschriebenes Register zeigt den Inhalt der jeweiligen Kapitel. Die Seiten stehen für den Kapitelinhalt. Durch Anklicken mit dem Zeigegerät auf eine bestimmte Rubrik, öffnet sich das betreffende Kapitel. Das virtuelle Buch wird über eine gewöhnliche Fernbedienung gesteuert. Mit der linken oder der rechten Navigationstaste werden die Kapitel ausgewählt und die Seiten umgeblättert. Diese Metapher hat den Vorteil, daß alle Rubriken des Hauptmenüs ständig sichtbar und verfügbar sind. Beim Wechsel zwischen verschiedenen Rubriken muß der Benutzer das Hauptmenü nicht explizit aufrufen, wie es bei anderen Alternativen der Fall ist (vgl. PREMIERE).

3.5 Die Menü-Metapher

Die Menü-Metapher zeigt eine Menüliste. Die einzelnen Menüeingänge werden durch entsprechende Felder und dazugehörigem Text dargestellt (Abb. 5).

Abb. 5: Die Menü-Metapher mit den Rubriken Freizeit, Kunst, Einkaufen, Fernsehen, Reisen, Nachrichten, Kommunikation und Finanzen.

Der Vorteil dieses Ansatzes ist die visuelle Darstellung der Punkte in einer sehr strukturierten Art und Weise. Der Bildschirm ist in zwei Spalten aufgeteilt, jede Spalte enthält vier Auswahlfelder. Diese Designalternative kann sowohl über die Navigations- und Bestätigungstaste als auch über ein Zeigegerät gesteuert werden. Neben der Akzeptanz dieser Metapher war das Verstehen der Aufschriften und Felder von besonderem Interesse für diese Studie.

3.6 Die Comic-Stadt

Diese farbenfrohe Alternative zeigt eine Stadt, die wie ein Comic gezeichnet ist (Abb. 6). Das Bild stammt von einem französischen Online-Anbieter. Die einzelnen Menüpunkte werden durch Gebäude dargestellt, die zusammen eine Fantasiestadt bilden. Auf jedem Gebäude steht der Titel der entsprechenden Rubrik geschrieben. Die Einkaufsstraße steht z. B. für all das, was im allgemeinen mit Einkaufen in Verbindung gebracht wird.

Abb. 6: Ein französischer Online-Dienst, basierend auf der Comic-Stadt-Metapher (Quelle: Multicable, Lyonnaise Communication und France Telecom, Frankreich)

3.7 Leute

Das Hauptmenü der Leute-Metapher zeigt verschiedene Personen, die für einzelne Menüeinträge stehen (Abb. 7).

Abb. 7: Persönliche Führung für verschiedene Informationsbereiche

Diese Metapher basiert auf der Idee, eine persönliche Führung mit einem Begleiter durch die ITV-Angebote zu bekommen. Den Benutzern wurde gesagt, daß sie eine virtuelle Begleitperson bekämen, die für eine bestimmte Rubrik steht und Informationen und weitere Erklärungen gibt.

3.8 Der Stuttgarter Stadtplan

Die Stadtplan-Metapher stellt die Stadt Stuttgart in Form eines Stadtplans sehr realistisch dar (Abb.8).

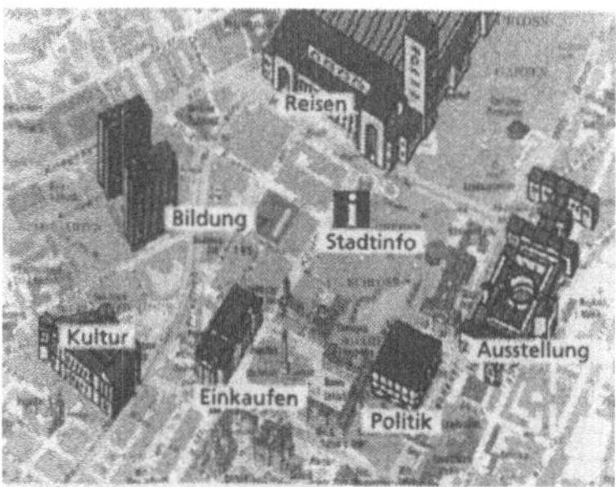

Abb. 8: Die Dienste werden auf dem Stuttgarter Stadtplan Gebäuden zugeordnet: Reisen dem Hauptbahnhof, Ausstellungen der Staatsgalerie, Politik dem Rathaus, Einkaufen einem Einkaufszentrum, Kultur einem bekannten Kulturgebäude und Bildung der Universität.

Die einzelnen Menüeinträge sind als verschiedenfarbige Gebäude auf grau-weißem Hintergrund gezeichnet. Die ITV-Inhalte verbergen sich hinter verschiedenen bekannten Gebäuden und Einrichtungen von Stuttgart. Jedes Gebäude trägt zusätzlich die Aufschrift des Rubriktitels, für die es steht.

Als Alternative zum realistischen Stadtplan wurde Testpersonen Freihandzeichnungen gezeigt, die die gleichen Gebäude darstellten. Diese Alternative zeigte allerdings keine Details wie z. B. die Angabe von Straßennamen, etc. Die Benutzungsoberfläche ist so entwickelt, daß sie hauptsächlich über ein Zeigegerät gesteuert werden kann.

3.9 Die VR-Stadt

Die VR-Stadt-Metapher stellt ebenfalls eine Stadt dar. Im Gegensatz zur vorherigen alternative, stellt sie dreidimensionale Elemente in futuristischem Design dar, die durch Linien miteinander verbunden sind (Abb. 9).

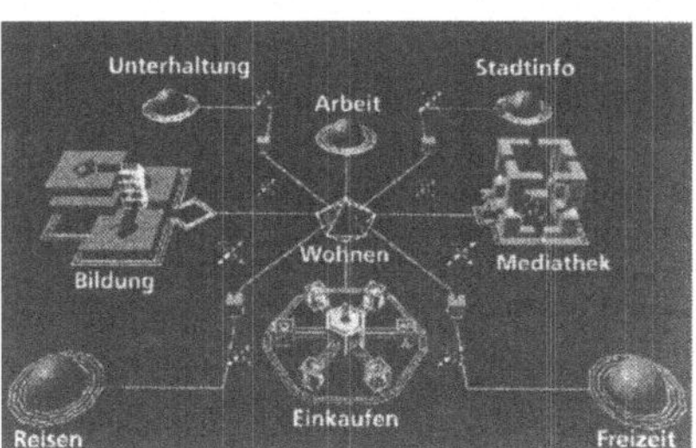

Abb. 9: Dreidimensionale, sehr abstrakt dargestellte Stadt mit persönlichem Lebensbereich im Zentrum. Andere Dienste sind Unterhaltung, Arbeit, Stadtinformationen, Mediathek, Freizeit, Einkaufen und Reisen (Quelle: Comenius Project, DeTeBerkom GmbH, Berlin).

Dieses futuristische und farbenfrohe Bild wurde dem COMENIUS-Projekt entnommen, das mehrere Berliner Schulen verbindet. Die Aufschriften wurden zu den grafischen Elementen der VR-Stadt zugefügt. Im Zentrum der Darstellung steht die Rubrik „Wohnen". Demzufolge stehen der Benutzer und seine Belange im Mittelpunkt dieser Alternative.

Die Metapher ist sowohl über ein Zeigegerät wie auch über die Navigationstasten einer gewöhnlichen Fernbedienung steuerbar.

4 Methodischer Ansatz

Für die Auswertung wurden die beschriebenen neun Alternativen insgesamt 19 Testpersonen im Alter von 18 bis 63 (12 männlichen und 7 weiblichen), die acht Zweiergruppen und eine Dreiergruppe bildeten, vorgelegt. Die Zweiergruppen wurden dazu ermuntert, über die unterschiedlichen Alternativen zu diskutieren. Generell führt diese Methode zu Diskussionen, die die Meinungen und Ideen der Benutzer zum Vorschein bringen.

Bei der Berücksichtigung der Forschungsfragen war es notwendig, eine Auswertungsmethode zu verwenden, die viele qualitativen Daten über Attribute, mentale Modelle und Ideen des Benutzers erzeugt. Die Codiscovery-Methode [1] ist gut geeignet, um viele qualitativen und subjektiven Daten in einer frühen Phase des Designprozesses zu erhalten. Die Idee dieser Technik ist die, daß zwei Testpersonen, die sich sehr gut kennen (z. B. Freunde, Familienmitglieder) an der Bewertung eines Produktes oder einer Benutzungsoberfläche teilnehmen.

Diese eher familiäre Situation hat die folgenden Vorteile:

- eine vertraute Person bietet Schutz in einer ungewohnten Situation,
- bei einer vertrauten Person ist es leichter, Gedanken spontan zu äußern, selbst in Gegenwart der Evaluatoren,
- die Unterhaltung zweier vertrauter Personen können die Gedanken und Ideen des Einzelnen stimulieren.

Um das freie Äußern ihrer Gedanken zu stimulieren, wurden die Benutzer wiederholt aufgefordert, laut zu denken [3] und wurden ermuntert, jeden Gedanken, der ihnen in den Kopf kam, auszusprechen. Um kreatives Denken zu fördern, wurden den Benutzern gesagt, daß alles erlaubt sei. Sie sollten auch untereinander ihre Meinungen austauschen. Die Benutzer wurden dabei immer wieder daraufhingewiesen, daß nicht sie, sondern die Designalternativen getestet würden, und daß ihre Meinungen entscheidend für die Verbesserung der Designalternativen wären.

Je Sitzung nahmen zwei Evaluatoren teil, einer übernahm die Rolle des Moderators, während der andere die Designalternativen vorstellte. Die Diskussion war halb strukturiert. Der Moderator hielt sich an eine vorbereitete Diskussionsstruktur und eine Reihe von Fragen, mit der Freiheit, den Ideen und Vorschlägen der Diskussion zu folgen. Der Evaluator nahm die Benutzerreaktionen schriftlich und auf Video auf. Für jeden Benutzer wurde ein separates Protokoll angefertigt, in dem die Meinungen beider Evaluatoren einflossen.

Zu Beginn einer jeden Sitzung wurde von den Testpersonen ein Fragebogen ausgefüllt, mit dem biografische Daten (Alter, Geschlecht, Beruf), Vorerfahrungen mit interaktiven Geräten (PC, Geldautomat) und der Grad der Vertrautheit mit Heimgeräten und elektronischen Dienstleistungen festgehalten wurden.

Die beiden vollfunktionsfähigen Designalternativen (Apple-Kaufhaus, PREMIERE) wurden in wechselnder Reihenfolge zu Beginn einer jeden Sitzung gezeigt, daraufhin wurden die sieben statischen Alternativen vorgestellt. Die Benutzer durften diese beiden Prototypen nicht benutzen, um sicherzustellen, daß alle Benutzer die gleichen Informationen erhielten, und um zu vermeiden, daß die Benutzer eine Zurückhaltung gegenüben den nichtinteraktiven Designalternativen entwickelten.

Um mögliche Effekte einer Reihenfolge zu vermeiden, wurde die Folge, in welcher diese Alternativen vorgeführt wurden, jedes Mal willkürlich festgelegt, was bedeutet, daß die Alternativen in jeder neuen Sitzung in wechselnder Reihenfolge gezeigt wurden. Die Demonstration erfolgte auf dem PC. Die Benutzer wurden jedoch aufgefordert, sich die Situation so vorzustellen, als würden sie zu Hause vor Ihrem Fernseher sitzen. Jede Alternative wurde zuerst vorgeführt und danach besprochen. Während der Diskussion konnten die Designalternativen immer wieder auf dem Bildschirm angeschaut werden.

Die Interview endete mit einer Evaluation der persönlichen Vorlieben der Testpersonen. Jede Person wurde gebeten - diesmal ohne Partner - die Alternativen auf eine Likert-Skala mit Stufen, die zwischen „sehr positiv" (Stufe 15) und „sehr negativ" (Stufe 1) lagen, anzuordnen, wobei mehrere Alternativen dieselbe Stufe besetzen durften. Jede Designalternative wurde auf eine Karte (Format einer Postkarte) gedruckt, die auf einer auf 2 Meter langem Papierband gedruckten Skala direkt plaziert werden konnte.

5 Qualitative Ergebnisse

In der Studie wurden 925 mündliche Äußerungen aufgezeichnet. Die Zahl der Äußerungen für jede Metapher varierte zwischen 59 und 139 (durschnittlich 103). Alle Kommentare, Ideen und Vorschläge der Benutzer wurden während der Analyse der Studie gesammelt und klassifiziert. Daraus entstand der folgende Klassifizierungsplan der mündlichen Äußerungen (Abb. 10):

	Äußerungen über den Prototypen		
Inhalt	positiv	negativ	Verbesserungen
Informationsdarstellung			
Dialog- und Navigationsdesign			
Allgemeine und Metaphern Akzeptanz			

Abb. 10: Klassifikationsschema für die verbalen Protokolle

Die folgende Abbildung 11 zeigt den Anteil der positiven und negativen Kommentare sowie den Anteil der vorgeschlagenen Verbesserungen für jede Metapher. Es zeigte sich, daß die Baum-Metapher die am schlechtesten beurteilte Metaphar war, auf die keine positiven Kommentare, eine großer Anteil von negativen Kommentaren und ein kleiner Anteil von möglichen Verbesserungen entfiel. Die PREMIERE Oberfläche wurde am besten angenommen und erhielt einen großen Anteil positiver und einen kleinen Anteil negativer Kommentare.

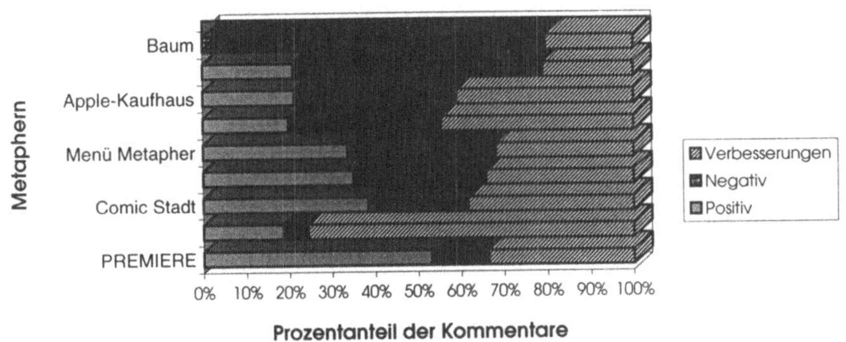

Abb. 11: Verhältnis der positiven und negativen Kommentare sowie der Verbesserungen

5.1   Das Apple-Kaufhaus

Die Benutzer bemängelten, daß das Apple-Kaufhaus zu viele Informationen zur Verfügung stelle, daß die Artikel in dem virtuellen Laden zu klein seien, und daß die Orientierung zwischen den verschiedenen Waren schwierig sei.

Die dreidimensionale Darstellung der Produkte und die Möglichkeit, sie zu manipulieren (Drehen der Produkte in alle Richtungen), wurde als sehr nützlich kommentiert. Die Benutzer waren durch diese Möglichkeiten beeindruckt und erachteten es als eine gute Methode sei, sich über die Produkte zu informieren.

Die Tatsache, daß keine anderen Personen in dem Laden anwesend waren, erzeugte den Eindruck, daß sie allein im Kaufhaus wären, was als unangenehm empfunden wurde. Deshalb schlugen die Benutzer vor, virtuelle Verkäufer oder virtuelle Kunden hinzuzufügen.

5.2 PREMIERE

Im allgemeinen bekam der elektronische Programmguide von PREMIERE die positivsten Kommentare. Vor allem die Informationsdarstellung wurde als einfach zu lesen, klar strukturiert und attraktiv betrachtet. Fast alle Benutzer waren sich darüber einig, daß dieses System einfach zu verwenden sei. Die Fernbedienung, die nur wenige Funktionstasten hat, wurde ebenfalls positiv beurteilt.

Der Premiere EPG wurde am wenigsten kritisiert. Als Verbesserungen schlugen 8 von 19 Benutzer eine „Bild im Bild"-Funktion vor, damit das laufende Programm während der Navigation im EPG weitergesehen werden könnte.

5.3 Baum-Metapher

Diese Metapher schnitt in der Beurteilung der Benutzer extrem schlecht ab. Die Benutzer konnten keine positiven Argumente für diese Art von Benutzungsoberfläche finden.

9 von 19 Benutzer äußerten die Kritik, daß ein Baum gar nichts mit interaktivem Fernsehen (semantisch zu weit entfernt von ITV) zu tun habe, und daß der Vorteil der Verwendung eines Baumes zur Auswahl elektronischer Dienste nicht klar sei. Außerdem wurde kritisiert, daß die auf den Äpfeln plazierten Icons nicht deutlich erkennbar waren. Auch die Anordnung der Äpfel wurde als verwirrend und unregelmäßig betrachtet.

Nur wenige Kommentare bezüglich Verbesserungen (3,2 % aller Verbesserungen, die in dieser Studie aufgezeichnet wurden) wurden hierzu gemacht. Die Tatsache, daß die Benutzer keine Ideen hatten, wie diese Metapher nützlicher zu machen sei, zeigt, daß generell die Idee eines Baumes als Auswahldienst nicht akzeptiert wird.

5.4 Buch-Metapher

Die einfache und klare Vorstellung dieser Metapher und die Vertrautheit mit dem Buch als Informationsmedium wurden gut akzeptiert; die Attraktivität eines Buches als Möglichkeit, ITV-Dienste auszuwählen, wurde jedoch für ITV als ungeeignet und „langweilig" (9 von 19) betrachtet. Es ist zweifelhaft, ob ITV versuchen sollte, eine solche Atmosphere zu schaffen. Die vertikale Darstellung der Dienste-Beschriftungen wurde als schwer lesbar bemängelt.

Um die Attraktivität eines Buches als Metapher zu verbessern, schlugen einige Benutzer vor, die Metapher durch die Verwendung von realistischen Blätter-Animationen oder von speziellen Effekten (z.B. Musik) zur Identifizierung von jedem Kapitel zu erweitern.

5.5 Menü-Metapher

Was die Darstellung und den Inhalt betrifft, wurde die Menü-Metapher ähnlich wie die Buch-Metapher bewertet. Die beiden Metaphern wurden oft als „langweilig" bezeichnet (6 von 19 Benutzern der Menü-Metapher). 5 von diesen 19 Personen sagten, sie erwarteten von ITV, daß es „mehr" darstelle.

Positive Kommentare wurden gemacht in bezug auf die klare Anordnung (13 von 19 Benutzer) und auf die Funktionalität der Menü-Metapher (12 von 19 Benutzer). Die Menü-Metapher vermittelte den Eindruck, man könne sie gut steuern. Diese Tatsache wurde von vielen Benutzern, die dies als einen wichtigen Faktor für die Vereinfachung der Auswahl eines Dienstes betrachten, betont.

Kritisiert wurden die undeutlichen Ränder der Menüs. Die Benutzer wünschten sich große und deutliche Symbole. Die Pictogramme für TV, Tourismus und Freizeit wurden als undeutlich bewertet.

5.6 Comic-Stadt

Insgesamt wurde die Comic-Stadt gut akzeptiert (14 von 19 Benutzern; nur 4 lehnten die Metapher ausdrücklich ab). Von 9 von 19 Benutzern wurde das Farbdesign gebilligt. Auch der Stil eines Zeichentricks wurde gutgeheißen.

Im Gegensatz dazu, wurde die Metapher als undeutlich und unstrukturiert bewertet. Außerdem wurde kritisiert, daß die Gebäude oder Icons nicht das darstellten, was auf ihnen stand. Einige Benutzer schlugen vor, eine Animation einzuführen, die bei der Wahl eines Gebäudes (eines Dienstes) das Gebäude im Stil des Apple-Kaufhauses öffnen würde.

5.7 Leute

Nachdem die Metapher vorgestellt und die Benutzer damit vertraut waren, wichen die Vorstellungen von der Grundidee der virtuellen Führer schnell ab. Sprachausgabe als Informationsquelle wurde bald ein wichtiges Thema der Interviews (9 von 19 Benutzer). Die Benutzer wünschten sich einen sprechenden virtuellen Führer, entweder indem er den Benutzer durch das System führt, oder indem er ihnen zusätzliche Informationen zur Verfügung stellt.

Die Vorstellung, daß bekannte Persönlichkeiten als virtuelle Führer verwendet werden sollten, wurde stark abgelehnt (10 von 19 Benutzer). 11 von 19 Benutzer schlugen vor, daß es möglich sein sollte, die einzelnen Führer aus einer Reihe von Führern zu wählen. Jedoch sollten die virtuellen Führer angenehm (8 Benutzer), ernst und in ihrem Bereich kompetent (6 Benutzer) sein. Eine Zeichentrickfigur wurde gut akzeptiert (5 Benutzer). Weiterhin hatten die Benutzer erwartet, den Führer ausschalten zu können (dies wurde von 8 von 19 Benutzer erwähnt).

In der Studie erhielt die Leute-Metapher die meisten Verbesserungsvorschläge zum Design. Was die Interaktion mit virtuellen Führern betrifft, wurden die folgenden Vorschläge gemacht:

- die Führer können sich bewegen und sprechen,
- die Informationen, die die Führer vermitteln, könnten als Kommentare, Hilfeinformationen oder als eine mündliche Produktvorstellung dargestellt werden, oder auch in direkter Interaktion mit dem Benutzer stehen,
- nur ein Führer sollte die gesamten Informationen vorstellen,
- junge Menschen ziehen Führer vor, die durch ihr Outfit die Lebensart der Rubrik, wofür sie stehen, darstellen, und ältere Menschen ziehen einen gutgekleideten Führer vor, der den Bereich, wofür er steht nicht durch seine Bekleidung darstellt,
- ein Zeichentrick-Führer scheint akzeptabel.

5.8 Der Stuttgarter Stadtplan

Alle Testpersonen kannten Stuttgart sehr gut. Deshalb wurde die Stuttgarter Stadtplan-Metapher (9 von 19 Benutzern) sowie die Idee, lokale Dienste vorzustellen (7 von 19 Benutzern) im allgemeinen sehr gut akzeptiert. Die bekannten Gebäude wurden gut erkannt.

Die Benutzer kommentierten, daß der Stuttgarter Stadtplan nur für lokale Dienste und nicht für landesweite Dienste nützlich sei (11 von 19 Benutzern). Als Haupteingangs-Metapher wurde diese Designalternative abgelehnt (6 von 19 Benutzern).

5.9 Die VR Stadt

Diese Metapher wurde als verwirrend, unklar und fremd betrachtet (10 von 19 Benutzern). Zusätzlich stellen die Symbole den Dienstinhalt nicht semantisch dar. Die Tatsache, daß der „lebendige" Bereich mitten im Bild war, so daß der Benutzer immer im Zentrum war, wurde positiv kommentiert. (10 von 19 Benutzern)

6 Quantitative Ergebnisse

Die folgenden Grafiken geben einen kurzen Überblick über die Benutzerakzeptanz der verschiedenen Metaphern. Abbildung 12 zeigt den meistverwendeten Wert auf der 15-stufigen Likert-Skala (modal). Dies zeigt eine klare Präferenz für PREMIERE, das Apple-Kaufhaus, die Comic Stadt, den Stuttgarter Stadtplan und den Menü-Ansatz (Menü-Metapher). Weniger akzeptiert sind die Designalternativen Leute, die VR-Stadt, und die Buch- und Baummetaphern. Das gleiche Bild erhält man, wenn der Medien berechnet wird.

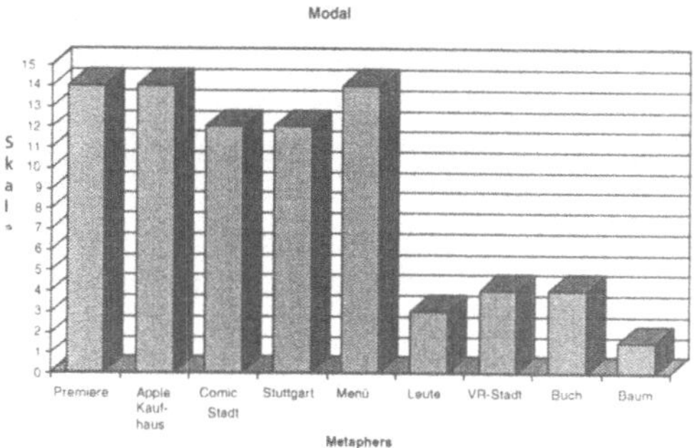

Abb. 12: Plazierung der jeweiligen Metaphern auf der Skala, die von 15 (außerordentlich gut akzeptiert) bis zu 1 (nicht akzeptiert) reicht.

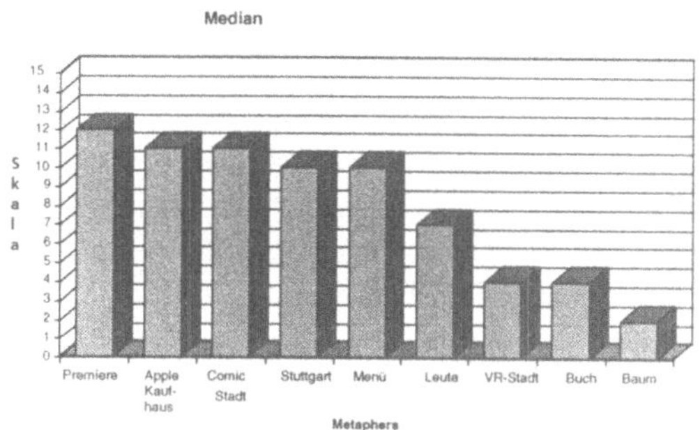

Abb. 13: Median der Plazierung der jeweiligen Metaphern

7 Schlußfolgerungen

Von den Reaktionen und Beobachtungen, die während der Benutzertests gemacht wurden, kann man bezüglich der Informationsdarstellung und der ITV-Benutzungsoberflächen die folgenden Anforderungen ableiten:

Es kann angenommen werden, daß die Punkte, die von den Benutzern häufiger erwähnt wurden, entscheidende Faktoren für den Design von ITV-Benutzungsoberflächen sind. Diese Anforderungen sind von den positiven Kommentaren, die die Benutzer über die verschiedenen Alternativen gemacht haben, abgeleitet worden. Außerdem stellen die Bemerkungen, die Mängel oder fehlende Merkmale der einzelnen Alternativen kritisierten, eine Hilfe dar, weitere Anforderungen zu formulieren.

Die Anforderungen, die von den Benutzertests abgeleitet werden konnten, sind die folgenden:

- die Metaphern für Benutzungsoberflächen sollten semantisch nah am ITV sein. Wenn dies nicht der Fall ist, wird die Metapher abgelehnt (Baum, die Comic Stadt, die VR-Stadt),
- die grafischen Elemente der Metapher dürfen nicht zu spielerisch, undeutlich oder klein sein,

- das Farbdesign der Metapher sollte zu ihrer Attraktivität beitragen,
- die Metapher muß einfach, aber nicht langweilig sein (die Buch- und Menümetaphern wurden als sehr klar strukturiert, aber zu langweilig bewertet),
- je realistischer das Design der Metapher, desto besser ihre Akzeptanz (ein abstraktes Design ist zu vermeiden),
- die Anordnung der einzelnen Komponenten einer Metapher sollte eine klare visuelle Struktur haben (lieber in der Form einer Liste oder einer Tabelle als in der Form eines Stadtplans),
- Symbole und Piktogramme müssen die Inhalte, für die sie stehen, darstellen,
- die Symbole müssen sich vom Hintergrund in Farbe und Form klar unterscheiden lassen,
- die Benutzer sollten ein unmittelbares, visuelles/akkustisches Feedback bekommen,
- das System sollte schnell sein; Wartezeiten sind nicht akzeptabel,
- Sprachausgabe wird unter der Bedingung angenommen, daß sie abgeschaltet werden kann,
- falls virtuelle Führer verwendet werden, sollten diese keine bekannten Persönlichkeiten sein. Die Möglichkeit, einen Führer aus Verschiedenen zu wählen, ist vonnöten. Die Vorstellung, daß ein persönlicher Führer den Benutzer direkt anreden sollte, wurde als positiv empfunden,
- in Fällen, wo die Benutzer eine Auswahl treffen müssen, sollten Animationen vermieden werden. Es muß besonders auf die Effizienz der Auswahl geachtet werden,
- zur Darstellung von Produkten sind Animationen und Videos angenommen,
- in Fällen, wo sich die Benutzer zwischen verschiedenen Diensten oder Produkten entscheiden müssen, ist die Effizienz wichtiger als die Attraktivität der Darstellung.

Aus den Ergebnissen der Interviews kann die folgende Hierarchie bezüglich Auswahl und Interaktion der Dienste abgeleitet werden:

Menü-Hierarchie	Anforderungen	Beispiele von Darstellungen und Interaktionsmodi
Wahl der Dienste	• technische Metaphern sind bevorzugt (Menüs, Buttons) • klare, visuelle Strukturen • attraktives Farbdesign • Kombinationen von Überschriften und Piktogramm für Buttons- oder Menü-Einträge • die Effizienz ist wichtiger als die visuelle Attraktivität • eine zwei-dimensionale Darstellung ist ausreichend	• Liste der Menüoptionen als Button mit Piktogramm und Beschriftung • Auswahl mit Navigations- und numerischen Tasten der Fernbedienung oder Zeigegerät
Wahl der Produkte	• 3D-Metaphern oder virtuelle Führer können verwendet werden • attraktives Farbdesign • Kombinationen von Beschriftung und Piktogramm • die Effizienz ist wichtiger als die visuelle Attraktivität	• Kaufhaus-Metapher • Auswahl der Produkte aus den Regalen (Reihenfolge der Produkte in Form von Tabellen) • Auswahl mit Navigations- und numerischen Tasten der Fernbedienung oder Zeigegerät
Betrachten von Produkten	• realistische Darstellung • verschiedene Medien sollten verwendet werden • verschiedene Informationsquellen sollten angeboten werden • die Information über ein Produkt ist wichtiger als die Effizienz • 3D-Darstellungen sind bevorzugt	• das Produkt (z. B. ein Staubsauger) ist auf einem Bild dargestellt • ein Video mit der Bedienungsanleitung und den wichtigsten Merkmalen des Produktes kann ausgewählt werden • das Produkt kann in 3D betrachtet werden (z.B. Quicktime VR) • eine Zoom-Funktion ermöglicht, kritische Merkmale des Produktes zu untersuchen (z. B. Schweißnähte)

Abb. 14: Benutzeranforderungen für ITV-Benutzungsoberflächen (n=19)

8 Literatur

[1] Kemp, J.A.M. and van Gelderen, T.(1993). The Co-Discovery Method: an informal method for iteratively designing consumer products. In: Institute for Perception Research, IPO Annual Progress Report 28; The Netherlands.

[2] Mack, R.L. & Nielsen, J. (1994): Executive Summary. In J. Nielsen & R. Mack (Eds.): Usability Inspection Methods (pp. 1-24). New York: John Wiley.

[3] Nielsen, J. (1994). Estimating the number of subjects needed for a thinking aloud test. Int. J. Human-Computer Studies,41, 385-397.

[4] Wozencroft, J. (1992). Die Grafiksprache des Neville Brody (pp. 144f). Bangert Verlag, München.

Adresse der Autoren

Dipl.-Psych. Michael Burmester
Dipl.-Inform. Franz Koller
Fraunhofer-Institut für Arbeitswirtschaft und Organisation
Nobelstr. 12
70569 Stuttgart
Tel.: +49 711 970 2311 (-2321)
Fax: +49 711 970 2300
Email: Michael.Burmester @iao.fhg.de
Email: Franz.Koller@iao.fhg.de

Benutzerorientiertes Design - PREMIERE Benutzungsoberfläche für interaktives Pay - TV

Michael Burmester, Franz Koller und Thilo König

Fraunhofer-Institut für Arbeitswirtschaft und Organisation, Stuttgart

PREMIERE Medien GmbH, Hamburg

Zusammenfassung

Die Verbreitung des Fernsehens in den 50er und 60er Jahren hat unseren Tagesablauf stark verändert. Für die kommenden Jahren zeichnen sich durch die starke Zunahme der Programmzahl und das erweiterte Service-Angebot weitere Entwicklungen ab. Daher werden sinnvolle Konzepte erforderlich, die einfach zu bedienen sind und bei denen der Fernsehzuschauer trotz der Informationsvielfalt nicht den Überblick verliert.

Ein solches Konzept soll hier vorgestellt werden. Der im Rahmen von PREMIERE entwickelte Programmguide soll die klassische Programmzeitschrift ergänzen.

Der Prototyp wurde in zwei Iterationen von potentiellen Benutzern evaluiert. Dazu wurden insbesondere die Methoden der Benutzerbeobachtung und des Lauten Denkens eingesetzt. Im Mittelpunkt der ersten Iteration stand die Navigation. Es zeigte sich, daß die Orientierung innerhalb des Systems noch unzureichend war. Als Konsequenz daraus wurde die Bedienstruktur stark vereinfacht. Die Benutzer zogen offenbar eine einfachere Struktur einer komplexeren Variante vor, auch wenn diese weniger Bedienschritte verlangte. In der zweiten Evaluation wurde der Prototyp weiter optimiert.

Das Projekt zeigte, daß iteratives Design in Verbindung mit Benutzerbeteiligung während des Designprozesses ein Schlüsselfaktor ist, um eine breite Benutzerakzeptanz zu sichern.

1   Einleitung

Die Vorstellung, ein schwarzer, viereckiger, flimmernder Kasten könne unser Leben so stark beeinflussen, daß sich täglich im Schnitt 90 % der Bundesbürger mehrere Stunden damit beschäftigen, hätte den Erfindern des Fernsehers in den dreißiger Jahren wohl durchaus gemischte Gefühle beschert. Der Fernseher ist ein bedeutender Teil unserer Gesellschaft geworden und hat wie keine andere technische Errungenschaft dieses Jahrhunderts in fast allen Bereichen des privaten und öffentlichen Lebens gravierende Veränderungen bewirkt. Die Fernsehgewohnheiten werden sich weiter ändern und neue Herausforderungen an die technische Ausstattung der Geräte und Sender, an die Produzenten und Fernsehschaffenden und nicht zuletzt an den Konsumenten stellen.

Die Umsetzung neuer Technologien ermöglicht faszinierende und interaktive Formen der elektronischen Kommunikation. Heute noch isolierte Medien wie Video, Fernsehen, Telefon und Zeitung wachsen zusammen, es entstehen neue, multimediale „Mehrwertdienste". In

Zukunft werden Fernsehzuschauer über ihren Fernseher auf zahlreiche Informationen zugreifen können, z.B. auf Videofilme, Lernprogramme, Spiele, Einkaufskataloge usw.

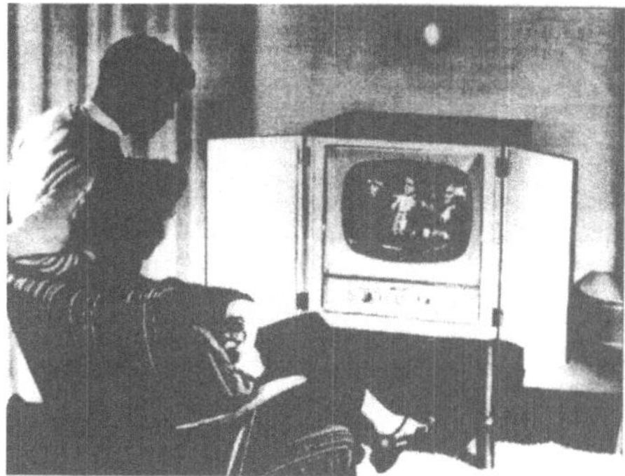

Abb. 1:

Fernsehen anno 1956

Gerade im technischen Bereich ergeben sich gravierende Veränderungen: Umfangreiche dynamische Datenmengen werden digitalisiert, neue Datenkomprimierungsmethoden angewandt und der Informationshighway zunehmend ausgebaut. Die interaktive Nutzung von Fernsehdiensten mittels spezieller Kanäle zur Rückkopplung und zum Informationsaustausch zwischen Konsument und Sender wird möglich. Set-Top-Boxen erlauben die Steuerung und Selektion von Daten in beiden Richtungen: Die Sendeanstalt bestimmt das Angebot; die Benutzer entscheiden, welche Serviceleistungen sie in Anspruch nehmen (z.B. Teleshopping, Pay-Radio, Video-on-demand).

2 Anforderungen an das Interaktive Fernsehen

Neben der entsprechenden Hardware und Infrastruktur erfordert der Zugriff auf die großen Informationsmengen vor allem tragfähige Bedienkonzepte für interaktive Dienste. Bei einer kritischen Betrachtung gegenwärtiger Pilotprojekte fällt auf, daß die Anforderungen an die ergonomische Gestaltung und Bedienbarkeit interaktiver Dienste bislang nicht ausreichend diskutiert und berücksichtigt werden. Insbesondere können grafische Oberflächen, wie man sie von Computern kennt, nicht einfach übertragen werden. Bedienkonzepte für interaktives Fernsehen müssen den folgenden Aspekten gerecht werden:

- Aus dem Bereich der Unterhaltungselektronik ist bekannt, daß umfangreiche Bedienmanuale von Benutzern selten benutzt werden [7]. Selbstbeschreibungsfähigkeit von Systemen ist somit eine zentrale Anforderung.

- Die Benutzer haben teilweise keinerlei Vorerfahrung mit Computern und interaktiven Systemen. Die Bedienkonzepte müssen deshalb an Bedienstrukturen bereits bekannter Geräte anknüpfen [3].

- Der räumlichen Situation des Benutzers muß Rechnung getragen werden (Entfernung zum Bildschirm/TV, Sitzposition, Lichtverhältnisse).

- Wenn herkömmliche Fernbedienungen zur Steuerung interaktiven Fernsehens eingesetzt werden, so sollten diese nur eine geringe Zahl von Bedienelemente aufweisen [13].

- Die Anzahl der Fernsehkanäle wird in absehbarer Zeit (3-5 Jahre) von derzeit 30 auf bis zu 100 - 500 anwachsen, d.h. mit einfachen Mitteln sollte eine Auswahl von Programmen/Diensten aus einer großen Anzahl von Alternativen möglich sein.

Durch das wachsende Programmangebot und die zusätzlichen Servicedienstleistungen müssen klassische Elemente der Orientierung und Navigation überdacht und durch neue Konzepte ergänzt oder ersetzt werden.

3 Gestaltung und Realisierung einer Benutzungsschnittstelle für Interaktives Pay TV: Iterativer Entwicklungsansatz

Für Design und Realisierung interaktiver Systeme eignet sich ein iterativer Entwicklungsansatz. Der Vorteil dieses Entwicklungsansatzes besteht darin, daß die Qualität des Systems bezüglich der Funktionalität, der formalästhetischen Gestaltung sowie der Informations- und Dialoggestaltung in mehreren Entwicklungszyklen (Analyse, Design, Prototyping, Evaluation) optimiert werden kann [2].

3.1 Analyse

Die Analyse beginnt mit einer detaillierten Spezifikation der Anforderungen. Zu klärende Fragen betreffen das Profil der Zielgruppe, den Umfang und Detailgrad der darzustellenden Information sowie die Festlegung der wesentlichen Ziele, die durch die Anwendung erreicht werden sollen.

3.2 Design

Der wichtigste Faktor für das Design ist die Benutzbarkeit (Usability). Folglich muß die Navigationsstruktur einfach und unmißverständlich sein. Der Benutzer sollte immer in der Lage sein, die vier folgenden Fragen zu beantworten: Wo bin ich? Warum bin ich hier? Wie bin ich dahin gekommen? Wohin kann ich gehen? In diesem Zusammenhang ist ein intuitives Verständnis der Navigationsstruktur und der Informationsdarstellung wünschenswert.

Die relevanten Kriterien können wie folgt zusammengefaßt werden:

- Konsistenz in der Navigationsstruktur und der Benutzungsschnittstelle

- intuitives Verständnis der Navigationsstruktur und der Informationsdarstellung

- emotionale Kriterien (wie z.B. Spaß, Attraktivität des Systems, etc.) und motivationale Aspekte (z.B. Interesse, das Informationsangebot zu erforschen)

- software-ergonomische Kriterien: Aufgabenangemessenheit, Selbstbeschreibungsfähigkeit, Steuerbarkeit, Fehlertoleranz, Erlernbarkeit.

3.3 Prototyping

Die Methode des Rapid Prototyping ermöglicht die Prüfung des Systems bereits in frühen Entwicklungsphasen. Um Fehler rechtzeitig beheben zu können und damit die Entwicklung effektiv und kostengünstig zu gestalten, ist eine frühe prototypische Realisierung des Systems unentbehrlich.

3.4 Evaluation

Um die Bedienqualität und die generelle Akzeptanz des im Entwicklungsprozeß befindlichen Produktes zu gewähren, ist es notwendig, einen Evaluationsschritt nach jeder Prototypenversion einzuführen. Dieser Schritt soll unter Einbezug von Endbenutzern stattfinden [1]. Nur so ist es möglich, Form- und Dialogentwürfe zu testen. Zum Test der Dialogentwürfe reichen bereits einfache Computersimulationen der Benutzungsoberfläche aus [10].

Nach Studien von Nielsen [9] können bereits mit drei bis vier Testpersonen bis zu 75 % der möglichen Bedienschwierigkeiten ermittelt werden. Virzi [12] konnte zeigen, daß bereits mit ein bis zwei Personen die größten Bedienprobleme - sogenannte „user interface disasters" [8] - ermittelt werden können.

Bei der Untersuchung von Prototypen werden die Methoden des „Lauten Denkens", der Beobachtung von Bedienschwierigkeiten durch Evaluatoren, der Messung von Aufgabenbearbeitungszeiten sowie der Befragung eingesetzt. Für Re-Analysen und zur Präsentation kritischer Bediensituationen werden Videoaufzeichnungen eingesetzt. Dabei empfiehlt sich der Einsatz von zwei Videokameras, wobei eine auf den Bildschirm und die zweite Kamera auf den Probanden gerichtet ist. Die Bilder werden über einen Videomischer zusammengefügt. Auf diese Weise können Reaktionen und spontane Äußerungen analysiert und dokumentiert werden.

Zum Abschluß eines Prototypentests erhalten die Benutzer die Möglichkeit, eine Gesamtbewertung zur Bedienbarkeit, Verständlichkeit und Akzeptanz des Prototypen abzugeben. Dies geschieht mit einem standardisierten Fragebogen zu software-ergonomischer Qualität und durch ein strukturiertes Interview, bei dem beispielsweise Fragen nach der Zufriedenheit des Benutzers mit dem System, nach Eigenschaften des Systems, die besonders positiv oder besonders negativ aufgefallen sind, gestellt werden. Ferner werden die Benutzer aufgefordert, Ideen zur Verbesserung des Prototypen zu äußern.

4 Der elektronische Programme Guide (EPG) von PREMIERE

4.1 Das Entwicklungsteam

Im Rahmen des digitalen Fernsehens wurde für den privaten deutschen Pay-TV Sender PREMIERE ein Konzept zur On-Screen-Navigation bei der Programmauswahl entwickelt. Als Ausgangsbasis hatte PREMIERE Konzeptionsentwürfe und Designvorschläge vorgelegt. In Kooperation mit dem Fraunhofer-Institut für Arbeitswirtschaft und Organisation in Stuttgart wurden in einem interdisziplinären Team bestehend aus Designern, Psychologen, Informatikern und Software-Ergonomen die verschiedenen Design-Alternativen diskutiert. Entsprechend der iterativen Vorgehensweise wurde das Produkt in verschiedenen Prototypen vorgefertigt und Tests mit Benutzern unterzogen.

4.2 Anforderungen an die Benutzungsschnittstelle

Von Anfang an wurde darauf geachtet, daß der Benutzer im Mittelpunkt der Aufmerksamkeit stand. Dieses oberste Ziel führte zur Formulierung der folgenden Anforderungen:

- Der Raum für die verschiedenen Arten von Informationen (Pay-TV Kanäle, Programmübersicht für den gegenwärtigen Tag und die darauffolgenden sieben Tage, verschiedene Filmgenres, Tele-Shopping, Informations-Service, Videospiele und Pay-Radio) mußte derart strukturiert sein, daß der Benutzer sehr schnell in der Lage war, sich ein mentales Modell [4] zu bilden.

- Die Dialoge zwischen Benutzern und dem PREMIERE Programmguide mußten effizient und selbstbeschreibungsfähig sein sowie dem Kriterium der Erwartungskonformität entsprechen.

- Das zur Firmenkultur (Corporate Identity) von PREMIERE gehörige Corporate Design mußte in die Benutzungsschnittstelle integriert werden. Das Produkt sollte einfach erkennbar sein, und sein visuelles Design sollte potentielle Benutzer ansprechen. Zu diesem Zweck wurden moderne grafische Trends, das PREMIERE Firmendesign und Key Visuals mit einbezogen [14].

- Die Motivation für die Benutzung des PREMIERE Programmguides sollte durch ein fehlertolerantes, ästhetisches, interessantes und steuerbares System gefördert werden. Es wurde darauf geachtet, Informationen so attraktiv darzustellen, daß der Benutzer sich eingeladen fühlt, nach weiteren Informationen zu suchen.

- Des weiteren mußten Marketing Kriterien berücksichtigt werden. Das bedeutete, daß sich dieses System von anderen existierenden Systemen unterscheiden mußte.

4.3 Die erste Iteration: Navigation im Informationsraum

4.3.1 Design und Prototyping

Abb. 2: Erste Version des Programmüberblicks: Der obere Bildteil fungiert als eine Art Werbefläche. Dort wird der Tip des Tages (ein aktueller Film) angeboten. Im unteren Bildbereich befindet sich die nach Genres geordnete Übersicht der aktuellen Sendungen. Im Fußbereich des Bildschirms befinden sich die sogenannten Shortcuts, mit denen in andere Informationsbereiche verzeigt werden kann (z.B. weitere Sender oder elektronische Dienste). In mit Links- und Rechtspfeilen gekennzeichneten Feldern kann in die Information „Tip des Tages" und der aktuell laufenden Sendungen "geblättert" werden. (Überblick Navigationsstruktur in Abb. 3)

Eine erste Version der Benutzungsschnittstelle des Programmguides wurde mit dem Rapid Prototyping Werkzeug Director 4.0 produziert. Die darzustellenden Informationselemente (Titel des Films, Genre, Sendezeit, Sendedauer, Darsteller usw.) waren zum größten Teil von PREMIERE vorgegeben worden (vgl. Kap. 4.1). Die visuelle Gestaltung der Oberfläche sollte die Benutzer durch die Kombination von bildlicher und textueller Information sowie die Darstellung der einzelnen Kanäle in parallel verlaufenden Sparten an die Struktur gängiger Programmzeitschriften erinnern und damit die einzelnen Funktionalitäten des Programmguides intuitiv verständlich machen. Bei der genauen Wahl der Farben, der

Bestimmung der Fontgröße usw. wurden software-ergonomische Kriterien wie Leserlichkeit und Übersichtlichkeit sowie die PREMIERE Corporate Identity als Kriterien herangezogen.

Die grundlegenden Merkmale des Systems, d.h. die Basis-Navigationselemente und das Interaktionskonzept, wurden implementiert. Die Funktionsfähigkeit wurde so weit entwickelt, daß die Benutzer sich einen ersten Eindruck über die Dialog- und Informationskonzepte verschaffen konnten. Ferner wurden Shortcuts in Form von zusätzlichen Schaltflächen unterhalb der Programmübersicht realisiert (s. Abb. 2). Über sie sollte es den Benutzern ermöglicht werden, direkt von einem Informationsbereich zum anderen zu navigieren, ohne das Hauptmenü aufzurufen. Die Shortcuts wurden eingeführt, um das System effizienter zu gestalten, da bei dieser Art der Navigation weniger Tastenanschläge notwendig sind.

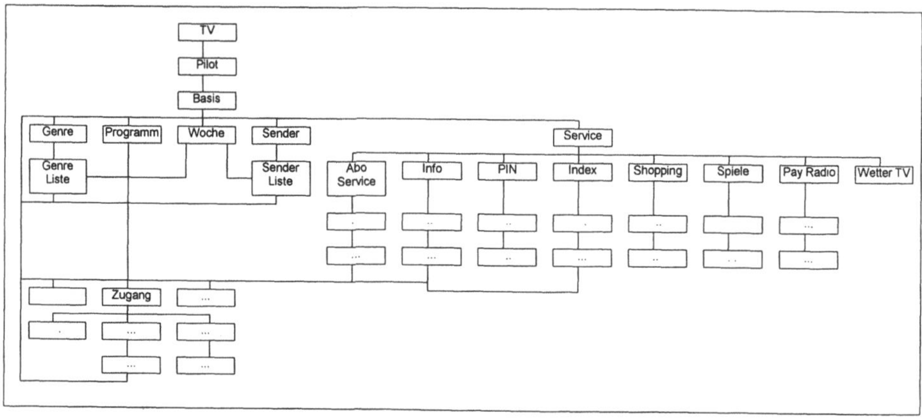

Abb. 3: Vorläufiges Navigationskonzept bei erster Iteration (vgl. auch Bildschirmdarstellung in Abb. 2)

4.3.2 Evaluation

Der erste Prototyp wurde in einer Versuchsreihe mit fünf Testpersonen getestet. Die Benutzer wurden aufgefordert, typische Aufgaben hinsichtlich der Navigation in verschiedenen Informationsbereichen zu lösen (z.B. feststellen, welcher Film zur gleichen Uhrzeit in einem anderen Kanal läuft; welche weiteren Filme innerhalb eines bestimmten Genres angeboten werden, usw.) und den Prototypen anschließend auf einer Rating-Skala hinsichtlich Attraktivität, Verständlichkeit, Übersichtlichkeit und ähnlicher Kriterien zu bewerten. Bei der Durchführung der Aufgaben wurden die Methoden des Lauten Denkens sowie der Beobachtung durch die Evaluatoren bzw. Dokumentation auf Videoband angewandt (vgl. Kap. 3.4).

Die Evaluatoren waren in erster Linie daran interessiert, zu erfahren, wie eine große Datenmenge übersichtlich dargestellt und über die Cursortasten effektiv gehandhabt werden kann. Zentrale Fragestellungen waren dabei: Behalten die Benutzer bei der Bedienung des Prototypen den Überblick oder verlieren sie die Orientierung? Werden alle angebotenen Informationen optimal verstanden und sämtliche Navigationsmöglichkeiten genutzt? Welche Verbesserungen schlagen die Benutzer vor?

4.3.3 Ergebnisse der ersten Iteration

Schon zu Beginn der Tests zeigte es sich, daß die Erwartungshaltung interaktivem Fernsehen gegenüber sehr hoch lag. Die Benutzer erwarteten bspw. Teledienste wie Teleshopping sowie verschiedene Kameraeinstellungen bei Fußballspielen. Als evtl. anfallende Kosten angesprochen wurden, nahmen die Benutzer schnell eine ablehnende Haltung ein.

Im allgemeinen hoben die Testpersonen die optisch ansprechende Gestaltung des Prototypen hervor. Auch wurde er - im Vergleich zu BTX - als einfach zu bedienen eingestuft. Kritisiert wurde die Struktur der Informationsdarstellung. Die Benutzer wünschten sich eine deutlichere Hervorhebung der einzelnen Bereiche über entsprechende Farbgebung, Veränderung der Fontgröße usw. Alle Testpersonen sprachen sich außerdem für zusätzliche Informationen zur Uhrzeit und dem laufenden Fernsehkanal im oberen Bereich des Fernsehbildes aus.

Die Orientierung in der Navigationsstruktur war bei allen Testpersonen gering. Dafür spricht vor allem auch die Tatsache, daß keine der Testpersonen Shortcuts benutzte. Anstatt direkt von einem Informationsbereich in den anderen zu wechseln (z.B. aus dem aktuellen Fensehprogramm in den Bereich Genre), verließen die Benutzer den Programmguide durch Betätigung der Exit-Taste der Fernbedienung, kehrten damit in das laufende Fernsehprogramm zurück und starteten den Programmguide erneut über die Menü-Taste der Fernbedienung. Von dort aus wurde dann der gewünschte Informationsbereich angewählt.

Aus der Beobachtung des Navigationsverhaltens wurde die Schlußfolgerung gezogen, dem Benutzer lediglich die Möglichkeit zum Navigieren über das Hauptmenü zu geben (vgl. Abb. 4). Infolgedessen wurde die Navigationsstruktur zwischen den verschiedenen Informationsbereichen extrem vereinfacht. Diese verbesserte Navigationsstruktur mit dem Hauptmenü als zentralem Punkt des Systems wurde für die zweite Iteration implementiert.

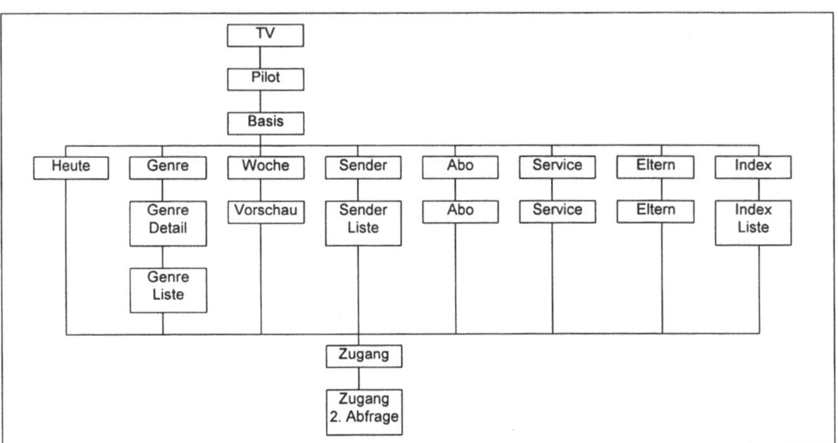

Abb. 4: Verbessertes Navigationsdesign in der zweiten Iteration

Die mit dieser Struktur verbundenen Screens sind im folgenden Kapitel dargestellt.

4.4 Die zweite Iteration: Dialog- und Informationsgestaltung

4.4.1 Design und Prototyping

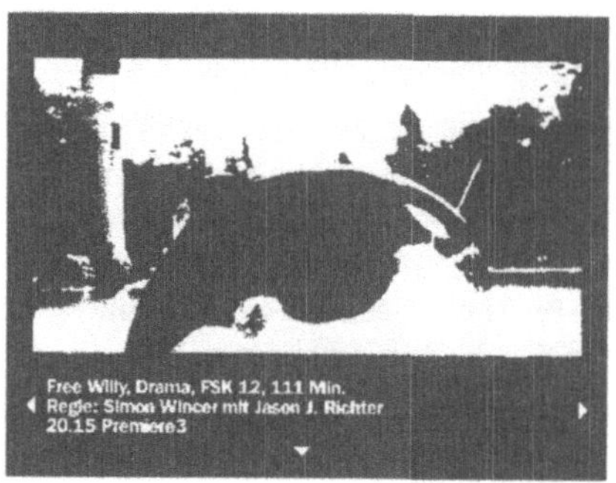

Abb. 5: „Pilot" bezeichnet das Einstiegsbild in den Premiere Programmguide. Der Fernsehzuschauer öffnet den Pilot, um sich Informationen über alternative Programme (die zur gleichen Zeit bzw. zu einem späteren Zeitpunkt laufen) zu verschaffen. Das Bild legt sich dabei vor das laufende Fernsehprogramm. Das hat den Vorteil, das der Benutzer nicht in einen anderen Kanal wechseln muß.

Der Pilot besteht aus einem on-screen Bild, das vor dem laufenden Fernsehprogramm erscheint. Er gibt Textinformationen über andere Programme. Am oberen Bildschirmrand sind Informationen über das laufende Fernsehprogramm und die Zeit eingeblendet. Diese Informationen verändern sich nicht während der Suche.

Wird die Pfeiltaste der Fernbedienung gedrückt, kann der Benutzer Textinformationen über andere Fernsehkanäle abrufen, ohne in diese Kanäle wechseln zu müssen. Es stehen sowohl Textinformationen über laufende Programme verschiedener Kanäle (horizontale links/rechts Tasten) ebenso wie über spätere Programme dieser Kanäle (vertikale Tasten) zur Verfügung. Ist der Benutzer an einem dieser Programme interessiert, drückt er die OK-Taste seiner Fernbedienung und wechselt damit zum gewünschten Programm.

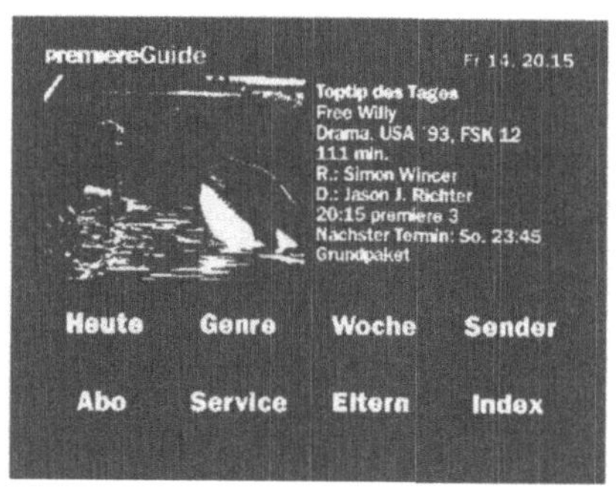

Abb. 6: Über das Hauptmenü können alle Bereiche des Programmguides erreicht werden. Das hierfür verwendete Layout (Bild in oberer linker Bildhälfte, rechts davon Textinformation, Buttons für bestimmte Funktionalitäten in unterer Bildhälfte) sowie die entsprechenden Funktionen wiederholt sich mehrmals innerhalb des Systems.

Die Kopfzeile des Basismenüs zeigt den Menütitel an, um den Benutzer über die akutelle Position innerhalb des Systems zu informieren. Darauf folgt ein Informationsblock, der in einen Bildabschnitt auf der linken Seite und einen Textabschnitt auf der rechten Seite aufgeteilt ist. Auf dem unteren Drittel des Bildschirms befinden sich acht Auswahlfelder, die es dem Benutzer ermöglichen, zwischen verschiedenen Bereichen zu wählen. Durch die Betätigung der Pfeiltaste auf der Fernbedienung kann der Benutzer die verschiedenen Auswahlfelder aktivieren und dadurch Bild und Text des oberen Bildschirmteils verändern.

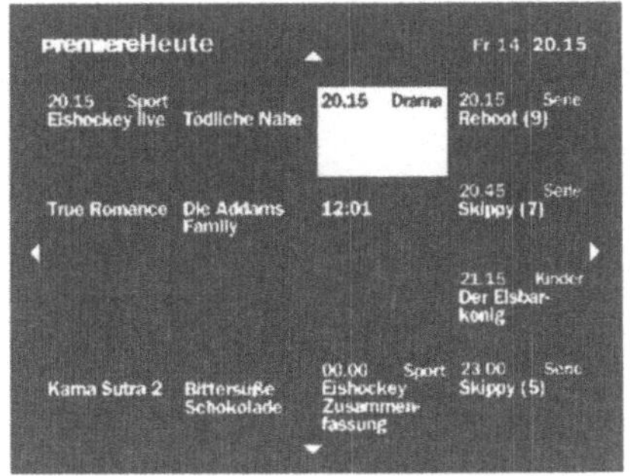

Abb. 7: Die Programmübersicht erinnert an ein Fernsehmagazin. Sie bietet eine Übersicht über laufende und zukünftige Programme. Generell listet die Programmübersicht alle Sendungen des aktuellen Tages auf. Die Programme der folgenden sieben Tage werden auf Verlangen gezeigt.

Die Programmübersicht zeigt Sendezeit, Genre und Titel der Programme an, wobei die Titel durch eine entsprechende Farbcodierung hervorgehoben werden. Der Text kann in vier Richtungen bewegt werden. Bei horizontalem Scrollen können die Kanäle gewechselt werden, während bei vertikalem Scrollen die Programme eines Kanals von 6:00 Uhr des aktuellen Tages bis 5:59 Uhr des folgenden Tages gezeigt werden.

Abb. 8: Um einen Film innerhalb eines bestimmten Genres auszuwählen, kann der Fernsehzuschauer den Informationsbereich „Genre" aktivieren. Er erhält dann eine Liste der entsprechenden Filme.

Das Layout der Genreauswahl ist mit dem des Hauptmenüs identisch. Wird eine große Rubrik wie z.B. Kino ausgewählt, erscheint ein Untermenü, mit identischem Layout und einer detaillierteren Inhaltsstruktur. Sobald der Benutzer ein Feld auswählt, das zu einem bestimmten Genre gehört, wechseln Bild und Text im oberen Bildschirmabschnitt.

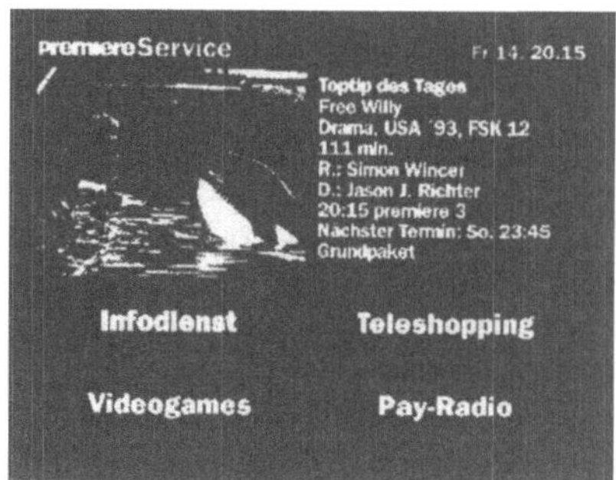

Abb. 9: Das Dienstleistungsangebot bietet Zugang zu verschiedenen Online-Diensten.

Zusätzlich zu den üblichen Dienstleistungen hat der Benutzer Zugang zu Online-Diensten, z.B. Teleshopping, Wetter TV oder Videospiele.

4.4.2 Evaluation

Die Evaluationsziele der zweiten Iteration waren folgende:

- Bewertung des ersten Eindrucks: Der erste Eindruck beeinflußt die Kaufmotivation des Benutzers (Kemp & van Gelderen, 1993);

- Exploration: Überprüfen, wie einfach den Benutzern die Exploration des Systems fällt;

- Navigation: Evaluation der Navigation innerhalb der Programmübersicht;

- Verständlichkeit: Verständnis der Textinformationen (z. B. Hintergrundinformationen zum Film).

Zusätzlich zu den oben aufgeführten Evaluationszielen leitete sich eine weitere Fragestellung aus der ersten Iteration ab. Wichtigstes Ergebnis der ersten Iteration war, daß die Benutzer die Shortcuts weder verstanden noch benutzten. Als Konsequenz wurden die Shortcuts entfernt. Es blieb jedoch die Frage offen, ob die Benutzer Shortcuts benutzen würden, wenn sie mit dem PREMIERE Programmguide besser vertraut wären.

Um diesen Gedankengang zu analysieren, wurde eine kleine Gruppe von Benutzern gebeten, an drei aufeinanderfolgenden Tagen an einem Test teilzunehmen. In jeder Sitzung mußten diese Benutzer Bedienaufgaben gleicher Struktur mit unterschiedlichen Inhalten bearbeiten. Es wurde folgende Hypothese aufgestellt: Wenn Benutzer Shortcuts brauchen, nachdem sie Erfahrungen mit dem System gesammelt haben, so würden sich entsprechende Äußerungen in den Protokollen des „lauten Denkens" finden.

4.4.3 Ergebnisse der zweiten Iteration

Es wurden einige Schwachstellen in der Navigation und in der Informationsdarstellung entdeckt. Die Mehrzahl der Benutzer wollten detaillierte und offensichtliche Informationen bezüglich der Kosten haben, die bei der Auswahl von Filmen oder Diensten entstehen können. Sie befürchteten, daß bei der versehentlichen Auswahl eines Filmes oder Dienstes Kosten entstehen, die nicht bemerkt werden. Diese Befürchtung wirkte demotivierend auf den Wunsch, das System zu explorieren.

Der Test zeigte, daß der Navigationsmechanismus so einfach und verständlich wie möglich gehalten werden muß. Die OK-Taste sollte bspw. nur zur Bestätigung einer Eingabe benutzt werden und nicht mit zusätzlichen Funktionen belegt sein. Des weiteren bereiteten die Fließtexte in den Informationsfeldern Schwierigkeiten beim Lesen. Die angebotenen Informationen sollten auf das absolut Notwendige beschränkt und visuell gut strukturiert werden (z.B. durch Schlagworte oder räumliche Gruppierung).

Der Pilot wurde als hilfreiche und nützliche Systemkomponente bewertet. Er wurde effektiv und effizient genutzt. Für den Piloten schlugen die Benutzer die folgende Reihenfolge der Informationdarbietung vor:

- Erste Zeile: Name des Pay-TV Kanals und Sendezeit der Programme,

- Zweite Zeile: Titel des Programms,

- Dritte Zeile: zusätzliche Informationen (z. B. Regisseur des Films).

Die zusätzliche Zeitangabenzeile am unteren Rand der Programmübersicht wurde nicht als Informationsquelle genutzt, sondern von den Testpersonen als irritierend empfunden.

Die Informationsdarstellung wurde generell als verständlich und gut strukturiert befunden. Infolge seiner visuellen Attraktivität und seiner leichten Handhabung wurde das System insgesamt sehr positiv bewertet.

Die einfache Navigationsstruktur mit dem zentralem Hauptmenü reduzierte die Schwierigkeiten, Systemkomponenten zu finden, auf ein Minimum. Selbst die Benutzer, die das System an drei aufeinanderfolgenden Tagen getestet hatten, vermißten die Shortcuts nicht. Es hat den Anschein, daß Benutzer eine leicht verständliche Menüstruktur ohne Shortcuts der Möglichkeit, durch Shortcuts weniger Tasten betätigen zu müssen, bevorzugen.

5 Ausblick

Zukünftig werden immer mehr Leute die Möglichkeit haben, vom interaktiven Fernsehen Gebrauch zu machen. Gleichzeitig wird die Zahl der angebotenen Dienstleistungen wachsen. Seitdem jeder Anbieter die Benutzungsschnittstelle der von ihm angebotenen Dienstleistungen entsprechend eigener Prinzipien selbst entwirft, wird der Benutzer mit einer Vielzahl unterschiedlich gestalteter Benutzungsschnittstellen konfrontiert. Diese Entwicklung birgt die Gefahr, daß Benutzer bei jedem neuen multimedialen Dienst neue Bedienprinzipien und -elemente erlernen müssen. Der zusätzliche Lernaufwand kann zu einer verminderten Akzeptanz multimedialer Dienstleistungen führen. Um eine breite Akzeptanz der Dienstleistungen zu sichern, sollten standardisierte und benutzerorientierte Bedienkonzepte und -elemente für interaktives Fernsehen entwickelt werden.

6 Literaturverzeichnis

[1] Bullinger, H.-J. (1993). Benutzergerechte Gestaltung von Software - Eine Herausforderung an den Industriestandort Deutschland. In W. Coy et al. (Hrsg.), Menschengerechte Software als Wettbewerbsfaktor. Stuttgart: Teubner Verlag.

[2] Bullinger, H.-J., Burmester, M., Mangol, P. & Vossen, P.H. (1995), Design interaktiver Produkte - Dialog zwischen Mensch und Produkt. In H.-J. Bullinger (Hrsg.), Design interaktiver Produkte - Dialog zwischen Mensch und Produkt. Stuttgart: IRB, S. 13-28.

[3]   Burmester, M. & Machate, J. (1994), "Common User Access" for Electronic Home Devices or 20 Ways to Set the Clock? In R. Oppermann, S. Bagnara & D. Benyon (Hrsg.), ECCE7 Seventh European Conference on Cognitive Ergonomics. Human-Computer Interaction: From Individuals to Groups in Work, Leisure, and Everyday Life. Proc. GMD-Studien Nr. 233. Sankt Augustin: GMD, S. 97-110.

[4]   Dutke, S. (1994), Mentale Modelle: Konstrukte des Wissens und Verstehens. Göttingen: Verlag für Angewandte Psychologie.

[5]   Hermanns, A. (1993), Multimedia im Marketing, Kongressunterlagen Werbung und Illusion, Frankfurt Oktober 93.

[6]   Koller, F. (1992), "Gestaltung von Multimedia-Systemen". In: Ergonomie und Informatik

[7]   Maguire, M.C., Butters, L.M. & McKnight, C. (1994), Usability Issues for Buyers and Users of Home Electronic Products. In: R. Oppermann, S. Bagnara & D. Benyon (Hrsg.), Human-Computer Interaction: From Individuals to Groups in Work, Leisure, and Everyday Life,. Proceedings ECCE7 Seventh European Conference on Cognitve Ergonomics, GMD-Studien Nr. 233, S. 117-133.

[8]   Molich, R. (1994), Preventing user interface disasters. Behaviour & Information Technology. Vol. 13, Nos. 1 and 2, S. 154-159.

[9]   Nielsen, J. (1994), Estimating the number of subjects needed for a thinking aloud test. Int. J. Human Computer Studies, 41, S. 385-397.

[10] Prümper, J., Heinbokel, T. & Käting, H.J. (1993), Virtuelle Prototypen als Werkzeuge zur benutzerorientierten Produktentwicklung; Anwendung einer handlungstheoretischen Fehlertaxonomie auf reale und virtuelle Oberflächen von Waschmaschinen, Prümper & Partner Produktergonomie, München.

[11] SPIEGEL special (1995),TV total, Macht und Magie des Fernsehens; Spiegel-Verlag, Rudolf Uhlstein GmbH & Co KG, Hamburg, Ausgabe 8, S. 47ff.

[12] Virzi, R.A. (1990), Streamlining the design process: Running fewer subjects. Proceedings of Human Factors Society, 34th Annual Meeting, Orlando, FL, 8-12 October, S. 291ff.

[13] Vorst, L.M.T., Kanis, H. & Marinissen, A.H. (1992), User involved design of a remote control. In H. Bouma & J.A.M. Graafmans (Hrsg.), Gerontechnology. Amsterdam: IOS Press, S. 343-348.

[14] Wozencroft, J. (1992), Die Grafiksprache des Neville Brody. München: Bangert Verlag.

Adresse der Autoren

Dipl.-Psych. Michael Burmester
Dipl.-Inform. Franz Koller
Fraunhofer-Institut für Arbeitswirtschaft u. Organisation
Nobelstr. 12
70569 Stuttgart
Tel.: +49 711 970 2311 (-2321)
Fax: +49 711 970 2300
Email: Michael.Burmester @iao.fhg.de
Email: Franz.Koller@iao.fhg.de

Thilo König
PREMIERE Medien GmbH
Tonndorfer Hauptstraße 90
22045 Hamburg
Tel.: +49 40 66 80 12 40
Fax: +49 40 66 80 16 94
E.Mail koen@premiere.bertelsmann.de

Ein Beitrag zur Gestaltung von Benutzungsschnittstellen in Terminal-Informationssystemen am Beispiel des ÖPNV & City Informationssystems Dresden

Kamen Danowski

Fraunhofer-Institut für Informations- und
Datenverarbeitung, EPS Dresden

Falk Fünfstück

Technische Universität Dresden
Fakultät Informatik

Zusammenfassung

Für die Entwicklung der Benutzungsschnittstellen von Informationssystemen existiert eine Vielzahl von allgemeinen Gestaltungsregeln und Normen. Die Besonderheiten der Terminal-Informationssysteme erfordern eine Spezifizierung dieser in Abhängigkeit vom Informationsinhalt und Nutzerkreis. Es werden einige allgemeine Prinzipien der Dialog- und Maskengestaltung diskutiert. Diese sollen aber nicht als Dogmen oder Einschränkungen, sondern als unterstützende Hinweise verstanden werden. In Abhängigkeit von der Aufgabenstellung soll am Ende jedes Entwicklungsprozesses ein Produkt entstehen, das vom Nutzer akzeptiert wird. Dies kann, wie die Entwicklung des ÖPNV & City Informationssystems Dresden zeigt, nur in einem mehrstufigen, iterativen Prozeß erarbeitet werden, indem ausgehend von den Anforderungen und Erfahrungen die Lösung schrittweise konstruiert und verbessert wird.

1 Einleitung

In einer Informationsgesellschaft ist die Art und Weise des Informationszugangs von entscheidender Bedeutung. Neben solchen etablierten Varianten wie die Standardsoftware auf PC oder Workstation und dem WWW, stellen Terminal-Informationssysteme eine weiteres wichtiges Mittel zur Informationsgewinnung dar. Diese existieren meist in Form von rechnergestützten Selbstbedienungsautomaten und sind mittlerweile unverzichtbarer Bestandteil des öffentlichen Lebens geworden. Selbstbedienungsautomaten können grob nach ihren Funktionen unterteilt werden:

- Service Automaten, wie z.B. Banken-, Fahrscheinautomaten, Visitenkartendrucker usw.
- Informationsautomaten, wie z.B. Kiosksysteme [9], Auskunftsautomaten für Städte, ÖPNV (Öffentlicher Personennahverkehr) [5], Behörden und Ämter, Produkte usw.
- Spielautomaten.

Der vorliegende Beitrag beschäftigt sich schwerpunktmäßig mit der Gestaltung von Benutzungsschnittstellen von Terminal-Informationssystemen unter ergonomischen Gesichtspunkten. Für die Konzeption solcher Informationssysteme sind maßgebend:

- das Ziel des Informationssystems,
- der potentielle Nutzerkreis und
- die Plazierung des Automaten.

Um diese Merkmale zu bestimmen, sind Voruntersuchungen nötig [4, 9, 10]. Das System sollte auf den späteren Nutzerkreis abgestimmt sein, und es müssen Vorüberlegungen zu seiner Begriffswelt, einer für ihn geeigneten Informationsdarstellung bis hin zur Unterstützung verschiedener Sprachen gemacht werden. Die Standortwahl richtet sich sehr stark nach dem Einsatzzweck. Die so gewonnen Erkenntnisse sind entscheidend bei der Konstruktion eines bedarfsgerechten, benutzerfreundlichen Informationssystems. Die Akzeptanz und damit der Erfolg hängen entscheidend von der Grundkonzeption ab.

Viele Untersuchungen konzentrieren sich auf isolierte Aspekte der Systemgestaltung. Andere bieten nur abstrakte Hinweise ohne konkrete Richtlinien für die Umsetzung. In diesem Beitrag werden einige Gestaltungsprinzipien diskutiert, die aus praktischen Erfahrungen beim Erstellen von Bildschirmdialogen in Informationssystemen resultieren.

2 Allgemeine Anforderungen

Auf Abb. 1 werden einige wichtige Einflußfaktoren bei der Konzeption eines Terminal-Informationssystems dargestellt. Bei öffentlichen Auskunftssystemen ist davon auszugehen, daß die Benutzer den unterschiedlichsten Erfahrungsklassen beim Umgang mit Computern zuzuordnen sind. Für jede dieser Gruppen muß das System einfach, schnell und möglichst ohne Einarbeitungszeit zu bedienen sein. Im allgemeinen wird davon ausgegangen, daß der Benutzer keine oder wenig Erfahrung hat [8]. Deshalb muß das Informationssystem in besonderer Weise an dessen mentales Modell angepaßt sein. Dem wird Rechnung getragen, indem die Informationen direkt und auf eine leicht zugängliche Art und Weise vermittelt werden, die der Benutzer auch bei anderen Gelegenheiten seiner Erfahrungswelt findet. Des weiteren sollte ihm nach Möglichkeit kein bestimmtes Vorgehen aufgezwungen werden, um an die gewünschte Information zu gelangen.

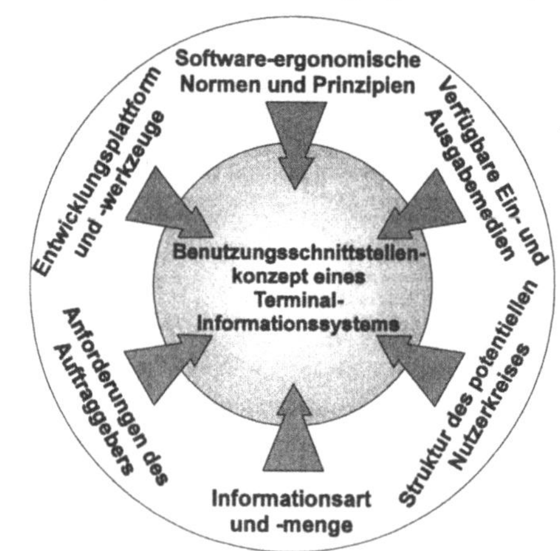

Abb.1: Einflußfaktoren bei der Konzeption eines Terminal-Informationssystems

Ein wichtiger Gesichtspunkt ist die Beachtung der Wahrnehmungsprinzipien und Gestaltgesetze [7]. Damit soll die leichte Erlernbarkeit beim Umgang mit dem System unterstützt werden. Der Benutzer muß sofort und ohne einen Einführungskurs das Informationsterminal intuitiv bedienen können [8]. Dazu gehören selbsterklärende Beschriftungen, die Verwendung eindeutiger Metaphern usw. [10] Um den Wiedererkennungseffekt zu erleichtern und Benutzungsfehler zu vermeiden, ist auf eine konsistente Verwendung der Oberflächenelemente zu achten, da bestimmte Handlungen schon nach wenigen Schritten automatisiert erfolgen [4].

Die Attraktivität eines Auskunftssystems hängt hauptsächlich von dem Informationsinhalt und der Präsentationsart ab. Ein öffentliches Informationssystem fungiert auch als Werbeträger für diejenige Instanz, die es einsetzt. Deshalb ist man bemüht, zunehmend multimediale Elemente zur Informationscodierung zu integrieren, womit die Aufmerksamkeit auf das System gelenkt und zu seiner Nutzung angeregt werden soll. Die Verwendung von Ton, farbigen Grafiken, Animationen und Videos erhöht entscheidend die Attraktivität des Informationssystems, wenn diese zur Steigerung der Aussagekraft und nicht als Selbstzweck eingesetzt werden. Eine übermäßige Benutzung von Farbenvielfalt und Animationen kann ermüdend und für den Nutzer sogar verwirrend wirken [7]. Um eine ästhetische Wirkung und visuelle Klarheit zu erzielen, ist nach einem annehmbaren Kompromiß zwischen dem technisch Machbaren und dem software-ergonomisch Günstigen zu suchen.

3 Grundprinzipien für die Gestaltung der Benutzungsschnittstelle von Terminal-Informationssystemen

3.1 Übersicht der Ein- und Ausgabemedien

Bevor die Benutzungsschnittstelle selbst gestaltet wird, muß Klarheit über die verwendeten Ein- und Ausgabemedien herrschen. Diese beeinflussen stark das Layout der Oberfläche. Als Eingabemedien bieten sich bei Selbstbedienungsautomaten die (eingeschränkte) Tastatur, der berührungsempfindliche Bildschirm (Touchscreen), das Touchpad, die Rollkugel (Trackball) bzw. die Rollwalze und der Steuerknüppel (Joystick) an. Ungeeignet sind die volle Standardtastatur, wegen ihrer Größe, die Maus und der Lichtgriffel (Light Pen), weil diese nicht in das System integriert sind und deshalb leicht beschädigt bzw. entwendet werden können. Alternativ dazu können sich die im Entwicklungsstadium befindenden Spracheingabe und das Verfolgen von Augenbewegungen (Eye Tracking) betrachtet werden. In der Praxis durchgesetzt haben sich inzwischen die eingeschränkte Tastatur, etwa zur Zifferneingabe, der Touchscreen und die Rollkugel bzw. -walze. Sie benötigen beim Umgang die wenigste Übung und entsprechen am ehesten den Erwartungen unerfahrener Benutzer.

Der Touchscreen ist das am einfachsten zu bedienende Gerät; die Eingabe erfolgt direkt über den Bildschirm, indem die angezeigten Schaltflächen berührt werden. Das entspricht weitgehend den Erwartungen des Benutzers, der hier eine Metapher zu realen Geräten und Automaten erkennt. Im Gegensatz dazu erfolgt die Eingabe beim Touchpad nicht über den Bildschirm, sondern lediglich über eine gesonderte Bedienfläche. Mit dem Trackball wird der Cursor indirekt über den Bildschirm bewegt, und eine Schaltfläche kann erst über eine spezielle Taste betätigt werden. Bei der oft anzutreffenden Rollwalze erfolgt die Bewegung des Cursors nur in einer Dimension, was die für die Bedienung zur Verfügung stehende Fläche stark einschränkt und eine lineare Anordnung der Oberflächenelemente erfordert. Die Rollkugel bzw. -walze setzt voraus, daß die Schaltflächen nicht in zu großem Abstand voneinander plaziert sind.

Das primäre Interaktionsmedium des Terminal-Informationssystems ist der Bildschirm. Einerseits dient er, wenn er als Touchschreen vorliegt, zur Eingabe, andererseits erfolgt über ihn auch die Ausgabe der meisten Informationen. Um aber Redundanz bezüglich der wichtigen Angaben zu schaffen und eine Benutzung des Informationssystems zu erleichtern, ist es vorteilhaft, zusätzlich eine Sprachausgabe vorzusehen [8]. Bestimmte Informationen, die umfangreich und schwer zu behalten sind, wie beispielsweise Adressen, Telefonnummern, Fahrplanauskunft etc., sollten des weiteren ausgedruckt werden können.

3.2 Dialoggestaltung

Für die Gestaltung von Benutzungsschnittstellen gelten die Normen der DIN 66234 T.8 bzw. ISO 9241 T.10. Diese müssen gleichermaßen bei der Gestaltung von Terminal-Informationssystemen berücksichtigt werden. Aufgrund der Besonderheiten von Informations- und Auskunftssystemen, die oft in Form von Selbstbedienungsterminals für einen unerfahrenen Benutzerkreis vorgesehen sind, können aber nicht alle Regeln und Grundsätze ohne Spezifizierung übernommen werden. Einerseits kann nicht davon ausgegangen werden, daß der Nutzer bereits über grundlegende Vorkenntnisse beim Umgang mit Software verfügt, andererseits stehen im Gegensatz zum PC nur sehr eingeschränkte Eingabemöglichkeiten zur Verfügung, und die Bildschirmgröße ist oft beschränkt. Besonderes Augenmerk muß deswegen auf folgende Grundsätze bei der Dialoggestaltung in Terminal-Informationssystemen gelegt werden:

- Jeder einzelne Dialog sollte nur ein Thema, einen Grundgedanken haben, um das Kurzzeitgedächtnis nicht zu überlasten. Dazu sollte der Benutzer möglichst nur eine Frage beantworten müssen. (Aufgabenangemessenheit)

- Die Steuerung der Dialogabläufe sollten konsistent gestaltet und den Erwartungen des Nutzers angepaßt sein, indem wechselnde Durchführungsbedingungen vermieden werden. (Erwartungskonformität)

- Der Benutzer muß sofort erkennen können, an welcher Stelle der Dialogführung er gerade ist, welche Eingabemöglichkeiten gegeben sind und wie er weiter vorgehen soll, um sein Ziel zu erreichen. (Selbstbeschreibungsfähigkeit)

- Die Eingabemöglichkeiten sollten auf sinnvolle, logische Dinge beschränkt sein und durch Auswahl von Objekten erfolgen. (Objekt-Aktion-Modell) Auf die Anzeige unnötiger und unpassender Informationen, wie z.B. explizite Werbung, ist zu verzichten.

- Zwangsfolgen bei der Dialogabarbeitung sind zu vermeiden. Es muß immer möglich sein, den Dialog zu beenden bzw. zu unterbrechen oder einen Schritt zurückzugehen, ohne daß bereits vorgenommene Eingaben wiederholt werden müssen. (Steuerbarkeit)

- Die Anzahl der Dialogschritte zur Erreichung eines Ziels ist zu minimieren. Die Obergrenze hierfür liegt bei drei bis vier einzelnen Dialogen. Deswegen sollten Dialogverzweigungen durch Gruppenbildungen realisiert werden.

- Der Nutzer sollte nicht während der Dialogabarbeitung durch die Verwendung von Zeitschranken „gehetzt" werden. Erst bei längerer Nichtbenutzung ist der Ursprungszustand wiederherzustellen.

- Die Antwortzeiten des Systems sind kurz zu halten. Bei Bedarf ist eine Meldung sowie eine Fortschrittsanzeige auszugeben.

- Potentielle Fehlerquellen sind zu minimieren. Der Nutzer sollte auf fehlerhafte oder unlogische Eingaben in verständlicher Art und Weise hingewiesen werden. (Fehlerrobustheit)

Dies sind einige der wichtigsten Grundprinzipien bei der Dialoggestaltung. Es können sicherlich nicht immer alle dieser Regeln eingehalten werden. Im Einzelfall ist daher zu überprüfen, welches der Kriterien den Vorrang hat und dem mentalen Modell des Benutzers weitestgehend entspricht. Dazu muß oft nach tragfähigen Lösungen in möglichen Widerspruchsfeldern, wie z.B. der Minimierung der Anzahl der Dialogschritte und dem Prinzip, daß jeder Dialog nur einen Grundgedanken haben sollte, gesucht werden.

3.3 Maskengestaltung

Bei der Maskengestaltung ist auf die Anordnung der einzelnen Informationsflächen zu achten:

- Es ist eine Ortscodierung vorzunehmen, indem die Status-, Verarbeitungs- und Steuerungsinformationen bei allen Dialogmasken konsistent angeordnet werden.
- Statusinformationen sollten in Form einer Titelzeile am oberen Rand des Bildschirms untergebracht werden. Wird ein Touchscreen verwendet, sollten die Eingabemenüs mit den Schaltflächen zur Dialogsteuerung am besten am unteren oder, wenn dies nicht möglich ist, am rechten Rand – die meisten Benutzer sind Rechtshänder – angeordnet werden, damit bei deren Betätigung keine Informationen von der Hand verdeckt werden. (Abb. 2)

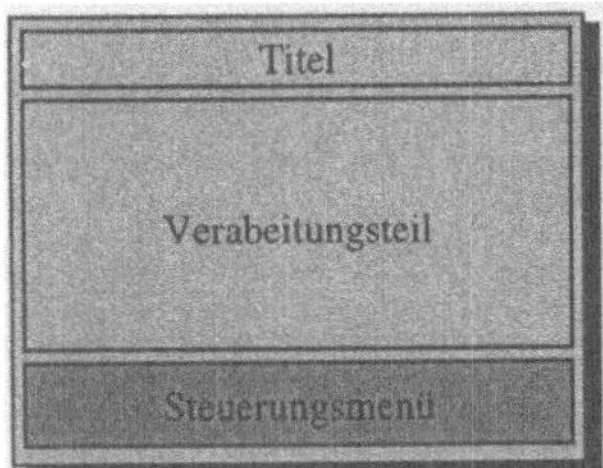

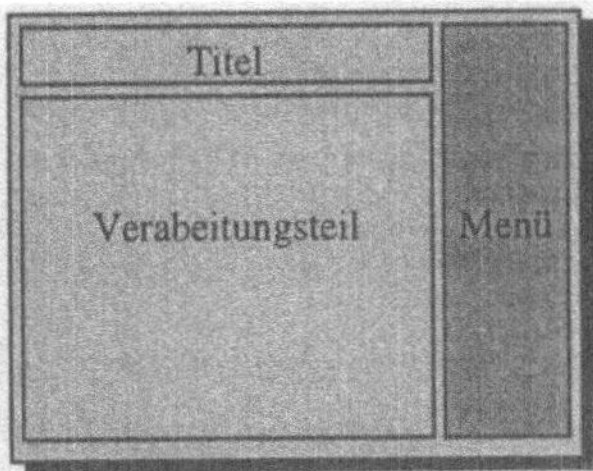

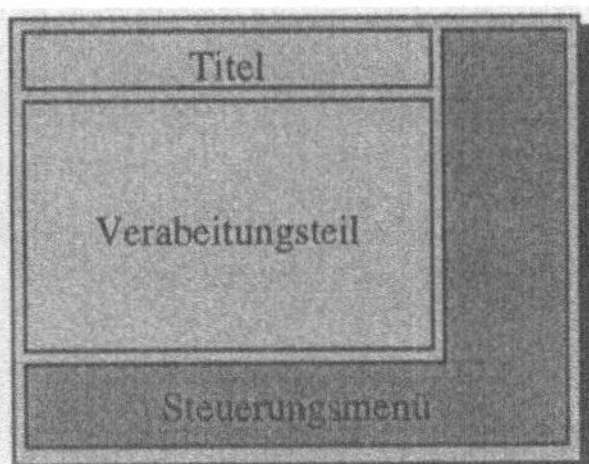

Abb. 2: Drei Varianten der Maskengestaltung

- Funktionstasten, die die gleichen Aufgaben erfüllen, sollten immer am gleichen Ort zu finden sein und das gleiche Aussehen haben. (Abb. 3) Ist eine Schaltfläche nicht auf allen Dialogmasken verfügbar, sollte der entsprechende Platz möglichst freigelassen werden. Dem Nutzer ist speziell bei Verwendung eines Touchscreens eine optische und / oder akustische Rückkopplung zu geben.

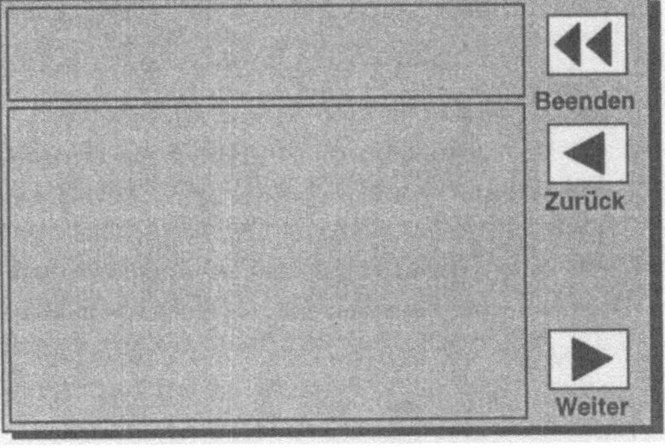

Abb. 3: Beispiel für die Aufteilung einer Steuerleiste

Bei der Informationscodierung sind folgende Prinzipien zu beachten:

- Themenverwandte Informationen sollten in räumlicher Nähe angeordnet werden, um die Zusammengehörigkeit visuell deutlich zu machen. Dazu können Gruppen gebildet werden. Im Gegensatz dazu sollte eine Hervorhebung zuerst durch eine räumliche Isolierung realisiert werden.
- Die Informationscodierung muß auf jeden Fall redundant vorgenommen werden, so daß möglichst alle Benutzer gut damit zurechtkommen. Es bietet sich die gleichzeitige Verwendung von aussagekräftigen Symbolen und Schrift an. Es sind Symbole zu verwenden, die auf Metaphern aus der realen Welt basieren.
- Zur Informationsausgabe sind kurze und prägnante Sätze zu wählen, die zur besseren Lesbarkeit in natürlicher Groß- und Kleinschreibung gehalten sind. Auf Abkürzungen ist möglichst zu verzichten.

- Farben dürfen niemals alleiniger Informationsträger sein. Sie sollten zur besonderen Hervorhebung sparsam eingesetzt werden. Hintergrundfarben können wegen ihrer neutralen Wirkung Grau- und Blautöne sein.

Diese Forderungen dürfen aber kein starres Dogma sein, sondern sollen grundlegende Hinweise liefern, die bei jeder einzelnen Anwendung zu überprüfen sind, und helfen, Fehler zu vermeiden.

4 Realisierung der Benutzungsschnittstelle im ÖPNV & City Informationssystem Dresden

4.1 Systemkonzept und Funktionalität

Das ÖPNV & City Informationssystem Dresden stellt ein multifunktionelles Softwareprodukt zur Fahrplan- und Stadtauskunft dar. Die Hauptfunktionen des Systems können in fünf Gruppen eingegliedert werden:

- Verbindungssuche und optimale Fahrweggenerierung im Netz der öffentlichen Verkehrsmittel (Straßenbahn, Bus, S-Bahn, Airport-City-Liner, Fähre, Bergbahn) im Raum Dresden unter Berücksichtigung der tageszeit- und tagesartabhängigen Fahrplan- und Liniennetzbesonderheiten. Als Start- und Zielpunkte können neben den Haltepunkten der Dresdner Verkehrsbetriebe über 5 000 verschiedene andere Objekte, wie z.B. Straßennamen und Adressen, Hotels, Bildungseinrichtungen, etc., angegeben werden.
- Informationen über mehr als 4 000 markante Objekte in der Stadt und der Umgebung, wie beispielsweise zu öffentlichen Einrichtungen, Sehenswürdigkeiten, Behörden, Einkaufszentren, kulturellen Einrichtungen, Veranstaltungen, etc. Dem Nutzer werden je nach Art des Objektes Kurzbeschreibungen, aktuelle Auskünfte, Termine, Geschichtliches, Adressen und Telefonnummern, Öffnungszeiten und Eintrittspreise angeboten.
- Ein digitalisierter Stadtplan in zwei Vergrößerungsstufen steht dem Nutzer zur Verfügung. In diesem können Adressen gesucht und angezeigt werden.
- Eine Abfrage der direkten Fernverkehrsverbindungen von und nach Dresden mit Eisenbahn, Schiff und Flugzeug ist auf der Grundlage der aktuell gültigen Saisonfahrpläne ebenfalls möglich.
- Aktuelle Mitteilungen der Verkehrsbetriebe, wie z.B. Informationen über Tarife, über die Bus- und Straßenbahntechnik, über kurz- oder längerfristige Baumaßnahmen und Umleitungen einzelner Linien etc., können in einer separaten Rubrik abgerufen werden [1, 2].

Bei der Gestaltung des Datenhaltungskonzeptes für das ÖPNV & City Informationssystem Dresden wurde ein Schritt vom statischen Auskunftssystem mit starren Informationen zum dynamischen System mit bedarfsabhängiger Datenaktualisierung gemacht. Es wird von einer Konzentration und Konvertierung unterschiedlicher Rohdaten

- der ÖPNV-Fahrpläne (Fahr- und Dienstplanungssystem EPON der Dresdner Verkehrsbetriebe AG),
- der Stadtinformationen (Datenbank der Dresdner Werbung und Tourismus GmbH),
- der Flugpläne (Flughafen Dresden GmbH) und
- des Bahnfahrplans (Deutsche Bahn AG)

auf einem Datenserver (UNIX-Workstation) ausgegangen, welcher ohne Benutzereingriffe die Informationen von den Quellen akquiriert, die Daten konvertiert und die Datenbestände der Auskunftsrechner aktualisiert [3].

4.2 Gestaltung der Benutzungsschnittstelle

4.2.1 Nutzergruppen und Anforderungsprofil

Die relevanten Schnittstellen eines Systems mit der Informationspalette, wie die vom ÖPNV & City Informationssystem Dresden, können aus Benutzersicht in folgende drei Gruppen eingeteilt werden:

- Benutzung durch den Fahrgast an Selbstbedienungsterminals,
- Benutzung durch Auskunftspersonal in Servicezentren oder per Telefon und
- Informationsabfrage über öffentliche Kommunikationsnetze. (Abb. 4)

Einer der Hauptschwerpunkte der Untersuchungen zur Gestaltung der Benutzungsoberfläche des ÖPNV & City Informationssystems Dresden lag auf der Anwendung in Form eines Auskunftsterminals, an dem der Fahrgast im Dialogbetrieb Informationen selbständig abfragen kann [11]. Diese Variante, die an wichtigen Verkehrsknotenpunkten installiert wird, steht sowohl Einheimischen als auch in- und ausländischen Besuchern der Stadt zur Verfügung. Deshalb muß die Benutzungsoberfläche einerseits für erfahrene und Gelegenheitsnutzer und andererseits für solche, die mit dem System zum ersten mal in Kontakt kommen, geeignet sein. Eine Dialogführung in unterschiedlichen Sprachen muß verfügbar sein.

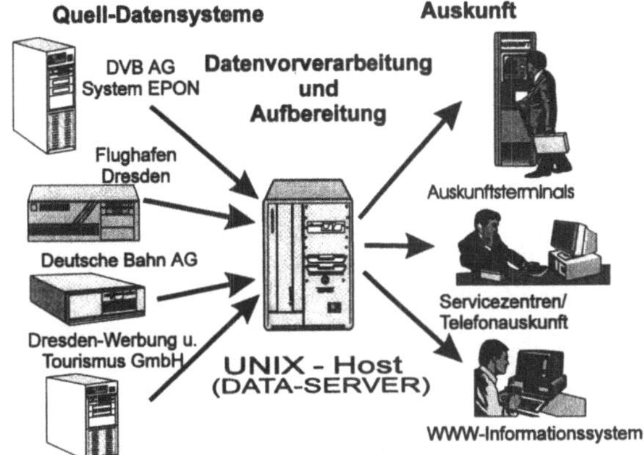

Abb. 4: Systemkonzept des ÖPNV & City Informationssystems Dresden [3]

Als Eingabemedien wurden seitens der Hardwarehersteller Touchscreen und Rollwalze vorgeschlagen. Für die Realisierungsvariante wurde durch den Auftraggeber der Touchscreen als Eingabemedium gewählt. Als Ausgabemedien stehen Bildschirm, Drucker und Lautsprecher zu Verfügung.

Bei der Erarbeitung eines Anforderungsprofils für die Benutzungsoberfläche des ÖPNV & City Informationssystems Dresden wurden mehrere aus der Literatur bekannte Grundsätze und Gestaltungsprinzipien ausgewertet [6, 7, 8]. Diese weitgehend anwendungsneutralen Richtlinien wurden auf die konkrete Systemlösung angewendet, wobei sich die Untersuchungen auf zwei Hauptgebiete konzentrierten:

- Dialoggestaltung (Festlegung der Reihenfolge und des Inhaltes der einzelnen Dialogschritte) und
- Dialogmasken-Layout (Auswahl, Gestaltung, Positionierung einzelner Dialogelemente).

Mit besonderem Augenmerk auf diese zwei Schwerpunkte und auf die im Abschnitt 3 definierten Anforderungen wurde die Benutzungsschnittstelle des ÖPNV & City Informationssystems Dresden konzipiert und programmtechnisch umgesetzt.

4.2.2 Dialoggestaltung

Ausgehend von der Struktur des potentiellen Nutzerkreises für das Informationssystem wurde eine Dialogstruktur konstruiert, die es erlaubt, in wenigen Schritten den Informationswunsch eindeutig zu ermitteln, um anschließend die gewünschte Information auszugeben. Trotz des breiten Informationsspektrums wurde versucht, im gesamten System weitgehend gleichartige Dialogschritte zu gestalten. Der Informationswunsch wird vom System schrittweise ermittelt, wobei dem Nutzer Freiheiten über die Reihenfolge und Art der Eingaben eingeräumt werden. Solche Freiheiten werden im allgemeinen vom Benutzer erwartet. Die Vielzahl der Objekte, über die im System Informationen generiert werden können (Fahrtverbindungen) und Informationen vorliegen (Stadtinformation), erfordert eine mehrstufige Detailierung der Angaben.

Das mentale Modell, das die Erwartungshaltung des Benutzers gegenüber dem System reflektiert, spielte eine primäre Rolle im Gestaltungsprozeß. Aufgrund umfangreicher Untersuchungen (Beobachtungen, Befragungen, etc.) wurde festgestellt, daß schon vor der Informationsabfrage bestimmte Vorstellungen über die Vorgehensweise und die Art der Auskunft beim Benutzer bestehen. Will er beispielsweise eine Fahrplanauskunft einholen, geht er davon aus, ähnlich wie in einem Kursbuch nach dem Startort, dem Zielort und der Abfahrts- bzw. Ankunftszeit gefragt zu werden. Entsprechend diesen Vorstellungen wurde die Dialogführung genau in dieser Reihenfolge implementiert.

Die Bestimmung des Startpunktes erfolgt, indem als erstes die Gruppe, zu der er gehört (z.B. Sehenswürdigkeiten), ausgewählt wird (Abb. 5) und im nächsten Dialogschritt innerhalb dieser Gruppe das konkrete Objekt (z.B. Semperoper) angegeben wird. Auf dieselbe Art und Weise wird der Zielpunkt bestimmt. Eine Auswahl aus der Objektgruppe entfällt, wenn der Start bzw. Zielpunkt der aktuelle Ort ist, da dieser in der Systemkonfiguration eindeutig bestimmt ist. Auch hier findet sich ein Bezug zum mentalen Modell des Benutzers, der eine Fahrplanauskunft sehr häufig für den aktuellen Standort

Abb. 5: Gruppenauswahl: Startort

wünscht. Die Zeitangaben werden in einem separaten Schritt gemacht, indem zwischen Sofortabfahrt, gewünschter späterer Abfahrtszeit vom Start und gewünschter Ankunftszeit am Ziel alternativ ausgewählt werden kann. Als Ergebnis der Routenberechnung wird dann eine Liste von einer oder mehreren Teilverbindungen angeboten, die durch Start, Ziel, Verkehrsmittelart, Liniennummer, Fahrtrichtung, Abfahrts- und Ankunftszeit beschrieben wird. Fußwege und Umsteigevorgänge werden ebenfalls detailliert beschrieben. Nach der Informationsausgabe kann der Nutzer durch das Aktivieren der entsprechenden Buttons eine frühere oder eine spätere Verbindung abfragen bzw. sich die Lage des Start- und des Zielortes im Stadtplan und die Linienführung anzeigen lassen. Sollten Alternativverbindungen existieren, welche weniger Umsteigerelationen enthalten oder kostengünstiger sind, so werden diese ebenfalls angeboten. Textinformationen können auch auf dem Drucker ausgegeben werden.

Bei der Dialogführung wurde eine eindeutige Feststellung des Informationswunsches und klare Informationsausgabe durch eine minimale Anzahl von Dialogschritten angestrebt. Dabei wurde darauf geachtet, daß im ÖPNV & City Informationssystem Dresden jeder Dialogschritt ein Thema realisiert und eindeutig klassifiziert werden kann. Durch die flexible, teilweise vom Nutzer bestimmbare Eingabereihenfolge wird den Erwartungen des Nutzers entsprochen, und Überraschungseffekte werden vermieden.

4.2.3 Dialogmasken-Layout

Bei der Gestaltung der Dialogmasken im ÖPNV & City Informationssystem Dresden wurde trotz der Vielfalt der Informationsthemen versucht, daß diese ein weitestgehend einheitliches Erscheinungsbild haben und der Wiedererkennungseffekt durch Ortscodierung gewährleistet ist. Es wurde weiterhin beachtet, daß bei der Bedienung des Touchscreens immer ein Teil des Bildschirms durch die Hand verdeckt wird. Um so wenig wie möglich wichtige Informationen im Bedienprozeß zu verdecken, wurden beim Design die Steuerungselemente am rechten Bildschirmrand angeordnet. Der Bildschirm wurde demzufolge in drei Bereiche [6] unterteilt, die eindeutig durch Panels abgegrenzt sind:

- Steuerungsmenü – „Übersicht", etc.,
- Verarbeitungsbereich – Direktauswahl und Informationsausgabe und
- Titelleiste – Überschrift der Bildschirmmaske. (Abb. 6)

Durch die Titelleiste und die Elemente im Verarbeitungsbereich wird eine sofortige Erkennung des Themas der Bildschirmmaske erreicht. Alle Bildschirmmasken wurden mit neutral wirkendem, grauem Hintergrund erstellt. Die Bedienelemente wurden mit farbigen Symbolen versehen, um durch die Metaphern die Wahrnehmbarkeit zu verbessern. Dazu wurden Piktogramme erstellt, welche die entsprechende Funktion oder Funktionsgruppe eindeutig kennzeichnen. Ihre Bedeutung wird zusätzlich mit einem entsprechend prägnanten, allgemeinverständlichen Begriff verdeutlicht.

Abb 6: Maskenlayout am Beispiel der Stadtplanausgabe

In dem ÖPNV & City Informationssystem Dresden wurden die Textinformationen und -hinweise qualitativ und quantitativ an das menschliche Kurzzeitgedächtnis angepaßt, indem diese möglichst kurz und zugleich allgemeinverständlich dargestellt wurden. So werden bei Eingabeaufforderungen ausschließlich kurze Sätze bzw. Begriffe verwendet, die leichter behalten und deren Bedeutungen schneller erkannt werden können.

In einem Informationssystem sollte nach Möglichkeit ein einheitliches Schriftbild genutzt werden. Abweichungen in Schriftart und -größe müssen eine definierte Bedeutung haben. Als Schriftart wurde mit Helvetica eine Schriftart ohne Serifen gewählt. Diese Schriftart zeichnet sich durch relativ wenig Pixel pro Buchstabe aus und ermöglicht somit durch Kontrast und Geschlossenheit eine schnellere Wahrnehmung [6]. Eine vergrößerte Schriftart wurde verwendet, um die Titel der einzelnen Bildschirmmasken darzustellen und somit die Aufmerksamkeit zu lenken.

4.3 Praktische Erfahrungen und Akzeptanz

Das ÖPNV & City Informationssystem Dresden befindet sich seit Mitte 1994 im täglichen Einsatz in den Servicezentren der Dresdner Verkehrsbetriebe und als Selbstbedienungsterminals. (Abb. 7) Über diesen Zeitraum hinweg werden automatisch Statistiken der Nutzungsart und -häufigkeit geführt. Die Eingabedialogschritte, an denen die Nutzer am längsten verbleiben oder abbrechen, wurden analysiert und überarbeitet. Die Anzahl unvollendeter – das System wird durch ein Timeout zurückgesetzt – oder abgebrochener Dialoge am Selbstbedienungsterminal liegt unter 10%. Über die Qualität der ausgegebenen Informationen wurden umfangreiche Befragungen durchgeführt, und praktische Hinweise wurden ausgewertet und eingearbeitet. Die Statistik der Benutzungshäufigkeit zeigt, daß am Tag zwischen 260 und 480 Informationen ausgegeben werden. Es wurden Unterschiede in der Nutzungsfrequenz zwischen der Einführungsphase und dem normalen Betrieb beobachtet. Am häufigsten werden Fahrplaninformationen (ca. 36%) und Stadtinformationen (ca. 24%) abgefragt. Ein starker Anstieg der Informationsausgaben wird beim Fahrplanwechsel beobachtet.

Abb. 7: Auskunft im Servicezentrum und am Selbstbedienungsterminal

5 Schlußfolgerungen

Die Akzeptanz von Terminal-Informationssystemen hängt weitgehend von der Gestaltung der Benutzungsschnittstelle ab. Die Untersuchungen im Vorfeld der Entwicklung des ÖPNV & City Informationssystems Dresden zeigten, daß trotz einer Reihe bestehender Terminal-Informationssysteme wenige Ansätze und Richtlinien zur Gestaltung nutzerfreundlicher, ergonomischer Benutzungsschnittstellen für diese Systemklasse existieren. Informationsterminals zeichnen sich einerseits durch die Verwendung besonderer Eingabemedien aus und sprechen andererseits einen breiten, zum Teil rechentechnisch unerfahrenen Nutzerkreis an. Um den Spezifika der Auskunfts- und Informationssysteme gerecht zu werden, wurde versucht, ausgehend von den Normen und Gesetzen der Software-Ergonomie einige Prinzipien und Grundsätze für die Gestaltung von Benutzungsoberflächen in Informationsterminals zu definieren.

Entsprechend diesen Erkenntnissen und dem speziellen Anforderungsprofil eines Fahrplan- und Stadtauskunftssystems für die breite Nutzung wurde das Konzept für die Benutzungsschnittstelle des ÖPNV & City Informationssystems Dresden entwickelt und seine Tragfähigkeit praktisch erprobt.

Literaturverzeichnis

[1] K. Danowski: Ein Beitrag zur wissensbasierten Modellierung von Entscheidungsprozessen in Verkehrsleit- und Informationssystemen. In: atp - Automatisierungstechnische Praxis, Oldenbourg Verlag, München, Heft 12/35 (1993), 677-682.

[2] K. Danowski: Ein wissensbasiertes, multimediales Fahrplan- und City Auskunftssystem. In: Mitteilungen aus dem Fraunhofer-Institut für Informations- und Datenverarbeitung '94, Fraunhofer-Institut für Informations- und Datenverarbeitung, Karlsruhe, 1994.

[3] K. Danowski, U. Jung: Zur Anwendung wissensbasierter Modellierungsmethoden in Verkehrsinformationssystemen. In: Wissenschaftliche Beiträge zur Informatik, Technische Universität Dresden, Fakultät Informatik, Progreßmedia, Dresden, Heft 1/8 (1995), 105-116.

[4] E. Eberleh et al. (Hg.): Einführung in die Software-Ergonomie. DeGruyter Verlag, Berlin, 1994.

[5] G. A. Giannopoulos, I. Retuerto: Rview of existing advanced passenger information systems and presentation of the POPINS project, Computer-aided operating systems (AVL-Systems) and traffic in the cities. International conference 30.11-2.12 Paris, 1992.

[6] T. Hoffmann: Handbuch zur softwareergonomischen Gestaltung von Bildschirmmasken. VDI-Verlag, Düsseldorf, 1989.

[7] G. Geiser: Mensch-Maschine-Kommunikation. Oldenbourg Verlag, München, 1990.

[8] O. Klose: Weiterentwicklung der Bedienoberfläche des wissensbasierten multimedialen ÖPNV & City Informationssystems Dresden unter Berücksichtigung von Anforderungen unterschiedlicher Anwendungsbereiche, Diplomarbeit, Fakultät für Verkehrswissenschaften, TU Dresden, 1994.

[9] L. Leventhal et al.: Hypertext-Based Kiosk Systems: Seven Challenges and an Empirical Study, Human-Computer Interaction. In: Lecture Notes in Computer Science EWHCI '94, Springer Verlag, Berlin, Heidelberg, 1994, 226-237.

[10] S. Musil et al.: Der Reisesupermarkt: Design und Entwicklung einer Interaktionsphilosophie für die Beratung in Reisebüros. In: Software-Ergonomie '95, Berichte des German Chaper of the ACM, Teubner Verlag, Stuttgart, 1995, 281-292.

[11] H. Strobel, M. Ritschel: Von rechnergestützten Betriebsleitsystemen zum computerintegrierten Stadtverkehr: Eine Übersicht über Szenarien, Integrationsmethoden und Managementkonzepte. Das öffentliche Verkehrswesen - Orientierungen, Berichte des Internationalen UITP Ausschusses an den 50. UITP Kongreß in Sydney, 02. bis 07. Mai 1993, herausgegeben durch UITP, Brüssel, 1993.

Adressen der Autoren

Dipl.-Ing. Kamen Danowski
Fraunhofer-Institut für Informations- und Datenverarbeitung
Außenstelle für Prozeßsteuerung
Zeunerstraße 38
D-01069 Dresden
Tel. 0351/4640 660
Email: danowski@eps.iitb.fhg.de

Falk Fünfstück
Technische Universität Dresden
Fakultät Informatik
D-01062 Dresden
Email: ff1@inf.tu-dresden.de

Benutzerorientierte Adaptivität und Adaptierbarkeit im Projekt AVANTI

Josef Fink, Alfred Kobsa und Andreas Nill

GMD - Forschungszentrum Informationstechnik GmbH
Institut für Angewandte Informationstechnik (FIT)
Forschungsbereich Mensch-Maschine Kommunikation (MMK)

Zusammenfassung

Ziel des Projekts AVANTI (AdaptiVe and Adaptable INteractions for Multimedia Telecommunications ApplIcations) ist es, multimediale Informationen über eine Region einem Benutzerkreis mit verschiedenen Vorkenntnissen und Interessenlagen, z.B. Touristen, Einwohnern, Reisekaufleuten, aber auch bestimmten Behindertengruppen wie Blinden und Rollstuhlfahrern adäquat anzubieten. Der Heterogenität der Benutzerbedürfnisse wird durch eine weitgehende Individualisierung des Informationsangebots auf der Basis modellierter Annahmen über die Interessen, Kenntnisse und Fähigkeiten von Benutzern Rechnung getragen. Der Zugriff auf das AVANTI-System kann sowohl von öffentlich zugänglichen Informationskiosken, Büros, vom heimischen Wohnzimmer als auch von entsprechenden mobilen Geräten (z.B. Laptops, Palmtops) aus erfolgen.

Die GMD ist innerhalb des Gesamtprojekts an der Ermittlung der Benutzerbedürfnisse, der Konzeption und Implementierung der für die Anpassungsleistungen zuständigen Komponenten und an der Evaluierung der Gesamtsysteme beteiligt. Darüber hinaus wirkt sie auch bei der Erstellung der Informationssysteme für die geplanten Feldtests mit. In diesem Papier stellen wir die wichtigsten Ergebnisse der ersten Projektphasen vor und geben einen Überblick über die Architektur des Gesamtsystems.

1   Einleitung

Das von der Europäischen Kommission im Rahmen des ACTS-Programms geförderte Projekt AVANTI wurde im März 1995 gestartet [1]. Elf Partner aus sieben europäischen Ländern sind derzeit an dem auf vier Jahre angelegten Projekt beteiligt. Adaptive und adaptierbare Systemleistungen[1] sollen dabei sowohl an der Benutzerschnittstelle als auch in den zu erstellenden Informationssystemen eingesetzt werden, um die *Benutzbarkeit* des Gesamtsystems für einen großen Personenkreis einschließlich bestimmter Behindertengruppen sicherzustellen und die individuelle *Nützlichkeit* der angebotenen Informationen zu gewährleisten [18]. Die Evaluierung der erstellten Informationssysteme hinsichtlich der Nutzenkriterien Effektivität, Effizienz und Akzeptanz mit dezidierten Benutzergruppen in Rom, Siena (Italien) und Kuusamo (Finnland) wird darüber Aufschluß geben, inwieweit diese Ziele erreicht werden.

Zur Zeit befinden sich die Projektarbeiten im Übergang von der Konzeptions- zur Implementierungsphase.

---

[1] Der Definition in [23] folgend bezeichnen wir die Anpassungsleistungen des Systems, die der Benutzer über Systemparameter selbst einstellen kann, als adaptierbar und die Anpassungsleistungen des Systems, die aus der Sicht des Benutzers vom System automatisch initiiert werden, als adaptiv.

Dementsprechend möchten wir in diesem Papier auch unsere Schwerpunkte setzen:

- Charakterisierung der im Projekt beteiligten Benutzer und ihrer Bedürfnisse,
- Überblick über die durchgeführte Benutzermodellierung,
- Szenarien adaptiver und adaptierbarer Systemleistungen, die der Heterogenität der Anforderungen an das AVANTI-System Rechnung tragen und
- Darstellung der zugrundeliegenden Software-Architektur auf der Basis des World-Wide Web.

2   Benutzercharakteristika

In der ersten Phase von AVANTI wurden in Italien, Finnland und Deutschland u.a. die Bedürfnisse der folgenden Benutzergruppen untersucht [2]:

- Benutzer mit speziellen Behinderungen:
  - Rollstuhlfahrer,
  - Blinde,
  - Personen, die an einer leichten Form von Muskeldystrophie (Muskelschwund) leiden und damit vor allem durch Störungen in der Feinmotorik gehandicapt sind,
- Reisekaufleute,
- Touristen,
- Einwohner und
- ältere Personen, meist ohne jede Computererfahrung.

Die durch obige Aufzählung beschriebene Benutzerpopulation ist sehr heterogen. Die Unterschiede betreffen u.a.:

- *Physische und sensorische Fähigkeiten*, die bei Benutzern mit speziellen Behinderungen und älteren Personen eingeschränkt sind [2].
- *Interessen und Präferenzen*, die u.a. vom Alter der Benutzergruppe abhängen [30]. So bevorzugen ältere Reisende i.d.R. eher kulturelle und geschichtliche Informationen, wohingegen jüngere Reisende Informationen über Einkaufs- und Unterhaltungsmöglichkeiten sowie über Sportereignisse präferieren. Auch wurden deutliche Unterschiede zwischen den Interessen und Präferenzen von Touristen und Einwohnern festgestellt [2].
- *Wissen*, das bei Benutzern z.B. auf die im AVANTI-Informationssystem beschriebene Stadt oder Region bezogen unterschiedlich sein kann. Reisekaufleute setzen sich in dieser Hinsicht als professionelle Experten gegenüber den anderen Benutzergruppen ab [2].
- *Kognitive Fähigkeiten*, die z.B. bei älteren Personen durch eine niedrigere Geschwindigkeit der Informationsverarbeitung sowie durch eine nachlassende Kapazität des Arbeitsgedächtnisses und der Konzentrationsfähigkeit beeinträchtigt sein können [17].

Die in dieser Phase erhobenen Benutzerbeschreibungen wurden in einem zweiten Schritt um die spezifischen Charakteristika der älteren Personen und der Einwohner reduziert, da diese Benutzergruppen nicht an den Feldtests teilnehmen. Anschließend konnte eine weitere Eingrenzung unter dem Bickwinkel Mobilität in einem urbanen Raum und Kompetenz im Umgang mit Computern bzw. dem AVANTI-System erfolgen. Dadurch konnte nicht zuletzt sichergestellt werden, daß eine handhabbare Menge an relevanten Benutzercharakteristika übrigblieb:

- *Physische und sensorische Fähigkeiten*, wie die Fähigkeit Treppen steigen zu können, ein Element des Benutzeroberfläche mit der Maus selektieren zu können oder grafische Information rezipieren zu können.
- *Interessen und Präferenzen*, z.B. das Interesse an Zugangsinformationen zu öffentlichen Gebäuden wie Angaben über das Vorhandensein von Aufzügen, Rampen, Leitsystemen, Telekommunikations- und sanitären Einrichtungen, an Detailinformationen über Kirchen oder die Präferenz eines Benutzers für bestimmte Modalitäten einer Information, z.B. Bilder oder Videos.
- *Wissen über die Domäne*, vornehmlich über prominente Orientierungspunkte wie öffentliche Plätze, Kirchen, Museen, Parks, etc. und die Infrastruktur in den Städten Siena, Kuusamo und Rom.
- Kognitive Fähigkeiten:
  - *Kompetenz im Umgang mit Computern*, z.B. die Fähigkeit, Elemente der Benutzeroberfläche eines Computers zielgerichtet manipulieren zu können.
  - *Kompetenz im Umgang mit dem AVANTI-System*, z.B. mit den im AVANTI-System angebotenen Orientierungs- und Navigationshilfen oder Kenntnisse über die angebotenen adaptiven und adaptierbaren Systemleistungen.

Die Relevanz dieser Benutzercharakteristika zur Steuerung der angebotenen Anpassungsleistungen wird nicht zuletzt dadurch betont, daß auch in der Literatur das Domänenwissen und die benannten Kompetenzfelder als Schlüsselfaktoren für die Lösung potentieller Ergonomieprobleme in Hypertext-Systemen beschrieben werden [21, 22].

Neben Benutzercharakteristika werden in AVANTI ein Domänenmodell und technische Kenngrößen, die die Leistungsfähigkeit der zur Verfügung stehenden Telekommunikations-Infrastruktur beschreiben, zur Steuerung von Anpassungsleistungen verwendet (vgl. das vierte Kapitel). Eine weitere Motivation für Anpassungsleistungen wäre der aktuelle Nutzungskontext des Systems [2], ob z.B. der Zugriff von einem Informationskiosk oder vom heimischen PC aus erfolgt. Dieses und weitere Motivationsfelder werden jedoch in den ersten Prototypen des AVANTI-Systems nicht berücksichtigt werden können.

3 Benutzermodellierung

Die Mehrzahl der Anpassungsleistungen in AVANTI stützen sich auf modellierte Annahmen[2] über Charakteristika eines Benutzers, die in einem sog. „Benutzermodell" gespeichert werden. Die Bildung dieser Annahmen zum Aufbau und zur Pflege des Benutzermodells wird in der einschlägigen Literatur oft als problematisch beschrieben [9, 23].

---

[2] Durch die Verwendung des Begriffs Annahme in diesem Zusammenhang betonen wir die prinzipielle Unsicherheit, mit der die Repräsentation von Benutzercharakteristika in Benutzermodellen behaftet ist. Dies schließt natürlich die Existenz sicherer Benutzermodellinhalte keineswegs aus.

In AVANTI werden Annahmen aus den folgenden Informationsquellen gebildet:

- Die Eingaben des Benutzers mittels eines kurzen Eingangsfragebogens oder in einem Klärungsdialog führen i.d.R. direkt zum Eintrag entsprechender Annahmen ins Benutzermodell. Deshalb werden sie aus Benutzerperspektive bisweilen auch als explizite Annahmen bezeichnet, im Gegensatz zu den nachfolgenden, aus Benutzerperspektive als implizit bezeichneten Annahmen.
- Ausgewählte Dialogakte3 des Benutzers können zur Bildung von Annahmen und deren Eintrag ins Benutzermodell verwendet werden. Fordert ein Benutzer z.B. weitere Detailinformationen zu einer bestimmten Kirche an, dann kann angenommen werden, daß er Interesse an Informationen zu dieser Kirche hat.
- Die Teile eines Benutzermodells, die Annahmen über relevante Eigenschaften von Benutzergruppen (vgl. das zweite Kapitel) enthalten, werden als Stereotype [25] bezeichnet. Unter bestimmten Voraussetzungen, wie z.B. einer einschlägigen Antwort des Benutzers im Eingangsfragebogen, kann ein Stereotyp für einen Benutzer aktiviert (bzw. deaktiviert) werden, was bedeutet, daß die Annahmen, die in dem Stereotyp enthalten sind, das Modell des Benutzers anreichern (bzw. aus diesem entfernt werden).
- Basierend auf Annahmen über den Benutzer und Zusatzinformationen über die Domäne, kann das System durch Inferenzen weitere Annahmen über den Benutzer bilden. Ein Beispiel dafür wäre das inferierte Interesse eines Benutzers an Gemälden aufgrund der Tatsache, daß der Benutzer sich im Laufe der Interaktion für mehrere Einzelgemälde interessiert hat. Interesse an Einzelgemälden kann vom System angenommen werden, wenn sich der Benutzer z.B. Detailinformationen dazu angefordert hat.

Die Unsicherheit mit der eine Annahme im Benutzermodell behaftet ist, hängt maßgeblich von deren Herkunft ab. Explizite und über Dialogakte gebildete Annahmen haben in dieser Hinsicht eine höhere Sicherheit als Annahmen die über Stereotype zugesprochen, oder von Inferenzen gebildet, wurden. Deshalb stehen Annahmen aus Stereotypen und Inferenzen in der Folge nur noch für Abfragen zur Verfügung, nicht aber für die weitere Aktivierung/ Deaktivierung von Stereotypen und für weitere Inferenzen. Dadurch wird die Bildung stetig unsicher werdender Annahmen im Benutzermodell vermieden.

4 Anpassungsleistungen

Die in AVANTI angebotenen Anpassungsleistungen betreffen sowohl die Benutzerschnittstelle, als auch die Hypermedia-Seiten an sich. *Anpassungen der Benutzerschnittstelle* umfassen die Integration spezieller Hardware, wie Braille-Displays, Switches und/oder komplementäre bzw. alternative Software, wie Screen-Reader, Eingabehilfen oder „Dual Browser" [3, 26]. *Anpassungen der Hypermedia-Seiten* [6] erfolgen sowohl bzgl. der Inhalte, wie beispielsweise die Prominenz, Detailliertheit, Ausführlichkeit und Modalität von Informationen über Objekte bzw. Links, als auch bzgl. der Struktur des Hyperraums, wie beispielsweise Anpassungen bzgl. der angebotenen Orientierungs- und Navigations-

---

[3] Dialogakte repräsentieren Aktionen des Benutzers auf der Benutzeroberfläche (z.B. Anforderung von Hilfe in einem bestimmten Kontext), die zur Bildung von Annahmen über den Benutzer verwendet werden können [24]. Dieser Prozeß der Annahmenbildung kann insofern auch als eine Verallgemeinerung einer Präsuppositionsanalyse von Sprechakten in natürlich-sprachlichen Systemen interpretiert werden [14].

möglichkeiten. Letztere orientieren sich dabei an den kognitiven Fähigkeiten des Benutzers [28, 27] und bei blinden Personen an deren zusätzlichem Bedarf an Orientierungs- und Navigationshilfen, wie seitenbezogene Inhaltsverzeichnisse und Übersichten mit Querverweisen [13].

Die folgenden Szenarien stellen einige der im System angebotenen adaptiven und adaptierbaren Systemleistungen im Kontext einer Systemnutzung vor. In unseren Beispielen beschränken wir uns dabei auf Anpassungsleistungen an Hypermedia-Seiten [6], da diese in der Literatur im Gegensatz zu Anpassungen der Benutzerschnittstelle weitaus seltener anzutreffen sind.

Für einen Benutzer, der in einem Eingangsfragebogen sein Interesse an Informationen für Rollstuhlfahrer bekundet hat, wird sein Benutzermodell mit dem Stereotyp „Rollstuhlfahrer" verbunden. Dadurch wird sein individuelles Benutzermodell durch typische Charakteristika von Rollstuhlfahrern, z.B. einschlägiges Interesse an bestimmten Informationen, angereichert. Während bei Gebäuden die Eignung für Rollstuhlfahrer z.B. von bestimmten Türbreiten oder dem Vorhandensein geeigneter sanitärer Einrichtungen abhängt, ist für die Auswahl von Routen in einer Stadt die Beschaffenheit des Straßenpflasters, des Bordsteins, die Steigung einer Straße und/oder das Attribut „Fußgängerzone" wichtig. Bei der Auswahl eines Buses sind dies Informationen wie vorhandene Einstiegshilfen oder Preisermäßigungen für behinderte Kunden. Diese relevanten Zusatzinformationen werden in der Folge auf den entsprechenden Hypermedia-Seiten automatisch vom System hinzugefügt. Ändert der Benutzer die vom System angebotene Informationsauswahl, dann wird das daraus gefolgerte Interesse an zusätzlichen Informationen bzw. Desinteresse an zuvor angebotenen Informationen im Benutzermodell vermerkt, wodurch ggf. aus dem Stereotyp „Rollstuhlfahrer" stammende Interessen ganz oder in Teilen überschrieben werden. Ändert der Benutzer die angebotene Informationsauswahl nicht, dann wird in diesem Fall angenommen, daß das Benutzermodell den aktuellen Stand des Benutzerinteresses widerspiegelt.

Informiert sich ein Benutzer z.B. über den Dom „Santa Maria" und fordert im Überblick weiterführende Informationen zur Baugeschichte an, dann kann auf ein diesbzgl. Interesse am Dom „Santa Maria" geschlossen werden, was zur Bildung und zum Eintrag einer entsprechenden Annahme ins Benutzermodell führt. Fordert der Benutzer im weiteren Verlauf der Sitzung baugeschichtliche Detailinformationen zur Basilika „San Domenico" an, so führt auch dies zur Bildung und zum Eintrag einer entsprechenden Annahme ins Benutzermodell. Da der Benutzer wiederholt Detailinformationen zu Sakralbauten angefordert hat, kann nun eine Inferenzregel innerhalb des Benutzermodells auf ein Benutzerinteresse an Detailinformationen zu Sakralbauten schliessen. Diese im Benutzermodell gezogene Inferenz würde dazu führen, daß der Benutzer bei weiteren sakralen Bauwerken entsprechende Zusatzinformationen automatisch angeboten bekommt, solange er dieses Informationsangebot nicht weiter qualifiziert (siehe oben).

Bei jeder Anforderung einer Hypermedia-Seite überwacht das AVANTI-System die Zeiten, die zur Anpassung, Übertragung und zum vollständigem Aufbau einer Hypermedia-Seite auf der Benutzeroberfläche notwendig sind. Die Verbindung der Übertragungszeit mit Größeninformation zu einer Hypermedia-Seite erlaubt zudem eine Berechnung der verfügbaren Bandbreite im Netzwerk. Die Anwendung einfacher statistischer Verfahren auf diesen Informationen erlaubt nun eine Prognose der verfügbaren Bandbreite und der Antwortzeit für zukünftige Anforderungen von Hypermedia-Seiten. Damit können bei der Zusammenstellung einer Hypermedia-Seite technisch aufwendige Modalitäten einer Information zu einem Objekt, wie Videos oder große Bilder, im Falle einer langsamen Modem-Verbindung durch inhaltlich gleichwertige textuelle Beschreibungen und einen Link auf die substituierte Modalität ersetzt werden. Zusätzlich können noch die Originalgröße und die voraussichtliche Ladezeit für die substituierte Modalität mit angegeben werden. Aufgrund der angebotenen Informationen kann der Benutzer dann entscheiden, ob er die substituierte Modalität anfordern möchte oder nicht.

5 Systemarchitektur

Die folgende Grafik gibt einen Überblick über die Architektur von AVANTI:

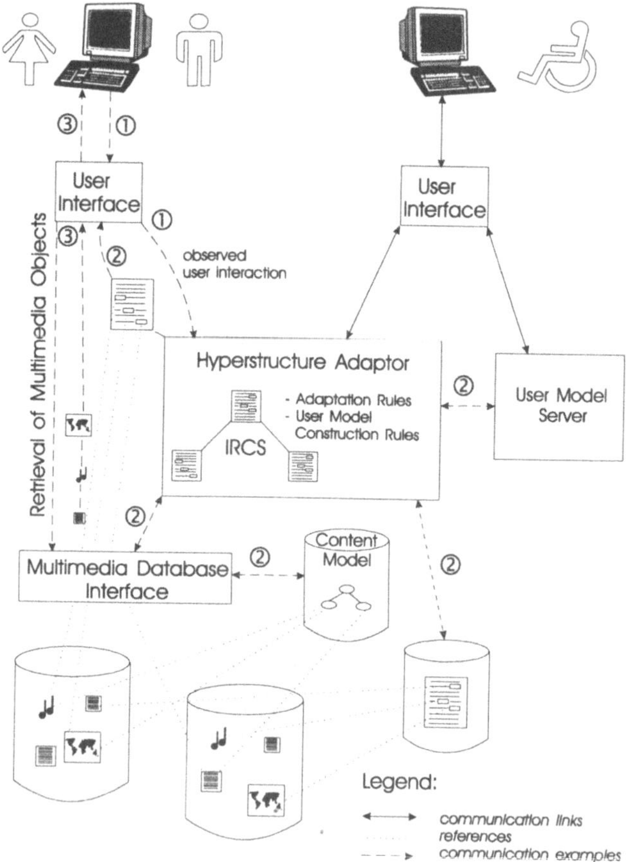

Abbildung 1: Architektur des AVANTI Systems

Die Aufgaben und die Interaktion zwischen den Hauptkomponenten, dem *User Interface* (UI), dem *Hyperstructure Adaptor* (HSA), dem *User Model Server* (UMS) und dem *Multimedia Database Interface* (MDI), können dabei entlang des in der obigen Abbildung angedeuteten Beispiels einer Seitenanforderung skizziert werden (①-③ finden sich dabei in der Grafik wieder):

① Ein Benutzer fordert über das UI eine neue Hypermedia-Seite vom HSA an.

② Der HSA lädt die Hypermedia-Seite im *IRCS*-Format *(Information Resource Control Structure)* [4], das eine Untermenge und eine Erweiterung von HTML [29] darstellt, und beginnt mit der inhaltlichen Aufbereitung. Dazu enthält die *IRCS*-Seite neben festen Bestandteilen Verweise auf Anpassungsregeln *(Adaptation Rules)*, die das Aufbereiten der adaptiven und/oder adaptierbaren Teile einer Hypermedia-Seite steuern. Anpassungsregeln können Informationen von anderen AVANTI-Komponenten einholen, z.B.:

- Annahmen über Fähigkeiten, Interessen oder Wissen des Benutzers vom UMS z.B. der Benutzer bevorzugt Bilder gegenüber textuellen Erklärungen, und/oder
- inhaltliche Meta-Informationen zu Multimedia-Objekten über das MDI aus dem *Content Model*, z.B. ein Bild ist eine Erklärung zu einem assoziierten Begriff.

Zusätzlich stehen innerhalb des HSA Informationen über die aktuelle Sitzung des Benutzers, z.B. frühere Eingaben oder Angaben über bereits angeforderte Seiten, zur Verfügung. Akquisitionsregeln *(User Model Construction Rules)* ermöglichen die Bildung von Annahmen aus der Interaktion des Benutzers mit dem System, z.B. kann angenommen werden, daß ein Benutzer an einem Objekt interessiert ist, wenn er Detailinformationen dazu anfordert. Die vom HSA solchermaßen aufbereitete *IRCS*-Seite wird dann an das UI übergeben.

③ Das UI fordert über das MDI die auf der *IRCS*-Seite referenzierten Objekte aus den AVANTI-Datenbanken an und präsentiert dem Benutzer die angeforderte Hypermedia-Seite[4].

Die Architektur von AVANTI baut auf dem World-Wide Web (WWW) bzw. den damit verbundenen Standards der IETF (Internet Engineering Task Force) auf [12], d.h.:

- URLs (Universal Resource Locators) als universellen Adressierungsmechanismus für jegliche Art von Objekten, z.B. *IRCS*-Seiten, Multimedia-Objekte,
- HTTP (HyperText Transfer Protocol) bzw. MIME (Multipurpose Internet Mail Extensions) als universelles Kommunikationsprotokoll zwischen allen Softwarekomponenten[5], und
- HTML (HyperText Markup Language) als Grundlage der *IRCS*-Beschreibungssprache für Hypermedia-Seiten.

---

[4] Auch das *User Interface* von AVANTI kann ggf. noch weitere Anpassungen durchführen. Dies wurde hier jedoch aus Gründen der Übersichtlichkeit weggelassen.

[5] Ausnahme: Die Kommunikation mit dem *User Model Server* wird mittels KQML (Knowledge Query and Manipulation Language [10, 14] durchgeführt.

142

Die wichtigsten Vorteile dieser Architektur sind:

- Der Zugriff auf das AVANTI-System kann über die gängigsten der heute im Umfeld des WWW verfügbaren Browser erfolgen.
- Bereits existierende Software im Umfeld des WWW, z.B. Browser, Server, Gateways, Proxies, Kommunikationsbibliotheken, kann ganz oder in Teilen als Entwicklungsgrundlage oder -umgebung für AVANTI-Komponenten verwendet werden.
- Komponenten von AVANTI können entsprechend organisatorischer und/oder technischer Anforderungen flexibel verteilt werden. Die Mobilität eines Benutzers wird dabei insofern unterstützt, als sein Benutzermodell und der korrespondierende UMS jederzeit ermittelt werden können. Ein HTTP-basierter Name Service bildet durch die Auflösung symbolischer Referenzen die hierfür notwendige Grundlage [11].
- Der HSA kann zur Laufzeit in der Programmiersprache Java [8] geschriebene Funktionalität aus dem *Hyperstructure Adaptor* in das *User Interface* verlagern, insofern dort die entsprechenden Voraussetzungen zur Ausführung von Java-Code gegeben sind.
- AVANTI unterstützt in Zusammenarbeit mit verbreiteten HTTP-Servern, wie z.B. den Servern der Netscape Corporation [19], gängige Sicherheitsstandards [7], wie z.B. SSL (Secure Sockets Layer) und SHTTP (Secure HTTP).

Die zentralen Komponenten in der AVANTI-Architektur sind der *Hyperstucture Adaptor* und der *User Model Server*. Die Grundlage für deren Entwicklung bilden das Produkt „WebObjects" [20] für den HSA und das Benutzermodellierungs-Shell-System „BGP-MS" (Belief, Goal and Plan Maintenance System) [5, 15] für den UMS.

Danksagung

Wir bedanken uns bei Gaby Nordbrock, Reinhard Oppermann und Christoph Thomas für ihre wertvollen Kommentare und Anregungen zu früheren Versionen dieses Papiers.

Literaturverzeichnis

[1] AVANTI: Offizielle AVANTI-Seite der Europäischen Kommission,
URL: http://drogo.cselt.stet.it/sonah/AVANTI/index.html;
GMD-Projektseite, URL: http://www.gmd.de/fit/projects/avanti.html, 1996.

[2] AVANTI Deliverable DE001 - User groups' needs and requirements. Project Report, Brüssel, 1996.

[3] AVANTI Deliverable DE004 - Selection of terminals and terminal adaptations. Project Report, Brüssel, 1996.

[4] AVANTI Deliverable DE005 - Adaptable Multimedia Document Structure. Project Report, Brüssel, 1996.

[5] BGP-MS: Projektseite, URL: http://zeus.gmd.de/projects/bgp-ms.html, 1996.

[6] P. Brusilovsky: Methods and Techniques of Adaptive Hypermedia. In: User Modeling and User-Adapted Interaction 6 (2-3), 1996, 87-129.

[7] A. Cain: Web Security - Technologies for Security, Authentication, and Privacy on the World-Wide Web. In: Tutorial Notes, Fifth International World-Wide Web Conference. Paris. O'Reilly & Associates, Inc., Sebastopol, California, 1996, 1-31.

[8] M. Campione, K. Walrath: The Java Tutorial. URL: http://www.javasoft.com/tutorial/index.html, 1996.

[9] D. N. Chin: Acquiring User Models. Artificial Intelligence Review 7, 1993, 185-197.

[10] T. W. Finin, J. Weber, G. Widerhold, M. Genesereth, R. Fritzson, D. McKay, J. McGuire, R. Pelavin, S. Shapiro und C. Beck: Specification of the KQML Agent-Communication Language. URL: http://www.cs.umbc.edu/kqml/papers/kqmlspec.ps, 1993.

[11] KAPI: KAPI Software Overview. URL: http://hitchhiker.space.lockheed.com:80/aic/shade/software/KAPI/README.html, 1996.

[12] D. Keller: Creating Web Pages: Introduction to HTML und HTML Authoring for the Web, Part II. In: Tutorial Notes, Fifth International World-Wide Web Conference, Paris. O'Reilly & Associates, Inc., Sebastopol, California, 1996, 46-58 und 59-100.

[13] A. Kennel, L. Perrochon, A. Darvishi: WAB: World-Wide Web Access for Blind And Visually Impaired Computer Users. New Technologies in the Education of the Visually Handicapped, Paris, Juni 1996 und ACM SIGCAPH Bulletin, Juni 1996. URL: http://www.inf.ethz.ch/department/IS/ea/blinds/.

[14] A. Kobsa: Benutzermodellierung in Dialogsystemen. Springer Verlag, Berlin u.a., 1985.

[15] A. Kobsa, W. Pohl: The User Modeling Shell System BGP-MS. In: User Modeling and User-Adapted Interaction 4(2), 1995, 59-106.

[16] A. Kobsa, J. Fink, W. Pohl: A Standard for the Performatives in the Communication between Applications and User Modeling Systems (draft). URL: ftp://ftp.informatik.uni-essen.de/pub/UMUAI/others/rfc.ps, 1996.

[17] A. Kruse, U. Lehr: Reife Leistung - Psychologische Aspekte des Alterns. In: Funkkolleg Altern, Studienbrief 2, Studieneinheit 5, DIFT (Deutsches Institut für Fernstudien an der Universität Tübingen), 1996.

[18] R. Kuhlen: Pragmatischer Mehrwert von Information - Sprachspiele mit informationswissenschaftlichen Grundbegriffen. Bericht 1/89, Informationswissenschaft, Universität Konstanz, 1989.

[19] Netscape Corporation: Introducing the next Generation of Open Servers. URL: http://home.netscape.com/comprod/server_central/index.html, 1996.

[20] NeXT Corporation. WebObjects. URL: http://www.next.com/WebObjects/Products.html, 1996.

[21] J. Nielsen: The matters that really matter for hypertext usability. In: Proc. ACM Hypertext '89, Pittsburg, 1989.

[22] J. Nielsen: Usability Engineering.Academic Press Inc., San Diego, 1993.

[23] R. Oppermann : Adaptively supported Adaptability. In: International Journal of Human-Computer Studies. 40, 1994, 455 - 472.

[24] W. Pohl, A. Kobsa, O. Kutter: User Model Acquisition Heuristics Based on Dialogue Acts. Bericht Nr. 56/94 (WIS-Bericht 6), Informationswissenschaft, Universität Konstanz, 1994.

[25] E. Rich: User Modeling via Stereotypes. In: Cognitive Science, 3, 1979, 329-354.

[26] A. Savidis, C. Stephanidis: Developing Dual User Interfaces for Integrating Blind and Sighted Users: the HOMER UIMS. In: Proceedings of the CHI'95 Conference on Human Factors in Computing Systems, Denver, Colorado, 1995.

[27] H. Schaumburg, L. J. Issing: Lernen mit Hypermedia: Verloren im Hyperraum? HMD - Theorie und Praxis der Wirtschaftsinformatik. Nr. 190, 1996, 108-121.

[28] B. White: Web Document Engineering. In: Tutorial Notes, Fifth International World-Wide Web Conference, Paris. O'Reilly & Associates, Inc., Sebastopol, California, 1996, 101-113.

[29] World Wide Web Consortium: HyperText Markup Language (HTML). URL: http://www.w3.org/pub/WWW/ MarkUp/, 1996.

[30] M. Zeiner, B. Harrer, L. Bengsch: Städtetourismus in Deutschland. Deutscher Fremdenverkehrsverband, Bonn, 1995.

Adressen der Autoren

Josef Fink
GMD - Forschungszentrum
Informationstechnik GmbH
Institut für Angewandte
Informationstechnik (FIT)
Forschungsbereich Mensch-
Maschine Kommunikation (MMK)
D-53754 Sankt Augustin
Tel.: 02241/14-2729
E-mail: Josef.Fink@gmd.de

Prof. Dr. Alfred Kobsa
GMD - Forschungszentrum
Informationstechnik GmbH
Institut für Angewandte
Informationstechnik (FIT)
Forschungsbereich Mensch-
Maschine Kommunikation (MMK)
D-53754 Sankt Augustin
Tel.: 02241/14-2315
E-mail: Alfred.Kobsa@gmd.de

Andreas Nill
GMD - Forschungszentrum
Informationstechnik GmbH
Institut für Angewandte
Informationstechnik (FIT)
Forschungsbereich Mensch-
Maschine Kommunikation (MMK)
D-53754 Sankt Augustin
Tel.: 02241/14-2859
E-mail: Andreas.Nill@gmd.de

Heuristische Evaluation und IsoMetrics:
Ein Vergleich

Günther Gediga und Kai-Christoph Hamborg

Fachbereich Psychologie, Universität Osnabrück

Zusammenfassung

Es werden die Ergebnisse aus zwei formativen Evaluationsstudien dargestellt, in denen parallel mit einer expertenbezogenen (heuristische Evaluation) und einer nutzerbezogenen (IsoMetrics) Methode gearbeitet wurde. Grundsätzliche Probleme in Bezug auf die Auswertung von Daten formativer Evaluationsuntersuchungen und ein Lösungsansatz für die Klassifizierung gewonnener Daten werden dargestellt. Ausgehend von der Klassifizierung der durch die Evaluationsmethoden evozierten Bemerkungen, werden die Methoden nach Qualität und Quantität der Evaluationsergebnisse verglichen. Anwendungszeitpunkt und ein möglicher kombinierter Einsatz der vorgestellten Evaluationsmethoden werden vor dem Hintergrund der Ergebnisse diskutiert.

1 Die Problemstellung

Um Benutzbarkeits-Schwächen eines Softwaresystems festzustellen gibt es eine Reihe unterschiedlicher Methoden, die im Extremfall „expertenbasiert" oder „rein nutzerbasiert" arbeiten. In diesem Beitrag wollen wir zwei Techniken zur Evozierung von Benutzbarkeits-Schwächen kontrastieren: Die „heuristische Evaluation" [5], [6], die „Usability"-Experten benötigt, und das IsoMetrics-Verfahren [11], [12], das mittels einer Benutzerbefragung versucht, die Schwächen eines Softwaresystems festzustellen.

In [4], [7] und [9] wird diskutiert, daß aus Experten- und Benutzerbefragungen oft wenig überlappende Ergebnisse resultieren. Dies führt zu der Frage, ob sich die Ergebnisse aus Experten- und Benutzerbefragungen ergänzen oder vielleicht doch substituieren. Ein Ziel der Untersuchung ist es, zu klären, welche unterschiedlichen gestaltungs-unterstützende Informationen mit den beiden Verfahren erhoben werden und worin die Stärken und Schwächen der beiden Befragungstypen bestehen. Erst wenn hier substantielle Unterschiede feststellbar sind, können beide Verfahren als sich ergänzend angesehen werden.

In [7] wird behauptet, daß ca. 5 Experten ausreichen, um eine heuristische Evaluation durchzuführen, allerdings ohne eine methodisch befriedigende Erklärung für die Berechnung dieser Anzahl anzugeben. Es soll deshalb eine Methode für die Bestimmung der Evaluatoren-Anzahl vorgestellt werden, die dann auch angewandt und diskutiert wird. Insbesondere wird man mit Hilfe der vorgestellten Methode die „ca. 5-Evaluatoren"-Regel bewerten können.

Bei der Durchführung der Datenauswertung tauchen eine Reihe von Detailproblemen auf: So muß zunächst festgelegt werden, was als „Bemerkung" oder „Schwäche" gezählt werden soll. Desweiteren ergibt sich als Schwierigkeit die unterschiedliche Qualität von Bemerkungen. Letztlich muß geklärt werden, wann Bemerkungen als gleich angesehen werden können. Erst nach der Lösung dieser Detailprobleme auf der Ebene gegebener qualitativer Daten kann man versuchen, quantitative Aussagen - wie z.B. den Verlauf der Anzahl von evozierten Bemerkungen in Abhängigkeit von der Anzahl der im Beurteilungsprozeß tätigen Personen - zur Gütebeurteilung der Verfahren sinnvoll zu bestimmen.

2 Methoden zur formativen Software-Evaluation

2.1 Aufgaben und Schwierigkeiten der formativen Software-Evaluation

Als Voraussetzung für die ergonomische Gestaltung von Softwaresystemen wird neben der Berücksichtigung von Arbeitsabläufen und der Mensch-Computer Arbeitsteilung die frühzeitige Evaluation erster Konzepte und Gestaltungslösungen der eigentlichen Software betrachtet [2], [3], [7], [8].

Alle Evaluationsmaßnahmen, die auf die Gestaltung und Verbesserung der Software im Entwicklungszyklus ausgerichtet sind, werden als formative Evaluation bezeichnet [7], [10]. Die Gestaltung und Verbesserung von Software soll durch formative Evaluationsverfahren i.d.R. dadurch ermöglicht werden, daß Bemerkungen und Aussagen über Schwachpunkte z.B. eines Prototypen bei Benutzern oder ausgewählten Experten evoziert werden.

Prinzipielle Probleme bei der formativen Software-Evaluation bestehen darin, zu welchem Zeitpunkt die Evaluation wie in den Entwicklungsprozeß integriert wird, wie Bemerkungen zu einem Softwaresystem evoziert und erhoben werden können und wie sich diese in Bezug auf das (Re-) Design eines Systems auswerten und umsetzen lassen. Für die formative Evaluation von Softwaresystemen liegen eine Vielzahl unterstützender Methoden vor (s. [7]), die sich größtenteils entweder auf den direkten Einbezug prospektiver Nutzer oder von Ergonomieexperten stützen. Zu diesen Methoden zählen auch die heuristische Evaluation und das IsoMetrics[L] Verfahren.

Simple and natural dialog
Speak the user's language
Minimize user memory load
Be consistent
Provide feedback
Provide clearly marked exits
Provide shortcuts
Good error messages
Prevent errors

Tab. 1: Heuristiken nach Nielsen und Molich ([5])

2.2 Die heuristische Evaluation

Die heuristische Evaluation [5], [6] ist eine informelle Methode zum Aufdecken möglicher Benutzbarkeits-Probleme, deren Einsatz in frühen Designphasen empfohlen wird [7, S. 224]. Bei der heuristischen Evaluation wird eine Menge von Richtlinien (*Heuristiken*) für die Untersuchung der vorgegebenen Software durch Experten zugrunde gelegt. Während es eine Reihe von Systemen gibt, die eine große Anzahl von Richtlinien zugrunde legen, wird bei der heuristischen Evaluation nur eine geringe Anzahl von globalen Heuristiken festgelegt, die vom einzelnen Untersucher interpretiert werden müssen. Nielsen und Molich [5] schlagen vor, die in der Tab. 1 dargestellten folgenden Heuristiken zu benutzen.

Nach Nielsen [6] soll ein Team von 3 Evaluatoren bereits zwischen 50% („novice evaluator'') und 90% („double specialist'') der Probleme finden können, die in der Software auftauchen, wobei allerdings die Aufdeckungsrate eine große Abhängigkeit von der spezifischen Software und der Fähigkeiten der Evaluatoren aufweist [4, S. 121].

2.3 IsoMetrics

Der Fragebogen IsoMetrics[L] umfaßt sieben Subskalen mit insgesamt 90 Items. Die Fragen wurden systematisch aus einem Fragenpool von über 600 Items ausgefiltert [11]. Die Konstruktion der Skalen und erste Untersuchungen, die die Validität und Reliabilität des Verfahrens belegen, sind detailliert an anderer Stelle beschrieben [11].

Jede Subskala von IsoMetrics[L] repräsentiert einen Gestaltungsgrundsatz gemäß ISO 9241/10. Die Subskalen sind daher wie folgt benannt (Anzahl Items je Skala in Klammern):

Skala 'A': Aufgabenangemessenheit	(17)
Skala 'S': Selbstbeschreibungsfähigkeit	(14)
Skala 'T': Steuerbarkeit	(14)
Skala 'E': Erwartungskonformität	(9)
Skala 'F': Fehlerrobustheit	(17)
Skala 'I': Individualisierbarkeit	(11)
Skala 'L': Erlernbarkeit	(8)

Tab. 2: Skalen des IsoMetrics-Fragebogens mit der Anzahl der zugehörigen Items

Je Item werden drei unterschiedliche Daten erhoben:

- die Bewertung der Software nach ISO 9241/10, wie durch das Item operationalisiert, auf einer 5-stufigen Ratingskala. Die Skalierung der Skala reicht von "stimme überwiegend nicht zu (Rating 1) über "teils/teils" (Rating 3) bis "stimme überwiegend zu" (Rating 5).

- die Gewichtung des Items als Ausdruck der Bedeutung des durch das Item operationalisierten Aspekts der ISO 9241/10 in Bezug auf den Gesamteindruck der Software. Die Gewichtung wird ebenfalls auf einer 5-stufigen Ratingskala, von "nicht wichtig" (Rating 1) über "teils/teils" (Rating 3) bis "wichtig" (Rating 5) vorgenommen.

- eine oder mehrere auf das Item bezogene Problemmeldungen, die der oder die Befragte in freiem Text selbst formuliert.

In Abb. 1 wird ein Item des Fragebogens zur Veranschaulichung dargestellt.

Die Gebrauchstauglichkeit der von den Benutzern erzeugten Bemerkungen für die Softwareentwicklung wurde in [12] dokumentiert.

	stimmt nicht	stimmt wenig	stimmt mittelmäßig	stimmt ziemlich	stimmt sehr		Keine Angabe
Wenn Menü-Optionen in bestimmten Bearbeitungsschritten nicht zur Verfügung stehen, wird mir die Sperrung sichtbar gemacht.	1	2	3	4	5		

	stimmt nicht	stimmt wenig	stimmt mittelmäßig	stimmt ziemlich	stimmt sehr		Keine Angabe
Wie wichtig ist dieser Aspekt für Ihren Gesamteindruck von der Software?	1	2	3	4	5		
Können Sie konkrete Beispiele nennen, bei denen Sie dieser Aussage nicht zustimmen können?							

Abb. 1: Ein Fragebogenitem aus IsoMetrics[L] (Erläuterungen im Text)

3 Versuchsplan und Methodik

3.1 Die Untersuchungen

In der Untersuchung 1 wurde der Prototyp eines Sprach-Lehr-Systems im Rahmen eines Forschungsprojekts (RECALL; [1]) untersucht. Es handelte sich hierbei um die Version 1.04a des Systems - die gröbsten Fehler des Systems waren damit durch mehrfache Entwicklertests und Expertenbegutachtung in den Vorversionen behoben worden. Der Prototyp wurde mit IsoMetrics[L] und dem Vorgehen zur heuristischen Evaluation [6] evaluiert. An der Evaluationsstudie mit IsoMetrics[L] nahmen 8 Teilnehmer teil. Die heuristische Evaluation wurde durch drei Experten durchgeführt. Die IsoMetrics-Rater (IM-Gruppe) arbeiteten 4 Stunden mit dem Softwareprototyp. Die IM-Gruppe bestand aus am Spracherwerb interessierten Laien, wobei Sprachlehrer, Linguisten, Softwareentwickler, Psychologen und CALL-Spezialisten ausgeschlossen waren. Für die geleistete Arbeit erhielten die Teilnehmer in der IM-Gruppe eine Aufwandsentschädigung von 50DM. Die heuristische Evaluation (HE-Gruppe) wurde von drei Usability-Experten (1 Informatiker, 1 Linguist, 1 Psychologe, die mindestens seit 0,5 Jahren im usability-engineering-Bereich arbeiten) durchgeführt, die mindestens einmal eine heuristische Evaluation durchgeführt hatten. Die drei Experten arbeiteten unabhängig voneinander jeweils einen Tag mit dem Softwareprototypen und führten begleitend die heuristische Evaluation durch. Die Aufwandsentschädigung betrug 100DM.

Gegenstand der Untersuchung 2 war der Prototyp eines multimedialen Informationssystems zur Studienberatung, das in einem Studienprojekt des Fachgebiets Arbeits- und Organisationspsychologie der Universität Osnabrück entwickelt wurde. Bei der Gestaltung des Informationssystems wurde ein iteratives Entwicklungskonzept verfolgt. Der nach Anforderungsanalyse und Spezifikation erstellte erste Prototyp wurde ebenfalls mit IsoMetrics[L] und dem Vorgehen zu der heuristischen Evaluation evaluiert. Die Evaluationsstudie mit IsoMetrics[L] wurde mit 16 Teilnehmern, die heuristische Evaluation mit fünf Experten durchgeführt. Die IM-Gruppe bearbeiteten mit dem zu evaluierenden System Aufgaben mit einem durchschnittlichen Zeitaufwand von ca. 1 Stunde. Daraufhin wurde die Software mit IsoMetrics[L] bewertet. Die IM-Gruppe bestand aus Personen, die im Fachgebiet als Haupt- oder Nebenfachstudierende studierten oder zu studieren beabsichtigten. Die Gruppenmitglieder wurden für die geleistete Arbeit mit 50 DM entlohnt. Die heuristische Evaluation (HE- Gruppe) wurde von fünf Usability-Experten (drei aus Untersuchung 1 und zwei weitere mit vergleichbarem Profil) durchgeführt. Die Evaluatoren der HE-Gruppe arbeiteten mit dem System ca. zwei Stunden und führten begleitend die Heuristische Evaluation durch. Die Aufwandsentschädigung für die Evaluatoren betrug 50 DM.

3.2 Die Datenauswertungs-Strategie

Während in der Literatur (z.B. [7]) als Resultat formativer Evaluation von identifizierten Problempunkten („problems", [7, S. 156]) geredet wird, ist die genaue Bewertung, was einen Problempunkt darstellt, nicht so einfach, wie es suggeriert wird. Zunächst werden bei der formativen Evaluation Bemerkungen zu einem System erhoben. Bei der Auswertung dieser Bemerkungen handelt es sich um ein *inhalts-analytisches* Problem, wobei die zu vermittelnde Information auf den Empfänger (Softwareentwickler) abgestimmt werden muß und die generierte Information erheblich von den gewünschten Ergebnissen abweichen kann, da selbst ein Usability-Experte kaum in der Lage sein dürfte, für den Entwickler leicht interpretierbare und umsetzbare Informationen zu liefern. Verschärft wird dieses Problem, wenn ein Nutzer mit einer anderen Qualifikation Bemerkungen über die Software generiert. Um den Transformationsprozeß transparenter zu gestalten, wurden die von den Untersuchungsteilnehmern erzeugten Phrasen zunächst auf für (wie wir denken) Entwickler mehr oder weniger wichtige Kategorien klassifiziert (Tab. 3).

Bei der Durchführung der Klassifikation stellte sich heraus, daß zunächst die „leicht" klassifizierbaren Kategorien „irrelevant" (Irr), „positive Bemerkung" (Pos) und „generelles Unverständnis" (X) ausgefiltert werden sollten. Die Bemerkungen in der Kategorie X sind oft nicht direkt verwertbar, geben aber häufig ebenfalls Aufschluß über die Schwächen eines Systems, wie die Beispielphrase in Tab. 3 belegt. Aus diesem Grunde werden Phrasen aus der Kategorie X als berichtenswert festgehalten. Nach der Bestimmung der X-Phrasen können die S und S+ Kategorien relativ leicht gefunden werden. Der Rest ergibt sich zu den G, bzw. G+ Kategorien. Da die S+ und G+ Phrasen in Abgrenzung zu den S und G Phrasen nicht unbedingt neue Probleme oder Fehler mit der Software behandeln, wurde in einem zweiten Schritt eine gröbere Kategorisierung gebildet, wobei die Veränderungsvorschläge ignoriert wurden. Man erhält statt der Kategorien S+, S eine gröbere Kategorie S', bzw. statt G+ und G die gröbere Kategorie G'. Die Elemente innerhalb des feinen Kategoriensystems nennen wir im folgenden *Bemerkungen*, während die Elemente des groben Kategoriensystems *Schwächen* (des Softwaresystems) genannt werden.

Im dritten Analyseschritt wurden die redundanten Bemerkungen bzw. Schwächen festgestellt. Aus naheliegenden Gründen gilt nach Aussonderung der Redundanzen:

$$|S'| \leq |S| + |S^+|$$
$$|G'| \leq |G| + |G^+|.$$, bzw.

Als vierter Analyseschritt wurden diejenigen - redundanzfreien - Bemerkungen markiert, die sowohl in der heuristischen Evaluation als auch bei der Untersuchung mit dem IsoMetrics-Instrument genannt wurden (s. Tab. 4 und 5).

Kategorie	Art der Bemerkung
Irr	Die von der Person abgegebene Bemerkung bezieht sich auf etwas anderes als die Interaktion der Person mit der Software (z.B. Bemerkungen zum Befragungsinstrument).
Pos	Die Bemerkung ist positiver Art. Klassischerweise werden diese Bemerkungen - wie auch die irrelevanten Anmerkungen - nicht dokumentiert.
X	Die Bemerkung drückt ein generelles Unverständnis aus. Sie handelt mehr vom Fühlen, Denken und Erleben der Person als von den auslösenden Schwächen der Software . Es muß aus der Bemerkung ersichtlich sein, daß es sich bei der Software um den auslösenden Faktor handelt. Beispiel: *Die Drag- und Drop-Funktion war nicht hilfreich. Habe lieber die Tastatur benutzt.*
S	Es ist nachvollziehbar, auf welchen Teil der Software sich die Bemerkung bezieht. Beispiel: *Ausführung von „Load Resources" führt zur Nicht-Benutzbarkeit des Systems. Ein Neustart des Systems war bei allen Versuchen erforderlich*
S+	Wie S, es wird jedoch zusätzlich mindestens ein Veränderungsvorschlag geäußert. Beispiel: *Um in der History einen Schritt zurückzugehen, sollte man nicht immer über „ZURÜCK" gehen müssen. Abhilfe z.B. über rechte Maustaste oder linkes Eselsohr auf der ersten Seite.*
G	Die Bemerkung ist genereller Natur. Sie bezieht sich zwar auf die Software, es ist aber nicht genau angegeben, welche Teile betroffen sind. Beispiel: *Die Page-up/down-Tasten werden nicht unterstützt.*
G+	Wie G, es wird jedoch zusätzlich mindestens ein Veränderungsvorschlag geäußert. Beispiel: *Online-Hilfen, Online-Lexikon ist dringend erforderlich. F1-Taste mit Help belegen.*

Tab. 3: Kategoriendefinition für die Analyse der Bemerkungen

Ein wichtiger Parameter für formative Evaluationsvorhaben ist die für die Evozierung eines bestimmten Potentials von Bemerkungen bzw. Schwächen notwendige Personenanzahl. Da man nicht beliebig viele Personen untersuchen kann, muß ein Modell herhalten, das das Verhalten bei größeren Personenstichproben extrapoliert, bzw. asymptotische Aussagen macht. Da die Anzahl der evozierten Bemerkungen eine negativ-beschleunigte Kurve bildet, ist das einfachste Modell eine Lernkurve mit

$$B(N) = \lambda(1 - \exp(-\alpha N)).$$

B(N) bezeichnet hierbei die Anzahl der redundanzfreien evozierten Bemerkungen (Schwächen) bei N Personen, die durch die Asymptote λ und den Lernparameter α modelliert werden. Lassen sich die Daten an dieses Modell anpassen, dann erhält man mit der Asymptote sowohl eine Schätzung der insgesamt zu evozierenden Bemerkungen (Schwächen) und mit α eine Schätzung wie schnell diese asymptotische Anzahl erreicht wird. Wenn man mindestens 50% der möglichen Bemerkungen mit einem Verfahren evozieren möchte, so ist man mit N>-ln(0,5)/α auf der sicheren Seite.

4 Ergebnisse

4.1 Untersuchung 1

Die Häufigkeiten der nach der Prozedur aus Kap. 3.2 klassifizierten nicht-redundanten Bemerkungen sind in Tab. 4 dargestellt. Es zeigt sich, daß in der ersten Untersuchung in der IM-Gruppe erheblich mehr Bemerkungen evoziert wurden, als in der HE-Gruppe, wobei bei beiden Methoden im Durchschnitt ca. 16 (redundanzfreie) Bemerkungen pro Teilnehmer entfallen.

Kategorie	H.E. gesamt	Schnittmenge	IsoMetrics gesamt
S	20	14	35
S+	16	7	18
G	9	6	42
G+	4	3	6
X	0	0	26
Gesamtsummen	49	30	127

Tab. 4: Analyse der Bemerkungen zu Untersuchung 1

Die Angaben aus der HE-Gruppe sind erheblich präziser, der überwiegende Teil der Anmerkungen befindet sich hier in den Klassen S und S+. In der IM-Gruppe sind die Häufigkeiten breiter verteilt. Es gibt hier auch einen hohen Anteil der Klasse X (ca. 25%), die in der HE-Gruppe nicht besetzt ist. Betrachtet man die Schnittmenge der Bemerkungen aus beiden Untersuchungsgruppen, so zeigt sich, daß bei beiden Methoden ein eigenständiger Rest vorhanden ist. Dieser Effekt ist nicht neu, so konnte in [9] gezeigt werden, daß Benutzer-orientierte und Experten-orientierte usability Verfahren in den meisten Fällen teilweise unterschiedliche Ergebnisse erbringen. Es zeigt sich in dieser Untersuchung aber auch, daß die Bemerkungen aus der IM-Gruppe erheblich mehr Bemerkungen aus der HE-Gruppe überdecken als umgekehrt.

Bei der Untersuchung, wieviele Teilnehmer für eine usability-Testung benötigt werden, beschränken wir uns in diesem Abschnitt auf das IsoMetrics-Verfahren, da an der heuristischen Evaluation in Untersuchung 1 nur 3 Experten teilnahmen. Bei der Berechnung der Vertrauensbereiche zeigt sich, daß zwischen 3 und 6 Teilnehmern ein Konfidenzband mit einer Breite von ca. 15 Bemerkungen/Schwächen zu erwarten ist. Die meisten Untersuchungsteilnehmer haben relativ viele Bemerkungen (zwischen 20 und 30) erzeugt, es gab aber auch einige wenige Teilnehmer, die wenige Bemerkungen erzeugt haben (im Extremfall eine auswertbare Bemerkung). Die Dynamik der berechneten Mittelwerte wird durch das Modell (Abb. 2) $B(N)=\lambda(1-\exp(-\alpha N))$ eingefangen.

Abb. 2 zeigt die gute Übereinstimmung zwischen dem Modell $B(k)=\lambda(1-\exp(-\alpha k))$ und den empirisch ermittelten Daten. Für die Bemerkungen erhalten wir $\alpha=0{,}122$ mit einer Asymptote $\lambda=202$, während die Schwächen mit den Parametern $\alpha=0{,}179$ und der Asymptote $\lambda=110$ beschrieben werden können. Bei $N>3{,}87$ ($=-\ln(0{,}5)/\alpha$) Untersuchungsteilnehmern kann man hier mit dem IsoMetrics-Verfahren mehr 50% der feststellbaren Schwächen abschöpfen, und mit $N>5{,}68$ könnten der Bemerkungen eingefangen werden.

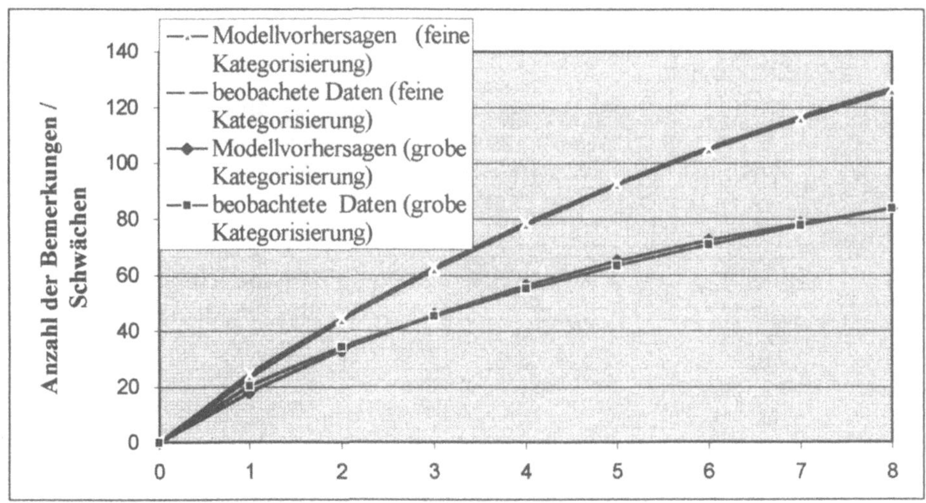

Abb. 2: Vergleich von empirischen und Modell-Erwartungswerten bei der Beschreibung von evozierten Bemerkungen in Abhängigkeit von der Rateranzahl (IsoMetrics / Untersuchung 1)

4.3 Untersuchung 2

Aus der in 3.2 beschriebenen Datenauswertungsstrategie resultieren die in Tab. 5 dargestellten Häufigkeiten nicht redundanter Anmerkungen zu dem evaluierten Prototypen.

Es zeigt sich, daß in der HE-Gruppe deutlich mehr Anmerkungen evoziert wurden als in der IM-Gruppe. Der Anteil spezifischer Bemerkungen ist, wie in der ersten Untersuchung, in der HE-Gruppe höher als in der IM-Gruppe (80% der Bemerkungen in der HE-Gruppe fallen in die Kategorien S und S+, in der IM-Gruppe nur 20%). In der HE- Gruppe wurden bei einem Range von 8 bis 25 Anmerkungen im Durchschnitt ca. 13 redundanzfreie Anmerkungen gemacht. In der IM-Gruppe liegt der Durchschnitt redundanzfreier Anmerkungen unter 3, bei einem Range von 0 bis 9 Anmerkungen.

Kategorie	H.E. gesamt	Schnittmenge	IsoMetrics gesamt
S	38	6	15
S+	17	0	5
G	12	2	17
G+	3	0	1
X	1	0	5
Gesamtsummen	68	8	43

Tab. 5: Analyse der Bemerkungen zu Untersuchung 2

Die Schnittmenge der Bemerkungen zwischen den Gruppen ist in den Kategorien S+, G, G+ und X äußerst gering bzw. gleich Null. Die größte Anzahl gemeinsamer Anmerkungen findet sich in Kategorie S (N=6).

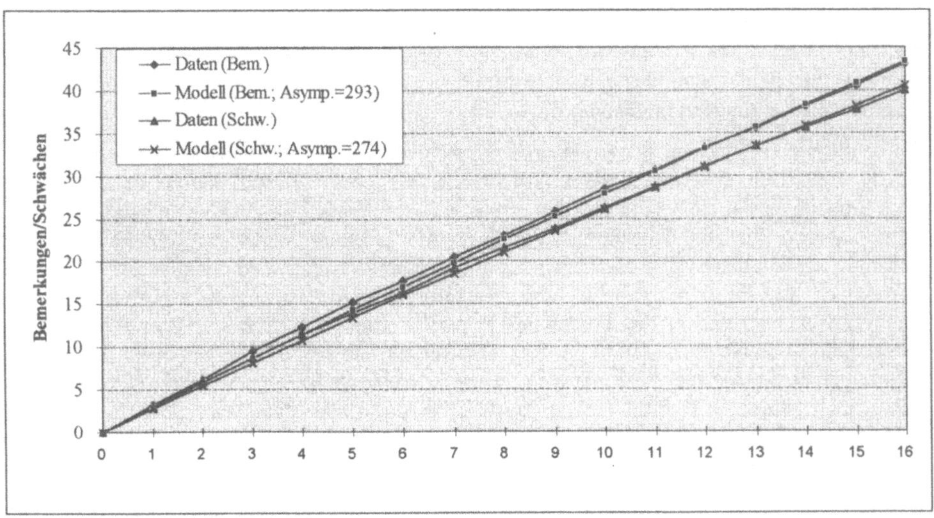

Abb. 3: Vergleich von empirischen und Modell-Erwartungswerten bei der Beschreibung von evozierten Bemerkungen in Abhängigkeit von der Rateranzahl (IsoMetrics / Untersuchung 2)

Für die Bemerkungen erhalten wir bei der IM-Gruppe einen schwach ausgeprägten Lernparameter von $\alpha = 0.01$, der sich in den empirischen Daten als ein nahezu linearer Zuwachs an Bemerkungen in Abhängigkeit von der Rateranzahl darstellt. Die Menge zu erwartender Bemerkungen liegt bei $\lambda = 293$. Der Lernparameter ist bei den „Schwächen" nicht unterschiedlich zu dem Lernparameter bei den Bemerkungen ($\alpha = 0.01$), erwartungsgemäß liegt die Anzahl zu erwartender genannter Schwächen unter der der Bemerkungen, jedoch nur schwach abweichend von diesen ($\lambda = 274$; s. Abb. 3). Abb. 3 zeigt eine recht gute Übereinstimmung zwischen der modellierten Lernkurve und den empirisch ermittelten Daten. Bei N > 69,31 Untersuchungsteilnehmern können bei $\alpha = 0.01$ mit dem Iso-Metrics Verfahren mehr als 50% der feststellbaren Bemerkungen und Schwächen abschöpft werden.

Die Modellanpassung für die HE-Gruppe ist sowohl für die Bemerkungen als auch für die Schwächen schlecht. Nach den Modellparametern findet praktisch kein Lernen statt ($\alpha=0.0002$, Bem.; $\alpha=0.0005$, Schw.) und die Schätzungen der Asymptoten sind infolgedessen extrem hoch ($\lambda=72.500$, Bem.; $\lambda=26.000$, Schw.). Selbst wenn man den Vorhersagen des Modells nicht folgen will, ist es zumindest einleuchtend, daß die Anzahl der benötigten Evaluatoren immens sein dürfte, um eine befriedigende Abdeckung der gesamten Schwächen des Systems zu erhalten.

5 Diskussion

Die dargestellten Untersuchungen haben gezeigt, daß sowohl durch den Einsatz der Heuristischen Evaluation als auch durch das IsoMetrics^L Verfahren ein Spektrum qualitativ unterschiedlicher Bemerkungen evoziert werden. Um die verschiedenen Bemerkungsqualitäten zu berücksichtigen erscheint eine Klassifizierung der Ergebnisdaten als sinnvoll. Die in den präsentierten Untersuchungen vorgenommene Lösung muß noch auf Reliabilität überprüft werden und ihre Praxistauglichkeit unter Beweis stellen.

Bei der Unterscheidung der Bemerkungen nach ihrer Spezifität zeigte sich, daß die expertenbezogene Heuristische Evaluation eine größere Anzahl spezifischer Bemerkungen evozierte, während mit dem nutzerbezogenen IsoMetrics^L Verfahren eine relativ größere Anzahl genereller Bemerkungen erhoben wurde.

Ein deutlicher Unterschied der Ergebnisse aus den beiden Untersuchungen besteht in Bezug auf die empirisch erhobene und zu erwartende Menge an Bemerkungen. In der ersten Untersuchung, in der ein bereits von den gröbsten Systemfehlern bereinigter Prototyp evaluiert wurde, wurde für das IsoMetrics^L-Verfahren ein deutlich stärkerer „Lernzuwachs" in Abhängigkeit von der Evaluatorenanzahl als in der zweiten Untersuchung registriert, in der die „Lernkurve" fast linear mit zunehmender Evaluatorenanzahl anstieg. Dieser Effekt kann mit dem Entwicklungsstadium der Prototypen erklärt werden. Während der Prototyp aus der ersten Untersuchung sich schon in einem ausgereiften Stadium befand und um einige grundsätzliche Schwächen bereinigt war, war der Prototyp aus der zweiten Untersuchung bisher noch keiner Evaluation unterzogen worden und daher noch mit einer Vielzahl von Schwächen versehen, was die Wahrscheinlichkeit vergrößerte, daß jeder neue Evaluator auch neue Schwachpunkte entdeckt.

Während sich in der ersten Untersuchung die Heuristische Evaluation und IsoMetrics^L in Bezug auf die Menge der Anmerkungen nicht wesentlich unterscheiden und sich gegenseitig ergänzen, ist in der zweiten Untersuchung die Heuristische Evaluation ergiebiger in Bezug auf die Menge evozierter Bemerkungen und Schwächen. Dies bestätigt zwar die Aussage Nielsens, daß die HE besonders in frühen Entwicklungsphasen sinnvoll ist, aber in Betracht der eher niedrigen Frequenz (im Vergleich zur Asymptote) von Bemerkungen/Schwächen in Untersuchung 2 ist noch nachzuweisen, ob in einem sehr frühen Entwicklungsstadium diese Ergebnisse wirklich besser sind als solche, die mittels Inspection- oder Walkthrough-Techniken zu erreichen wären.

Letztlich ist zu bemerken, daß in keiner der dargestellten Untersuchungen sich die in [5] präsentierte „Lernkurve" wiedergefunden hat. Insbesondere dürfte es ein eher seltenes Ereignis sein, daß bereits bei fünf Evaluatoren mit einer Annäherung an die Asymptote zu erhebender Bemerkungen/Schwächen zu rechnen ist.

Literaturverzeichnis

[1] Bosch, P.: Repairing Errors in Computer-Assisted Language Learning. Telematics Programme LE1-1615 (1995). [HTML-Document] URL: http://www.infj.ulst.ac.uk/~recall/

[2] Gould, J.D.: How to Design Usable Systems. In: M. Helander (ed.). *Handbook of Human-Computer Interaction*. Amsterdam: Elsevier, 1988.

[3] Gould, J.D. & Lewis, C. : Designing for Usability: Key Principles and What Designers Think. *Communications of the ACM*, 28(1985), 300-311.

[4] Lindgaard, G.: *Usability testing and System Evaluation*. Chapman & Hall, 1994.

[5] Nielsen, J. & Molich, R.: Heuristic evaluation of user interfaces. In: *Proceedings of the CHI '90 Conference on Human Factors in Computing Systems*. ACM Press, 1990, 249-256.

[6] Nielsen, J.: Finding usability problems through heuristic evaluation. In: *Proceedings of the CHI '92 Conference on Human Factors in Computing Systems*. ACM Press, 1992, 373-380.

[7] Nielsen, J.: *Usability Engineering*. Boston: AP Professional, 1993.

[8] Oppermann, R. & Reiterer, H.: Software-ergonomische Evaluation. In: E. Eberleh, H. Oberquelle & R. Oppermann (Hrsg.). *Einführung in die Software-Ergonomie*. Berlin: de Gruyter, 1994.

[9] Smilowitz, E.D., Darnell, M.J. & Benson, A.E.: Are we overlooking some usability testing methods? A comparison of lab, beta, and forum tests. *Behaviour & Information Technology*, 13 (1993), 183-190.

[10] Williges, R.C., Williges, B.H. & Elkerton, J.: Software Interface Design. In: G. Salvendy. *Handbook of Human Factors*. New York: Wiley, 1987.

[11] Willumeit, H., Gediga, G. & Hamborg, K.-C. : *Validation of the IsoMetrics usability inventory.* Universität Osnabrück, Fachbereich 8, Forschungsberichte Nr. 105, 1995, s.a. URL http://www.psycho.uni-osnabrueck.de/ggediga/www/isometr1.html

[12] Willumeit, H., Gediga, G. & Hamborg, K.-C.: IsoMetrics[L] Ein Verfahren zur formativen Evaluation von Software nach ISO 9241/10. *Ergonomie & Informatik*, 27 (1996), 5 - 12.

Adressen der Autoren

PD Dr. rer. nat. Günther Gediga
Universität Osnabrück
Fachbereich Psychologie
Abt. Methodenlehre
49069 Osnabrück
Email: ggediga@luce.psycho.uni-osnabrueck.de

Dr. rer. nat. Kai-Christoph Hamborg
Universität Osnabrück
Fachbereich Psychologie
Abt. Arbeits- und Organisationspsychologie
49069 Osnabrück
Email: hamborg@luce.psycho.uni-osnabrueck.de

Dialogbausteine
Ein Konzept zur Verbesserung der Konformität von Benutzungsschnittstellen mit internationalen Standards

Claus Görner

Gesellschaft für Software
Management mbH

Michael Burmester

Fraunhofer-Institut für
Arbeitswirtschaft und
Organisation

Manfred Kaja

ISA GmbH

Zusammenfassung

Die Entwicklung von Benutzungsschnittstellen wird künftig im wesentlichen von zwei Standards beeinflußt werden: EU-Richtlinie 90/270 [1] und DIN EN ISO 9241 [2]. Für den Umgang mit den mehr als 400 Gestaltungsempfehlungen wird folgender Lösungsvorschlag diskutiert: Für viele der Regeln lassen sich Implementierungsbeispiele erstellen, die von den Entwicklern weiter verwendet werden können. Es ist wesentlich einfacher, solche Einzelbeispiele mit den Empfehlungen der Standards abzugleichen. Weiterhin kann sich der Prüfungsaufwand für eine bestimmte Benutzungsschnittstelle später dadurch reduzieren, daß solche geprüften Beispiele wiederverwendet werden.

Wird eine Benutzungsschnittstelle hauptsächlich aus solchen, auf Ergonomie geprüften Bausteinen aufgebaut, können viele Designfehler vermieden werden und es besteht die berechtigte Hoffnung, daß so eine Benutzungsschnittstelle wesentlich besser den Standards entspricht, als wenn sie vollständig neu entwickelt wird. Über diese Vorgehensweise bleibt einerseits Software-Designern erspart, dicke Dokumente mit Gestaltungsregeln zu lesen, andererseits entsteht der Qualitätssicherung ein wesentlich geringerer Prüfungsaufwand. Nicht zuletzt lassen sich durch diesen Ansatz Iterationszyklen beim Prototyping verkürzen. [1]

1 Einleitung

Die Entwicklung von Benutzungsschnittstellen wird künftig im wesentlichen von den beiden Standards EU-Richtlinie 90/270 [1] und DIN EN ISO 9241 [2] beeinflußt werden.

Im Rahmen des europäischen Einigungsprozesses entstand die Richtlinie 90/270 [1], die seit Januar 1993 in Kraft ist und in allen europäischen Ländern verbindlich gilt. Darin werden Arbeitgeber verpflichtet, nur Hard- und Software einzusetzen, die bestimmte Mindestanforderungen erfüllt. Erstmals werden damit software-ergonomische Forderungen in einem gesetzlichen Rahmen verankert und es ist zu erwarten, daß daraus mittelbar auch für Softwarehersteller Handlungsbedarf gegeben ist [3]. Obwohl in der EU-Richtlinie keine Hinweise auf andere Normen und Standards gegeben werden, tendieren die unterschiedlichen

[1] Die vorgestellte Arbeit entstand im Projektverbund SANUS: Sicherheit und Gesundheitsschutz bei der Arbeit an Bildschirmen auf der Basis internationaler Normen und Standards.(Fördernummer 01HP214/1) und wird gefördert vom Projektträger Arbeit, Technik, Umwelt.

nationalen Umsetzungsaktivitäten eindeutig dahin, die DIN EN ISO 9241 als Konkretisierungshilfe heranzuziehen [4].

DIN EN ISO 9241 [2] befaßt sich mit ergonomischen Fragen zum Einsatz von Bildschirmgeräten und deckt neben hardware-ergonomischen Aspekten vor allem software-ergonomische Fragen ab. Die Teile 10 bis 17 dieses Standards beinhalten mehr als 400 einzelne Empfehlungen zur Informationspräsentation, geeigneten Benutzerführung, Menügestaltung, Interaktion über Kommandos, direkte Manipulation und Formulare bis hin zur Beurteilung von Gebrauchstauglichkeit (usability). Obwohl noch nicht alle Teile endgültig verabschiedet sind, ist dieser Standard eine aktuelle Sammlung ergonomischen Wissens, die in weltweiten Abstimmungsprozeduren breite Akzeptanz gefunden hat [5].

Dennoch scheinen bis heute zwei wesentliche Aspekte ungelöst:

- Wie können die Gestaltungsempfehlungen von Software-Designern effizient bei der Entwicklung einer bestimmten Benutzungsschnittstelle angewendet werden.
- Wie können Qualitätssicherer auf effiziente Weise die korrekte Einhaltung der Gestaltungsempfehlungen überprüfen?

Beide Ziele sind heute nur mit hohem Zeitaufwand zu erreichen, und erfordern umfangreiche Kenntnisse zur Gestaltung von Benutzungsschnittstellen. Software-Designer, die die beiden Standards berücksichtigen möchten, müssen zunächst einige Hürden überwinden: Typische Hürden bei der Umsetzung der EU-Richtlinie:

- Die Richtlinie nennt Gestaltungsziele. Sie gibt keinen konkreten Weg zur Erreichung dieser Ziele vor.
- Die aufgeführten Ziele und Prinzipien zur Mensch-Maschine-Schnittstelle erscheinen als relativ unsystematische Auszüge aus verschiedenen, nicht genannten Quellen.
- Ein allgemeiner Hinweis auf „Grundsätze der Ergonomie" wurde aufgenommen, ohne zu erläutern, wo diese niedergelegt sind.

Typische Hürden bei der Anwendung von DIN EN ISO 9241:

- Mehr als 400 einzelne Regeln müssen gekannt und angewandt werden.
- Alle Regeln sind technologie-neutral formuliert und müssen daher für die jeweilige Entwicklungsumgebung (z.B. grafische, textuelle oder sprachliche Benutzungsschnittstelle) interpretiert werden.
- Jede Regel muß daraufhin untersucht werden, ob sie in dem konkreten Designkontext relevant ist oder ob sie ignoriert werden kann.
- Es ist nicht ausreichend, jede Regel einmal zu prüfen. Viele Regeln müssen an mehreren Stellen der Benutzungsschnittstelle wiederholt berücksichtigt werden (z.B. in jedem Fenster von neuem).

Aufgrund dieser Schwierigkeiten ist es nahezu unmöglich, eine Benutzungsschnittstelle zu entwickeln, die jede relevante Gestaltungsregel umsetzt und so voll den Anforderungen der beiden Standards entspricht. Zumindest wäre es schwierig, einen solchen Nachweis zu führen.

Zusätzlich zu diesen Werken haben Software-Designer Quellen mit Gestaltungsregeln zu berücksichtigen, die sich auf eine konkrete Entwicklungsumgebung beziehen [6], [7], [8].

Da es die Aufgabe solcher hersteller-spezifischen Styleguides ist, Gestaltungshinweise für ein einheitliches Aussehen und Verhalten der einzelnen Elemente zu geben, stellt sich unmittelbar die Frage nach einem Zusammenhang mit den genannten Standards.

- Sind Herstellerstyleguides mit den Forderungen aus EU-Richtlinie und DIN EN ISO 9241 vereinbar?
- Können Styleguideregeln als Untermenge der Standards betrachtet werden oder umgekehrt?

Bedauerlicherweise liegt kein so eindeutiger Zusammenhang vor. Wenn eine Software den Regeln eines bestimmten Styleguides entspricht, braucht das nicht zu bedeuten, daß sie auch die ergonomischen Anforderungen der genannten Standards erfüllt.

Allerdings haben Analysen gezeigt, daß es aufgrund des hohen Detaillierungsgrades von Styleguides möglich ist, viele Software-Muster zu definieren. Solche Muster können wiederum leichter auf Kompatibilität zu den genannten Standards getestet werden. Es ist anzunehmen, daß eine Benutzungsschnittstelle wesentlich standardkonformer ausfällt, wenn sie aus solchen Softwaremustern zusammengesetzt wird, als wenn sie vollständig neu entwickelt wird.

2 Das Konzept ergonomischer Dialogbausteine

Unter Dialogbausteinen sind vordefinierte Interaktionselemente grafischer Benutzungsschnittstellen zu verstehen [9], [10]. Im einfachsten Fall besteht ein Dialogbaustein aus einem einfachen Bedienelement (z.B. ein Radiobutton, Pushbutton, etc.), für das Aussehen und Verhalten bereits festgelegt sind. Einfache Bedienelemente können zu größeren Bedieneinheiten kombiniert werden, wie z.B. Tabellen oder Formulare, die sich aus vielen Eingabeelementen zusammensetzen, Notebooks, die sich u.a. aus Reitern, Karteikarten, Eingabefeldern zusammensetzen. Schließlich können solche komplexeren Bedienelemente zu standardisierten Operationen mit vordefinierten Bedienabläufen zusammengesetzt werden.

Daraus ergibt sich, daß Dialogbausteine

- ein für eine bestimmte Systemplattform vordefiniertes Aussehen und Verhalten besitzen sollen,
- nachweislich über eine ausreichende Gebrauchstauglichkeit verfügen sollen, die an den genannten Standards gemessen ist,
- sinnvolle Gestaltungslösungen für häufig benötigte Arbeitsabläufe und Handlungsziele von Benutzern bereitstellen sollen.

Das folgende Beispiel zeigt, wie Dialogbausteine Software-Designern helfen, Standards und Styleguides korrekt zu interpretieren und effizient anzuwenden.

1. Empfehlung aus DIN EN ISO 9241 (sinngemäß)	Benutzer sollten informiert werden, wenn die Gefahr eines unbeabsichtigten Datenverlusts besteht.
2. Korrespondierende Regel aus einem Styleguide (sinngemäß)	Wenn ein Fenster geschlossen wird, das noch ungesicherte Daten enthält, muß eine Sicherheitsabfrage erfolgen.
3. Beispielhaft lassen sich daraus folgende Implementierungen ableiten:	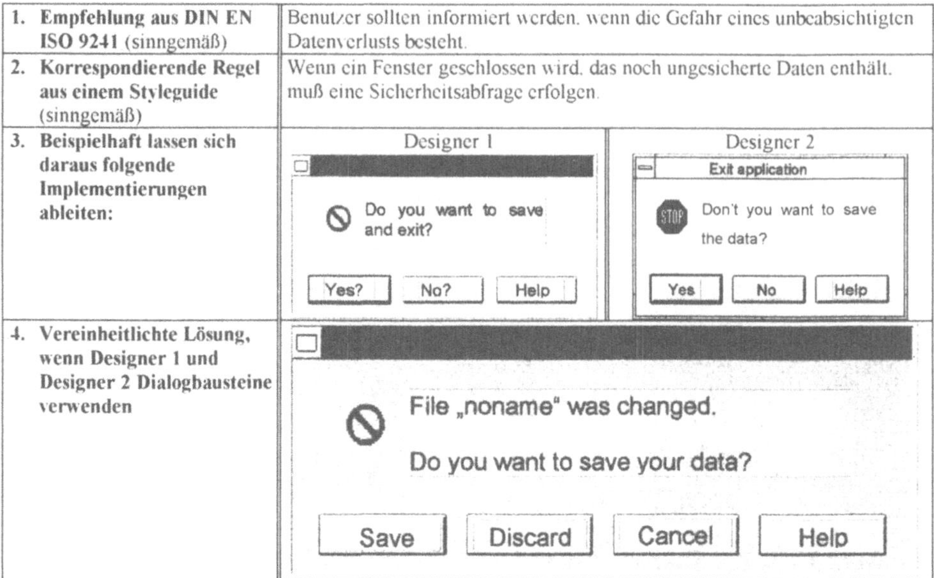
4. Vereinheitlichte Lösung, wenn Designer 1 und Designer 2 Dialogbausteine verwenden	

Abb. 1: Gestaltungsregeln aus Standards und Styleguides sind interpretationsbedürftig. Unterschiedliche und fehlerhafte Designlösungen können vermieden werden, wenn vordefinierte und auf Ergonomie geprüfte Dialogbausteine verwendet werden.

3 Verwendung von Dialogbausteinen

Mit Dialogbausteinen läßt sich nicht nur das Design einer Benutzungsschnittstelle verbessern. Vielmehr kann auch der Prototypingprozeß beschleunigt und der Implementierungsaufwand reduziert werden [11]. Dies soll wiederum an einem kurzen Szenarium verdeutlicht werden.

Szenarium:	Ein Software-Designer möchte in einer Benutzungsschnittstelle die Möglichkeit einbauen, nach Adressen zu suchen.
Schritt 1:	Der Designer benötigt ein Fenster für die Anwendung Auswahlhilfen und Spezifikationen zu verfügbaren Dialogbausteinen unterstützen bei der Wahl des richtigen Bausteins „Hauptfenster"
	Folgende Gestaltungsregeln sind darin bereits implementiert: • Ein Hauptfenster enthält bereits eine Menüleiste; die Menüleiste enthält bereits bestimmte. • vordefinierte Pulldown-Menüs; • vordefinierte Pulldown-Menüs enthalten standardisierte Menüeinträge (z.B. Beenden der Anwendung mit Sicherheitsabfrage)
Schritt 2:	Der Designer benötigt im Hauptfenster eine Tabelle, um alle verfügbaren Adressen anzuzeigen. Aus den angebotenen Bausteinvarianten wählt er eine Tabelle mit editierbaren Feldern, Spalten und Zeilen, die der Benutzer auch sortieren kann.
	Folgende Gestaltungsregeln sind darin bereits implementiert: • Die Tabelle kann sortiert werden, in dem auf die jeweilige Spaltenüberschrift geklickt wird. • Die Tabelle ist mit einem Pulldown-Menü verknüpft, über das standardisierte Bearbeitungsfunktionen (Zeile einfügen, löschen, etc.) verfügbar sind.
Schritt 3:	Der Designer benötigt eine Funktion, mit der ein Benutzer nach einer bestimmten Adresse suchen kann. Aus den angebotenen Bausteinvarianten wählt er eine Suchfunktion, die das erste gefundene Element in einem separaten Fenster anzeigt.
	Folgende Gestaltungsregeln aus einem denkbaren Styleguide sind darin bereits implementiert: • Die Suchfunktion trägt die Bezeichnung „Suchen..." • Es gibt einen entsprechenden Menüeintrag im Pulldown-Menü „Bearbeiten" • Nach Aufruf der Funktion wird ein Dialogfenster eingeblendet, in das Suchkriterien eingegeben werden können. • Die Pushbuttons in diesem Fenster sind vordefiniert. • Wenn die Suche mit dem Pushbutton OK gestartet wird, erscheint in einem vordefinierten Fenster eine Fortschrittsanzeige.

Abb. 2: Iterative Entwicklung einer Benutzungsschnittstelle mit Hilfe vorgefertigter Dialogbausteine am Beispiel einer Anwendung mit Suchfunktion.

Selbstverständlich können für jeden Arbeitsablauf alternative Dialogsequenzen vordefiniert werden. In der Regel gibt es nicht eine beste Lösung, aber mehrere äquivalente Lösungen [12]. Aus diesem Grund sollte das Ziel dahingehen, solche Bausteine anzubieten, die sich nachweislich bewährt haben und als normkonform gelten.

4 Instrumente für den Umgang mit Dialogbausteinen

In der ersten Phase des SANUS Verbundvorhabens wurden bereits mehr als 100 potentielle Dialogbausteine aus verschiedensten Anwendungsbereichen identifiziert (vgl. auch [13], [14], [15]). Dazu gehören z.B. typische Datensichten, Strukturierungshilfen für große Datenmengen, Tabellen mit unterschiedlichstem Verhalten, Fenstertypen. Zahlreiche dieser Bausteine wurden als Prototypen implementiert.

Aufgrund der großen Zahl von Bausteinen wurden Konzepte und Hilfsmittel entwickelt, die den Umgang mit den Bausteinen innerhalb des Software-Entwicklungsprozesses [16] vereinfachen und die Auswahl des jeweils passenden Bausteins erleichtern

- Entscheidungsbaum
  Ein Hilfsmittel, das den Designer über eine Reihe von Fragen auf einen oder mehrere passende Bausteine hinweist. Wesentlich ist dabei, daß die Designentscheidungen an Handlungszielen des Benutzers und seinen Arbeitsaufgaben orientiert werden.
- Bausteinmanager
  Eine Softwareumgebung, in der die Bausteine auf anschauliche und übersichtliche Weise angezeigt und aufgerufen werden können. Hier können Beispiele, neutralisierte und individualisierbare Muster oder Software-Codes eingesehen werden.

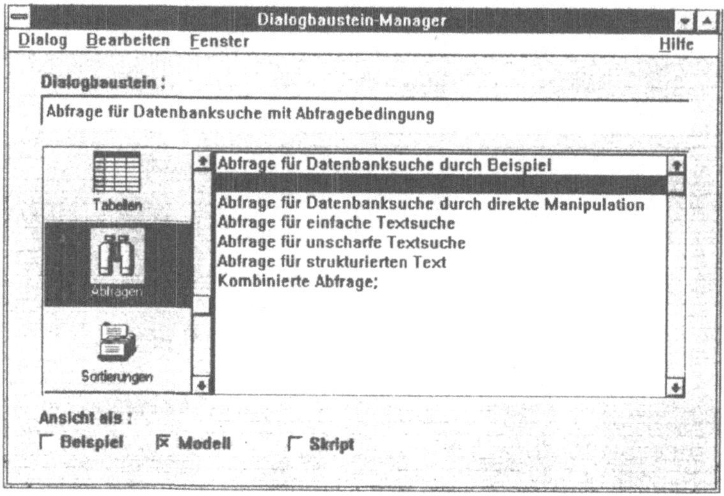

Abb. 3: Software zur Verwaltung der Dialogbausteine.

- Editor zur Parametrisierung eines Bausteins
  Ein Werkzeug, über das ähnliche Bausteine zu Varianten zusammengefaßt werden und situationsspezifisch konkretisiert werden können (z.B. welche Spalten und Zeilen für einen Tabellenbaustein, editierbare Zellen oder nur Zeilenselektion).
- Bausteinspezifikation
  Eine Online-Dokumentation, die Auskunft gibt über Inhalt und Verwendungszweck jeden Bausteins. Über elektronische Links werden die im Baustein berücksichtigten Regeln in einem korrespondierenden Styleguide nachlesbar.

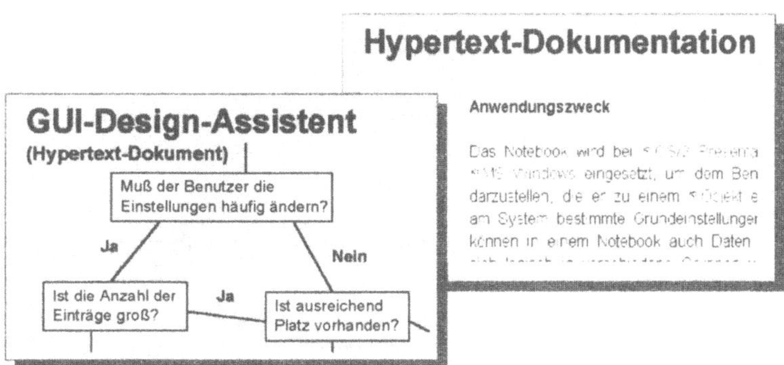

Abb. 4: Der Umgang mit Dialogbausteinen erfordert einerseits Entscheidungs- und Auswahlhilfen, andererseits die Anbindung an einen Styleguide, um das relevante Gestaltungswissen im Einzelfall nachlesen zu können.

5 Absicherung der Konformität der Dialogbausteine zur Europäischen Bildschirmrichtlinie 90/270 und zur ISO 9241

Die Dialogbausteine sollen auf Konformität zur Europäischen Bildschirmrichtlinie 90/270 überprüft werden. Dies ist möglich in dem die Einhaltung der Gestaltungsprinzipien und -regeln der ISO 9241 Teil 10 und 12 bis 17 bei jedem Dialogbaustein überprüft wird (vgl. Abb. 5) [17], [18].

Die Überprüfung jedes Dialogbausteins anhand der über 400 Regeln in der ISO-Norm 9241 wäre extrem aufwendig. Der Aufwand wird dadurch reduziert, daß nicht jede Regel in den Teilen 12 bis 17 auf jeden Dialogbaustein zutrifft.

Eine ökonomische Überprüfung der Dialogbausteine wird mit dem im SANUS-Projekt entwickelten, rechnergestützten Evaluationswerkzeug SHIVA (Structured Human Interface Validation technique) [19] (vgl. auch [20]) vorgenommen.

Das Evaluationswerkzeug SHIVA bietet die Möglichkeit, Eigenschaften des Dialogbausteins (z.B. die Dialogelemente, wie Push Buttons oder Menüs, aus denen ein Dialogbaustein besteht) als Suchbegriffe zum Auffinden der relevanten Gestaltungsregeln zu nutzen. Die Gestaltungsregeln liegen in der SHIVA-Datenbank sowohl als Prüffragen als auch als Originalregeln vor. Die Prüffragen, die während der Evaluation auf einen Dialogbaustein angewendet wurden, werden automatisch protokolliert. SHIVA generiert am Ende der Evaluation einen Prüfbericht, der die auf den zu prüfenden Dialogbaustein angewandten Gestaltungsregeln und deren Grad der Einhaltung sowie Verbesserungsvorschläge des Evaluators enthält. Anhand dieses Berichtes werden die Dialogbausteine in der nachfolgenden Entwicklungsphase optimiert.

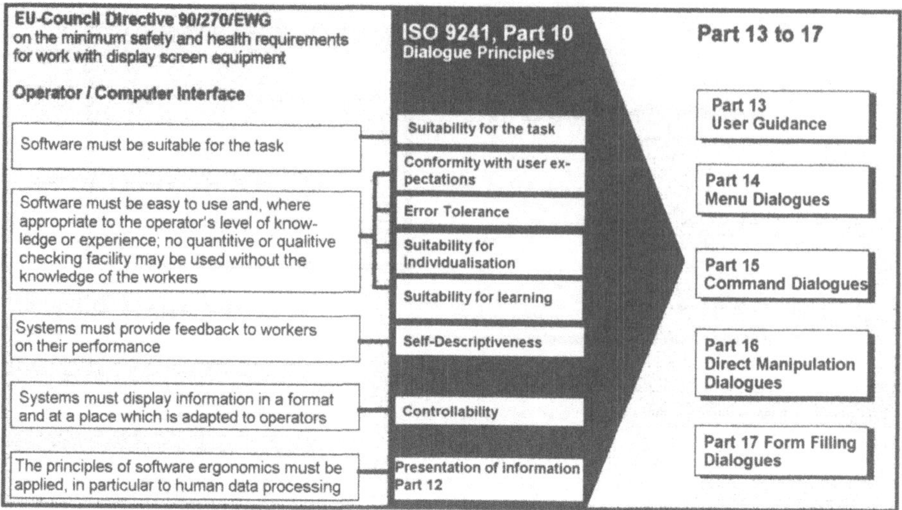

Abb. 5: Ableitung der ISO 9241 aus den Anforderungen an die Mensch-Maschine-Schnittstelle aus dem Anhang 3 der Europäischen Bildschirmrichtlinie 90/270. Der Teil 10 der ISO 9241 ist der erste Schritt der Konkretisierung der allgemeinen Richtlinienanforderungen. Die über 400 Gestaltungsregeln der Teile 12 bis 17 konkretisieren die Dialogprinzipien des Teil 10 für Informationsgestaltung und verschiedene Dialogtechniken.

Die anhand der ISO 9241 überprüften Dialogbausteine sichern bei der Entwicklung von Benutzungsschnittstellen eine ergonomische Grundqualität. Die Evaluation der Benutzungsschnittstelle während ihrer Entwicklung kann somit auf die Prüfung der Aufgabenangemessenheit sowie auf die Berücksichtigung der spezifischen Benutzeranforderungen beschränkt werden.

Literatur

[1] EU-Directive 90/270/EWG: Minimum safety and health requirements for work with display screen equipment. Official Journal of the European Communities No. L 156/14 (1990).

[2] ISO 9241: Ergonomic requirements for office work with visual display terminals (VDTs).

[3] Görner, C., Bullinger, H.-J.: Leitfaden Bildschirmarbeit. Sicherheit und Gesundheitsschutz. Universum Verlagsanstalt: Wiesbaden, 1995.

[4] Görner, C.: Software-Design. Wege zu belastungsfreier Arbeit am Bildschirm. In: Die EU-Richtlinie zur Bildschirmarbeit und ihre Umsetzung in deutsches Recht. EUROFORUM Konferenz, 1995.

[5] Beimel, J., Schindler, R., Wandke, H.: Ergonomic requirements for office work with visual display terminals. Understanding and acceptance of the first Draft International Standard ISO 9241 Part 10: Dialogue Principles. Humboldt University Berlin, Institute of Applied Psychology, 1992.

[6] IBM - Object-Oriented Interface Design: IBM Common User Access Guidelines. Que Corp., 11711 N. College Avenue, Carmel, IN 46032; 1992.

[7] OSF/Motif Styleguide (Open Software Foundation) Revision 2.0. Open Software Foundation, 11 Cambridge Center, Cambridge, MA 02142; 1994.

[8] Microsoft: The Windows Interface Guidelines. A Guide for Designing Software. Microsoft Press, Redmond, Washington 98052 - 6399; February 1995.

[9] Bodart, F., Vanderdonckt, J.: Widget standardisation through abstract interaction objects (pp. 300-305). In A.F. Özok & G. Salvendy (Eds.), Advances in Applied Ergonomics. Proceedings of the 1st International Conference on Applied Ergonomics. Istambul: USA Publishing, 1996.

[10] Vanderdonckt, J., Bodart, F.: Encapsulating Knowledge for intelligent automatic interaction objects selection. In S. Ashlund, K. Mullet, A. Henderson, E. Hollnagel, T. White (Eds.), Procedings of INTERCHI '93, New York: ACM Press, 1993.

[11] Reiterer, H.: The development of design aid tools for human factor based user interface design. In: IEEE (Ed.): International Conference on Systems, Man Cybernetics. Piscataway: IEEE, 1993.

[12] Vanderdockt, J.: A corpus of selection rules for choosing interaction objects. Technical Report 93/3. Institute d'Informatique: Namür, 1993.

[13] Vanderdockt, J.: Guide ergonomiquedes interface homme-machine. Presses Universitaires de Namur: Namür, 1994.

[14] Reiterer, H.: User interface design assistants. CHI '94 Conference. Special Interest Group (SIG) on tools for working with guidelines. Boston 1994. ACM SIGCHI Bulletin Vol. 27, No. 2 April 1995.

[15] Ilg, R., Weisbecker, A.: Basic Dialogue Building Blocks. Styleguide for graphical user interfaces of transaction processing applications. IAO Technical Report, 4/1993.

[16] Beck, A., Janssen, C., Weisbecker, A. & Ziegler, J. (1994). Integrating Object-Oriented and Graphical User Interface Design. Proceeding of the SE/HCI Workshop, Sorrento, Italy, May 16-17.

[17] Fähnrich, K.-P.: Special Session „Standardisation in Human-Computer Interaction and European Directive 90/270 on VDT" - Introduction: "Standardisation in Human-Computer Interaction and European Directive 90/270 on VDT". In A.F. Özok & G. Salvendy (Eds.), Advances in Applied Ergonomics. Proceedings of the 1st International Conference on Applied Ergonomics. Istambul: USA Publishing, 1996.

[18] Burmester, M.: Tools for Conformance Testing and Software Evaluation (pp. 306-311). In A.F. Özok & G. Salvendy (Eds.), Advances in Applied Ergonomics. Proceedings of the 1st International Conference on Applied Ergonomics. Istambul: USA Publishing, 1996.

[19] Ziegler, J., Burmester, M.: Structured Human Interface Validation Technique - SHIVA. In Y. Anai, K. Ogawa & H. Mori (Hrsg.), Symbiosis of Human and Artifact. Proceedings of the 6th International Conference on Human-Computer-Interaction, Tokio, Japan, 9.-14. July 1995, Volume 2 (pp. 899-906). Amsterdam: Elsevier, 1995.

[20] Oppermann, R., Wick, C., Geis, T., Koch, M., Lutz, P., Prümper, J., Reiterer, H., Strapetz, W.-W.: Der ISO 9241-Evaluator. Ergonomie & Informatik, 27, 13-17, 1996.

Adressen der Autoren

Dr. Claus Görner
Gesellschaft für Software
Management mbH
Azenbergstraße 35
70174 Stuttgart
Email: goerner@gsm.de

Dipl.-Psych. Michael Burmester
Fraunhofer-Institut für Arbeitswirtschaft
und Organisation
Nobelstraße 12
70569 Stuttgart
Email:michael.burmester@iao.fhg.de

Manfred Kaja
ISA GmbH
Azenbergstraße 35
70174 Stuttgart
Email: kaja@isa.de

Die Gestaltung der Benutzungsschnittstelle von Prozeßleitsystemen nach der Leitstandsmetapher

Udo Griem und Horst Oberquelle

Werum GmbH, Universität Hamburg

Zusammenfassung

Eine der wesentlichen Aufgaben von Prozeßleitsystemen besteht darin, Betriebsleitpersonal bei der Führung ihrer Anlagen und bei der Steuerung und Überwachung ihrer Produktionsprozesse zu unterstützen. Die Frage, wie Leitsysteme zu entwickeln sind, die diesen Anforderungen genügen, steht im Mittelpunkt dieses Beitrags. Ausgehend von der Betrachtung einer speziellen Anwendungssituation wird ein Oberflächenkonzept vorgestellt, das, an die Begriffswelt des Leitpersonals anknüpfend, unter Verwendung einer „Leitstandsmetapher" entwickelt wurde.

1   Einleitung

Die Anwendungsgebiete von Prozeßleitsystemen sind in den unterschiedlichsten Bereichen zu finden: Pharma-, Nahrungs- und Genußmittel-, Stahl- und chemische Industrie ebenso wie Erdölraffinerien, Automobilfertigung und Verkehrsleittechnik sind beispielhaft als Einsatzbranchen zu nennen. Zusätzlich zu der bereits hieraus resultierenden Anforderungsvielfalt weisen auch die unterschiedlichen Anwendungssituationen jedes einzelnen Bereichs zum Teil erhebliche Divergenzen auf. Verschiedenartige Organisationsstrukturen und Automatisierungsgrade der Produktion sind als herausragende Unterscheidungsfaktoren zu nennen. Hieraus ist zu erklären, daß es sich bei Prozeßleitsystemen in der Regel nicht um Standardsoftware handelt, sondern, daß sie, basierend auf einem Satz von Basisfunktionalitäten, in Form von Anwendungssoftware für die jeweilige Situation „maßgeschneidert" werden (vgl. [1]). Ausgehend von der Betrachtung einer beispielhaften Anwendungssituation werden in diesem Artikel Probleme, die mit einer Gestaltung der Benutzungsoberfläche verbunden sind, aufgezeigt und zu Anforderungen verdichtet. Darauf basierend wird ein Oberflächenkonzept vorgestellt, welches in besonderem Maße auf die Erfordernisse des Echtzeitbetriebs innerhalb von Leitwarten ausgerichtet ist.

Die Grundlage dieses Artikels bildet eine Untersuchung, die in Kooperation zwischen der Firma Werum GmbH und der Universität Hamburg durchgeführt wurde [9]. Als Untersuchungsgrundlage diente das von Werum entwickelte und bei vielen namhaften Produktionsbetrieben im Einsatz befindliche Prozeßleitsystem PAS-PLS.

2   Ausgangssituation und Zielsetzung

Zu Beginn der hier dargestellten Untersuchung stand eine Analyse des Entwicklungsprozesses des Prozeßleitsystems PAS-PLS. Als ein wesentliches Problemfeld dabei wurde die bereits angedeutete, im Bereich der Entwicklung von Prozeßleitsystemen übliche Trennung zwischen

Standard- und Anwendungssoftwareentwicklung erkannt. Eine Übernahme von erfolgreichen Konzepten und Lösungen aus den unterschiedlichen Anwendungssituationen in die Entwicklung der Standardkomponenten ist dabei als wesentliche Aufgabe anzusehen. Die hierfür unumgängliche Bewertung der unterschiedlichen Spezialfälle erfordert aus softwareergonomischer Sicht stets die Einbeziehung der jeweiligen Anwender in deren realen Arbeitssituationen und Organisationsstrukturen (s.[12]). Aus diesem Grund wurde eine konkrete Einsatzsituation des Leitsystems ausgewählt, anhand derer versucht werden sollte, wesentliche, im Umgang mit dem Leitsystem auftretende Probleme zu erkennen sowie Lösungsmöglichkeiten hierfür zu entwerfen. Falls möglich, sollte hieraus eine allgemeinere Oberflächenkonzeption entwickelt werden.

3 Auswahl einer Methode

Aufgrund dieser Zielsetzung galt es zunächst ein Verfahren zur effektiven Erfassung der Probleme einer speziellen Arbeitssituation zu finden. Als Grundlage hierfür wurden die Prinzipien der ethnographischen Methoden der Softwareentwicklung (s. [2]) und der Methode „Contextual Inquiry" (s. [11]) verwendet. Da die zur Verfügung stehende Zeit innerhalb der Anwendungssituation aufgrund von diversen organisatorischen Randbedingungen auf den Zeitraum eines Tages beschränkt bleiben mußte, konnte keine umfassende ethnographische Untersuchung durchgeführt werden. Daher wurde entschieden, ein Interview im Arbeitskontext durchzuführen, bei dem lediglich die allgemeinen Leitlinien der erwähnten Methoden Anwendung finden sollten.

4 Ein Interview im Arbeitskontext

Als Anwendungssituation wurde die Leitwarte eines Betriebes der Aluminium produzierenden Industrie ausgewählt. Die Leitwarte dient zur Überwachung der Produktion innerhalb einer etwa 200 m langen und 80 m breiten Werkshalle. Hier befinden sich 8 Schmelzöfen mit einem Fassungsvermögen von jeweils etwa 70 Tonnen. In neben den Öfen positionierten Gießanlagen werden die ca. 700 Grad heißen Aluminiumlegierungen zu festen Blöcken unterschiedlicher Länge gegossen. Das Aufgabenfeld der Anlagenfahrer setzt sich aus einer Vielzahl von Einzelaufgaben von unterschiedlichem Anforderungsniveau zusammen. Dazu gehört die Beobachtung der Prozeßverläufe, das Einstellen von Sollwerten, die Koordination von Arbeiten an den Öfen (befüllen, entleeren, warten usw.), Personaleinsatzplanung und -überwachung sowie die Behebung von Störungen und eine Reaktion auf diverse unvorhersehbare Ereignisse.

Zwei unterschiedliche Gruppen von Anlagenfahrern wurden über den Zeitraum von insgesamt etwa 14 Stunden bei ihrer Arbeit begleitet und befragt. Hauptinterviewpartner jeder Gruppe war ein Vorarbeiter. Andere Personen wurden nur dann berücksichtigt, wenn sie sich im gleichen Raum aufhielten.

Nach einer kurzen Vorstellung der Beteiligten und einer Darstellung der allgemeinen Ziele der Befragung folgte eine etwa einstündige, in Verlauf und Charakter von den Befragten bestimmte Phase. Die Vorarbeiter erläuterten darin die Grundlagen des Produktionsprozesses und ihre spezielle Arbeitsweise.

Für den weiteren Ablauf war ein detaillierter Fragebogen als Arbeitsgrundlage und zur Festlegung der Untersuchungsschwerpunkte vorbereitet worden. Die Fragen waren aufgeteilt in allgemeine Fragen zur Anwendungssituation, Fragen zur Benutzung des Leitsystems und solche zu dessen Bewertung (vgl. [9]). Sie dienten als Anhaltspunkte, um das Interview aus der jeweiligen Situation heraus gestalten zu können. D.h., es wurden nur jene Fragen gestellt, die in der jeweiligen Situation als angemessen bewertet wurden. Im wesentlichen orientierte sich der Verlauf des gesamten Interviews an den Handlungen der Beobachteten.

Während der gesamten Befragung wurden Feldnotizen angefertigt. Hier wurden räumliche Gegebenheiten, wichtige Ereignisse und Antworten für eine nachträgliche Auswertung protokolliert.

Es fand eine engagierte Zusammenarbeit zwischen Beobachter und Anlagenfahrern statt, wobei ein großes Interesse der Anlagenfahrer deutlich wurde, ihre Probleme im Umgang mit dem Leitsystem darzustellen. Motivationsprobleme seitens der Anlagenfahrer traten nicht auf, obwohl allen Beteiligten deutlich war, daß die Befragung keine unmittelbare Verbesserung der beobachteten Arbeitssituation mit sich bringen würde.

Bei der Einschätzung des vorhandenen Systems unterschieden sämtliche Anlagenfahrer zwischen zwei Teilkomponenten, die sehr unterschiedlich bewertet wurden. Die erste Teilkomponente enthielt im wesentlichen die zentrale Anzeige von Prozeßwerten, die vor der Einführung des Leitsystem ausschließlich dezentral an den einzelnen Schmelzöfen abzulesen waren. Sie wurde durchweg positiv bewertet und im Arbeitsgeschehen häufig verwendet.

Die Bewertung der zweiten Teilkomponente fiel erheblich kritischer aus. Sie sollte dazu dienen, die Zusammensetzungen der unterschiedlichen Ofeninhalte zu protokollieren, darzustellen und für Qualitätssicherungsnachweise verfügbar zu machen. Dafür wurde ein PC bereitgestellt, auf dem diverse manuelle Eingaben vorzunehmen waren. Dieser Systemteil wurde allerdings nicht verwendet und innerhalb der ersten Beobachtungsphasen von seiten der Anlagenfahrer auch nicht erwähnt. Erst auf Nachfrage des Beobachters wurde die Funktionalität erläutert, die sich hinter der an Bildschirmmasken orientierten Benutzungsoberfläche verbarg. Die eigentlich am Computer einzugebenden Daten wurden, wie vor der Systemeinführung, auf acht nebeneinander liegenden DIN A3-Zetteln protokolliert und die notwendigen einfachen Berechnungen mit Hilfe eines Taschenrechners durchgeführt. Als Begründung hierfür wurden unter anderem folgende Mängel des Computersystems genannt:

- umständliches Wechseln zwischen den unterschiedlichen Bildschirmdialogen,

- undurchsichtige Gliederung der Benutzungsoberfläche,

- mangelnde Korrekturmöglichkeiten falscher Eingaben und

- eine mangelnde Übersicht über die aktuelle Gesamtsituation.

Zusammenfassend ist festzustellen, daß das kurze Interview im Arbeitskontext eine Reihe von Problemen der Systemnutzung zutage treten ließ. Der Umgang der Anlagenfahrer mit dem System und dessen Bewertung wurde trotz der recht kurzen Zeitspanne deutlich, so daß Relevanz und Anwendbarkeit partizipativer Vorgehensweisen im Entwicklungsprozeß des Leitsystems PAS-PLS gezeigt werden konnten. Darüber hinaus warf die Beobachtung einige Fragen allgemeiner Art auf. Wodurch sind Akzeptanz und Ablehnung verursacht? Warum erschien den Benutzern ein Teil der Bildschirmdialoge so undurchsichtig, ein anderer jedoch klar und einleuchtend?

Versucht man diese Fragen zu beantworten, so ist zu bemerken, daß die Prozeßwerte der nicht verwendeten Systemkomponenten in einer Weise angezeigt wurden, die nur in geringem Bezug zum Erfahrungsalltag der Anlagenfahrer standen. Die vor dem Systemeinsatz verwendete Strukturierung wurde vollkommen aufgelöst und die Anzeigefläche auf einen Bruchteil verringert, indem die vorher auf 8 DIN A3-Zetteln notierten Werte auf einem 17 Zoll Bildschirm angezeigt wurden. Offensichtlich war es den Anlagenfahrern nicht möglich, ein inneres, sogenanntes mentales Modell der Struktur dieses Systemteils zu bilden. Die Struktur blieb unverstanden und das System unverwendet. Allgemein wird die Problematik der mentalen Modelle in [7] erläutert.

Eine Methode, die Bildung eines mentalen Modells zu unterstützen, besteht in der Bereitstellung einer Metapher, die es ermöglicht, in der täglichen Erfahrung erworbenes Wissen auf die Verwendung eines Computersystems zu übertragen (vgl. [3]). Als eine für den Bereich der Prozeßleitung naheliegende Metapher wurde die „Leitstandsmetapher" im weiteren Verlauf der hier dargestellten Arbeit verwendet, um ein Konzept zur Strukturierung der Benutzungsoberflächen von Prozeßleitsystemen auszuarbeiten. Dabei wurden insbesondere Anforderungen, die im Echtzeitbetrieb von Prozeßleitsystemen gestellt werden, berücksichtigt.

4 Anforderungen an die Benutzungsschnittstellen von Prozeßleitsystemen

Abb. 1 dient dazu, die speziellen Probleme, die bei der Verwendung von Leitsystemen im Echtzeitbetrieb auftreten, zu verdeutlichen. Dem Modell liegt die Grundstruktur heutiger Prozeßleitsysteme zugrunde (vgl. [8]).

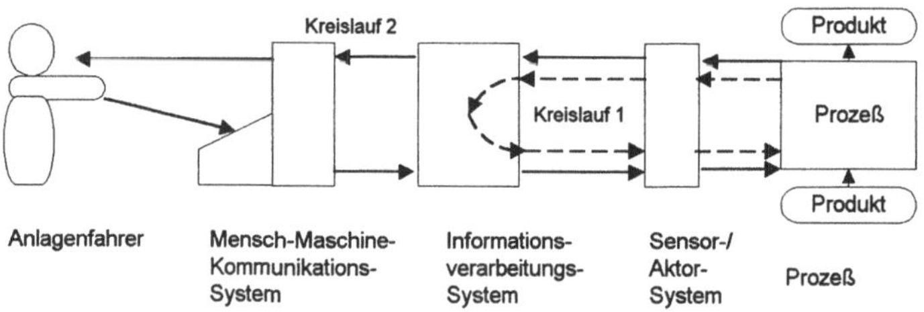

Abb. 1: Modell der Prozeßleitung

Den Ausgangspunkt bildet der Produktionsprozeß. Er ist durch sich dynamisch ändernde physikalische Größen wie Temperaturen, Längen, Gewichte und ähnliches gekennzeichnet. Diese werden unter Verwendung von Sensoren in elektrische Signale umgewandelt, die an die Verarbeitungskomponente des Leitsystems gesendet werden. Bei automatisierten Anlagen werden die Eingangssignale direkt in Ausgangssignale umgewandelt, welche ohne Eingriff des Bedienpersonals wieder auf den Prozeß zurückwirken (Abb. 1, Kreislauf 1).

Alle relevanten Prozeßzustandsänderungen werden an die Mensch-Maschine-Schnittstelle weitergereicht und dort visualisiert bzw. akustisch dargestellt. Die Anlagenfahrer müssen den Prozeßzustand wahrnehmen, bewerten und entscheiden, was zu tun ist sowie angemessen auf die jeweilige Situation reagieren (vgl. [10]). Die im allgemeinen erforderliche Bedienung von Eingabegeräten führt zur Ansprache von Stellgeräten (Aktoren), die wiederum den Prozeßverlauf beeinflussen (Abb. 1, Kreislauf 2).

Die hier beschriebenen Kreisläufe müssen jeweils beendet sein, bevor sich der Prozeßzustand derart signifikant ändert, daß der angestoßene Prozeßeingriff unangemessen wird. Diese Rechtzeitigkeit von Verarbeitungs- und Entscheidungsergebnissen wird als Echtzeitfähigkeit eines Systems bezeichnet und muß als das bedeutendste Unterscheidungskriterium zu Nicht-Echtzeit-Systemen angesehen werden. In diesem Zusammenhang sei auch auf die wichtige technische Frage der Ausfallsicherheit hingewiesen, die allerdings an dieser Stelle nicht weiter diskutiert werden kann.

Aus der Echtzeitfähigkeit des aus Leitpersonal und Prozeßleitsystem bestehenden Mensch-Maschine-Gesamtsystems ergibt sich direkt die Relevanz einer einfachen *Verständlichkeit* oder Selbstbeschreibungsfähigkeit im Bereich der Leitsysteme. Sie ist hier von größerer Bedeutung als bei „normalen" EDV-Systemen, da in kritischen Situationen der Prozeßleitung in der Regel keine Zeit zum Aufruf von „Help-Funktionen" oder zum Nachschlagen in Handbüchern besteht.

Eine weitere Anforderung an Leitsysteme besteht darin, daß *sowohl eine allgemeine Prozeß-übersicht als auch detaillierte Informationen* für die Reaktion der Anlagenfahrer auf diverse Ereignisse angezeigt werden müssen. So erfordert die Leitung eines Prozesses in unterschiedlichen Betriebsbereichen (Normalbetrieb, Anfahrbetrieb, Störbetrieb etc.) auch unterschiedliche Informationsmengen und Abstraktionsstufen. Im Störbetrieb sind ausschließlich wichtige Informationen prägnant darzustellen. Die Anzeige von vielen Einzelheiten ist dort im allgemeinen unangemessen, da Menschen dazu neigen, unter Streß die wesentlichen Ziele bzw. global kritische Zustände aus den Augen zu verlieren und sich stattdessen mit (unbewußt vorgeschobenen) Detailfragen zu beschäftigen [14]. Das Vorhandensein von Darstellungen unterschiedlicher Abstraktionsstufen ist auch für den Normalbetrieb relevant. Besonders in hochautomatisierten Anlagen können recht lange Zeitspannen auftreten, in denen kein manueller Eingriff in den Prozeß nötig ist. Um einem Nachlassen von Aufmerksamkeit und Konzentration entgegenzuwirken, ist es sinnvoll, für diesen Fall Darstellungen bereitzustellen, die ein Mitverfolgen des Prozeßgeschehens auf niedriger Abstraktionsstufe ermöglichen (vgl. [14]). Ein einfacher Wechsel zwischen Übersichts- und Detaildarstellungen ist daher als wichtige Forderung an Leitsysteme zu stellen.

Während Bürosysteme im allgemeinen von einzelnen Nutzern verwendet werden und somit das Konzept des „Personal Computers" sinnvoll eingesetzt werden kann, werden Leitsysteme in der Regel durch *mehrere Nutzer* in direkter Folge oder gleichzeitig verwendet. Diese gleichzeitige Verwendung bezieht sich nicht nur auf das Leitsystem als verteiltes System in einem lokalen Netzwerk, sondern betrifft häufig einzelne Bedienstationen. Spätestens beim Schichtwechsel wird das Bedienpersonal vollkommen ausgetauscht. Die ablösenden Anlagenfahrer müssen den Prozeßverlauf in kurzer Zeit nachvollziehen können, um den Erfordernissen des Prozesses entsprechend eingreifen zu können.

5 Die Leitstandsmetapher als Lösungsansatz

Moderne Benutzungsoberflächen sind geprägt durch die Verwendung von Oberflächenelementen, die auf Grundlage der Desktop-Metapher entworfen wurden. Frei verschiebbare, sich überlappende Fenster und aus dem Bürobereich entlehnte Icons (Aktenordner, Reißwölfe u.a.) sind Kennzeichen hierfür. Betrachtet man die oben dargestellten Anforderungen an Benutzungsoberflächen von Prozeßleitsystemen, so ist die Zweckmäßigkeit einer Übernahme der

Konzepte in den Bereich der Leitwarten in Frage zu stellen. Insbesondere stehen die Anforderungen des Mehrbenutzerbetriebs und die einer schnellen Auffindbarkeit von Informationen einer Verwendung der Fenstertechnik entgegen.

Mit der Leitstandsmetapher wird im folgenden ein Konzept vorgestellt, das als Alternative zu Oberflächen, die auf Basis der Desktop-Metapher realisiert sind, zu verstehen ist. Das Entwurfskonzept enthält eine Einschränkung der freien Verschiebbarkeit von Fensterbereichen und berücksichtigt in besonderem Maße die Anforderungen an die Benutzungsschnittstelle von Leitsystemen im Echtzeitbetrieb. Die dabei verwendete Metapher ist in Analogie zu realen Leitwarten entworfen, deren Komponenten zunächst dargestellt werden sollen.

5.1 Leitstände, Leitwarten

Die Zusammenfassung von Steuerungs- und Anzeigekomponenten in Instrumententafeln, Informationsborden und Pulten wird als Leitstand bezeichnet. Leitstände sind die wesentlichen Komponenten von Leitwarten, den zur Prozeßführung dienenden Räumen. Die wichtigsten Arbeitsmittel in Leitwarten werden im folgenden kurz vorgestellt (vgl. [4], [6]):

Instrumententafeln sind zumeist großflächig an den Wänden der Leitstände zu finden. Hinter der Frontplatte befindet sich ein Hohlraum, der zur Aufnahme von Anzeigern, Schreibern, Meldefeldern und Schaltelementen dient, deren Sichtfenster und Bedienelemente auf der Tafel sichtbar sind. Häufig ist ein Fließbild des Prozesses auf der Tafel angebracht, wobei Anzeige- und Bedienkomponenten in die Darstellung integriert sein können.

In *Informationsborden* werden Anzeiger, Schreiber und Bildschirme in einem ca. 0.5 m hohen Gehäuse so niedrig über dem Fußboden installiert, daß die Anzeigen von einem vorgelagerten Pult aus beobachtet werden können, ohne daß der Blick auf die Instrumententafel verstellt wird. Es werden vorwiegend Anzeigen installiert, die lediglich durch qualitatives Sehen zu beobachten sind, d.h. bei denen keine exakten Werte abgelesen werden müssen.

Das *Bedienpult* dient zum Betätigen von Anzeigeeinrichtungen und Stellteilen, mit deren Hilfe auf den Prozeß eingewirkt werden kann. Die Bauformen können erheblich variieren.

Neben den drei Komponenten Pult, Informationsbord und Tafel sind in den meisten Leitständen *Alarm- und Warnmelder* angebracht. Neben Lautsprechern zur akustischen Alarmmeldung werden hierfür Lampen verwendet, die in der Regel an einem gut einsehbaren Ort (z.B. der Decke der Leitwarte) montiert sind.

Erhebliche Unterschiede bei der Gestaltung von Leitwarten sind in deren *Gliederung* zu finden. Während in kleinen Produktionsanlagen sämtliche Funktionen von einem Leitstand zu bedienen sind, werden Leitstände großtechnischer Anlagen (z.B. Kraftwerke) im Sinne einer besseren Übersichtlichkeit gegliedert. Bei der Gliederung nach den Aufgaben des Personals erhält man einen Hauptleitstand, von dem aus die Prozeßführung erfolgt und Nebenleitstände, die nur bei der Inbetriebnahme, im Wartungsfall, bei bestimmten Störungen oder anderen speziellen Situationen besetzt sind.

5.2 Die Benutzungsoberfläche

Entsprechend den Hauptkomponenten von Leitwarten werden auf der Benutzungsoberfläche Bereiche definiert. So steht ein Bedienbereich in Analogie zum Bedienpult, ein Anzeigebereich zum Informationsbord und ein Tafelbereich zur Wandtafel realer Leitstände. Die Anordnung dieser Bereiche auf dem Bildschirm entspricht im Prinzip der Sichtweise der Anlagenfahrer auf die realen Leitstandskomponenten (Abb. 2):

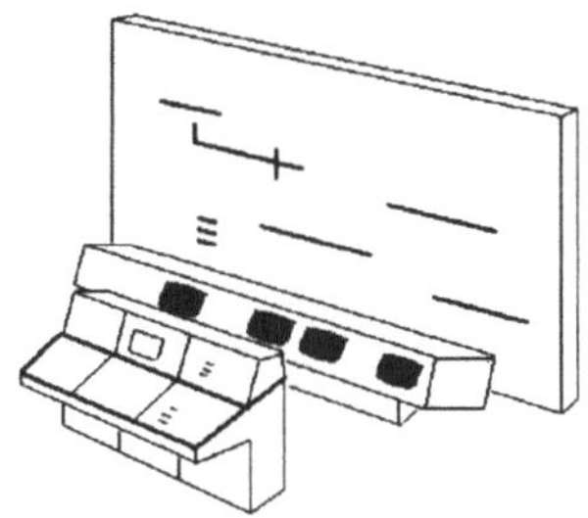 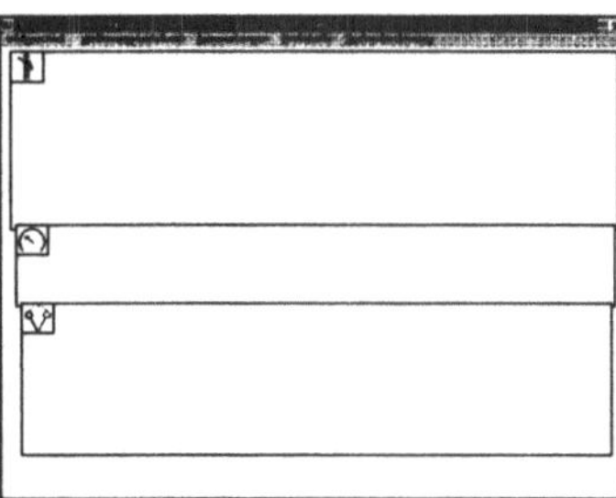

Abb. 2: Gegenüberstellung der Komponenten einer Leitwarte
(links, von unten nach oben: Bedienpult, Informationsboard und Instrumententafel)
mit den Bereichen des Oberflächenkonzeptes der Leitstandsmetapher
(rechts, von unten nach oben: Bedienbereich, Anzeigebereich und Tafelbereich)

Die möglichen Funktionen, die innerhalb der einzelnen Bereiche ausgelöst werden können, stimmen im wesentlichen mit den Funktionen der Leitstandskomponenten überein:

 Der *Bedienbereich* ist in Analogie zu den Bedienpulten realer Leitstände zu verstehen. Er soll dazu dienen, Funktionen auszulösen, die direkt auf den Prozeß einwirken. Hierzu können standardisierte Oberflächenelemente wie Buttons oder Auswahllisten verwendet werden. Das Icon des Bedienbereichs stellt ein simplifiziertes Bild eines Schalthebels dar, wobei der Vorgang des „Umlegens" durch die Verwendung eines grauen und eines schwarzen Teilsymbols visualisiert wird.

 Der *Anzeigebereich* dient entsprechend den Informationsborden ausschließlich zur qualitativen Anzeige von Informationen. Hier sollten vorwiegend Informationen angezeigt werden, die während der Prozeßleitung ständig sichtbar sein müssen. Verläufe aktueller Prozeßwerte sind beispielhaft zu nennen. Die Inhalte der Darstellungen des Anzeigebereichs lassen sich unter Verwendung von Oberflächenelementen des Bedienbereichs manipulieren (Skalieren, Daten anwählen usw.). Im Anzeigebereich selbst können keine Eingaben durchgeführt werden. Das zugehörige Icon zeigt ein konventionelles Anzeigeinstrument.

 Der *Tafelbereich* hat eine komplexe Funktionalität, die in Analogie zum Umgang mit Wandtafeln realer Leitstände zu verstehen ist. Vorwiegend dient er für die Aufnahme von Fließ- bzw. Anlagenbildern. Hiermit soll eine allgemeine Übersicht des Prozeßzustands gegeben werden. Der Tafelbereich ist durch ein Icon gekennzeichnet, das eine Person in Aktion zeigt. Diese Person steht stellvertretend für Anlagenfahrer, die an einer Wandtafel entlanggehen und die dort angebrachten Geräte bedienen.

Der Tafelbereich kann vertikal in beliebig viele Ausschnittsbereiche unterteilt sein. In Äquivalenz zum „Hingehen" zur Wandtafel kann jeder Ausschnitt des Tafelbereichs angewählt werden (z.B. Anklicken mit einer Maus). Der Ausschnitt überdeckt nach seiner Anwahl den Bildschirm komplett. Bedien- und Anzeigebereich werden dann nicht angezeigt. Stattdessen erscheint ein

Anwahlbutton am linken unteren Bildrand, der die Möglichkeit gibt, zur vorhergehenden Darstellung zurückzukehren (s. Abb. 3, rechts).

Wurde der Tafelbereich auf die dargestellte Weise angewählt, so lassen sich Bedienfunktionen durchführen, die direkt auf den Prozeß wirken können. Darüber hinaus ist in Äquivalenz zum „Entlanggehen am Wandtafelbild" ein Wechsel auf eine angrenzende Ausschnittsdarstellung möglich. Dieser Wechsel wird durchgeführt, wenn ein Randbereich des Ausschnittsbildes oder das Icon des Tafelbereichs angewählt wird.

Innerhalb des vorgeschlagenen Konzepts ist es möglich, unterschiedliche *Übersichtsdarstellungen* (Haupt- und Nebenleitbilder) festzulegen. Hierdurch können verschiedene Sichtweisen auf das Prozeßgeschehen visualisiert, unterschiedliche Anlagenteile betrachtet oder Darstellungen für spezielle Aufgaben integriert werden. Spezielle Störleitbilder sollten beispielsweise nur verdichtete Prozeßinformationen enthalten und mit Assistenz- und Erkärungskomponenten ausgestattet sein, unter deren Verwendung Hinweise auf mögliche Maßnahmen zur Fehlerkorrektur angezeigt werden können.

Die Übersichtsdarstellungen können unterschiedliche Anordnungen der Bedien-, Anzeige- und Tafelbereiche enthalten. Sie sind in Analogie zur Gliederung von Leitständen zu verstehen, wonach Leitwarten größerer Anlagen in Hauptleitstand und Nebenleitstände unterteilt sein können.

Die verschiedenen Übersichtsbilder können mit Hilfe einer am unteren Rand des Bildschirms angeordneten *Auswahlzeile* angewählt werden. Sie enthält Knöpfe, die als 1 aus n Auswahl realisiert und jeweils genau einer Übersichtsdarstellung zugeordnet sind. Hierdurch ist ein schneller Wechsel zwischen den unterschiedlichen Darstellungen möglich (siehe Abb. 3, links).

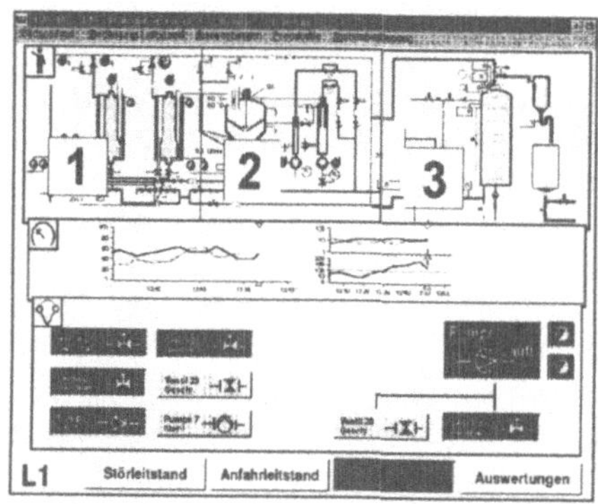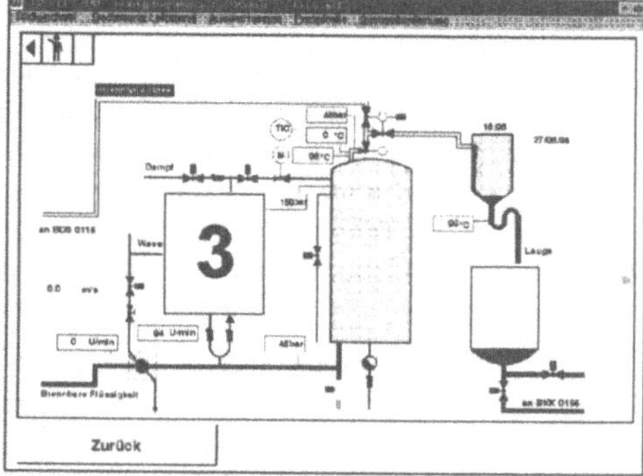

Abb. 3: Beispiel zur Verdeutlichung der Funktionalität des Tafelbereichs
Wird Bereich 3 (linkes Bild, oben rechts) angewählt, dann überdeckt dieser Ausschnitt die Bildschirmoberfläche (rechtes Bild). Danach besteht die Möglichkeit der Bedienung von Interaktionselementen zur Änderung von Prozeßparametern, des direkten Wechsels in den angrenzenden Ausschnitt 2 (Anklicken des Pfeilicons) oder des Rücksprungs zur Übersichtsdarstellung (Zurück-Button).

Ein *Alarmbereich* wird am oberen Rand des Bildschirms in Analogie zu Signallampen realer Leitstände dargestellt. In den Bildbereich hineinragende Alarmsymbole werden automatisch sichtbar, wenn mindestens ein aktueller Alarm ansteht (s. Abb 4). Dabei ist gleichgültig, ob ein Ausschnittsbild oder eine Übersichtsdarstellung angezeigt wird. Bei der Gestaltung der Übersichtsbilder ist darauf zu achten, daß die Alarmsymbole keine wichtigen Anzeige oder Bedienelemente überdecken. Der Alarm mit der höchsten Priorität (z.B. Sammelstörung) wird zusätzlich textuell angezeigt. Die Anzahl der anstehenden Alarme steht als weitere Information im Alarmbereich zur Verfügung. Jedes Alarmsysmbol ist einer Alarmgruppe zugeordnet, so daß aus der Position des angezeigten Symbols auf Herkunft und Bedeutung des Alarms geschlossen werden kann. Das Anklicken eines Alarmsymbols bewirkt die Anzeige eines der entsprechenden Alarmgruppe zugeordneten Übersichtsbildes. Dort werden Hinweise, wie mögliche Ursachen und Folgen der anstehenden Alarme, angezeigt.

Abb. 4 zeigt beispielhaft das Hauptleitbild des auf der Grundlage der Leitstandsmetapher realisierten Prototypen. Erkennbar sind der Bedienbereich mit Schaltfeldern zur Betätigung von häufig verwendeten Ventilen, der Anzeigebereich mit Kurvenverlaufsdarstellungen, der zur Übersicht dienende Tafelbereich und der Alarmbereich mit Alarmen aus zwei unterschiedlichen Alarmgruppen.

5.3 Erste Bewertung

Der vorgestellte Ansatz der Leitstandsmetapher wurde in erster Linie im Hinblick auf die in Abschnitt 4 dargestellten Gestaltungsprobleme von Prozeßleitsystemen entwickelt. Auf der Grundlage dieser Metapher realisierte Benutzungsoberflächen scheinen insbesondere geeignet zur Verwendung durch mehrere Nutzer und zur Unterstützung eines schnellen Wechsels zwischen Darstellungen unterschiedlichen Abstraktionsniveaus. Zudem scheint ein hoher Grad an Verständlichkeit gewährleistet.

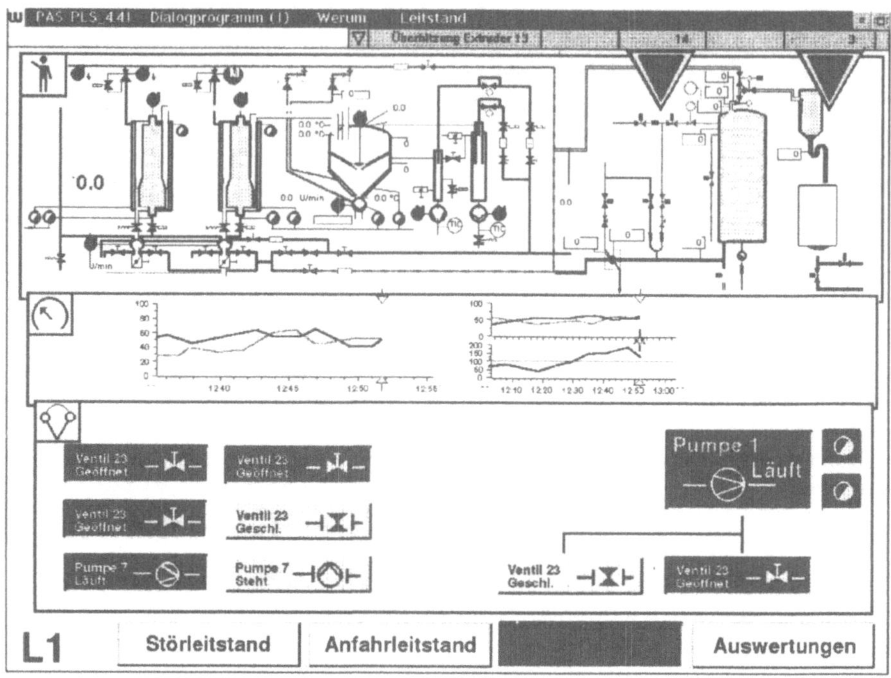

Abb. 4: Darstellung der Bereiche nach dem Oberflächenkonzept der Leitstandsmetapher:
(von oben nach unten:) Alarmbereich, Tafelbereich, Anzeigebereich, Bedienbereich, Auswahlzeile.

Insgesamt stellt die Leitstandsmetapher einen Gestaltungsrahmen für den Entwurf von Prozeß-leitsystemen sowie zur Systemeinführung und Schulung dar. Die wesentliche Aufgabe der Systementwickler besteht auch unter ihrer Verwendung darin, diesen Rahmen im Kontext der jeweiligen Anwendung neu auszufüllen. Dabei ist eine intensive Zusammenarbeit zwischen Systementwicklern und Anwendern, wie sie beispielsweise in [13] dargestellt wird, erforderlich. Partizipative Methoden der Systementwicklung sind als wichtige Hilfsmittel hierfür anzusehen. Überblicksartige Darstellungen dieser Arbeitsweisen sind in [5] und [15] zu finden.

Die Verwendung der Leitstandsmetapher sollte nicht bedeuten, daß die Anzeigen herkömmlicher Leitwarten lediglich in verkleinerter Form auf der Bildschirmoberfläche dargestellt werden oder prinzipiell ausschließlich eine Abbildung der realen Anlagenkomponenten stattfindet. Die einzelnen Übersichtsbilder sollten hingegen auch verwendet werden, um logisch zusammenhängende, aber räumlich in der Anlage verteilte Prozeßwerte zusammenzufassen und zu strukturieren. Der Gestaltungsspielraum unter Verwendung der Leitstandsmetapher erstreckt sich zudem auf neuere Konzepte der Visualisierung von Prozeßdaten, wie sie beispielsweise in [16] vorgestellt werden. Sinnvoll scheinen unter anderem die Integration von Video-Konferenz-Systemen oder Darstellungen wie Polardiagrammen, die Prozeßwerte graphisch, unter Berücksichtigung von Erkenntnissen der Mustererkennung und Gestaltpsychologie visualisieren.

6 Zusammenfassung

Die während eines Interviews im Arbeitskontext festgestellten Probleme von Anlagenfahrern beim Umgang mit einem konkreten Leitsystem bildeten die Grundlage für den Entwurf des hier dargestellten Oberflächenkonzeptes. Diese speziellen Probleme wurden bewertet, zusammengefaßt und zu allgemeinen Anforderungen verdichtet. Unter Verwendung einer Metapher aus dem Bereich der Prozeßleitung wurde ein Ansatz zur Gliederung von Benutzungsoberflächen entworfen, der in besonderer Weise auf diese Anforderungen ausgerichtet ist. Bei der Entwicklung eines Prototypen wurden die allgemeinen Konzepte konkretisiert und deren Realisierbarkeit nachgewiesen. Eine umfassende Bewertung des Ansatzes im Umfeld einer realen Anwendungssituation konnte allerdings bislang nicht geleistet werden, wäre aber als Ausgangspunkt weiterer Untersuchungen denkbar.

Literaturverzeichnis

[1]: H.-P. Bartenschlager et al.: Anforderungen an ein Leitsystem der Fabrik von morgen. Arbeitspapier der GI, Arbeitskreis „Fertigungsleitsysteme". 1994.

[2]: J. Blomberg et al.: Ethnographic Field Methods and Their Relation to Design. In: D. Schuler, A. Namioka: Participatory Design, Lawrence Erlbaum, New Jersey. 1993. 123-155.

[3]: J.M. Caroll, J.C. Thomas: Metaphor and the cognitive representation of computing systems. IEEE Transaction on Systems, Man, and Cybernetics. 12, 1982. S. 107-116.

[4]: H.J. Charwat: Lexikon der Mensch-Maschine-Kommunikation. Oldenbourg, München, 1992.

[5]: Communications of the ACM. Participatory Design, Vol 36, No. 4, June 1993.

[6]: DIN 33414: Ergonomische Gestaltung von Warten. Beuth, Berlin, 1985.

[7]: S. Dutke: Mentale Modelle: Konstrukte des Wissens und Verstehens. Kognitionspsychologische Grundlagen für die Software-Ergonomie. Verl. f. Angewandte Psychologie, Göttingen, Stuttgart, 1993.

[8]: U. Epple: Die leittechnische Anlage und ihre Elemente - Prozeßleitsysteme. In: M. Polke: Prozeßleittechnik, Oldenbourg, München, 1994, 453-533.

[9]: U. Griem: Zur benutzergerechten Gestaltung eines Prozeßleitsystems, Universität Hamburg Fachbereich Informatik, Diplomarbeit, 1995.

[10]: A.M. Heinecke: Visualisierung in Echtzeitsystemen - ergonomische und technische Fragen. In: PEARL94, Workshop über Realzeitsysteme, Springer, Berlin, 1994, 12-21.

[11]: K. Holtzblatt, S.J. Jones.: Contextual Inquiry: A Participatory Technique for System Design. In: D. Schuler, A. Namioka: Participatory Design, Lawrence Erlbaum, New Jersey, 1993, 177-210.

[12]: Oberquelle H.: MCI - Quo vadis? Perspektiven für die Gestaltung und Entwicklung der Mensch-Maschine-Interaktion. In: D. Ackermann, E. Ulrich (Hrsg): Software-Ergonomie '91. Benutzerorientierte Software-Entwicklung, Teubner, Stuttgart, 1991, S. 9 - 24.

[13]: J. Rasmussen: Information processing and human-machine interaction. Elsevier Scince Publishing Co., New York, 1986.

[14]: V. Risak: Mensch-Maschine-Schnittstelle in Echtzeitsystemen. Springer, Wien, 1986.

[15]: D. Schuler, A. Namioka: Participatory Design, Lawrence Erlbaum, New Jersey, 1993.

[16]: K. Zinser: Neue Formen und Medien der Prozeßvisualisierung. In: ATP-Automatiserungstechnische Praxis 35, Oldenbourg, München, Sept. 1993, S. 499-509.

Adressen der Autoren

Udo Griem,
Werum GmbH,
Erbstorfer Landstr. 14
21337 Lüneburg

Prof. Dr. Horst Oberquelle
Universität Hamburg
Fachbereich Informatik/ASI
Vogt-Kölln-Str. 30
22527 Hamburg
Email: oberquelle@informatik.uni-hamburg.de

Flexibilität bei Workflow-Management-Systemen[1]

Jens Hagemeyer[#],Thomas Herrmann[*], Katharina Just-Hahn[*], Rüdiger Striemer[§]

[#] IWi, Saarbrücken, [*] Universität Dortmund, [§] Fraunhofer ISST, Dortmund

Zusammenfassung

Daß Workflow-Management-Systeme (WMS), ebenso wie andere, die Organisation abbildende und definierende Software, flexibel sein müssen, ist weitgehend anerkannt. Allerdings berücksichtigen die heute verfügbaren Systeme nur in eingeschränkter Form Anpassungserfordernisse von seiten der Endbenutzer. Deshalb ist es sinnvoll, die in diesem Aufsatz vorgestellten Ansätze in WMS zu integrieren. (1) "late modeling" erlaubt es, unstrukturierte Anteile eines Geschäftsprozesses in das Modell zu integrieren. Die Festlegung des Ablaufs erfolgt erst fallorientiert und während der Ausführung. (2) Der zweite Ansatz bietet den Benutzern eine Vielzahl von Ausführungsvarianten zur Auswahl, wobei die Menge von Varianten durch die Benutzer, unter Verwendung gegebener Anpassungsoptionen, erweitert werden kann. (3) Der abschließend vorgestellte Step-by-Step-Ansatz bietet einen Lösungsvorschlag für die mit der Anpassung eines kooperationunterstützenden Systems auftretenden Probleme, z.B. "wer ist/wird von einer Anpassung betroffen" oder "wer ist über eine Modifikation zu informieren".

1 Einleitung

Adaptivität, Individualisierung, Adaptierbarkeit, Flexibilität etc. sind in der Software-Ergonomie bekannte Konzepte, um mit der Anforderung umzugehen, daß interaktive Systeme menschengerecht und aufgabenangemessen sein sollen, obwohl Menschen und Aufgaben eine Vielzahl unterschiedlicher, sich ständig ändernder Eigenschaften aufweisen. Lange Zeit konzentrierte sich die Diskussion auf die Anpassungserfordernisse für jeweils einzelne Individuen (daher auch die Gestaltungsanforderung „Individualization" nach ISO9241/10 [12]), während erst in den 90er Jahren der Anpassung von Groupware erhöhte Aufmerksamkeit gewidmet wurde (s. [16]). Die Übertragung der im Bereich Groupware diskutierten Anpassungskonzepte auf Workflow-Management-Systeme (WMS) bereitet jedoch aufgrund von deren Besonderheiten Schwierigkeiten. Da ein WMS proaktiv steuert, wie mehrere Beschäftigte bei der Erledigung von Geschäftsprozessen zusammenarbeiten, sind neben den verwendeten Dokumenten vor allem die implementierte Arbeitsverteilung und die koordinierende Ablaufsteuerung Gegenstand der Anpassung. Da die Anwender von WMS insbesondere daran interessiert sind, mit solchen Systemen sicherzustellen, daß die für ihre Geschäftsprozesse notwendigen Standards und Fristen eingehalten werden, kann eine Anpassung solcher Systeme in Widerspruch zu übergeordneten Interessen geraten. Anpassungsverfahren bei WMS und deren technische Unterstützung müssen darauf Rücksicht nehmen.

Der momentan üblicherweise praktizierte Ansatz zur Unterstützung von Geschäftsprozessen durch WMS verläuft von der Modellierung der Ist-Prozesse über Schwachstellenanalyse und anschließende Sollkonzeption bis hin zur Verfeinerung der Geschäftsprozeßmodelle (GPM) in Workflow-Modelle (WFM) [6]. Bei erforderlichen Anpassungen werden Rückmeldungen

---

[1] Diese Arbeit entstand im Rahmen des Vorhabens: „Verbesserung von Geschäftsprozessen mit flexiblen Workflow-Management-Systemen (MOVE)", gefördert vom BMBF unter der Fördernr. 01HB 9606/1.

aus der Führungs- oder Ausführungsebene an die Geschäftsprozeß- oder Workflow-Modellierer weitergegeben, ein erneuter Transformationsprozeß beginnt. An diesem Vorgang beteiligt sind Mitarbeiter aus den Fachabteilungen und der Datenverarbeitung und der Organisation, eventuell unterstützt von externen Beratern oder Softwarehäusern. Dieser Prozeß weist meistens Schnittstellen zwischen unterschiedlichen Modellierungsmethoden und -werkzeugen auf und ist verbunden mit einem großen Kommunikationsbedarf zwischen verschiedenen Spezialisten [9]. Diese Vorgehensweise kann nur in längeren Zeitintervallen wiederholt werden und verhindert eine kurzfristige Anpassung der durch WMS unterstützten Prozesse.

Eine solche kurzfristige Anpassung kann aber aus verschiedenen Gründen erforderlich sein: Es können nicht antizipierbare Ausnahmefälle auftreten (z.B. Fehlzeiten bei Mitarbeitern), Kundenwünsche oder andere Markt- und Umweltbedingungen können sich ad-hoc ändern, informelle Aspekte der internen Aufbau- und Ablauforganisation müssen integriert werden (s. [10]). Es ist daher sinnvoll, Möglichkeiten vorzusehen, durch die die Ausführenden der Geschäftsprozesse – also die Endbenutzer eines WMS – in die Lage versetzt werden, die Eigenschaften eines Geschäftsprozesses an die aktuellen Erfordernisse eines konkreten Falls anzupassen. Die Endbenutzer sollten auch festlegen können, ob die Änderungen nur für einen einzelnen Fall, für eine Menge sich ähnelnder Fälle oder für einen bestimmten Zeitraum gelten. Da Endbenutzer für solche Anpassungsaufgaben in der Regel unterschiedlich qualifiziert sind, sollten Anpassungsmöglichkeiten auf verschiedenen Komplexitätsstufen angeboten werden. Zum Beispiel: 1. Auswahl zwischen Alternativen (etwa zwischen verschiedenen, vorprogrammierten Workflows), 2. Veränderung vorgegebener Strukturen und Werte (etwa durch Überspringen oder Vertauschen von Aktivitäten im Geschäftsprozeß oder durch Änderung von Defaultwerten in Formularen), 3. Nachträgliche Programmierung oder Verfeinerung von Geschäftsprozessen. In der Literatur finden sich detaillierte Darstellungen sinnvoller Anpassungsmöglichkeiten bei WMS (s. [10]) und der verschiedenen Komplexitätsstufen (s. [16]). Darüber hinaus sind bei der Anpassung von Groupware und insbesondere von WMS immer mehrere kooperierende Personen betroffen, was zu Konflikten führen kann (s. [19]), zu deren Vermeidung Abstimmungsprozesse notwendig sind. Aus der Sicht der Benutzer wird sich der Aufwand für Anpassungen und Abstimmung um so mehr lohnen, je nachhaltiger sie von einer Anpassung der Geschäftsprozesse profitieren.

Im folgenden beschreiben wir verschiedene Ansätze, die eine angemessene Anpassung von WMS, die sich im Einsatz befinden, ermöglichen, um einen raschen und auch nachhaltigen Abgleich zwischen den sich ändernden ergonomischen und betriebswirtschaftlichen Anforderungen einerseits und den im WMS abgebildeten Sollmodellen andererseits zu erreichen. Ziel dieser Beschreibung ist es, diese Ansätze nebeneinander zu stellen, um in der Zusammenfassung Synergiemöglichkeiten aufzuzeigen. Ansätze zur Adaption von WMS können sich nicht auf technische Konzepte, etwa auf eine reine Erweiterung der Prozeßbeschreibungssprache, beschränken [1]. Vielmehr müssen auch Vorgehensmodelle und organisatorische Regelungen zur Anpassung betrieblicher WMS miteinbezogen werden.

Die verschiedenen Ansätze zur Flexibilisierung von WMS werden anhand eines einheitlichen Beispiels erläutert. Es handelt sich um einen realen Ausschnitt eines Geschäftsprozesses eines Industrieunternehmens, der sich mit der Bearbeitung von Lieferterminanfragen bei komplexen Produkten befaßt. Er kann grob in vier Aktivitäten zerlegt werden: 1) Zunächst wird geprüft, welche Teile zur Fertigung des Produkts benötigt werden und ob alle Teile vorhanden sind. 2) Falls der Bestand an benötigten Einzelteilen unvollständig ist, wird deren spätester Liefertermin für die fehlenden Teile festgestellt. 3) Es wird geklärt, welche Fertigungskapazitäten ab dem frühest möglichen Start der Fertigung verfügbar sind und welche alternativen Betriebsmittel ggf. benötigt werden. 4) In Abhängigkeit von den verfügbaren Kapazitäten wird der Liefertermin mit dem Montagemeister abgestimmt und weitergegeben. In der folgenden Beschreibung wird davon ausgegangen, daß dieser Prozeßausschnitt nicht in allen Details von vornherein festgelegt werden kann (2.1), daß festgelegte Abläufe häufiger geändert werden und neue Varianten des Prozesses entstehen (2.2) und daß bei der nachträglichen Festlegung von Details oder Varianten mehrere Teilnehmer des Prozesses in vordefinierter Weise miteinbezogen werden müssen (2.3).

2   Drei Ansätze zur Flexibilisierung

2.1   Modellierung nicht planbarer Teilprozesse zur Laufzeit

Vielfach kann der Ablauf einer Bearbeitung im Vorhinein nicht oder nur ungenau festgelegt werden. Erst zur Laufzeit der Workflow-Anwendung kann bestimmt werden, auf welche Art ein Prozeß bearbeitet werden soll. In [4] werden Prozeßklassen nach der Art der Unplanbarkeit der Geschäftsprozeß-Merkmale Lösungsweg, Kooperationspartner und Informationsbasis unterschieden. Unter Planbarkeit verstehen wir in diesem Zusammenhang die grundsätzliche Möglichkeit, bestimmte Entitäten eines Workflow-Modells in der Phase der Modellierung hinreichend detailliert festlegen zu können [4]. Nehmen wir an, daß Schritt 2 des beschriebenen Beispiels im Vorhinein nicht ausreichend geplant werden kann. Dieser Fall bezieht sich auf die Klasse semi-strukturierter Teilprozesse mit unplanbaren, da unbekannten Lösungswegen. Bei Prozessen dieser Art sind Teile des Geschäftsprozesses wohlstrukturiert und demnach modellierbar. Andere Teile - auch semi-strukturiert genannt - wiederum können nur ungenau modelliert werden. Für solche Teilprozesse wird eine nachträgliche Modellierung zur Laufzeit vorgesehen. Erst wenn ein bestimmtes Ereignis eintritt, wird die Nachmodellierung dieses Teilprozesses erforderlich. Abbildung 1 zeigt für das Beispiel der Lieferterminanfrage eine solche Situation. Der gesamte Geschäftsprozeß (in Abb. 1 nur in einem Ausschnitt dargestellt) ist bis auf den Teilprozeß, der sich ergibt, wenn der Bestand an Einzelteilen nicht ausreicht, modelliert. Für den nicht planbaren Teilprozeß wurde eine „black box" eingefügt, deren genaue Spezifikation zur Laufzeit nachzumodellieren ist, falls dieser Fall eintritt. Die Abbildung zeigt ein FUNSOFT-Netz. FUNSOFT-Netze [8] werden mit der Prozeß- und Workflow-Management-Umgebung CORMAN (früher: MELMAC, vgl. [2]) modelliert und ausgeführt, für die auch die Möglichkeit des „late modeling" realisiert wurde.

Das Konzept des „late modeling" sieht vor, daß sogenannte „modifikationsauslösende Ereignisse" (mit einem Fähnchen gekennzeichnet) existieren, die eine Modifikation von Teilprozessen während der Laufzeit der Workflow-Anwendung anstoßen. Im Beispiel handelt es sich dabei um das Ereignis „Bestand reicht nicht aus". Tritt dieses Ereignis ein, wird zunächst die Aktivität „Komponente ermitteln", die nicht näher spezifiziert ist, aus der Ausführung des Workflows entnommen. Dies bedeutet, daß alle Teile des Workflows, mit Ausnahme des zu modifizierenden Teils, weiterbearbeitet werden können.

Eine Modifikation kann für beliebig viele Teile eines Workflows gelten. Das bedeutet, daß alle modifizierbaren Teile nach Eintritt des modifikationsauslösenden Ereignisses durch eine im Vorfeld spezifizierte Personengruppe verändert werden können. Modifikationen können dabei das Neuanlegen, Ändern oder Löschen von Netzteilen beinhalten.

Nachdem nun die Aktivität „Komponente ermitteln" im Rahmen des late modeling spezifiziert wurde, kann mit der Bearbeitung fortgefahren werden. Die Modifikation würde im Beispiel bedeuten, daß die einzelnen Aktivitäten, ihre Reihenfolge, die bearbeitenden Personen sowie alle weiteren zur Ausführung notwendigen Informationen nachmodelliert werden. Das Konzept des „late modeling" sieht zu diesem Zweck vor, daß zunächst ein Vorschlag für die Gestaltung des Teilprozesses erarbeitet wird. Dieser Vorschlag ist Grundlage für die Aushandlung der endgültigen Version durch die beteiligten Personen. Im Rahmen der Aushandlung kann auch darüber entschieden werden, ob die durchgeführte Modifikation nur für den aktuellen Geschäftsprozeß gilt oder in das allgemeine GPM übernommen werden soll und somit auch für zukünftige Workflows dieses Typs gilt[2]. Denkbar wäre hier auch die Schaffung von Workflow-Varianten (s. Abschnitt 2.2). Wurde die Aushandlung erfolgreich abgeschlossen, gilt die Modifikation als beendet und das modifizierte Teilnetz kann ausgeführt werden. Während der Modellierung dieses Teilprozesses wurde der restliche Geschäftsprozeß durch das Workflow-System bearbeitet, falls er von der zu spezifizierenden Aktivität unabhängig ist, wenn also in unserem Beispiel die Kapazitätsprüfung parallel zur Ermittlung der Komponenten stattfände (s. Abschnitt 2.3).

Auf die dargestellte Art und Weise können auch solche Teilprozesse, deren genauer Ablauf zum Zeitpunkt der Modellierung nicht bekannt ist, in die Workflow-Anwendung integriert werden. Voraussetzung ist, daß zum Zeitpunkt der Modellierung die modifikations-auslösenden Ereignisse und die nachträglich zu spezifizierenden Teile des Workflows bestimmt werden können. Darüber hinaus sollte festgelegt werden, wie der Aushandlungsprozeß zwischen den Personen abläuft, die von der nachträglichen Spezifizierung betroffen sind (s. Abschnitt 2.3).

Das Vorgehensmodell für die Aushandlung ist dabei selbst als Prozeßmodell hinterlegt und somit jederzeit änderbar. Bei der Festlegung der Menge der modifizierbaren Elemente werden automatisch die möglichen Konsequenzen von Änderungen bestimmt. So muß beispielsweise vorgesehen werden, daß bei Änderungen an einem Dokumentenspeicher auch die auf diesen Speicher zugreifenden Aktivitäten änderbar sein müssen.

Das beschriebene Konzept wurde für die

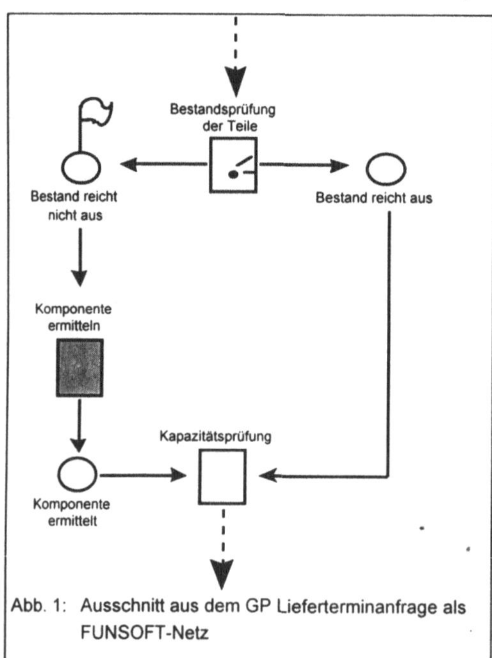

Abb. 1: Ausschnitt aus dem GP Lieferterminanfrage als FUNSOFT-Netz

---

[2] Eine weitergehende Differenzierung von Arten der Übernahme von Prozeßmodifikationen findet sich in [5].

Prozeßmanagementumgebung CORMAN, die am Fraunhofer-Institut für Software- und Systemtechnik entwickelt wurde und den FUNSOFT-Ansatz unterstützt [3], realisiert. Ziel war es, die Menge der mit CORMAN ausführbaren Geschäftsprozesse um solche Teilprozesse zu erweitern, die im Vorfeld nicht planbar sind und somit semi-strukturierten Charakter haben. Auf diese Weise konnte eine Methode gezeigt werden, mit der auch bestehende Workflow-Management-Systeme flexibler gestaltet und ihre Anwendungsmöglichkeiten erweitert werden können.

2.2 Prozeßadaption durch Variantenerzeugung

Anstelle des herkömmlichen top-down-Vorgehensmodells zur Workflow-Modellierung kann auch ein Ansatz verfolgt werden, der durch die Erzeugung und Auswertung von Varianten der Workflow-Modelle zu adaptiven Workflow-Management-Systemen führt. Dies ist eine Alternative zu der in Abschnitt 2.1 dargestellten Modellierung zur Laufzeit, mit dem Unterschied, daß keinerlei Modellierungskenntnisse seitens des Benutzers erforderlich sind.

Adaptivität von Workflow-Anwendungen an Änderungen der Aufbau- und Ablauf-organisation setzt eine Erweiterung der Geschäftsprozeßbeschreibungssprache voraus. So muß das Prozeßmodell um Informationen erweitert werden, wie sie üblicherweise in Organigrammen und Aufgabenbeschreibungen von Unternehmen zu finden sind. Auf diese Weise kann ein GPM nicht nur der Abbildung der betrieblichen Realität dienen, sondern zugleich der WMS-gestützten Steuerung der Arbeitsabläufe. Wird dies ergänzt durch die Trennung der festen Zuordnung von Organisationseinheiten zu Funktionen und Daten, kann eine flexible Architektur geschaffen werden, die evolutionäre, adaptive GPM und Organisationsmodelle ermöglicht.

In WMS werden WFM auf Typebene definiert, von denen bei Eintreten des auslösenden Ereignisses Instanzen gebildet werden. Innerhalb der Modelle existieren Alternativen für Teilprozesse. Bereits heute existieren Ausnahmebehandlungsmechanismen in WMS, die während der Bearbeitung ein Abweichen vom instantiierten GPM möglich machen (vgl. [15]). Beispiele sind die Wahl von alternativen Bearbeitungswegen und die Delegation von Aufgaben, d.h. die Weitergabe der Verantwortung für einen Prozeßschritt an eine andere Organisationseinheit.

Zur Flexibilisierung der Workflow-Bearbeitung sollten dem Benutzer darüber hinaus Anpassungsmöglichkeiten wie das Überspringen bestimmter Funktionen oder die Wahl eines anderen Anwendungsprogrammes zur Unterstützung der zu bearbeitenden Funktion gegeben werden. Zudem muß der Bearbeiter die Möglichkeit haben, bei Bedarf zusätzliche Funktionen oder Teilprozesse einzufügen (vgl. auch [15, S. 12]), möglichst ohne ihn zu zwingen, eine Modellierungssprache oder -notation zu erlernen. Diese Prozeßadaption kann nur bei Schritten durchgeführt werden, die bedingt modifizierbar oder flexibel sind (s. Abschnitt 2.3). Diese Erweiterung des WMS bewirkt ein Aufbrechen des Gegensatzes zwischen flexibler Groupware und starren Workflow-Systemen und erlaubt ein dynamisches, fallbezogenes ad-hoc-Anpassen der betrieblichen Informationssysteme an die Erfordernisse der Organisation.

Werden die Varianten der ausgeführten Workflow-Instanzen gespeichert (vgl. Abb. 2), stehen sie für Auswertungszwecke zur Verfügung. Eine bisher nicht genutzte Möglichkeit der Auswertung dieser Daten besteht in der Generalisierung dieser Workflows zu Workflow-

Typen, also Modellen. Zu diesem Zweck müssen die Runtime-Daten der Workflow-Engine, die die automatisch oder durch Benutzerauswahl getroffenen Entscheidungen bzgl. Verzweigungen, Einfügen und Auslassen von Funktionen oder Benennung einer anderen Organisationseinheit enthalten, als ein Ist-Modell abgebildet werden. Nach wiederholter Ausführung dieses Geschäftsprozesses liegen mehrere adaptierte Ist-Modelle als Varianten des GPM vor.

Diese Varianten können in zweierlei Hinsicht genutzt werden. Zum einen kann bei der nächsten Bearbeitung des Geschäftsvorfalls auf die Modifikationen früherer Bearbeiter zurückgegriffen werden. Die Benutzer können sich an Erfahrungen ihrer Kollegen orientieren und eine ihnen adäquat erscheinende Vorlage aus einer Variantenbibliothek nutzen. Die zweite Nutzungsmöglichkeit besteht in der Verarbeitung der Varianten durch eine Analysekomponente, die (z.B. durch statistische Auswertung oder Mustererkennung) Vorschläge zur Generalisierung der Varianten zu einem neuen Sollmodell entwickelt. Resultat ist die stete Synchronisation der Modellebene mit der Handhabung in der betrieblichen Realität. Modelle, die auf diese Art und Weise aktualisiert werden, werden auch als aktive Modelle bezeichnet [7]. Eine Systemarchitektur für adaptive WMS ist in Abb. 3 abgebildet. Sie erweitert das weithin akzeptierte Referenzmodell für WMS [18]. Der Workflow-Ausführungsservice erstellt ausgehend von einem Workflow-Modell eine Workflow-Instanz, welches von den Bearbeitern über den Task-Manager, wie oben dargestellt, modifiziert werden kann. Der Task-Manager stellt auch die Verbindung zu den operativen Anwendungen her und stellt dem Benutzer die erwähnten Ausnahmebehandlungsoptionen zur Verfügung.

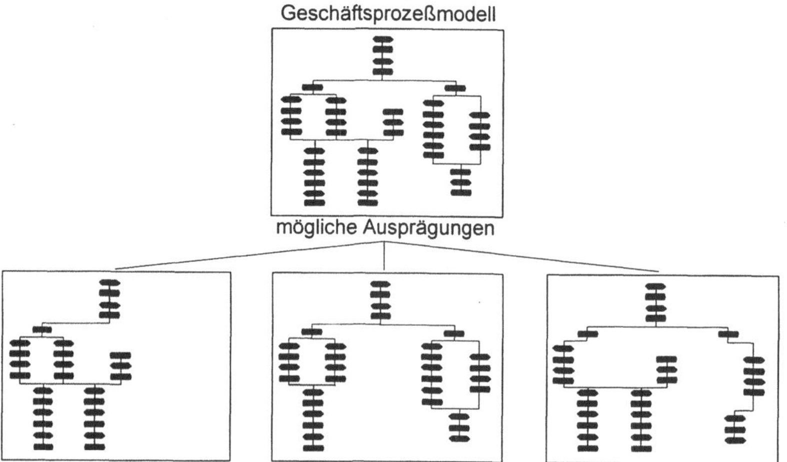

Abb. 2: Während der Prozeßbearbeitung entstandene Varianten von GPM

Der Einsatz des adaptiven WMS soll anhand des Beispielszenarios erläutert werden. Dabei wird davon ausgegangen, daß der eingangs beschriebene Ablauf als Prozeß modelliert sei und im Unternehmen durch ein WMS und angebundene Funktionen eines Produktionsplanungs- und Steuerungssystems unterstützt werde.

Entsprechend des vorgegebenen Modells wird vom System die Funktion zur Feststellung des Lieferzeitpunktes angestoßen, wenn sich bei der Bestandsprüfung ergibt, daß ein benötigtes

Teil nicht verfügbar ist. Nun könnte der Sachbearbeiter, der erfahren hat, daß es für eine bestimmte Klasse von Bauteilen seit kurzem Ersatzprodukte eines anderen Herstellers gibt, anstelle der möglicherweise langwierigen Beschaffung des ursprünglich vorgesehenen Bauteils jedoch auch den Einsatz des Alternativbauteils vorschlagen. Zu diesem Zweck fügt er eine Rückfrage an den Kundenbetreuer in den Prozeß ein, um herauszufinden, ob der Ersatz akzeptabel ist. Ist dies der Fall, wird die Verfügbarkeitsprüfung für dieses Teil durchgeführt, andernfalls die Feststellung des Lieferzeitpunktes des Originalteiles ausgelöst.

Diese Prozeßvariante wird vom System gespeichert und kann entsprechend der oben erwähnten zwei Möglichkeiten zur Adaption genutzt werden. Zum einen kann beim nächsten Fehlen eines Bauteiles dem jeweiligen Bearbeiter diese Alternative zur Bearbeitung des Falles angeboten werden. Zum anderen kann bei der nächsten Analyse der Ist-Prozesse ein Vorschlag zur Erweiterung des Sollmodells generiert werden, wenn diese Anpassung des Sollmodells häufig gewählt wurde.

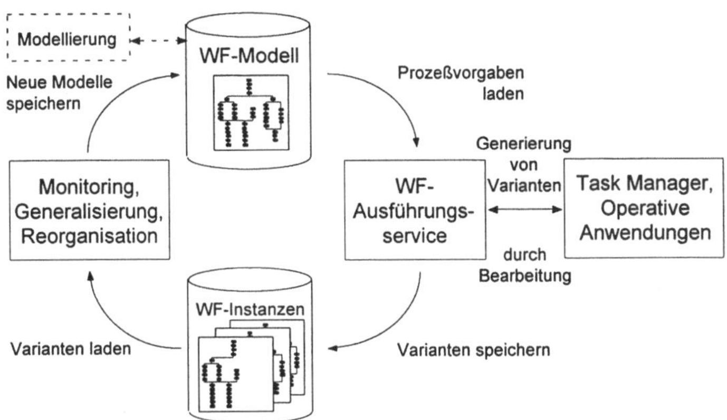

Abbildung 3: Architektur adaptiver WMS

2.3 Schrittweises Festlegen von Bedingungen zur Durchführung von Adaption

Mit der nachträglichen Adaption eines Workflow-Modells treten i.d.R. die folgenden Probleme auf: (p1) Wer wird/ist von dieser Modifikation betroffen? und (p2) Wer muß über diese Modifikation informiert werden. Sofern eine Information versendet wurde, ergeben sich weitere Probleme: (p3) Wer hat das Recht dieser Modifikation zu widersprechen? und (p4) Wie kann sicher gestellt werden, daß bereits vorhandenes Erfahrungswissen bzgl. des zu modifizierenden Workflows bei der Ausführung der Adaption berücksichtigt wird? Der im Weiteren vorzustellende Ansatz Step-by-Step [13] bietet eine Lösung für diese Probleme.

Step-by-Step versteht einen Workflow als eine Abfolge einzelner Schritte, wobei jeder Schritt genau zwei Aktivitäten miteinander verbindet (s. Abb. 4). Diese Verbindung beschreibt den Transfer von Arbeitsergebnissen zwischen diesen Aktivitäten. Step-by-Step klassifiziert einen Schritt eines Workflows danach, ob durch ihn verbundene Aktivitäten von einzelnen Personen (*elemental steps*, z.B. Abb. 4: C->D) oder von einer Gruppe von Personen (*compound steps*, durch einen Stern gekennzeichnet B*) ausgeführt werden und ob deren Ausführung *obligatorisch* (durchgezogene Linie) oder *abhängig* (gestrichelte Linie) ist. Die Sollkonzeption des Geschäftsprozesses bestimmt, ob ein Schritt obligatorisch ist oder abhängig. Ist

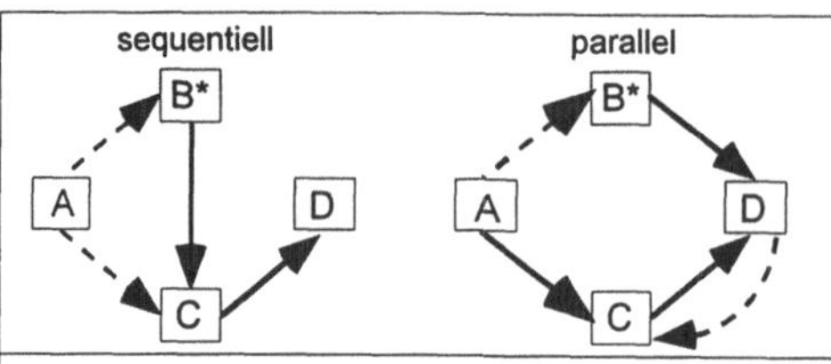

Abb. 4: Ausschnitt aus dem Beispiel Lieferterminanfrage als ein Step-by-Step Modell

ein Schritt abhängig, so wird er nur ausgeführt, wenn vorab festgelegte Bedingungen mit den spezifischen Eigenschaften des jeweils in Bearbeitung befindlichen Falls vereinbar sind. Die Blackbox, wie sie beim late modeling (s.o. Abschnitt 2.1) verstanden wird, wird im Rahmen von Step-by-Step als compound step repräsentiert, bei dem zwar die verantwortlichen organisatorischen Einheiten (z.B. eine Projektgruppe oder Abteilung) spezifiziert wurden, nicht jedoch die auszuführenden Aktivitäten oder deren Abfolge. Den Ausführenden, ermöglichen compound steps u.a. die fallorientierte Planung von Aktivitäten und Ressourcen. In einem Step-by-Step Modell kann auf die Darstellung von Ereignissen oder Dokumenten verzichtet werden, da vor allem die kooperativen Aspekte des Prozesses für die Abstimmung einer Anpassung relevant sind. Zur Handhabung der oben angeführten Probleme sind bei der Spezifikation eines Schrittes unter Step-by-Step verschiedene adaptionsrelevante Informationen zu spezifizieren:

Adaptionsrelevante Attribute zur Spezifizierung von Schritten

Die Bedingungen, unter denen ein Schritt modifiziert werden kann: Die Anpaßbarkeit eines Schritts oder eines Modells kann, ebenso wie dessen Ausführung, von verschiedenen Rahmenbedingungen abhängig sein oder bestimmten Einschränkungen unterliegen. Dies begründet eine Klassifizierung der Schritte nach: statisch, bedingt und flexibel. *Statische* Schritte dürfen von den Nutzenden nicht modifiziert werden, sondern lediglich von besonders ausgezeichneten Personen, z.B. einem Modellierer oder der Prozeßverantwortlichen. Z.B. könnte festgelegt sein, daß die Prüfung der Verfügbarkeit benötigter Teile bei der Lieferterminanfrage nie außer acht gelassen werden darf; dies würde bedeuten, daß die Schritte die zu dieser Aktivität hinführen nur bedingt veränderbar wären, da es immer mindestens einen Schritt geben muß, der diese Prüfung in den Teilprozeß miteinbezieht. Die Anpaßbarkeit eines *bedingt* modifizierbaren Schrittes ist im Einzelfall zu überprüfen. Bedingungen können dabei bspw. sein: die Rolle des Initiierenden (z.B. ein Mediator [17]), der Geltungsbereich der vorgeschlagenen Modifikation, die Zustimmung eines Vertreters des Managements oder die Einschränkung auf Variantengenerierung (s. Abschnitt 2.2) etc. Alle anderen Schritte gelten als *flexibel* und dürfen von allen Beteiligten zu jedem Zeitpunkt modifiziert werden. Diese Klassifizierung von Schritten bzgl. ihrer Adaptierbarkeit erfolgt im Verlauf der Modellierung oder zur Laufzeit des Systems. Damit ist auch sichergestellt, daß veränderte Bedingungen unmittelbar und gezielt abgebildet werden können. Diese Klassifizierung kann ebenfalls Objekt einer Anpassung sein.

Wer ist von einer Adaption betroffen? (s.o. p1): Jede Anpassung eines Workflows bedeutet eine organisatorische Veränderung des Geschäftsprozesses. D.h., daß neben denjenigen, die eine Anpassung vornehmen, andere, irgendwie in den Prozeß involvierte Personen von einer Änderung betroffen sind oder sein können, wobei das Maß der Betroffenheit unterschiedlich sein kann. Zur Erhaltung der Transparenz und der Handhabbarkeit des Geschäftsprozesses ist es notwendig, daß diejenigen, die von einer Modifikation betroffen sind, auch an dem Modifikationsprozeß selbst beteiligt werden. Aus diesem Grunde gilt es, die Betroffenen zu identifizieren. Die Liste derer, die betroffen sind wird zum einen bestimmt von Abhängigkeiten zwischen Aktivitäten und den zur Ausführung von Aktivitäten berechtigten Personen. Zum anderen können modellunabhängige Faktoren, wie das Erfahrungswissen (z.B. einer Person, die lange Zeit für den relevanten Bereich zuständig war, nun aber in einem anderen tätig ist), die organisatorische Kompetenz (z.B. als Abteilungsleiterin) oder auch das Management dafür ausschlaggebend sein, daß bestimmte Personen, Rollen oder Gruppen an bestimmten Modifikationsprozessen zu beteiligen sind. Step-by-Step sieht daher vor, daß für jeden Schritt, unabhängig von anderen, die Liste der Betroffenen definiert werden kann. Dieses Vorgehen erlaubt es den Nutzenden, aus ihrem Arbeitskontext heraus zu entscheiden, welche Schritte eines Geschäftsprozesses für sie so relevant sind, daß sie über deren Veränderungen informiert oder am Modifikationsprozeß aktiv beteiligt werden möchten oder zu beteiligen sind.

Auf welche Art eine betroffene Person einzubinden ist: Wie bereits angesprochen kann das Ausmaß der Betroffenheit sehr unterschiedlich sein. Je nachdem sind unterschiedliche Beteiligungsformen angemessen: Information (p2), Konsultation (p4) und Einspruchsrecht (p3). Die Beteiligungsform wird zusammen mit der Nennung als Betroffene in der Schrittdefinition festgelegt. Beteiligung bedeutet mindestens, daß Information darüber vermittelt wird, wer, was, warum geändert hat. Betroffene können als Berater oder indem sie das Einspruchsrecht wahrnehmen, aktiv eingebunden werden. Letzteres kann im Rahmen einer rechnergestützen Aushandlung [19] ausgeübt werden. Resultate einer Aushandlung sind: Das Fallenlassen einer Modifikation, deren Umsetzung auf Zeit, die Umsetzung einer Alternative bzw. die Zustimmung zur vorgeschlagenen Modifikation.

Was passiert im Falle einer Modifikation?

Abbildung 4 zeigt links den bereits erprobten sequentiellen Ablauf der Aktivitäten B* und C. Da es sich in der Vergangenheit herausgestellt hat, daß i.d.R. der späteste Liefertermin eines fehlenden Bauteils noch vor dem frühest möglichen Starttermin der Montage liegt (s. Bsp. Abschnitt 1), sollen B* und C aus Gründen der Effizienz parallelisiert werden. Diese Anpassung erfordert, Schritte zu verändern, zu ergänzen, bzw. zu löschen. Ein System, welches den Step-by-Step Ansatz integriert hat, unterstützt die Nutzenden dabei, z.B. indem die geplante Modifikation vorab simuliert werden kann. (Die Integration des Konzepts Step-by-Step in die Modellierungsumgebung CORMAN und 'late modeling' (s. Abschnitt 2.1) ist geplant.) Das System identifiziert dann aufgrund des existierenden und im Hinblick auf Step-by-Step attributierten Modells: 1. ob bzw. unter welchen Bedingungen Schritte modifiziert werden dürfen und 2. die Menge der Betroffenen sowie deren Beteiligungsform. Anhand dieser Information kann die ausführende Person die beabsichtigte Anpassung evaluieren. Soll die evaluierte Adaption ausgeführt werden, d.h. sofern die initiierende Person auch über die entsprechenden Kompetenzen verfügt, erfolgt die systemgestützte Einbeziehung der Betroffenen. Unser Bsp. (s.o.) enthält compound und elemental steps sowie abhängige und obligatorische Schritte. Der Schritt A->B* (abhängig) wird nur dann ausgeführt, wenn Bauteile fehlen. Die Aktivitäten unter C sind in jedem Falle erforderlich. Die linke Version (s. Abb. 4) stellt dies durch den obligatorischen Schritt B*->C sicher. Für die rechte Version wird dies erreicht durch den obligatorischen Schritt A->C. Damit erfolgt bei der neuen Version rechts der Transfer von Informationen und Kontrollfluß nach C sobald die Aktivität A abgeschlossen ist, sofern bestimmte Bauteile fehlen erfolgt nun parallel dazu ein Transfer zu B*. Der Schritt D->C wird dann ausgeführt, wenn der Liefertermin für fehlende Teile nach dem geplanten Montagetermin liegt, daher wird er abhängig klassifiziert.

Im weiteren wird fiktiv und beispielhaft dargestellt, wie sich die Menge betroffener Benutzer zusammensetzen könnte. Als Betroffene dieser Modifikationen mit Einspruchsrecht seien anhand der Attributierung des Modells identifiziert worden: Die an der Ausführung direkt beteiligten Personen (bei compound steps können dies natürlich auch sehr viele Personen sein) sowie die jeweils namentlich genannten Personen oder Rollen, in unserem Beispiel sind er die Meister der beteiligten Abteilungen sowie die Leiterin des Vertriebs, da dieser Geschäftsprozeß für die Kundenbetreuung von besonderem Interesse ist.

Im Rahmen der Aushandlung kann es auch dazu kommen, daß neben Veränderungen der Kooperation auch Veränderungen von Tätigkeitsbeschreibung, z.B. zu Gunsten einer gerechteren Arbeitsverteilung, diskutiert und vorgenommen werden. Gemäß dem Resultat der Aushandlung wird der initiierte Modifikationsprozeß abgeschlossen. Das Aushandlungsresultat sowie die Erläuterung der Modifikation werden abschließend dokumentiert, um die evolutionären Veränderungen des Systems nachvollziehen und ggf. in einem umfassenderen Re-Konfigurationsprozeß [11] berücksichtigen zu können.

3 Zusammenfassung

Die in 2.1 und 2.2 beschriebenen Verfahrensweisen zur Anpassung von WMS durch Endbenutzer basieren auf unterschiedlichen Ausgangssituationen. Die erste unterstellt, daß Teile einer Geschäftsprozeßbeschreibung zu Beginn der Bearbeitung konkreter Fälle noch nicht spezifizierbar sind oder - ganz gezielt - nicht spezifiziert werden sollten, um den Ausführenden Handlungs- und Entscheidungsspielräume zu eröffnen. Diese können dann genutzt werden, wenn durch "late modeling" eine fallbezogene Spezifizierung im nachhinein erfolgt. Der zweite Ansatz unterstellt, daß alle relevanten Aspekte eines Geschäftsprozesses im Vorhinein modelliert worden sind, daß sich aber während der Laufzeit Abweichungen vom vorgegebenen Weg aus der Sicht des Ausführenden als erforderlich erweisen. Gängige WMS und -Konzepte bieten durchaus Möglichkeiten, vom vorgegebenen Modell abzuweichen. Neu ist bei diesem Ansatz, die bei der Anpassung entstehenden Geschäftsprozeßvarianten abzuspeichern und ein Verfahren zu implementieren, mit dem diese Varianten in eigenständige alternative GPM umgesetzt werden. Dies hat den Vorteil, daß hier eine Anpassung der GPM *by doing* erfolgt, da eine einmal durchgeführte Abweichung der

Bearbeitung in einer Form dokumentiert wird, die deren spätere Wiederholung erlaubt und die auch Ansatzpunkt für ein diskursives Verfahren, sprich Aushandlung, zum Geltungsbereich der Abweichung sein kann. Mit Hilfe des Systems wird aus der 'Anpassung by doing' ein 'Modellieren by doing'. Dieses Verfahren kann auch für das "late modeling" verwendet werden, wobei die Vorgehensweise innerhalb der Blackbox nicht auf der Modellebene, sondern durch die Ausführung selbst definiert würde. Diese Ausführung wäre dann nicht von einem WMS, sondern durch nicht proaktive Groupware-Anwendungen zu unterstützen, bei denen die Koordination durch die Benutzer selbst gesteuert würde. Die Kombination von "late modeling" und "modeling by doing" gibt einen Hinweis, wie künftig ein fließender Übergang zwischen WMS und Groupware realisiert werden könnte: Mehrere Autoren könnten z.B. Joint-Editing in spontaner Koordination nutzen, um ein Dokument zu erstellen und zu überarbeiten. Danach wird ihnen dann ein Modell ihrer Vorgehensweise gezeigt und sie können entscheiden, ob sie in ähnlichen Fällen die alte Vorgehensweise wiederholen oder gezielt Änderungen vornehmen.

Immer wenn eine vorab festgelegte Vorgehensweise, die die Kooperation mehrerer Ausführender steuert, geändert wird, sind Abstimmungsprozesse zwischen den Kooperierenden bzw. den Prozeßverantwortlichen notwendig. Das Konzept Step-by-Step (s. Kap. 2.3) unterstützt diese Prozesse, indem aus dem GPM die für die Koordination essentiellen Aspekte herausgestellt werden, nämlich die Schritte, die die Aktivitäten potentiell unterschiedlicher Personen miteinander verbinden. Auf diese Weise werden die entscheidenden Entitäten sichtbar, mit denen die Bedingungen für die Gültigkeit einer Blackbox-Spezifizierung oder einer Variante zu verknüpfen sind. Hierzu gehört insbesondere die Festlegung, wer an einem Abstimmungsprozeß mit welchen Rechten zu beteiligen ist. Das durch Step-by-Step geregelte Beteiligungskonzept macht es erforderlich, daß die zu Beteiligenden vorab über die geplante Spezifizierung oder Anpassung informiert werden. Dazu können entweder geänderte Modelle oder auch die Simulation eines geänderten Ablaufs verwendet werden. Auch hier ist es hilfreich, wenn ein "modeling by doing" durch das System unterstützt wird, um ein Dokument zu erhalten, auf das sich der Abstimmungsprozeß der Beteiligten fokussieren kann.

Es zeigt sich, daß es derzeit unterschiedliche Konzepte zur Unterstützung der Anpaßbarkeit von WMS gibt, deren Potential erhöht werden kann, indem sie integriert werden und fließende Übergänge zwischen WMS und Groupware geschaffen werden.

Literaturverzeichnis

[1] T. Allweyer: Modellierung und Gestaltung adaptiver Geschäftsprozesse. (Veröffentlichungen des Instituts für Wirtschaftsinformatik, Heft 115), Saarbrücken 1995.

[2] W. Deiters, V. Gruhn, H. Weber: Software Process Evolution in MELMAC. In: The Impact of CASE on the Software Development Life Cycle, World Scientific, Singapore u.a., 1994, pp.301-323.

[3] W. Deiters, V. Gruhn, R. Striemer: Der FUNSOFT-Ansatz zum integrierten Geschäftsprozeßmanagement. Wirtschaftsinformatik (1995), Heft 5, S. 459-466

[4] W. Deiters, Th. Herrmann, T. Löffeler, R. Striemer: Identifikation, Klassifikation und Unterstützung semi-strukturierter Prozesse in prozeßorientierten Telekooperationssystemen. In: Tagungsband zur D-CSCW '96 In: H. Krcmar, H. Lewe, G. Schwab (Hg.): Herausforderungen der Telekooperation. Springer, Berlin u.a. 1996, S. 261-274.

[5] C. Ellis, K. Keddara, G. Rozenberg: Dynamic Change within Workflow Systems. In: Comstock, N. et al. (Hg.): Proc. Conference of Organizational Computing Systems (COOCS) '95. ACM-Press Milpitas, 1995, S. 10-21.

[6] J. Galler, A.-W. Scheer: Workflow-Projekte: Vom Geschäftsprozeßmodell zur unternehmensspezifischen Workflow-Anwendung. Information Management 10 (1995) 1, S. 20-35.

[7] R. Greenwood et al.: Active Models in Business. In: Proceedings of the 5th Conference on Business Information Technology. Manchester 1995.

[8] V. Gruhn: Validation and Verification of Software Process Models. Dissertation. Universität Dortmund 1991

[9] J. Hagemeyer, J. Galler, A.-W. Scheer: Koordination verteilter Workflow-Modellierung mit ContAct. In: J. Becker, M. Rosemann (Hg.): Workflowmanagement - State-of-the-Art aus Sicht von Theorie und Praxis, Proceedings zum Workshop der Gesellschaft für Informatik, Münster 1996.

[10] Th. Herrmann: Workflow Management Systems: Ensuring organizational Flexibility by Possibilities of Adaption and Negotiation. In: Proc. Conference on Organizational Computing Systems (COOCS) '95. ACM. New York S. 83 - 95.

[11] Th. Herrmann, K. Just: Anpaßbarkeit und Aushandelbarkeit als Brücke von der Software-Ergonomie zur Organisationsentwicklung. In: U. Hasenkamp (Hg.): Einführung von CSCW-Systemen in Organisationen. Tagungsband der D-CSCW '94. Vieweg, Braunschweig, 1994, S. 89-107.

[12] ISO/TC159/SC4/WG5: ISO 9241: Ergonomics requirements for office work with visual display terminals (VDTs): Part 10: Dialogue principles.

[13] K. Just, Th Herrmann: Step-by-Step: A concept for integrating self-organisation within workflow-mamagement-systems. E-CSCW '95 Workshop: Groupware for Self-Organising Units (to be published)

[14] B. Karbe, N. Ramsperger: Advanced Task Allocation in ProMInanD. In: HCI '91. Stuttgart 1991.

[15] S. Kirn: Organisatorische Flexibilität durch Workflow-Management-Systeme? (Arbeitsbericht des Instituts für Wirtschaftsinformatik Nr. 38.) Münster 1995.

[16] H. Oberquelle: Situationsbedingte und benutzerorientierte Anpaßbarkeit von Groupware. In: A. Hartmann et al. (Hg.): Menschengerechte Groupware - Software-ergonomische Gestaltung und partizipative Umsetzung. Teubner, Stuttgart, 1994, S. 31-50.

[17] K. Okamura, M. Fujimoto, W.J. Orlikowski: Helping CSCW Applications Succeed: The Role of Mediators in the Context of Use. In: Proc. Conference on Computer Supported Cooperative Work. ACM Press, New York, 1994, S. 55-67.

[18] Workflow Management Coalition Members: Glossary - A Workflow Management Coalition Specification. Brüssel, 1994.

[19] V. Wulf: Konfliktmanagement bei Groupware. Dissertation. Universität Dortmund. 1996

Adressen der Autoren

Jens Hagemeyer
Institut für Wirtschaftsinformatik
an der Universität des Saarlandes
Postfach 15 11 50
66041 Saarbrücken
Email: hagemeyer@iwi.uni-sb.de

Thomas Herrmann
Universität Dortmund, FB Informatik, LSVI
Fachgebiet Informatik und Gesellschaft
44221 Dortmund
Email: herrmann@iug.informatik.uni-dortmund.de

Katharina Just-Hahn
Universität Dortmund, FB Informatik, LSVI
Fachgebiet Informatik und Gesellschaft
44221 Dortmund
Email: just@iug.informatik.uni-dortmund.de

Rüdiger Striemer
Fraunhofer-Institut für Software- und
Systemtechnik
Postfach 52 01 30
44207 Dortmund
Email: striemer@do.isst.fhg.de

Die Priorisierung von Problemhinweisen in der software-ergonomischen Qualitätssicherung

Marc Hassenzahl

Technische Hochschule,
Darmstadt

Jochen Prümper

Fachhochschule für Technik
und Wirtschaft, Berlin

Uta Sailer

Ludwig-Maximilians-
Universität, München

Zusammenfassung

Im Rahmen software-ergonomischer Qualitätssicherung sind die Ressourcen für Gestaltung beschränkt. Um die effiziente Bearbeitung zu gewährleisten, wird die Priorisierung von Problembereichen eines Softwaresystems nötig. Es werden zwei Arten der Priorisierung von Problemhinweisen, nämlich *Priorisierung nach Fehlerkosten* und *Priorisierung nach Auftretensstabilität*, vorgestellt und empirisch analysiert. Hierbei steht der Aspekt der mentalen Beanspruchung im Zentrum der Überlegungen. Prinzipiell eignen sich beide Methoden zur Priorisierung und Auswahl der wichtigsten Problembereiche eines Softwaresystems. Ob man sich für die eine oder andere Methode entscheiden soll, hängt jedoch von verschiedenen Randbedingungen, wie Auftretenshäufigkeit und zeitökonomischen Überlegungen ab. In jedem Fall kann eine der hier vorgeschlagenen Priorisierungsmethoden den Prozeß der software-ergonomischen Qualitätssicherung effizienter gestalten und damit im Ergebnis die mentale Beanspruchung der Benutzer reduzieren.

1 Einleitung

Unserer Erfahrung nach ist es nicht schwer, im Laufe eines Softwareentwicklungsprozesses von Projektverantwortlichen, Endanwendern oder Entwicklern eine Erklärung zu bekommen, welche Mängel im Rahmen software-ergonomischer Qualitätssicherung am dringlichsten bearbeitet werden müssen. Leider stimmen diese Prioritäten in den seltensten Fällen miteinander überein: Entwickler denken stark in technischen Kategorien von Funktionalität und Realisierung, Benutzer in Kategorien der Gebrauchstauglichkeit und Projektverantwortliche nicht zuletzt in Kategorien der Verträglichkeit der Gestaltungsmaßnahmen mit dem Budget.

Damit stellt sich die Frage, wie Problembereiche einer Software priorisiert werden können, so daß sie auf der einen Seite möglichst unabhängig von den unterschiedlichen Wünschen und Sichtweisen einzelner Gruppen sind, es auf der anderen Seite aber zu einem Konsens der beteiligten Gruppen über die Rangfolge der Bearbeitung kommt. Dies ist besonders dann wichtig, wenn Benutzerpartizipation das Mittel der Erarbeitung von Lösungsvorschlägen für Problembereiche im Rahmen einer formativen Evaluation [5,15] ist.

Egal welche Art und Weise der Partizipation man bevorzugt, immer wird es eine Beschränkung der Ressourcen geben. Eine bedeutende Rolle spielen dabei Einschränkungen, die durch den - häufig sehr engen - zeitlichen und finanziellen Rahmen, in dem die Gestaltung beendet sein soll und durch die beschränkte Verfügbarkeit von beteiligten Benutzern entstehen [9].

Der erste Schritt zur Auswahl besonders schwerwiegender Problemursachen ist das Aufstellen einer Rangfolge von Problemhinweisen. Dies hat den Vorteil, daß ein beliebiger Prozentsatz der Problembereiche eines Softwaresystems systematisch abgearbeitet, d.h. umgestaltet werden

---

Dieser Beitrag entstand im Rahmen des PEGASUS-Projektes (Partizipative Evaluation und Gestaltung aufgabenorientierter Software unter Berücksichtigung europäischer Standards). Projektleitung: Jochen Prümper

kann. Dieser Prozentsatz muß zwar an die jeweiligen Ressourcen angepaßt werden; man ist sich aber immer sicher, wichtige Probleme zuerst bearbeitet zu haben.

Oft spielt bei der Auswahl der zu bearbeitenden Probleme die Antizipation der durch die Umgestaltung entstehenden Kosten eine große Rolle. Es werden also die Probleme ausgewählt, von denen man annimmt, daß sie leicht und damit kostengünstig zu beheben sind. In dieser Strategie spiegelt sich der Trugschluß wieder, daß die Art des Problems die Umgestaltungskosten festlegt. Dem ist nicht so. Vielmehr ist es der konkrete Gestaltungsvorschlag des software-ergonomischen Experten, der die Umgestaltungskosten bestimmt. Es liegt also in der Verantwortung des Gestalters, eine finanziell vertretbare, kreative und ggf. unvollständige Lösung eines Problems zu erarbeiten, anstatt das ganze Problem fallen zu lassen.

Was aber sind die wichtigsten Probleme?

Zwei Möglichkeiten der Priorisierung sollen im folgenden diskutiert und empirisch untersucht werden: *Priorisierung nach Fehlerkosten* und *Priorisierung nach Auftretensstabilität*.

2 Die Priorisierung von Problemhinweisen

Bevor diese beiden Methoden der Priorisierung vorgestellt werden, soll zunächst das Konzept der „Fehlereinheit" als Gegenstand der Priorisierung eingeführt werden, da die in der Fehlereinheit enthaltenen Daten die empirische Grundlage der Priorisierung bilden.

2.1 „Fehlereinheiten" als Gegenstand der Priorisierung

Personen haben oft ganz unterschiedliche Probleme im Umgang mit Softwaresystemen. Zur systematischen Erfassung derartiger Probleme hat sich in den letzten Jahren die „handlungsorientierte Fehlertaxonomie" [16] bewährt, da mit ihr Fehlerereignisse systematisch klassifiziert und verbessert werden können [6]. Hier werden *Fehlerereignisse*, die in Folge eines „Mismatches", d.h. einer Nicht-Passung von *Mensch* und *Computersystem* entstehen, als Nutzungsprobleme bezeichnet und nach Schritten im Handlungsprozeß und nach Ebenen der Handlungsregulation aufgeschlüsselt.

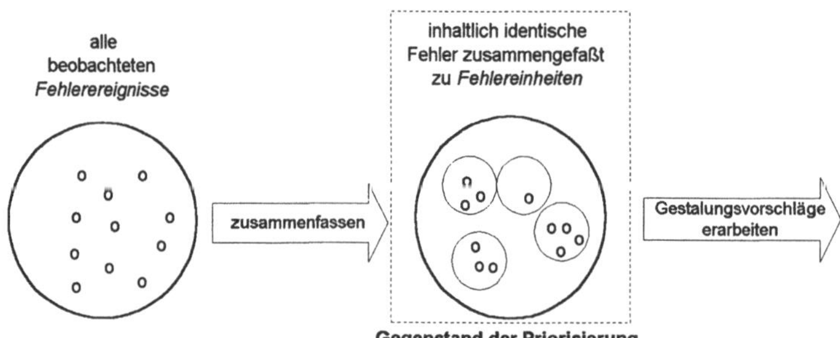

Abb. 1: Zusammenhang zwischen Fehlerereignissen und Fehlereinheiten

Nun sind viele Probleme im Umgang mit einem Softwaresystem keine singulären Ereignisse. Dies bedeutet, daß inhaltlich identische Probleme zusammengefaßt werden können. Diese zusammengefaßten Fehlerereignisse können als *Fehlereinheiten* bezeichnet werden. Abbildung 1 zeigt schematisch dieses Vorgehen. Beobachtete, inhaltlich identische *Fehlerereignisse* werden auf der theoretischen Grundlage der handlungsorientierten Fehlertaxonomie [16] zu einer

Fehlereinheit zusammengefaßt. Die Fehlertaxonomie mit ihren *Fehlerkategorien* liefert den Fehlerbegriff und ermöglicht durch die Zuordnung der *Fehlereinheit* zu einer Fehlerkategorie eine erste allgemeine Interpretation des zugrundeliegenden Problems. Gegenstand der Analyse und Gestaltung bleibt allerdings im ganzen Prozeß die *Fehlereinheit*.

Eine Fehlereinheit besteht aus einer *inhaltlichen Beschreibung*, der Zuordnung zu einer *Fehlerkategorie*, der *mittlere Fehlerbewältigungszeit* und der *Auftretenshäufigkeit* (vgl. Abb. 2). Die *inhaltliche Beschreibung* der Fehlereinheit hält den unmittelbaren Aufgabenzusammenhang und die Funktion des Systems fest, auf die sich der Fehler bezieht. Dies macht den Fehler für andere Personen nachvollziehbar und gibt Anhaltspunkte für konkrete Veränderungen.

Die Zuordnung der Fehlereinheit zu einer *Fehlerkategorie* gibt erste Hinweise, welche Prozesse am Zustandekommen der zugrundeliegenden Fehlerereignisse beteiligt sind und welche grundsätzlichen Gestaltungsmaßnahmen zu ergreifen sind.

Fehlerbewältigungszeit und *Auftretenshäufigkeit* sind Kennwerte, die Fehlereinheiten miteinander vergleichbar machen und eine Einschätzung der Konsequenzen einer Fehlereinheit erlauben. Die Fehlerbewältigungszeit ist die Zeit, die zur Korrektur eines Fehlers benötigt wird. Die Auftretenshäufigkeit ist die Anzahl der inhaltlich identischen Fehler, die zu einer Fehlereinheit gebündelt werden.

Inhaltliche Beschreibung: Der Benutzer möchte Antragsblätter aufrufen. Statt des Menüpunkts "Formantrag" wählt er "Haushalt".		
Fehlerkategorie:	*mittlere Fehlerbewältigungszeit:*	*Auftretenshäufigkeit:*
Wissensfehler	6 sec	2

Abb. 2: Bestandteile einer Fehlereinheit

2.2 Methoden der Priorisierung von Problemen

Nachdem *Fehlereinheiten* als Hinweise auf Problembereiche eines Softwaresystems vorliegen, stellt sich die Frage, wie diese Fehlereinheiten zur Bearbeitung priorisiert werden können. Im folgenden werden zwei Methoden vorgestellt: erstens nach *Fehlerkosten*, zweitens nach *Auftretensstabilität*.

2.2.1 Priorisierung nach Fehlerkosten

Zur Priorisierung kann man Fehlereinheiten im Hinblick auf ihre Kosten betrachten. Dabei lassen sich *ökonomische* Kosten eines Problems für ein Unternehmen (Institution) und *psychische* Kosten für den einzelnen Benutzer unterscheiden [16]. So wurde in einer Studie gefunden, daß 10% der gesamten Computerarbeitszeit auf Fehlerbewältigung entfällt. Diese Fehlerbewältigungszeit steht wiederum in signifikantem Zusammenhang zu negativen emotionalen Reaktionen [1,2]. Betrachtet man Computersysteme als Werkzeuge zur effizienten und effektiven Bearbeitung von Arbeitsaufgaben, ist dies bedenklich. Im Gegensatz zum Computer als Spielzeug muß der Computer als Werkzeug aus dem Aufmerksamkeitsbereich des Benutzers zurücktreten [8]. Er darf den Benutzer bei der Erledigung seiner Arbeitsaufgabe nicht behindern.

Die *Priorisierung nach Fehlerkosten* geht davon aus, daß die Probleme zuerst behoben werden müssen, die gleichzeitig hohe ökonomische und hohe psychische Kosten verursachen. Diese Fehlerkosten setzen sich zusammen aus der Häufigkeit des Auftretens des Fehlerereignisses und der Zeit, die benötigt wird, um das Fehlerereignis zu bewältigen. Mit anderen Worten: Fehlereinheiten kommt dann eine hohe Priorität zu, wenn sowohl die Auftretenshäufigkeit als auch die Bewältigungszeit hoch ist. Abbildung 3 veranschaulicht diesen Gedanken. Fehlereinheiten lassen sich in eine Rangfolge bringen, indem die Bewältigungszeiten aller Fehlerereignisse einer Fehlereinheit summiert werden.

	Bewältigungszeit	
Auftretenshäufigkeit	*niedrig*	*hoch*
hoch	mittlere Priorität	hohe Priorität
niedrig	niedrige Priorität	mittlere Priorität

Abb. 3: Priorität in Abhängigkeit von Auftretenshäufigkeit und Bewältigungszeit

2.2.2 Priorisierung nach Auftretensstabilität

In der handlungstheoretischen Fehlertaxonomie werden Nutzungsprobleme als „Mismatch" definiert. Das bedeutet, daß eine eindeutige Schuldzuweisungen in Richtung Computer oder Benutzer nicht möglich ist [12]. Aus praktischen Erwägungen bietet es sich jedoch an, den einen oder anderen Pol dieses „Mismatches" stärker zu berücksichtigen. Unterläuft mehreren Personen derselbe Fehler, dann ist es sinnvoll, die Software zu verändern; passiert es hingegen, daß eine einzelne Person Fehler macht, die anderen Personen nicht unterlaufen, dann sind entsprechende Qualifizierungsmaßnahmen sinnvoll [4].

Die Priorisierung nach Auftretensstabilität geht davon aus, daß „systembedingte" Fehlereinheiten mit einer höheren Priorität bearbeitet werden sollen, da hier die Chance, durch Veränderungen des Systems ein verringertes Auftreten des Problems bewirken zu können, höher eingeschätzt werden muß.

3 Methode

3.1 Versuchspersonen

Im Rahmen einer Studie zur partizipativen Evaluation und Gestaltung einer Bürosoftware wurden 12 Personen, sieben Frauen und fünf Männer untersucht. Das durchschnittliche Alter der Teilnehmer betrug 34 Jahre. Sie waren im Mittel neun Jahre in ihrem Beruf tätig, davon im Mittel 5 Jahre und 1 Monat in dem Tätigkeitsbereich, den die Funktionalität des Softwaresystems berührt. Die Teilnehmer arbeiteten im Durchschnitt seit rund 86 Monaten mit Computern und absolvierten im Schnitt 5½ Trainingstage für die untersuchte Software. Die Frage „Wie gut beherrschen Sie die getestete Software?" beantworteten die befragten Personen auf einer Skala von 1 (sehr schlecht) bis 7 (sehr gut) im Durchschnitt mit 2,8. Mit Computern arbeiteten die Personen im Mittel rund 17 Stunden pro Woche. Insgesamt arbeiteten die Befragten durchschnittlich mit 3,7 verschiedenen Softwareprogrammen. Die getestete Software war zum Zeitpunkt der Untersuchung noch nicht am Arbeitsplatz eingeführt.

3.2 Abhängige Variablen

Abhängige Variablen sind die *Aufgabenbearbeitungszeit*, die *Fehlereinheiten* und die *mentale Beanspruchung*. Die Aufgabenbearbeitungszeit muß registriert werden, da nur so der *Anteil der Fehlerbewältigungszeit an der Aufgabenbearbeitungszeit* bestimmt werden kann. Dies ist nötig, da individuelle Unterschiede in der Bearbeitungsgeschwindigkeit sich auch in längeren Fehlerbewältigungszeiten niederschlagen.

3.2.1 Aufgabenbearbeitungszeit

Die *Aufgabenbearbeitungszeit* ist die Zeit, die jeder Teilnehmer zur Erledigung einer Aufgabe benötigt. Eine maximale Aufgabenbearbeitungszeit wurde nicht vorgegeben. Vielmehr lag es in der Verantwortung der Teilnehmer, die Aufgabe dann zu beenden, wenn sie der Meinung waren, entweder das gewünschte Arbeitsziel erreicht zu haben oder aber keine Möglichkeit der Aufgabenlösung mehr zu sehen.

3.2.2 Fehlereinheiten

Fehlereinheiten lassen sich über Auftretenshäufigkeit und Fehlerbewältigungszeit charakterisieren. *Auftretenshäufigkeit* ist die Anzahl der inhaltlich identischen Fehlerereignisse, die zu einer Fehlereinheit gebündelt werden. *Fehlerbewältigungszeit* ist definiert als die Zeitspanne vom Beginn der fehlerhaften Aktion, bis zu dem Punkt, an dem eine neue Aktion ausgeführt wird. Diese Aktion kann dann entweder direkt der Fehlerbewältigung dienen, dann wird sie mitgemessen, oder aber wiederum ein neues Fehlerereignis sein, dann beginnt eine neue Messung.

3.2.3 Mentale Beanspruchung

Zur Annäherung an die psychische Belastung, im Sinne von subjektiv aversiv erlebter Beanspruchung [13], kam die BSMA-Skala [3,17] zum Einsatz. Diese eindimensionale Skala ermöglicht die Einschätzung der mentalen Beanspruchung von 0 bis 220. Neben den Zahlen der Skala erleichtern verbale Anker wie „kaum anstrengend" bei 20 oder „außerordentlich anstrengend" bei 204 die Einschätzung. Ein Vorteil dieses Verfahrens ist seine leichte Anwendbarkeit und schnelle Beantwortbarkeit. Dies ermöglicht, Beanspruchungsmessungen mehrmals während einer Testphase durchzuführen.

3.3 Versuchsablauf

Jeder Teilnehmer hatte in Einzelsitzungen nacheinander jeweils fünf Testaufgaben unterschiedlicher Komplexität in einer randomisierten Reihenfolge zu lösen. Bei der Aufgabenkonstruktion wurde hohen Wert auf Repräsentativität und unterschiedliche Schwierigkeitsgrade gelegt. Jede Aufgabe sprach andere Aufgabenbereiche der Tätigkeit und damit auch andere Funktionalitätsbereiche des Softwaresystems an.

Die Aufgabenbearbeitungszeit wurde mit Ergonom 1.0 [14] aufgezeichnet; Aktionen zur Aufgabenlösung sowie die verbalen Äußerungen des Benutzers mit ScreenCam 2.0 [7], einer Protokollsoftware, die die Aufzeichnung von Bildschirmaktionen als eine Art „Film" ermöglicht, der unabhängig von dem Softwaresystem abgespielt werden kann.

Die Beurteilung der Beanspruchung mittels BSMA wurde direkt nach jeder Aufgabenbearbeitung vorgenommen, d.h. im Laufe der Untersuchung wurden von jedem Teilnehmer fünf Beurteilungen abgegeben.

4 Ergebnisse

4.1 Fehleranalyse

Von 60 Untersuchungseinheiten (12 Teilnehmer mal 5 Testaufgaben) wurden 47 Einheiten (78%) analysiert. 13 Untersuchungseinheiten konnten aufgrund von Systemabstürzen, Netzwerkproblemen oder ähnlichem bei der Aufgabenbearbeitung nicht protokolliert werden.

Insgesamt wurden 353 Fehlerereignisse registriert. Faßt man diese zusammen, ergeben sich

117 verschiedene Fehlereinheiten. Diese Fehlereinheiten bündeln im Mittel rund 3 Fehlerereignisse (Auftretenshäufigkeit) mit einer mittleren Fehlerbewältigungszeit von 10,7 Sekunden.

Abbildung 4 zeigt alle 117 Fehlereinheiten mit ihren Kennwerten „mittlere Fehlerbewältigungszeit" und „Auftretenshäufigkeit". Es wird deutlich, daß der größte Anteil der Probleme relativ niedrige Auftretenshäufigkeiten und relativ kurze Fehlerbewältigungszeiten aufweist. Nur wenige Fehlereinheiten haben hohe Bewältigungszeiten oder hohe Auftretenshäufigkeiten und es finden sich keine, die hohe Werte auf beiden Kriterien erreichen. Die beiden Kennwerte korrelieren nicht miteinander.

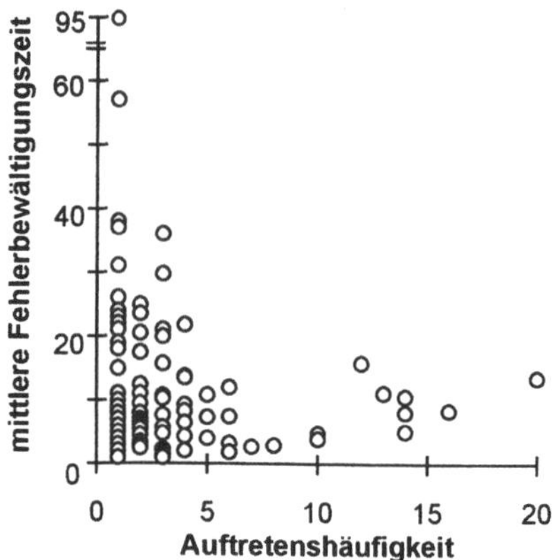

Abb. 4: Die Kombination mittlerer Fehlerbewältigungszeit und Auftretenshäufigkeit bei Fehlereinheiten

4.2 Ergebnisse zur Priorisierung nach Kosten

Die Priorisierung nach Kosten macht nur dann Sinn, wenn (1) der Anteil der Fehlerbewältigungszeit an der Aufgabenbearbeitungszeit so groß ist, daß überhaupt von *ökonomischen* Kosten im Sinne einer Beeinträchtigung der Produktivität ausgegangen werden kann und, wenn (2) die angenommenen *psychischen* Kosten der Probleme über die Fehlerbewältigungszeit eingeschätzt werden können.

Dazu wurde die Aufgabenbearbeitungszeit, die Fehlerbewältigungszeit, die Fehlerhäufigkeit und die Beanspruchung pro Teilnehmer errechnet, indem die Werte der von dem jeweiligen Teilnehmer einzeln bearbeiteten Untersuchungseinheiten gemittelt wurden.

4.2.1 Ökonomische Kosten: Anteil der Fehler- an der Aufgabenbearbeitungszeit

Betrachtet man die Aufgabenbearbeitungszeit gemittelt über die Teilnehmer ($N = 12$), so benötigten diese durchschnittlich 327 Sekunden für die Bearbeitung einer Untersuchungseinheit ($s = 307$, *min* = 65, *max* = 1264). Die mittlere Fehlerbewältigungszeit pro Untersuchungseinheit betrug 102 Sekunden ($s = 151$, *min* = 2, *max* = 574). Dies entspricht einem Anteil der mittleren Fehlerbewältigungszeit an der Aufgabenbearbeitungszeit von 31,2% und weist auf eine massive Beeinträchtigung der Produktivität hin.

4.2.2 Psychische Kosten: Zusammenhang von Fehlerbearbeitungszeit und Beanspruchung

Es ergibt sich über die Teilnehmer eine mittlere Beanspruchung von 49 laut BSMA-Skala ($s = 26$, *min* = 20, *max* = 120). Dieser Skalenwert deutet auf eine eher schwache mentale Beanspruchung hin. Er liegt 9 Skalenpunkte über dem verbalen Anker „etwas anstrengend" und 25 Skalenpunkte unter „einigermaßen anstrengend".

Mentale Beanspruchung und Fehlerbewältigungszeit korrelieren signifikant (r = .88, p < .01, N = 12). Personen, die länger zur Bewältigung ihrer Fehler benötigen, fühlen sich mental höher beansprucht.

Die Korrelation von mentaler Beanspruchung und prozentualem Anteil der Fehlerbewältigungszeit an der Aufgabenbearbeitungszeit beträgt r = .65 (p < .05, N = 12). Das bedeutet, daß Personen, die einen höheren Anteil der Aufgabenbearbeitungszeit mit Fehlerbewältigung verbringen, eine höhere Beanspruchung empfinden. Fehlerbewältigungszeit kann als Indikator für psychische Kosten betrachtet werden.

4.3 Ergebnisse zur Priorisierung nach Auftretensstabilität

Als Indikator für die Auftretensstabilität einer Fehlereinheit kann die Anzahl der unterschiedlichen Benutzer herangezogen werden, bei denen dieselbe Fehlereinheit beobachtet werden kann. Abbildung 5 zeigt diese Häufigkeiten. Von 117 verschiedenen Fehlereinheiten konnten 80 bei jeweils nur einer Person beobachtet werden. Dies sind idiosynkratische Fehlereinheiten, die den „Benutzer"-Pol des „Benutzer-Software" Kontinuums markieren. Betrachtet man das andere Ende des Kontinuums, so finden sich keine Fehlereinheiten, die bei allen Personen gleichermaßen beobachtet wurden. Die höchste Übereinstimmung ist dieselbe Fehlereinheit bei 10 von 12 Benutzern. Dieser Fehler kann als interindividuell stabil bezeichnet werden.

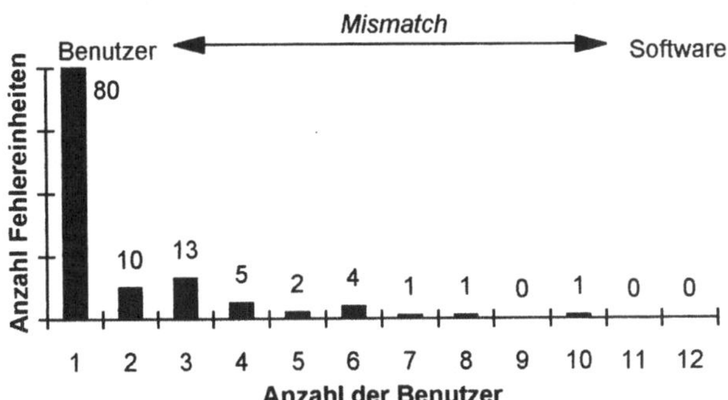

Abb. 5: Anzahl der Fehlereinheiten, die bei Benutzern gleichzeitig beobachtet wurden

4.4 Vergleich der Priorisierungen

Insgesamt ergeben sich aufgrund der hier vorgestellten Überlegungen zur Priorisierung zwei Möglichkeiten der Auswahl von Problemhinweisen, die im folgenden miteinander verglichen werden sollen: (1) Priorisierung nach Fehlerkosten und (2) Priorisierung nach Auftretensstabilität.

Auf Grundlage der Überlegungen zur Priorität in Abhängigkeit von Auftretenshäufigkeit und Bewältigungszeit (vgl. Abb. 3) schlagen wir vor, die 25% der Fehlereinheiten zur Bearbeitung auszuwählen, die den größten Anteil an der gesamten Fehlerbewältigungszeit ausmachen. In dem vorliegenden Fall sind dies 29 von insgesamt 117 Fehlereinheiten. Diese 25% der Fehlereinheiten sind für 69% der gesamten Fehlerbewältigungszeit verantwortlich. Könnten diese Fehlereinheiten beseitigt werden, würde sich der Anteil der Fehlerbewältigungszeit an der Aufgabenbearbeitungszeit von 31,2% auf 12,3% verringern und damit auch die assoziierten ökonomischen und psychischen Kosten. Obwohl dies ein idealisierter Wert ist, der davon ausgeht, daß keine neuen Probleme durch Umgestaltung erzeugt werden, wird deutlich, daß in der Bearbeitung weniger, aber sorgfältig ausgewählter Fehlereinheiten ein hohes Potential zur Verbesserung steckt.

Eine gröbere Priorisierung kann anhand der Auftretensstabilität vorgenommen werden. Hierbei kann die Anzahl der Benutzer, die dasselbe Problem haben, als Rangfolge betrachtet werden. Im vorliegenden Fall ist unser Vorschlag, auf die Bearbeitung der idiosynkratischen Fehler zu verzichten. Die 37 der 117 Fehlereinheiten (31,6%), auf die dieses Kriterium zutrifft, sind immer noch eine akzeptable Menge an zu bearbeitenden Problemhinweisen. Auf diese Art ausgewählte Problemhinweise sind für 63,2% der gesamten Fehlerbewältigungszeit verantwortlich. Die Beseitigung der zugrundeliegenden Probleme könnte zu einer Reduktion des Anteils der Fehlerbewältigungszeit an der Aufgabenbearbeitungszeit von 31,2% auf 14,3% führen.

Die Ergebnisse der beiden Priorisierungsmethoden sind ähnlich. Vergleicht man die Selektion aufgrund beider Priorisierungsmethoden, so findet man, daß von den 29 Problemhinweisen mit dem größten Anteil der Fehlerbewältigungszeit an der gesamten Fehlerbewältigungszeit (Priorisierung nach Fehlerkosten) 19 identisch mit den durch die zweite Methode ausgewählten sind (vgl. Abb. 6). Dies entspricht einer signifikanten Korrelation ($Phi = .42, p < .001, N = 117$ Fehlereinheiten). Auch der Anteil der erklärten Fehlerbewältigungszeit ist bei beiden in etwa vergleichbar. Allerdings müssen bei der Priorisierung nach Auftretensstabilität 8 Problemhinweise (6,8%) mehr bearbeitet werden.

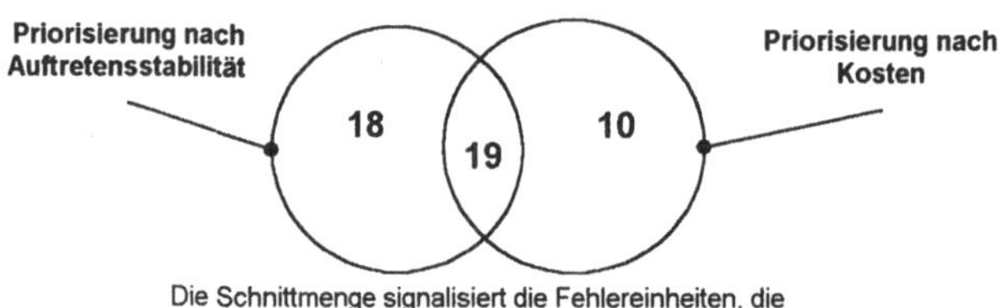

Abb. 6: Anzahl der über Priorisierungsmethoden ausgewählten Fehlereinheiten

5 Diskussion

In den Ergebnissen der Fehleranalyse dominieren Fehlereinheiten, die Fehlerereignisse mit relativ kurzen Fehlerbewältigungszeiten und geringen Auftretenshäufigkeiten zusammenfassen. Um detaillierte Hinweise auf die Schwachstellen eines Softwaresystems zu bekommen, muß mit einem hohen Detailliertheitsgrad in der Analyse gearbeitet werden. Je detaillierter die Analyse, desto weniger Fehlerereignissen werden als ähnlich klassifiziert und zu Fehlereinheiten gebündelt.

Ein Drittel der Aufgabenbearbeitungszeit wird auf die Bewältigung von Fehlern verwendet. Dieser Anteil ist hoch, selbst wenn man davon ausgeht, daß die Teilnehmer relativ ungeübt im Umgang mit dem System waren und dadurch der zeitliche Aufwand zur Behebung steigt [10, 11]. Eine Beseitigung der identifizierten Problembereiche wird zu einer deutlichen Steigerung der Produktivität des Benutzers führen.

Die Einschätzung der subjektiv empfundenen, mentalen Beanspruchung, die zwischen „etwas anstrengend" und „einigermaßen anstrengend" liegt, deutet auf eine eher schwache Beanspruchung hin. Selbst die Person, die die höchste Beanspruchung angab, konnotierte mit „ziemlich anstrengend" einen moderaten Wert. Hier wird der artifizielle Charakter einer experimentellen

Gebrauchstauglichkeitsstudie offensichtlich. Man kann davon ausgehen, daß ähnliche Probleme im Kontext der alltäglichen Aufgabenbearbeitung aufgrund ihrer Konsequenzen zu massiveren Beanspruchungen führen werden. Der Bedeutung der Erfüllung bzw. Nichterfüllung einer Aufgabe im Rahmen einer solchen Untersuchung wird ein geringerer Stellenwert als im Arbeitsalltag zugemessen.

Der signifikante Zusammenhang zwischen Fehlerbewältigungszeit bzw. dem Anteil der Fehlerbewältigungszeit an der Aufgabenbearbeitungszeit und subjektiv empfundener mentaler Belastung macht die Fehlerbewältigungszeit zu einem Indikator psychischer Fehlerkosten. Personen, die einen höheren Anteil ihrer Aufgabenbearbeitungszeit mit Fehlerbewältigung verbringen müssen, empfinden auch eine höhere Beanspruchung. Es liegt also nahe, Problemhinweise so zu priorisieren, daß die Fehlereinheiten, die den größten Anteil der Fehlerbewältigungszeit ausmachen, zuerst zu bearbeiten sind. So hat die Bearbeitung weniger Fehlereinheiten eine große Auswirkung auf die Reduktion der Fehlerbewältigungszeit und damit u.U. auch auf die Reduktion der Beanspruchung.

Priorisierung nach Auftretensstabilität ist eine einfachere Methode, da die Fehlerbewältigungszeit nicht erhoben werden muß. Im Vergleich zur Priorisierung nach Kosten führt sie zu ähnlich guten Ergebnissen. Von Nachteil ist, daß im Gegensatz zur Priorisierung nach Fehlerkosten die Zahl der ausgewählten Fehlereinheiten nicht von vornherein festliegt, da die Auftretenshäufigkeit idiosynkratischer Fehlerereignisse von Situation zu Situation unterschiedlich sein kann.

Die Übereinstimmung der anhand beider Methoden ausgewählten Fehlereinheiten ist abhängig von der Konstellation der Auftretenshäufigkeiten und der Bewältigungszeiten von Fehlerereignissen. Die Priorisierung nach Auftretensstabilität unterschätzt im Gegensatz zur Priorisierung nach Kosten systematisch die Einflußkraft von Fehlereignissen mit hohen Fehlerbewältigungszeiten und niedrigen Auftretenshäufigkeiten. In vorliegenden Fall wurde diese Kombination nicht oft beobachtet, was zu der hohen Übereinstimmung führte. Sollte es der Regelfall sein, daß es eher selten Probleme gibt, die nur vereinzelt auftreten und dann stets lange brauchen, bis sie bewältigt sind, dann ist die Methode der Priorisierung nach Auftretensstabilität der Priorisierung nach Kosten vorzuziehen: sie ist leichter durchführbar und erzielt trotzdem ähnlich gute Ergebnisse. Gibt es viele Fehlerereignisse mit hohen Fehlerbewältigungszeiten und niedrigen Auftretenshäufigkeiten, ist die Priorisierung nach Kosten die Methode der Wahl.

Im Zentrum der vorgeschlagenen Priorisierungsarten stehen Fehlerkosten, die über Fehlerbewältigungszeiten abgeschätzt werden. Dies geschieht vor dem Hintergrund, daß Beanspruchung im Arbeitsalltag zu Streß wird, und daß durch software-ergonomische Gestaltungsmaßnahmen die Beanspruchung reduziert werden kann. Deutlich wird, daß durch die Anwendung der vorgeschlagenen Priorisierungsmethoden zur Auswahl von Fehlereinheiten ein großer Teil der Fehlerbewältigungszeit - und damit der Fehlerkosten - reduziert werden könnte.

Offen ist, ob diese Kriterien zur Priorisierung von allen Personengruppen geteilt werden. Unter Umständen ziehen Personen Kriterien vor, die wenig mit objektiven Fehlerkosten, sondern eher mit ganz persönlichen Interessen, mit Vorstellungskraft, Gestaltungsspielraum oder finanziellen Aspekten zu tun haben.

Der Einsatz von Fehleranalysen und das Konzept der Fehlereinheit kann im Rahmen einer formativen Evaluation hilfreich sein. Besonders die Fehlereinheit bietet die Möglichkeit, qualitative mit quantitativen Daten zu verbinden. Aber sie darf nicht das einzige Mittel zur Analyse eines Softwaresystems sein. In einer empirischen Gebrauchstauglichkeitsstudie sind Merkmale des Systems und Merkmale der teilnehmenden Benutzer, z.B. Werkzeugwissen und Computerwissen, stark miteinander verschränkt. Es wird dabei nicht nur das System an sich getestet, sondern auch die Angemessenheit des Systems für die testende Benutzerpopulation. Verändert sich diese, werden sich auch die Ergebnisse der Untersuchungen ändern. Um diesen Einfluß zu kontrollieren, müssen auf der einen Seite Maßnahmen zur Erfassung der relevanten Benutzermerkmale ergriffen werden, auf der anderen Seite müssen beobachtete Nutzungsprobleme immer auf dem Hintergrund der Benutzerpopulation interpretiert und erst dann in Gestaltungsmaßnahmen umgesetzt werden.

In weiteren Arbeiten muß geprüft werden, inwiefern das Konzept der Fehlereinheit praxistauglich ist und in Verbindung mit der hier vorgeschlagenen Priorisierung in ein umfassendes Gestaltungskonzept eingepaßt werden kann. Da es immer auch die Möglichkeit gibt, Probleme durch beteiligte Personengruppen, wie Benutzer, Entwickler oder Auftraggeber, priorisieren zu lassen, müssen Untersuchungen folgen, die auch diese Methode weiter explorieren und bewerten.

Wie dem auch sei: die vorliegende Arbeit konnte zeigen, daß Priorisierung anhand empirischer Daten möglich ist und im Rahmen software-ergonomischer Qualitätssicherung einen Beitrag zur effizienten Bearbeitung von Problembereichen eines Softwaresystems zu leisten vermag.

Literaturverzeichnis

[1] F.C. Brodbeck: Fehlerbewältigungszeiten und die Nutzung von Unterstützungsmöglichkeiten. In: M. Frese & D. Zapf (Hg.): Fehler bei der Arbeit mit dem Computer - Ergebnisse von Beobachtungen und Befragungen im Bürobereich. Huber, Bern, 1991, 80-94.

[2] F.C. Brodbeck, D. Zapf, J. Prümper & M. Frese: Error handling in office work with computers: A field study. In: Journal of Occupational and Organizational Psychology 60(4) (1993), 303-317.

[3] K. Eilers, F. Nachreiner & K. Hänecke: Entwicklung und Überprüfung einer Skala zur Erfassung subjektiv erlebter Anstrengung. In: Zeitschrift für Arbeitswissenschaft 40 (1986). 215-224.

[4] M. Frese & D. Zapf: Fehlersystematik und Fehlerentstehung: Eine theoretische Einführung. In: M. Frese & D. Zapf (Hg.): Fehler bei der Arbeit mit dem Computer - Ergebnisse von Beobachtungen und Befragungen im Bürobereich. Huber, Bern, 1991, 14-31.

[5] L. Herrmann, L.L. Morris & C. Taylor Fitz-Gibbon: Evaluator's handbook. Beverly Hills, Sage, 1988.

[6] A. Kensik, J. Prümper & M. Frese: Ergonomische Gestaltung von Software auf Grundlage handlungsorientierter Fehleranalysen. In: H.D. Böcker (Hg.): Software-Ergonomie '95 - Anwendungsbereiche lernen voneinander. Stuttgart, Teubner, 1995, S. 217-232.

[7] Lotus: ScreenCam 2.0. Lotus Development Corporation, 1995.

[8] T.W. Malone: Toward a theory of intrinsically motivating instruction. In: Cognitive Science 4 (1981), 333-369.

[9] L. Müller-Böling: Akzeptanz und Partizipation - Sind Systemgestalter lernfähig? In: K.T. Schröder (Hg.): Arbeit und Informationstechnik. Springer, Berlin, 1986, 153-166.

[10] J. Prümper: Handlungsfehler und Expertise. In: M. Frese & D. Zapf (Hg.): Fehler bei der Arbeit mit dem Computer - Ergebnisse von Beobachtungen und Befragungen im Bürobereich. Huber, Bern, 1991, 118-130.

[11] J. Prümper, D. Zapf, F.C. Brodbeck & M. Frese: Some surprising differences between novice and expert errors in computerized office work. In: Behaviour & Information Technology 11(6) (1992), 319-328.

[12] J. Rasmussen: Human Error Data. Facts or Fiction. Riso National Laboratory, Roskilde, 1985.

[13] I. Udris & M. Frese: Belastung, Streß, Beanspruchung und ihre Folgen. In: D. Frey, C. Graf Hoyos & D. Stahlberg (Hg.): Angewandte Psychologie. Psychologie Verlags Union, München, 1988, 428-447.

[14] S. Willmeroth: Ergonom 1.0. Ziff Verlag GmbH, 1993.

[15] H. Wottawa & H. Thierau: Evaluation. Huber, Bern, 1990.

[16] D. Zapf, F.C. Brodbeck & J. Prümper : Handlungsorientierte Fehlertaxonomie in der Mensch-Computer Interaktion. In: Zeitschrift für Arbeits- und Organisationspsychologie 33 (1989), 178-187.

[17] R. Zijlstra & L. van Doorn: The construction of a scale to measure subjective effort. Universität Groningen: Unveröffentlichter Bericht, 1985.

Adressen der Autoren

Marc Hassenzahl
TH-Darmstadt
Fachbereich Psychologie
Hillebergstraße 42
64319 Pfungstadt

hassenzahl@aol.com
hassenzahl@hrz1.hrz.th-darmstadt.de

Prof. Dr. Jochen Prümper
FHTW-Berlin
Fachgebiet Wirtschaftspsychologie
Treskowallee 8
10313 Berlin

pruemper@aol.com
pruemper@rz.fhtw-berlin.de

Uta Sailer
LMU- München
Fachbereich Psychologie
Orleansstraße 39
81667 München

saileru@aol.com
sailer@mip.paed.uni-muenchen.de

Software, Ergonomie, Gestaltung?

Falk Höhn

Fachbereich Kunst & Design, FH-Hannover

Zusammenfassung

Der folgende Beitrag stellt den Versuch eines Formgestalters[1] dar, sich in die vielschichtigen Diskussionen von Software-Entwicklern einzumischen und seine Sicht auf das Problemfeld der Software-Entwicklung, besonders ihrer Ergonomie und Gestaltung, darzustellen. Gestalter verfügen über die Erfahrungen, mit komplexen, ihnen mehr oder weniger geläufigen Aufgabenstellungen umzugehen und diese zu lösen. Die gegenwärtige Gestaltungsdiskussion im Kontext der Software-Entwicklung zeigt, daß es Zeit für die Gestalter ist, in diese Diskussion einzugreifen und ihren spezifischen Beitrag einzubringen.

1. Formgestaltung

Die Formgestaltung (das Industriedesign) befaßt sich allgemein mit der Optimierung industriell gefertigter Erzeugnisse hinsichtlich

- utilitärer-   (...auf ihre Nutzung gerichteten...)
- faktibilitärer-   (...auf ihre Herstellung gerichteten...)
- operationaler-   (...mit ihrer Bedienung zusammenhängenden...)
- ökologischer-   (...auf ihre Umweltverträglichkeit gerichteten...) und
- ökonomischer-   (...auf ihre Verwertung gerichteten...)

Eigenschaften [1]. Dabei ist allerdings in den letzten Jahren ein Trend spürbar, der von der ausschließlichen Gestaltung technischer Produkte wegführt, hin zu einer stärkeren Orientierung auf die Gestaltung von komplexeren Strukturen (z. B. Informations- oder Verkehrssysteme). Der Gestalter übernimmt dabei zunehmend die Rolle des Organisators und Managers solcher Produktentwicklungen. Sein unschätzbarer Vorteil liegt insbesondere darin, daß er unvoreingenommen und ganzheitlich (alle Aspekte berücksichtigend und gegeneinander abwägend) an die Lösung der Probleme herangehen kann, ja muß. Die eigentlichen Fachleute (Konstrukteur, Technologe, Ergonom, Ökonom usw.) sind oftmals schon zu spezialisiert, um noch die Gesamtheit zu überblicken.

Da jedes zu gestaltende Produkt/System eine „Schnittstelle" zum Menschen besitzt, spielt die Ergonomie eine zentrale Rolle für die gestalterische Arbeit und liefert oftmals wesentliche Lösungsansätze bei der Verbesserung von Produkten, da es im Kern meistens um die Verbesserung der „Schnittstellen" geht. Bei der Betrachtung von Systemstrukturen spielt dann die Gestaltung von Arbeitsabläufen oder -prozessen eine wesentliche Rolle. Ergonomie und Gestaltung sollte man trotzdem nicht gleichsetzen!

Seit einiger Zeit rückt nun die Produktklasse „Software" ins Blickfeld der Gestalter. Dafür gibt es zwei wesentliche Gründe:

---

[1] **Formgestalter** - nicht **Designer**, weil der Terminus des „Design" im Zusammenhang mit „Software" zu Irritationen führen kann.

1. „Klassische" technische Produkte verschwinden immer mehr, da sich in rasanter Geschwindigkeit die Mensch-Maschine-Schnittstellen von „mechanischen" zu „elektronischen" Schnittstellen verwandeln. Ein Charakteristikum dabei ist, daß die Schnittstelle von der Maschine räumlich abkoppelbar wird. Zunehmend geht es also primär um die Gestalt der Schnittstelle (Hardware/Software) und nur noch sekundär um die Gestalt der Maschine bzw. des Systems.

2. In immer stärkerem Maße werden komplexe Arbeitsabläufe (, wie man sie beispielsweise im Büro oder in Schaltwarten vorfindet,) in Softwaresystemen nachgebildet. Diese Systeme werden dann eingesetzt, um die Effizienz der „klassischen" Arbeitsabläufe zu erhöhen bzw. zu rationalisieren. Software-Entwickler stoßen bei der Gestaltung aber sehr schnell an Grenzen, da ihr Repertoire an ergonomischen und besonders an gestalterischen Mitteln und ihr Vermögen, diese bei der Gestaltung der Schnittstelle und ihrer Software einzusetzen, begrenzt sind. Die Schlußfolgerung: Erkenntnisse der Ergonomie und Gestaltung müssen angewendet werden. [2]

Die Probleme bestehen nun darin,

- daß die Gestalter zur Software-Entwicklung recht schwer einen Zugang finden, da ihnen oftmals die sehr speziellen technologischen Grundlagen fehlen;

- daß die Gestalter mit den entwickelten und vorhandenen Methoden wenig anfangen können, weil sie für ihre Zwecke entweder zu allgemein (z. B. KADS) [3;4;5;6] oder zu speziell und eng (z. B. MIKE/KARL) [7; 8] sind;

- daß man sie zu spät zur Gestaltung eines Softwareproduktes hinzuzieht;

- daß den Gestaltern keine adäquaten Werkzeuge zur Verfügung stehen, um ihre Ideen in allen Phasen der Entwicklung einbringen zu können;

Die Software-Entwickler haben inzwischen ein reiches methodisches und technologisches Instrumentarium für sich selbst entwickelt, es jedoch versäumt, eine die Gestalter integrierende Methodik zu entwickeln (; die Gestalter haben bisher aber auch nicht entschieden genug danach verlangt). Verfolgt man die Diskussionen, kann man aber leicht den Eindruck gewinnen, daß die Software-Entwickler alles (Ergonomie und Gestaltung) selbst machen wollen bzw. können, vorausgesetzt man gibt ihnen die entsprechenden Guide-Lines. Das wird so aber nicht funktionieren!

2. Produktentwicklungsprozeß

Auf diesen Problemkreis soll an dieser Stelle nur in der für diesen Beitrag gebotenen Kürze eingegangen werden.
Der Autor geht grundsätzlich davon aus, daß es bei der Entwicklung eines technischen Produktes einen unerläßlichen „Kern"-Entwicklungsprozeß gibt, ohne den eine technische Entwicklung nicht möglich wäre. Daneben gibt es, diesen „Kern" umklammernde Entwicklungsprozesse, die vor allem die Qualität und die Effizienz der Entwicklung beeinflussen, aber für eine Produktentwicklung (scheinbar!) nicht unbedingt erforderlich sind.

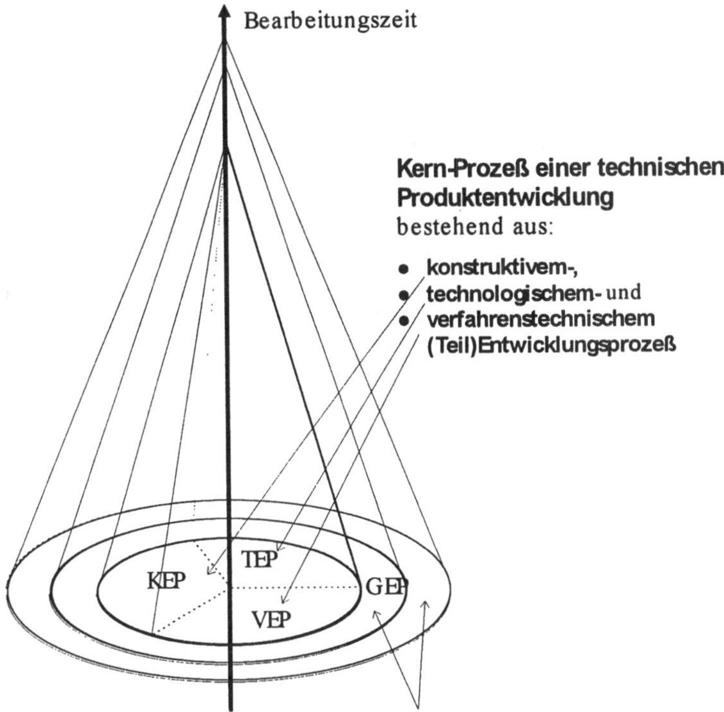

Überlagernde, aber nicht unbedingt erforderliche,
(Teil)Entwicklungsprozesse;
z. B. **Formgestaltung** (gest. Entwicklungsprozeß)

Bild 1   Am Produktentwicklungsprozeß Beteiligte

Bei der Entwicklung „klassischer" Produkte hat sich gezeigt, daß ohne die Integration des
gestalterischen Entwicklungsprozesses, sinnvollerweise keine Produktentwicklung stattfinden
sollte. Die Integration konnte aber nur gelingen, weil die grundsätzlichen Methoden von
Konstrukteuren, Technologen und Gestaltern sehr ähnlich sind und in der Ausbildung der
Gestalter entsprechende Inhalte und das notwendige Kontaktwissen vermittelt werden. Nach
Meinung des Autors stehen heute die Softwareentwickler vor der Grundsatzentscheidung,
Gestalter in ihren Entwicklungsprozeß *sinnvoll* zu integrieren. Bei der Ausbildung von
Gestaltern muß dem Rechnung getragen werden. Die Parallelen zur „klassischen" Technik
liegen dabei auf der Hand.

Zum gestalterischen Entwicklungsprozeß:

Vielleicht können einige grundsätzliche Ausführungen zur Struktur des gestalterischen
Entwicklungsprozesses helfen, den Software-Entwicklungsprozeß in Hinsicht auf die Gestaltung effektiver zu erweitern:

Inzwischen gibt es zahlreiche Vorschläge, wie der gestalterische Entwicklungsprozeß
modellhaft zu strukturieren ist. Darauf soll an dieser Stelle nicht näher eingegangen werden.

Man kann sich an der von FRICK [9] vorgeschlagenen Struktur des gestalterischen Entwicklungsprozesses orientieren:

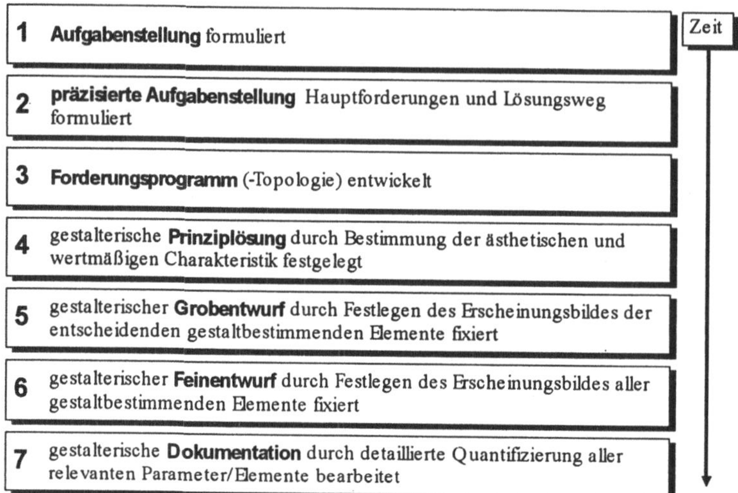

Bild 2: Grundstruktur des gestalterischen Entwicklungsprozesses

Der Gestalter wird mit unterschiedlichen Aufgabenklassen und mit unterschiedlichen Aufgabentypen konfrontiert. Die Aufgabenklassen ergeben sich aus der „Formgebundenheit" in Abhängigkeit von den technischen Freiheitsgraden

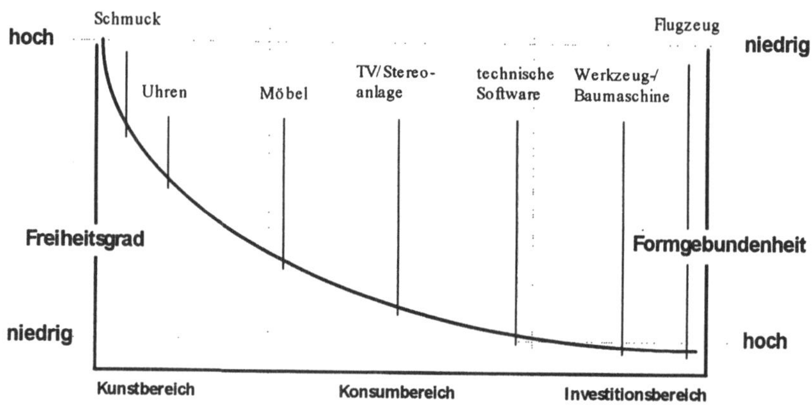

Bild 3: Freiheitsgrad in Abhängigkeit von der technisch bedingten Formgebundenheit (nach FRICK)

Folgt man diesem Gedanken, könnte man sich eine ähnliche Klassifizierung für Software-Produkte vorstellen:

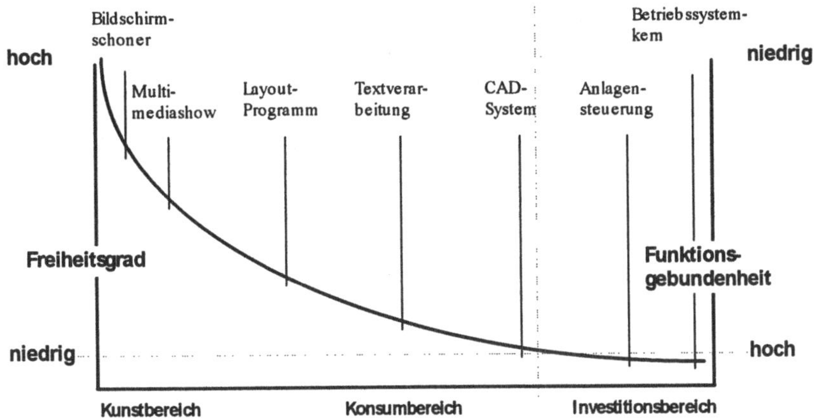

Bild 4: **Freiheitsgrad** in Abhängigkeit von der technisch bedingten **Funktionsgebundenheit**

Je nach Form-/Funktionsgebundenheit ist der für das jeweilige Produkt zu erbringende gestalterische Beitrag innerhalb des Produktentwicklungsprozesses unterschiedlich umfangreich. Aber auch die Geläufigkeit der Aufgabenstellung spielt hinsichtlich des Ergebnisses der gestalterischen Tätigkeit eine nicht zu unterschätzende Rolle. Im Gestaltungsbereich finden wir prinzipiell drei Aufgabentypen [10]:

1. Die **gestalterische Überarbeitung**

   Gesucht wird die Variante bekannter Lösungen.
   Fast alle relevanten Gestaltungsparameter sind bekannt; der Gestalter kennt sich in der Materie aus, seine Vorstellung von der Lösung - das Leitbild - und seine Strategie dazu sind sehr konkret; die zu erwartende Innovation der Lösung ist gering...

2. Die **gestalterische Bearbeitung**

   Gesucht wird eine innovative Lösung mit bekannter Technik (Erfahrungen liegen vor). Der Gestalter hat die Freiheit, grundsätzlich die Struktur, das Wirkprinzip etc. neu zu bearbeiten und mit neue Materialien ins Kalkül zu ziehen; die Aufgabe fordert den Gestalter heraus; die Innovation der Lösung hängt vom Können des Gestalters und seiner Fähigkeit, sich vom Leitbild zu lösen, ab...

3. Die **gestalterische Erfindung**

   Gesucht wird eine völlig neuartige Lösung.
   Der Gestalter entwirft in den vorhandenen objektiven Grenzen ein bis dahin nicht existierendes Objekt. Die Innovation ist garantiert - aber nicht die Optimalität der Gestalt.

   Dieser Aufgabentyp kommt sehr selten vor. Wirklich neue Produkte durchlaufen nach ihrer Erfindung schnelle Zyklen der Variantenbildung und Bearbeitung, bis sich eine optimierte Gestalt/Form, die dann zum Leitbild wird, herausgebildet hat.

Die gesamte Produktklasse „Software" ist im Sinne von (3) neu. Die gegenwärtige Phase der Diskussion dreht sich daher vor allem um die Optimierung der neuen Produkte, und zwar hinsichtlich Nutzungsfreundlichkeit, Komfort und Aufgabenangemessenheit (Ergonomie) sowie - im kommerziell interessanten Bereich - hinsichtlich Zielgruppenbezug, Differenzierung gegenüber den Konkurrenzprodukten und Ästhetik (Gestaltung).

Eine Ursache dafür, daß die Software-Entwickler beim Problem der Gestaltung der Software nicht vorankommen, liegt aus Sicht des Autors darin begründet, daß (Software)**Gestaltung** und (Software)**Ergonomie** in hohem Maße gleichgesetzt werden. Genau das führt aber letztendlich dazu, daß man eigentlich nicht über Gestaltung diskutiert.

Die (Software)Ergonomie ist für die (Software)Gestaltung unerläßlich, kann aber aus gestalterischer Sicht im Prinzip nur dabei helfen, die gröbsten Schnitzer zu verhindern. Folgt man all ihren Empfehlungen, hat man garantiert nichts falsch gemacht, aber ob das (Software)Produkt auch gut gestaltet ist, steht auf einem anderen Blatt. Genau darin liegt das Problem z. B. der Styleguids.

Daraus resultiert die Uniformität von Software-Oberflächen, die sich ihrerseits wieder aus der geringen Anzahl von Entwicklungswerkzeugen bzw. deren gestalterischer Beschränktheit ergibt. Das ist einerseits unter dem Aspekt der konsistenten Bedienung von Softwaresystemen innerhalb einer Softwareumgebung (angeblich) erwünscht. Andererseits verschenkt man bei der Gestaltung aber die Differenzierung/Optimierung der Gestalt, weil man einem engen Leitbild folgt (oder folgen muß). Aber würde es uns denn im täglichen Leben noch gefallen, wenn plötzlich alle Trinkbehälter (Tassen/ Gläser...) wie Schnabeltassen aussehen, bloß weil diese der Erfüllung der Trinkfunktion am nächsten kommen und für den Mitteleuropäer intuitiv bedienbar sind?

Nun hat sich inzwischen natürlich die Erkenntnis durchgesetzt, daß Software-Produkte auch „gestaltet" werden sollten. Da die Schnittstelle zum Menschen/Nutzer in fast 100 % der Fälle den Bildschirm benutzt und dieser ein grafisches Abbild unterstützt, lag es für die Software-Entwickler selbstverständlich nahe, mit der Gestaltung dieser Schnittstelle Grafik-Designer zu betrauen. Und genau an dieser Stelle liegt ein entscheidender Denkfehler: Grafik-Designer lernen (und tun dies dann auch), Informationen für den visuellen Konsum aufzubereiten. Deshalb liegen neben der Typografie und dem Layout ihre unbestrittenen Stärken z. B. in der Gestaltung von Informationssystemen oder der Informationsaufbereitung in Multimedia-Anwendungen. Sie lernen nicht, (technische) Systeme zu analysieren, nach Ansätzen für deren Verbesserung zu suchen und diese Ideen als Lösungsvorschlag für ein verbessertes technisches Produkt zu synthetisieren.
D. h., der grundlegende Ansatz des Formgestalters ist (auch bei der Gestaltung von Softwareprodukten) völlig anders als der eines Grafik-Designers. Das hat heute zur Folge, daß sich vermeintlich gut gestaltete Software im täglichen Nutzungsprozeß nicht bewährt oder immer noch als unhandlich erscheint. (Man denke nur an CAD-Systeme, für deren Benutzung ein „gestandener" Konstrukteur monatelang geschult werden muß, damit er dannach - zweifellos besser, schneller, schöner - das Gleiche machen kann wie vorher, nämlich Konstruieren; oder an Textverarbeitungssysteme aus dem useability-Labor, deren Funktionalität im täglichen Gebrauch überhaupt nur zu ca. 10 % genutzt - sprich: gebraucht - wird...)

3. Schlußfolgerungen

Die Qualität einer Software wird zukünftig wesentlich durch ihre Gestaltung bestimmt. Die Gestaltung muß deshalb möglichst frühzeitig in den Software-Entwicklungsprozeß einbezogen werden. Dazu kann man auf ein Erfahrungspotential der Gestalter zurückgreifen. Das anzuwendende Prinzip sollte dabei in Abhängigkeit vom Aufgabentyp gewählt werden. Da man davon ausgehen kann, daß viele der Software-Entwicklungen gegenwärtig noch ein hohes Maß an „Erfindung" darstellen, sollten die gestalterischen Spielräume weit gefaßt sein. Es empfiehlt sich, zeitlich davor gelagerte Gestaltungsstudien erstellen zu lassen, diese zu diskutieren und, sofern sie positiv beurteilt wurden, als Entwicklungsgrundlage im Sinne eines gestalterischen Leitbildes zu benutzen. Nach diesem sich bisher bewährenden Prinzip wird gegenwärtig am ERBUS-Projekt [11] gearbeitet.

Es ist erforderlich, (Software)Gestaltungswerkzeuge zu entwickeln, die einerseits den gestalterischen Spielraum nicht beschränken, andererseits aber auch kompatibel zu den Werkzeugen der Software-Entwickler sind. Damit könnte die Effektivität bei der Entwicklung insgesamt deutlich gesteigert werden. [12]
Notwendig sind aber auch Werkzeuge, mit denen Gestalter „Software-Dummys" generieren und flexibel und schnell bearbeiten können (Software-Modellbau). Zu diesem Zweck werden heute meist Systeme wie Toolbook oder Macromedia-Director benutzt, die dazu eigentlich nur als bedingt geeignet anzusehen sind.
Es fehlt an integrierten Werkzeugen zur Beobachtung des Nutzerverhaltens an funktionierender Software, um Rückschlüsse auf spezifische Eigenheiten der Nutzer ziehen zu können usw.

Weiterhin fehlt gestalterisches Methodenwissen über die Analyse von Softwarestrukturen mit dem Ziel, diese Strukturen - falls erforderlich - gestalterisch an den Nutzer und seinen Arbeitsprozeß anzupassen. Es fehlt an handhabbaren Methoden der Erprobung von Software-Prototypen unter realistischen Einsatzbedingungen.

Es besteht erhöhter Bedarf, die Arbeitsteilung im Software-Entwicklungsprozeß unter Berücksichtigung der Integration von Gestaltern zu untersuchen. Ein solcher Ansatz wird bei dem vom BMFT geförderten Forschungsprojekt WORKS/ERBUS verfolgt [13].

In der Folge müssen neue Ausbildungsinhalte für die Gestalter etabliert werden, um den Bedürfnissen bei der Gestaltung dieser neuen Produktklasse nachkommen zu können.

Literatur

[1]   OELKE, H.; FRICK, R.:
Bestimmung der Funktion von industriell hergestellten Produkten. Halle (1977), F/E-Bericht, HiF.

[2]   RÖDIGER, K.-H.:
Anwendungsbereiche lernen voneinander ... und woraus lernen wir? In:Software-Ergonomie '95, Hrsg.: H.-D. Böcker, S. 45 ff, B. G. Teubner, Stuttgart.

[3]   GEORGES, M. ET AL.:
KADS-II. An advanced and comprehensive methodology for integrated KBS development. Report ESPRIT Project Nr 5248. 1992.

[4]   PORTER, D.:
Overview of differences between KADS and CommonKADS. Report ESPRIT Project P5248 KADS-II. 1 June 1992.

[5]   WIELINGA, B. J.; SCHREIBER, A. TH.; BREUKER, J. A.:
KADS: A modelling approach to knowledge engineering.
Report ESPRIT Project P5248 KADS-II, Amsterdam, University of Amsterdam, 1991.

[6]   WIELINGA, B. J.; SCHREIBER, A. TH.; BREUKER, J. A.:
KADS: A modelling approach to knowledge engineering. Knowledge Acquisition. 4 (1992) 5-53.

[7]   ANGELE, J.; FENSEL, D. and STUDER, R.:
The Knowledge Acquisition and Representation Language KARL. Research report no. 316, Institut für Angewandte Informatik und Formale Beschreibungsverfahren, Universität Karlsruhe, May 1995.

[8]   FICHTNER, W.; LANDES, D.; RENTZ O.; RUCH, M.; SPENGLER, T. und STUDER, R.:
Der MIKE-Ansatz zur Modellierung von Expertenwissen im Umweltbereich - dargestellt am Beispiel des Bauschuttrecyclings. In Proceedings of the 9th International Symposium on Computer Science for Environment Protection (Berlin, September 27-29), 1995.

[9]   FRICK, R.:
Designmethodik - Eine Einführung für Studierende -. Halle (1982), Lehrmaterial an der HIF-Halle.

[10]   HÖHN, F.:
Informationstheoretische Untersuchungen zur Rationalisierung des gestalterischen Entwicklungsprozesses durch faktografische Informationssysteme und der daraus resultierende rechnerunterstützte Ergonomie-Sachverhaltspeicher ERGOFAKT. TH Ilmenau, Fakultät für Technische Wissenschaften, Diss. A, 1988.

[11/13]   WORKS/ERBUS - Verbundprojekt - Methodik der arbeitsorientierten Gestaltung von Wissenssystemen am Beispiel eines Wissenssystems für die Formgestaltung von Investitions- und Konsumgütern. BMBF-Projekt Nr. 01H5014

[12]   SCHWIGON, H.:
MAKEMENU - Ein Werkzeug für den Programmierer. Halle (1987), Forschungsbericht, HiF Halle.

Adresse des Autors

Prof. Dr.-Ing. Falk Höhn
FH Hannover, Fachbereich Kunst & Design
30419 Hannover
Herrenhäuser Str. 8
e-mail: falk@kd.fh-hannover.de

Kokonstruktive Weiterentwicklung eines Groupwareproduktes
Das Beispiel der Reimplementierung eines Suchtools

Helge Kahler und Volker Wulf

Projektbereich Software-Ergonomie und CSCW

Institut für Informatik III, Universität Bonn

Zusammenfassung

Die Entwicklung von Groupware ist durch ein produktorientiertes, von einzelnen Anwendungsfeldern abstrahierendes Vorgehen geprägt. Die dabei entstehende Software erfüllt aber häufig nicht die Anforderungen, die Nutzer in konkreten Anwendungsfeldern an sie richten. Dies soll hier gezeigt werden am Beispiel eines Suchtools, das in einem marktgängigen Groupwareprodukt zur Unterstützung räumlich verteilter telekooperativer Arbeit realisiert war. Die sich aus der unzulänglichen Realisierung dieses Suchtools ergebende kokonstruktive Reimplementierung basierend auf der Programmierschnittstelle des Groupwareproduktes wird beschrieben. Abschließend werden Konsequenzen diskutiert, die sich aus dieser Fallstudie für eine auftragsorientierte Weiterentwicklung von Groupwareprodukten ergeben.

1. Einleitung

Der Anwendungskontext von Groupware ist unter den gegebenen makroökonomischen Rahmenbedingungen durch Differenzierung und Dynamik gekennzeichnet. Organisationen und ihre Untereinheiten differenzieren sich aus, um so besser auf sich rasch wandelnde Markterfordernisse reagieren zu können. Aus diesen Merkmalen des Anwendungskontextes ergeben sich zwangsläufig Konsequenzen für den Software-Entwicklungsprozeß bei Groupware.

Aus der Differenziertheit verschiedener Anwendungsfelder zwischen und innerhalb von Organisationen ergibt sich die Notwendigkeit, Nutzer in den Software-Entwicklungsprozeß aktiv einzubeziehen. Nur sie kennen ihren Anwendungskontext hinreichend genau, um entsprechende Anforderungen an einen ihre kooperative Arbeit unterstützenden Artefakt zu formulieren (vgl. [8]). Ein solches Vorgehen soll im folgenden als kokonstruktives Vorgehen (zwischen Entwicklern und Nutzern) bezeichnet werden. Im Gegensatz zur skandinavischen Tradition der partizipativen Systementwicklung, bei der die emanzipatorische Wirkung von Nutzerbeteiligung im Vordergrund stand (vgl. [1]), wollen wir hier unter dem Begriff kokonstruktiver Entwicklung auch Ansätze subsummieren, die lediglich auf die funktionale Wirkung von Benutzerbeteiligung abstellen (vgl. [2]). Im Gegensatz zum Mainstream der US-amerikanischen Diskussion zum Thema "Usability-Engineering" (z.B. [13]) beschränken wir uns beim kokonstruktiven Vorgehen nicht ausschließlich auf die Gestaltung der Ein-/Ausgabe- und Dialogaspekte der Benutzungsschnittstelle gemäß dem IFIP-Modell. Vielmehr wird auch die Gestaltung der Funktionalität und die organisatorische Einbettung der Anwendungen thematisiert (vgl. [4]).

Aus der Dynamik des Anwendungskontextes läßt sich die Notwendigkeit zu einem evolutionären Software-Entwicklungsprozeß ableiten. Da sich die Arbeitsbedingungen während der Nutzung verändern, muß die Möglichkeit bestehen, die Software – wenn nötig – entsprechend den veränderten Anforderungen weiterzuentwickeln.

Um zu untersuchen, inwiefern Möglichkeiten zu einer kokonstruktiven und evolutionären Entwicklung von Groupware bestehen, soll auf Grudins Unterscheidung in Produkt- und Auftragsentwicklung zurückgegriffen werden (vgl. [9, S. 60ff]). In der Produktentwicklung werden aufeinanderfolgende Versionen einer Software für eine Vielzahl von Anwendungsfelder entwickelt. Die anwendende Organisation entscheidet sich in der Regel für ein "fertiges" Produkt. Die Funktionalität eines solchen Produktes wird vom Management und der Marketingabteilung des Herstellers vorgegeben und von den Entwicklern ausgearbeitet. Obwohl es je nach Produkt verschiedene Kommunikationsforen zwischen Hersteller- und Anwenderorganisation gibt (z. B. Anwenderarbeitskreise), findet in der Regel keine direkte Beteiligung von Nutzern im Entwicklungsprozeß statt. Demgegenüber wird bei der Auftragsentwicklung eine Software für genau ein Anwendungsfeld erstellt. Da die künftigen Nutzer bei Beginn der Entwicklung bekannt sind, ergibt sich die Funktionalität der Anwendung aus den Anforderungen des Anwendungsfeldes. Auch wenn dabei in vielen Fällen die Chance zur Etablierung eines kokonstruktiven Vorgehens nicht genutzt wird, so sind die Bedingungen hierfür in der Auftragsentwicklung viel günstiger als in der Produktentwicklung. Ein evolutionäres Moment ist in der Produktentwicklung durch die fortlaufende Erstellung neuer Versionen gegeben, es stellt sich für die Nutzer allerdings die Frage, inwiefern die neuen Versionen die Anforderungen aus den jeweiligen Anwendungsfeldern erfüllen. Bei der Auftragsentwicklung erfordert die Weiterentwicklung häufig hohen Aufwand, so daß dort ein evolutionäres Vorgehen nicht die Regel ist.

Trotz der hier diskutierten Probleme, in der Produktentwicklung auf die Differenziertheit und Dynamik des Anwendungskontextes eingehen zu können, ist Groupwareentwicklung bisher im wesentlichen Produktentwicklung (vgl. [10, S. 94ff]). Vor dem Hintergrund der sich aus der Organisationsumwelt ergebenden Anforderungen stellt sich allerdings die Frage, ob dies in Zukunft so bleiben kann. Da die Produktentwicklung den Anwendern den offensichtlichen Vorteil bietet, mit geringem Aufwand schnell über einsatzbereite Funktionalität zu verfügen, ist zu überlegen, wie ein Produktentwicklungsprozeß durch auftragsorientierte Elemente anzureichern ist. Möglichkeiten hierzu sind bei Produkten sowohl gegeben, wenn deren Funktionalität durch Eingriffe in den Programmcode verändert werden kann, als auch dann, wenn deren Funktionalität durch Einbindung externen Programmcodes mittels einer Programmierschnittstelle modifiziert werden kann. Die Groupware LinkWorks der Firma DEC ist ein Produkt, das beide Formen der Modifikation des Programmcodes ermöglicht.

Bisher finden sich in der einschlägigen Literatur wenig Studien, die kokonstruktives Vorgehen bei der Entwicklung von Groupware thematisieren (z.B. [3]; [5]). Berichte über die kokonstruktive Weiterentwicklung von bestehenden Groupwareprodukten sind bisher kaum veröffentlicht. Dies hat seinen primären Grund darin, daß in Forschungsprojekten Software in der Regel neuentwickelt wird. Nichtsdestotrotz ist dies für den erfolgreichen Einsatz von Groupware in betrieblichen Anwendungskontexten eine wichtige Fragestellung.

In diesem Aufsatz soll über die kokonstruktive Reimplementierung des Suchtools in LinkWorks berichtet werden. Dazu wollen wir im ersten Schritt zeigen, welche Schwächen das ursprünglich in LinkWorks implementierte Suchtool aufwies. Wir werden dann unsere Vorgehensweise beschreiben und darstellen, in welcher Weise sich das kokonstruktiv entwickelte Suchtool von dem ursprünglichen Design unterscheidet. Es wird diskutiert, inwiefern diese Reimplementierung durch die von LinkWorks vorgegebenen Modifikationsmöglichkeiten beeinflußt wurde. Abschließend werden die Vor- und Nachteile des hier gewählten Vorgehens diskutiert und untersucht, auf welche Weise die Gestaltung und die Entwicklungsmethodik des zu Grunde liegenden Groupwareproduktes verändert werden muß, um die Verwendung in einem spezifischen Anwendungskontext zu erleichtern.

2. Probleme des Originalsuchtools in POLITeam

Der Projektbereich Softwareergonomie und CSCW des Instituts für Informatik III der Universität Bonn ist Konsortialpartner des von Bundesministerium für Bildung, Wissenschaft, Forschung und Technologie geförderten POLITeam-Projektes. Dieses Projekt beschäftigt sich mit der Unterstützung von Telekooperation in großen räumlich verteilten Organisationen, wobei u.a. die Metapher von "gemeinsamen Schreibtischen" verwandt wird, auf die mehrere mit der POLITeam-Groupware arbeitende Benutzer Zugriff haben können. Das im Rahmen des POLI-

Team-Projektes genutzte Groupwareprodukt LinkWorks der Firma DEC enthielt ein Suchtool, das es erlaubte, prinzipiell jedes dem System bekannte Objekt auf allen "elektronischen Schreibtischen" zu suchen.

Beim Originalsuchtool konnte der Suchende eine Vielzahl von Suchkriterien im Eingabefenster angeben, darunter die Objektklasse (z. B. ”Text” oder ”Ordner”), das Datum der letzten Änderung und den Namen des Erstellers. Die entsprechend der Kriterien spezifizierten Objekte wurden dann im gesamten System gesucht und im Suchergebnisfenster des Aktivators, also derjenigen Person, die die Suche initiiert hat, angezeigt. Wählte der Suchende bestimmte Objekte zur Übernahme auf seinen Schreibtisch aus, so wurden für diese Dokumente Verweise erzeugt, so daß der Suchende auf sie zugreifen konnte. Um zu vermeiden, daß dem Aktivator private oder geheime Dokumente anderer Nutzer auf diese Weise zugänglich würden, hatten die Entwickler vorgesehen, daß für jedes Objekt mittels eines Attributes festgelegt werden konnte, ob es durch das Suchtool auffindbar sein oder ihm verborgen bleiben sollte. Dieses Attribut konnte nicht von den Nutzern direkt manipuliert werden, sondern war Bestandteil von komplexeren Zugriffsprofilen. In den Anwendungsfeldern von POLITeam wurden nach Absprache mit den Nutzern drei verschiedene Zugriffsprofile definiert. Wollte ein Nutzer sicher sein, daß sein Objekt anderen bei ihrer Suche verborgen blieb, mußte er diesem das Zugriffsprofil ”privat” geben, wohingegen die Zugriffsprofile ”zur Information” und ”öffentlich” ein Objekt mit dem Suchtool auffindbar machten. Da LinkWorks erlaubte, Dokumente mit dem Zugriffsprofil "privat" zu verschicken, dies aber dazu führte, daß die Dokumente beim Empfänger weder bearbeitet, gelesen oder auch nur zurückversandt werden konnten, wurde den Nutzern im Rahmen der Schulung zur Vermeidung von Fehlern empfohlen, lediglich für ausschließlich zum privaten Gebrauch bestimmte Dokumente das Zugriffsprofil "privat" zu vergeben. Insofern hätte der Gebrauch des Suchtools nahezu alle Dokumente dem Suchenden zugänglich gemacht.

Neben diesen Problemen der Gestaltung der Funktionalität des Suchtools bestanden auch Probleme hinsichtlich der Dialogaspekte der Benutzungsoberfläche. Durch die Entscheidung, eine Vielzahl von Suchkriterien mittels nur eines Eingabefenster abzufragen, war dieses Eingabefenster des Originalsuchtools überladen und unübersichtlich.

Diese Probleme führten zur Entscheidung, von dem in POLITeam-Projekt üblichen Vorgehen abzuweichen, Groupwareprodukte zunächst einzuführen und die dabei sich zeigenden Probleme für ein späteres Redesign zu nutzen (vgl. [15]). Statt dessen wurde das Suchtool in den Anwendungsfeldern deaktiviert und unmittelbar ein Reimplementierung des Suchtools angestoßen.

3. Vorgehen bei der kokonstruktiven Entwicklung

Gemäß dem für POLITeam gewählten kokonstruktiven und evolutionären Vorgehen erfolgte die Reimplementierung in enger Zusammenarbeit mit den POLITeam-Nutzern (vgl. [6]; [8]). Abb. 1 stellt das dabei verfolgte Vorgehen vor. Wie bereits dargestellt, begann der Entwicklungsprozeß mit der Evaluation des bestehenden Suchtools (vgl. Kap. 2). Außerdem nahmen wir die Ergebnisse der laufenden Diskussion im Bereich menschengerechter Gestaltung von Groupware zur Kenntnis. Von besonderer Bedeutung war die in Voruntersuchungen gewonnene Erkenntnis, daß der Gebrauch einzelner Groupwarefunktionen zu Konflikten zwischen einzelnen Nutzern führen kann (vgl. [14]; [17]).

Basierend hierauf wurde ein Interviewleitfaden bestehend aus insgesamt 29 offenen Fragen entwickelt, mit dem potentielle Nutzer von Groupware zu ihrer bisherigen Suchpraxis befragt wurden. Dabei wurde im ersten Teil des Fragebogens allgemein erhoben, welches grundlegende Suchverhalten bei den Befragten bei papierbasierter oder elektronischer Suche vorherrscht. Hier wurde nach der Art der gesuchten Objekte, dem Suchanlaß und dem Suchvorgang (u.a. exemplarische Beschreibung des Vorgehens, Dauer, verwandte Hilfsmittel wie Post-It-Zettel etc.) gefragt.

Um das Konfliktpotential, das Funktionen des Suchtools anhaftet, aufzudecken, wurden im zweiten Teil des Fragebogens die Rollen einer suchenden Person und derjenigen Person unterschieden, in deren Arbeitsbereich (z. B. auf oder in deren Schreibtisch oder auf deren Festplatte) jemand anderes etwas sucht. Hier wurde u. a. erfragt, in welchen Bereichen von Kollegen gesucht werden darf, ob der Suchende Einblick in Vorgänge erhält, nach denen er nicht gesucht hat, ob eine Suche die Befragten in der Rolle des Betroffenen stört und welche Bedeutung bei der Suche die Position der Beteiligten in der Organisation hat.

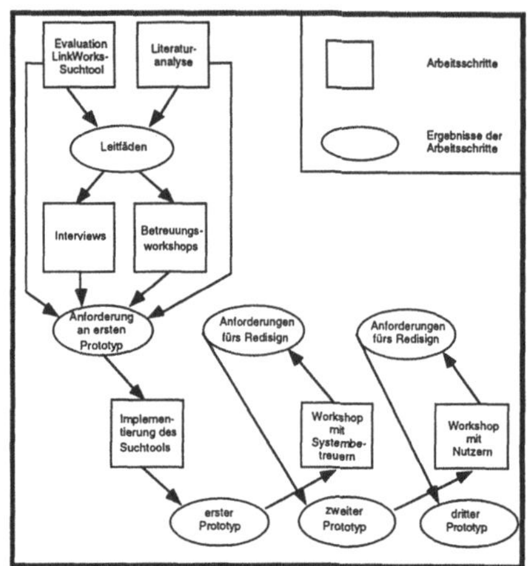

Abb. 1: Vorgehen bei der Reimplementierung des Suchtools

Um nicht ausschließlich die Besonderheiten eines Anwendungsfeldes zu erfragen, wurden Interviews in vier verschiedenen Organisationen aus dem Büro- und Verwaltungsbereichs durchgeführt. In diesen Anwendungsfeldern wurden insgesamt 10 etwa 30- bis 45-minütige Interviews geführt und ausgewertet. Außerdem wurden zwei zur Betreuung von POLITeam-Nutzern durchgeführte Workshops im Hinblick auf sich daraus an das Design des Suchtools ergebende Anforderungen hin beobachtet und ausgewertet. Es zeigte sich dabei, daß insbesondere die Auswertung der Interviews eine reiche Quelle der Inspiration war. Als Ergebnis der Voruntersuchungen ergaben sich die im folgenden Kapitel dargestellten Anforderungen an einen ersten Prototypen.

Nach dessen Implementierung führten wir zwei weitere Workshops mit dem Ziel durch, das Design der Prototypen weiterzuentwickeln. Dazu wurde ein lauffähiger Prototyp an Hand eines typischen Anwendungsszenarions den Workshop-Teilnehmern vorgestellt. Im ersten Workshop setzte sich die Teilnehmerschaft aus Anwendungsentwicklern und Nutzerbetreuern des POLITeam-Projektes zusammen. In diesem Workshop wurden insbesondere Verbesserungen der Handhabung und der Bedienung des Suchtools eingefordert. Am zweiten Workshop nahmen neben den Entwicklern des Suchtools drei Mitarbeiter aus einem der in der ersten Interviewphase bereits involvierten Anwendungsfelder teil. Die wichtigsten der in den Workshops für die Weiterentwicklung des Suchtools geäußerten Ideen sind in Kap. 6 dargestellt.

4. Anforderungen an den ersten Prototypen

Aus den Problemen mit dem Originalsuchtool, den Ergebnissen der Interviews und den Erfahrungen aus den Betreuungsworkshops ergaben sich die folgenden Anforderungen an das Design des ersten Prototypen.

Differenzierung verschiedener Suchbereiche – Durch die rollenorientierte Befragung in den Interviews wurde das Konfliktpotential der Suche außerhalb des eigenen Arbeitsbereichs deutlich. Befragte in der Rolle des Suchenden sprachen sich für eher uneingeschränkten Zugriff auf den gesamten Suchraum aus, während sie in der Rolle des von der Suche Betroffenen eher die Befürchtung äußerten, daß private elektronische Bereiche in nicht nachvollziehbarer Weise durchsucht werden. Im vortechnischen Zustand ließ sich in den meisten Anwendungsfeldern ein differenzierter Umgang mit verschiedenen Suchbereichen feststellen ("Die Büros werden grundsätzlich nicht abgeschlossen, die Kollegen haben in der Regel freien Zugang. Eigentlich kann überall gesucht werden, Schubladen werden allerdings als privat angesehen. Da sucht normalerweise keiner drin"). Andere Befragte unterschieden zwischen "erlaubter und unerlaubter Suche zum Zweck einer Überprüfung" und erwähnten bei einer Suche in den Büros Anderer existierende "Tabuzonen, die von der Stellung des Kollegen, aber auch seinem Charakter und seinem persönlichen Typ abhängen" (vgl. [12, S. 63ff]). Aus diesen Gründen sahen wir für den ersten Prototypen die Möglichkeit vor, drei Suchbereiche zu differenzieren: den eigenen Schreibtisch, den einer anderen Person und die Registratur bzw. das Archiv. Dem Aktivator sollte nach der Suche und vor der Übernahme der gefundenen Objekte auf den eigenen Schreibtisch anzeigen werden, wo ein gefundenes Objekt lokalisiert wurde. Die Interessen der Betroffenen sollten dann in einem weiteren Schritt durch unterschiedliche Konfliktregelungsmechanismen – wie z.B. der Unterdrückung der Anzeige gefundener Objekte oder der Anzeige des Suchvorgangs – gewahrt werden (vgl. [17]).

Beschränkung auf wichtige Suchkriterien – Die Interviews ergaben, daß der Eingabedialog des Originalsuchtools einige Suchkriterien enthielt, die für die Arbeit in den Anwendungsfeldern als nicht bedeutsam angesehen wurden. Als Suchkriterien waren unseren Interviewpartnern der Name des Objekts, das Datum der Erstellung und/oder letzten Veränderung, der Autor, der Eigentümer, ein Stichwort/Betreff, das Aktenzeichen und die Objektklasse des Objekts wichtig. Darüber hinaus wurde Möglichkeiten zur Volltextsuche gewünscht, weil Textdokumente laut Aussagen von Befragten ("Ich suche eigentlich nur nach Dokumenten") die mit Abstand meistgesuchten Objekte darstellten (vgl. [12, S. 52ff]).

Einbindung eines Kommunikationskanals – In den Interviews berichteten die Nutzer, daß sie während des Suchens häufig Kollegen befragen, um basierend auf deren Auskünften den Suchraum einschränken zu können. Aussagen wie "Suche ich ein Dokument, daß ich an einen Mitarbeiter ausgeliehen habe, so versuche ich ihn persönlich oder aber per Telefon zu erreichen. Ist der Mitarbeiter nicht anwesend, dann lege ich ihm einen Zettel in seinen Postfach oder, wenn es dringend ist, direkt auf seinen Stuhl" oder Fragen an Kollegen wie "Kannst Du Dich noch an die Bestellung vor ein oder zwei Jahren erinnern, wo zuviel Skonto abgezogen wurde?" können als typisch angesehen werden (vgl. [12, S. 59ff]). Da sich Groupwarenutzer nicht immer zur selben Zeit am selben Ort aufhalten, sollte das Suchtool Möglichkeiten zur Kommunikation während der Suche bereitstellen, also beispielsweise direkten Zugriff auf die Groupware-eigene Mailfunktionalität anbieten. Dabei erschien es sinnvoll, das Eingabefenster zur Spezifikation der Suchkriterien mitverschicken zu können.

5. Realisierungsalternativen

Um ein Suchtool zu implementieren, das diesen Anforderungen gerecht werden sollte, bietet LinkWorks prinzipiell zwei Vorgehensweisen an. Zum einen besteht die Möglichkeit, das Originalsuchtool durch Reprogrammierung der LinkWorks-eigenen Objektklassen zu modifizieren, zum anderen gibt es die Alternative, ein in einer gängigen Programmiersprache entwickeltes Suchtool in LinkWorks einzubinden. Dafür stellt LinkWorks eine spezielle Programmierschnittstelle zur Verfügung (APO für *Application Plus Objects*), die einen Zugang zu Objekten innerhalb von LinkWorks und die Veränderung und Ableitung von Methoden erlaubt. Zur APO gehört unter anderem eine Abfragesprache, die die Suche nach Objekten erlaubt. Für die Modifikation des Originalsuchtools sprachen neben dem geringeren Entwicklungsaufwand auch Performancegründe, da auf Grund der eingeschränkten Möglichkeiten der APO-Abfragesprache abzusehen war, daß mit einem extern angebundenen Suchtool die Geschwindigkeit des Originalsuchtools nicht zu erreichen war. Gegen die Modifikation des Originalsuchtools sprach die Tatsache, daß über die Programmierung von Objektklassen und die Änderung der Menüstruktur von LinkWorks lediglich Eingriffe bei der Behandlung der Suchergebnisse möglich gewesen wären. Dagegen ließen sich der Eingabedialog und der Suchvorgang selbst nicht ändern. Dies hätte insbesondere zur Folge gehabt, daß die Eingabemaske nicht hätte modifiziert werden können und daß eine Unterteilung der gefundenen Objekte nach den Orten, an denen sie gefunden wurden, *vor* der Übernahme in den Suchergebnisordner nicht möglich gewesen wären. Somit hätten wichtige der erhobenen Anforderungen nicht realisiert werden können.

Unter Berücksichtigung dieser Einschränkungen bei der Modifikation des Originalsuchtools entschieden wir uns für die Realisierung einer externen Lösung, die über die LinkWorks Programmierschnittstelle an die LinkWorkseigene Datenbank angebunden wurde. Da bei den

LinkWorks Anwendungspartnern fast ausschließlich Microsoft Windows Clients im Einsatz sind, entschieden wir uns zudem, auf eine aufwendige plattformübergreifende Lösung zu verzichten und realisierten das prototypische Suchtool in Microsoft Visual Basic for Windows. Bei der Realisierung ergab sich gegenüber dem Originalsuchtool eine Performanceeinbuße um den Faktor drei bis fünf je nach Suchanwendung für das noch nicht optimierte externe Suchtool. Diese Performanceschwäche läßt sich auf die Unzulänglichkeit der APO zurückführen, die keinen direkten Zugriff auf *alle* im Orginalsuchtool verwandte Funktionalität zuläßt. So stehen beispielsweise Systemattribute wie "Betreff" und "Schlüsselwörter" zwar dem Orginalsuchtool als Suchkriterien zur Verfügung, werden aber von der APO-Abfragesprache nicht unterstützt, was eine laufzeitkritische Nachprogrammierung der Funktionalität in Visual Basic erfordert.

Neben dem eigentlichen Suchvorgang erfolgt bei der Verwendung des externen Suchtools innerhalb von LinkWorks noch die Bereichsüberprüfung, die für jedes gefundene Objekt feststellt, ob es auf dem Schreibtisch der suchenden Person, einer anderen Person oder im Archiv / in der Registratur zu finden ist. Diese Bereichsüberprüfung ist zur Zeit noch zeitintensiv, da die Information kein Attribut des gefundenen Objekts ist.

6. Neues Suchtool

Das nach Abwägung der Realisierungsalternativen erstellte neue Suchtool erlaubt die Umsetzung der oben genannten Designanforderungen. Nach dem Start der Suche öffnet sich der Eingabedialog, der gemäß den Maßgaben der Voruntersuchung neu gestaltet ist (s. Abb. 2).

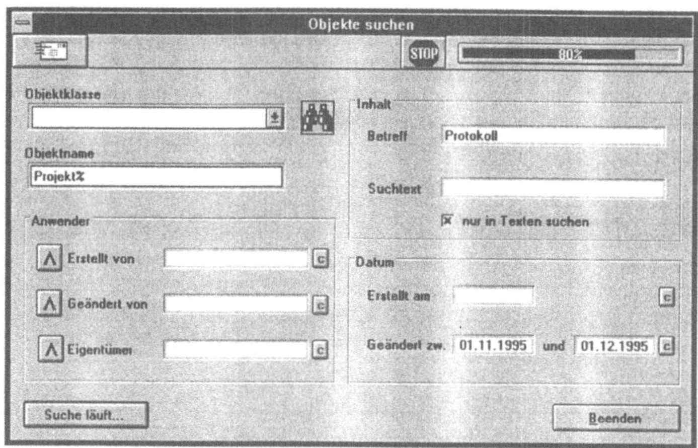

Abb. 2: Eingabedialog des Suchtools

Im Eingabedialog lassen sich die Suchkriterien für gesuchte Objekte spezifizieren. Die Suche verläuft um so schneller, je stärker die Suchkriterien, die bei einer Suchanfrage mit einem logischen UND verknüpft werden, den Suchraum einschränken. Grundlegende objektbezogene Suchkriterien sind die Objektklasse, die sich mit Hilfe einer Listbox auswählen läßt, und der Objektname, bei dessen Angabe auch Platzhalter verwandt werden können. Im Bereich *Inhalt* des Eingabedialogs läßt sich ein objekteigenes Betreff sowie ein Suchtext spezifizieren, der dann bei einer Volltextsuche gefunden werden kann. Im Bereich *Datum* kann als Erstellungs- und Änderungsdatum sowohl ein Zeitpunkt (Tag) als auch ein Zeitraum eingegeben werden. Der Bereich *Anwender* schließlich ermöglicht eine Einschränkung der Suche insofern, als daß bei Kenntnis der Person, die das Objekt erstellt oder geändert hat oder sein Eigentümer ist, sich diese noch angeben läßt. Im zweiten Workshop mit Nutzern von POLITeam wurde vorgeschlagen, die Eingabe für den Bereich *Anwender* gegenüber dem hier vorgestellten zweiten Prototypen noch dahingehend zu ändern, daß hier die Angabe einer Person auch ohne ihr Verhältnis zum Objekt reicht, da oft nicht klar sei, ob diese Person Ersteller, Änderer oder Eigentümer des Objekts ist.

Oberhalb der Eingabefelder für die Suchkriterien im Eingabedialog befinden sich noch zwei Buttons und ein Anzeigeelement. Einer der Buttons aktiviert den Eingabedialog für Kurznachrichten von POLITeam und leistet so einen Beitrag zu Kommunikation mit Kollegen bezüglich der Suche. Für eine geplante Einbindung von Audio- und Videotechnik in POLITeam ist hier auch die Öffnung eines entsprechenden Kommunikationskanals zu einer Person denkbar, die zur Suche angesprochen werden soll. Der andere Button ermöglicht den Abbruch des Suchvorgangs. Das Anzeigeelement gibt Auskunft über den Fortschritt bei der Suche.

Nach einer erfolgreichen Suche und Überprüfung, auf welchem Schreibtisch ein gefundenes Objekt liegt, werden die gefundenen Objekte in den drei Gruppen "eigener Schreibtisch", Schreibtisch von Kollegen", "Archiv" angezeigt (s. Abb. 3). Von dort aus können sie ausgewählt und in ein Ergebnisfenster auf dem elektronischen Schreibtisch übernommen werden. Diese Unterteilung in die drei Bereiche dient der ersten Orientierung des Suchenden darüber, wann die Übernahme welcher Objekte eventuell die Interessen Anderer berührt (vgl. Kap. 4). Hinsichtlich des Umgangs mit Konflikten bezüglich des Suchens äußerten die Nutzer im zweiten Workshop, daß sie als von einer Suche Betroffene für einzelne Dokumente individuell festlegen wollten, ob diese von anderen Nutzern gefunden werden können. Außerdem wünschten sie sich eine Anzeige dann, wenn Andere auf ihrem elektronischen Schreibtisch suchen. Eine weitere von den Nutzern geäußerte Idee betraf die Art, in der bei Anderen gefundene Objekte dem Suchenden zugänglich gemacht werden. Im Orginalsuchtool ebenso wie im ersten Prototypen führte eine erfolgreiche Suche dazu, daß das beim Suchenden ein Verweis auf das entsprechenden Objekt erzeugt wird, so daß der Suchende darauf Zugriff hat, ohne daß andere Personen davon wissen. Um das Risiko unerwünschter Manipulationen am Original auszuschließen, verlangten die Nutzer, daß im Normalfall dem Suchenden lediglich eine Kopie des Objektes bereitzustellen ist und nur dann, wenn dieser explizit zugestimmt hat, ein Verweis zu erstellen ist.

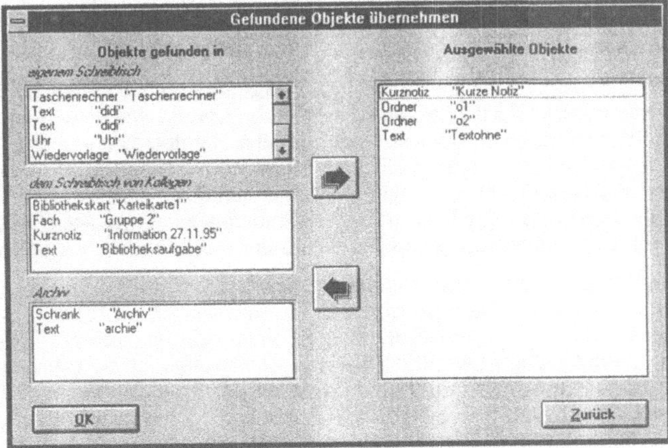

Abb. 3: Übernahmefenster des Suchtools

Die ausgewählten Objekte werden in einen Suchergebnisordner übernommen, von dem aus mit ihnen wie mit jedem anderen LinkWorks Objekt gearbeitet werden kann.

7. Zusammenfassung und Ausblick

In der hier beschriebenen Fallstudie wurden Schwächen eines produktorientierten Entwicklungsansatzes für Groupware deutlich. Die Gestaltung des ursprünglichen Suchtools deutet darauf hin, daß die Entwickler die bei Groupware im Vergleich zu einzelplatz-orientierten Systemen zu berücksichtigenden Probleme nur unzureichend wahrgenommen hatten. Für die hier involvierten Anwendungsfelder ließ sich durch interview- und prototypbasierte Kokonstruktion ein die Nutzeranforderungen befriedigenderes Design erzielen. Der im weiteren geplante Einsatz der ersten Version des neuentwickelten Suchtools muß allerdings noch erweisen, inwiefern damit alle augenblicklichen Gebrauchsanforderungen des im Workshop repräsentierten Anwendungsfeldes abgedeckt sind. Dabei ist insbesondere auf Einflüsse zu achten, die sich aus der Dynamik dieses Anwendungskontextes ergeben. Darüber hinaus ist zu untersuchen, inwiefern auch weitere – über die bisher einbezogenen – Anwendungsfelder die existierenden Gestaltungsanforderungen teilen oder ob und inwieweit diese zu modifizieren sind. Insofern stellt die präsentierte Version des Prototypen lediglich ein "Zwischenprodukt" eines kokonstruktiven und evolutionären Softwareentwicklungsprozesses dar.

Bei der Reflexion des Vorgehens zeigte sich, daß die Auswertung der interview-orientierten Exploration von nicht technisch unterstützen Suchvorgängen wichtige Designideen für das zu entwickelnde Suchtool erbrachte. Die aus der Voruntersuchung abgeleitete rollenorientierte Form der Befragung deckte das einzelnen Funktionen des Suchtools immanente groupwarespezifische Konfliktpotential auf. Techniken, die dazu beitragen, daß Nutzer im Rahmen einer kokonstruktiven Vorgehensweise sich in verschiedene technisch induzierte Rollen versetzen können, sollten weiterentwickelt werden. Solche Techniken stellen eine groupware-spezifische Erweiterung bisher bekannter Ansätze kokonstruktiver Systementwicklung dar.

Die hier dokumentierte Fallstudie macht deutlich, daß sich die kokonstruktive Weiterentwicklung eines Groupwareproduktes nicht auf die Ein-/Ausgabe- und Dialogschnittstelle beschränken darf. Vielmehr erfordern gerade die Verschiedenheit der potentiellen Anwenderorganisationen und die ihnen inhärente Entwicklungsdynamik, daß auch Möglichkeiten für eine Anpassung der Funktionalität bereitgestellt werden. Hier müssen klassische software-ergonomische Erkenntnisse daraufhin untersucht werden, inwieweit sie Hilfestellung für den Prozeß der Anpassung einer generischen Groupware für eine spezielle Organisation bieten können. Dabei spielen sowohl technische Voraussetzungen der Groupware eine Rolle als auch das methodische Vorgehen bei der Anpassung sowie die Fragen, wer auf welche Weise die Anpassungen vornehmen kann.

Hinsichtlich der Integration des neuentwickelten Suchtools in das Produkt LinkWorks zeigten sich deutliche Grenzen der Möglichkeit, das bestehende Produkt weiterzuentwickeln (vgl. auch [11]). Ähnliche Schwierigkeiten zeigen sich auch bei der Reimplementierung eines den Nutzeranforderungen entsprechenden Ereignisdienstes in LinkWorks (vgl. [7]). Diese Probleme scheinen typisch für augenblicklich auf dem Markt erhältliche Groupwareprodukte zu sein. LinkWorks bietet durch die Möglichkeit der Modifikationen bestimmter Methoden und einer leistungsfähigen Programmierschnittstelle vergleichsweise eher günstige Voraussetzungen.

Es stellt sich abschließend die Frage, welche Hinweise sich aus der hier dokumentierten Fallstudie für den Softwareentwicklungsprozeß bei Groupware gewinnen lassen. Die Fallstudie zeigt, daß ein ausschließlich produktorientierter Entwicklungsprozeß zu unbefriedigenden Ergebnissen führt. Andererseits spricht die Höhe des Aufwandes und die Dauer der Implementierung gegen eine in jedem Anwendungsfeld erfolgende Auftragsentwicklung. Insofern scheint eine modifizierte Kombination der beiden Ansätze ein aussichtsreiches Konzept. Ein "Rohprodukt" könnte produktorientiert allerdings unter erheblich stärkerer Betonung kokonstruktiver Vorgehensweise entwickelt werden. Die Entwickler sollten die zu beteiligenden Anwendungsfelder so auswählen, daß eine möglichst weitgehende Heterogenität der Nutzeranforderungen berücksichtigt wird. Sich dabei widersprechende Anforderungen sollten entweder alternativ umgesetzt werden, so daß bei der Einführung des Produktes im Anwendungskontext eine einfache diesbezügliche Konfiguration vorgenommen werden kann. Ansonsten sollte an solchen Stellen Möglichkeiten zur Reimplementierung in einer der Produktentwicklung zeitlich nachfolgenden Auftragsentwicklung ermöglicht werden. Solche Reimplementierungen werden bisher durch die vorherrschende "Produktsicht" der Hersteller nicht hinreichend unterstützt. Im

Falle des Suchtools bot die Möglichkeiten zur Klassenprogrammierung und die APO-Abfragesprache zu wenig Möglichkeiten zu einer suchzeiteffizienten Implementierung (vgl. Kap. 5). Ein weiteres Problem stellt der Umgang mit zeitlicher Dynamik der Nutzeranforderungen dar. Zwar sieht produktorientiertes Vorgehen eine iterative Produktversionen vor. Augenblicklich bestehen in der Regel nur schwache Informations- und keinerlei Mitbestimmungsmöglichkeiten der Anwender und Nutzer bei der Definition der zukünftigen Gestaltungsziele des Herstellers. Um eine nutzerorientierte Auftragsentwicklung auf Basis eines solchen Produktes über einen längeren Zeitraum betreiben zu können, müßten deshalb die Ziele der Versionsentwicklung Gegenstand intensiver Kommunikation zwischen Hersteller, Auftragsentwicklern und Nutzern werden.

Insofern stimmen wir mit Sumner und Stolze ([16]) überein, die als Ergebnis einer empirischen Untersuchung der Nutzung einzelplatzorientierter Softwareprodukte dynamisch sich entwikkelnde Anforderungen feststellten und zu deren Umsetzung eine kokonstruktive Entwicklung basierend auf den Möglichkeiten von in den Produkten implementierten Programmierschnittstellen vorschlugen. Zur Unterstützung von Auftragsentwicklungen basierend auf Groupwareprodukten glauben wir jedoch, daß bei der Festlegung der Eigenschaften einer Programmierschnittstelle bereits erhebliches Wissen über mögliche Anwendungsfelder vorliegen sollte, um die hier dargestellten Probleme vermeiden zu können. Auf diese Weise könnte aus einer Kombination von produkt- und auftragsorientiertem Vorgehen eine Perspektive für ein nutzergerechtes Software-Engineering bei Groupware erwachsen.

Literaturverzeichnis

[1] Bjerknes, G.; Ehn, P.; Kyng, M.: Computers and Democracy: A Skandinavian Challenge, Avebury 1987

[2] Clement, A.: Computing at Work: Empowering Action by "Low-level Users", in: Communications of the ACM, Vol. 37, No. 1, 1994, S. 53 - 63

[3] Cool, C.; Fish, R.S.; Kraut, R.E.; Lowery, C.M.: Interactive Design of Video Communication Systems, in: CSCW '92. Sharing Perspectives. Proceedings of the Conference on Computer-Supported Cooperative Work, ACM Press, New York 1992, S. 25 - 32

[4] Dzida, W.: Das IFIP-Modell der Benutzerschnittstelle, in: Office Management, Sonderheft Software-Ergonomie, Vol. 31, 1983, S. 6 - 8

[5] Egger, E.; Wagner, I.: Time Management: A Case for CSCW, in: CSCW '92. Sharing Perspectives. Proceedings of the Conference on Computer-Supported Cooperative Work, New York 1992, S. 249 - 256

[6] Floyd, Ch; Reisin, F.-M. and Schmidt, G. 1989: *STEPS* to software development with users, in: Ghezzi, C.; McDermid, J.A. (eds.): ESEC'89 - 2nd European Software Engineering Conference, University of Warwick, Coventry. Lecture Notes in Computer Science No. 387, Heidelberg: Springer, S. 48 - 64

[7] Fuchs, Ludwin; Sohlenkamp, Markus; Genau, Andreas; Kahler, Helge; Pfeifer, Andreas; Wulf, Volker: Transparenz in kooperativen Prozessen: Der Ereignisdienst in POLITeam. In: Krcmar, Helmut; Lewe, Henrik, Schwabe, Gerhard (Hrsg.): Herausforderung Telekooperation (Proceedings of the DCSCW 1996), Springer, Berlin u.a. 1996, S. 3-16

[8] Grønbæk, K.; Kyng, M.; Mogensen, P.: Cooperative Experimental System Development – Cooperative Techniques Beyond Initial Design and Analysis, in: Proceeding of the Third Decennial Conference: Computers in Context: Joining Forces in Design, Aarhus, DK, August 14 - 18, 1995, S. 20 - 29

[9] Grudin, J.: Interactive Systems: Bridging the Gaps between Developers and Users, in: IEEE Computers, No. 4, 4/1991, S. 59 - 69

[10] Grudin, J.: Groupware and Social Dynamics: Eight Challenges for Developers, in: Communications of the ACM, Vol. 37, No. 1, 1/1994, S. 92 - 105

[11] Kahler, Helge: Developing Groupware with Evolution and Participation - A Case Study. In Proceedings of the Participatory Design Conference 1996, Cambridge, MA, November 1996, S. 173-182
[12] Krüdenscheidt, G.: Partizipative Entwicklung eines Suchtools für Groupware, Diplomarbeit am Institut für Informatik der Universität Bonn, Bonn 1996
[13] Nielsen, J.: Usability Engineering, Academic Press, Boston et al. 1993
[14] Rohde, M.; Pfeifer, A.; Wulf, V.: Konfliktmanagement bei Vorgangsbearbeitungssystemen; in: WIRTSCHAFTSINFORMATIK, 38. Jg., Nr.2, S. 199 - 209
[15] Sohlenkamp, M.; Mambrey, P.; Fuchs, L.; Prinz, W.; Syri, A.; Kolvenbach, S.; Klöckner, K.; Pankoke-Babatz, U.: Unterstützung verteilter Regierungsarbeit mit POLITeam, in: Augsburger, W.; Ludwig, H.; Schwab, K. (Hrsg.): Tagungsband des Workshops "Koordinationsmethoden und Werkzeuge bei der computergestützten kooperativen Arbeit" vom 7.7.95 im Bamberg, Universität Bamberg, Bamberg 1995, S. 2 - 14
[16] Sumner, T; Stolze, M.: Evolution, not Revolution: PD in the Toolbelt Era, in: Proceeding of the Third Decennial Conference: Computers in Context: Joining Forces in Design, Aarhus, DK, August 14 - 18, 1995, S. 30 - 39
[17] Wulf, V.: Konfliktmanagement bei Groupware, Vieweg, Braunschweig 1997 (in Vorbereitung)

Adressen der Autoren

Helge Kahler
Universität Bonn
Institut für Informatik III
Projektbereich Software-Ergonomie und CSCW
Römerstr. 164
53117 Bonn
Email: kahler@informatik.uni-bonn.de

Dr. Volker Wulf
Universität Bonn
Institut für Informatik III
Projektbereich Software-Ergonomie und CSCW
Römerstr. 164
53117 Bonn
Email: volker@informatik.uni-bonn.de

Interaktion Blinder mit virtuellen Welten auf der Basis von zweidimensionalen taktilen Darstellungen

Martin Kurze

Institut für Informatik, Freie Universität Berlin

Zusammenfassung

Dieser Beitrag betrachtet zunächst das Verhältnis von grafischer Darstellung und Dargestelltem mit dem Schwerpunkt auf computergenerierten Bildern räumlicher Szenen. Für die tastende Wahrnehmung solcher Bilder müssen diese schon im Entwurfsprozeß auf die nicht-visuelle Rezeptionsmodalität abgestimmt werden. Das heißt, daß nicht Licht(-reflexion) das Aussehen des Bildes bestimmt, sondern Kraft (Reibung, Deformation). Den Kern dieses Beitrags bildet die Beschreibung eines haptischen Renderers, der Bilder räumlicher Gegenstände für die tastende Wahrnehmung - insbesondere durch Blinde - erstellt. Da Bilder tastend anders als sehend wahrgenommen werden, müssen Interaktionsmechanismen und -Geräte entwickelt werden, die anschließend dargestellt werden. Schließlich werden die gefundenen Resultate bewertet und ihr Nutzen auch für die Mensch-Computer-Interaktion Nicht-Blinder gezeigt.

1 Einleitung

Unsere Welt (einschließlich unserer Mensch-Computer-Interaktion) scheint immer stärker visuell geprägt zu sein. Bei genauerer Betrachtung ist nicht die Welt selbst visuell, sondern unsere Wahrnehmung der Welt basiert nur immer mehr auf dem visuellen Sinn. Dies mag auch daran liegen, daß unsere Wahrnehmung der Welt sich mehr und mehr indirekt über das visuelle Medium "Bildschirm" abspielt (in Form von Fernseh- oder Computerbildschirmen).

Die physische Welt ist nach wie vor so aufgebaut wie bisher. Sie kann auch nach wie vor wahrgenommen werden wie bisher, nämlich visuell, akustisch, über den Tastsinn und über Geruchs- und Geschmackssinn. Mit den Händen, dem Langstock, dem Gehör und einer guten Ausbildung sind auch Blinde durchaus in der Lage, sich in der physischen Welt zu bewegen.

Anders verhält es sich mit der abgebildeten Welt. Ist die physische Welt erst einmal unter Verwendung einer Kamera (oder eines Kameraersatzes) auf ihre optischen Eigenschaften reduziert und diese wiederum auf die zwei Dimensionen der Abbildungsfläche, so haben diese Informationen für Blinde keinen Wert. Bisher blieben hör- und tastbare Eigenschaften in den Medien bisher weitgehend unberücksichtigt. Während der akustische Kanal und akustische Eigenschaften der abgebildeten Welt in modernen "Multimedia"-Anwendungen immer häufiger genutzt werden, führt der Tastsinn nach wie vor ein Schattendasein.

Daß dies so ist, liegt nicht nur an den noch recht wenigen und teueren Ausgabegeräten für diesen Sinn, sondern auch daran, daß es noch kein Konzept für die Abbildung tastbarer Eigenschaften realer Gegenstände auf die vorhandenen oder möglichen tastbaren Medien gibt. Dabei spielt der Tastsinn nicht nur für Blinde eine besondere Rolle. Wie schon gezeigt wurde [8], heißt Tasten auch immer Handeln und hat damit neben der wahrnehmenden auch immer eine aktive Komponente. Nicht nur in virtuellen Welten vermittelt das Fehlen von Tasteindrücken das Gefühl, sich in einer Geisterwelt zu befinden, in der Objekte (wie Gespenster) zwar gesehen (und evtl. gehört), aber nicht ertastet werden können.

Mein Beitrag versucht, einen Weg zum Verständnis der Rolle des Tastens in der MCK aufzuzeigen und gleichzeitig mit den heute zur Verfügung stehenden Techniken Möglichkeiten des Tastsinns zu nutzen.

Nach einigen Begriffsbestimmungen am Ende dieser Einleitung wird eine Methode vorgestellt, Abbildungen speziell für die tastende Wahrnehmung zu entwerfen. Eine Beschreibung der entsprechenden Implementierung rundet diesen Teil ab. Schließlich werden Interaktionsmechanismen eingeführt, die den Tastsinn nicht nur nutzen, um Bilder zu erkennen. Eine Diskussion der gefundenen Ergebnisse schließt den Beitrag ab.

1.1 Begriffsbestimmungen

In diesem Text werden Begriffe verwendet, die in der Mensch-Computer-Interaktion bisher wenig benutzt wurden. Andere, wie der Begriff des "Renderns", sind im Bereich der visuellen Computergrafik wohl definiert, werden hier jedoch in nicht-visuellem Zusammenhang verwendet und bedürfen einer kurzen Erläuterung.

In der visuellen Computergrafik wird unter *Rendering* der Prozeß verstanden, "mit dem Beleuchtung, Textur und Lichteffekte an 3D-Computermodellen angebracht werden, um klare, scharfe Bilder mit fotorealistischen Details zu generieren" [1]. Allgemeiner wird der Begriff hier für den Prozeß, mit dem aus 3D-Computermodellen 2D-Bilder generiert werden, verwendet. Hierbei kann es sich bei dem 2D-Bild durchaus um ein tastbares Bild handeln.

Als *taktil* wird ein Reiz bezeichnet, der die Mechanorezeptoren der Haut, insbesondere der Finger, stimuliert. Man kann verschiedene Rezeptoren und Stimuli unterscheiden [3].

Kinästhetisch ist ein Reiz, der von den Mechanorezeptoren in Muskeln und Gelenken wahrgenommen wird. Über kinästhetische Reize wird auch die Position der Körperteile und die Widerstandskraft von Hindernissen eingeschätzt.

Als *haptische* Wahrnehmung wird hier die Summe der taktilen und kinästhetischen Reize verstanden. Da beide Reiztypen praktisch bei jeder Berührung zusammenwirken, ist eine Beschränkung auf die Betrachtung des taktilen Reizes allein wenig sinnvoll.

Erfassen wird hier als Überbegriff für visuelle und haptische Bildwahrnehmung benutzt. Für Sehende ist Erfassen gleichbedeutend mit Ansehen, Blinde ertasten (haptische) Zeichnungen. Der Wahrnehmungskanal ist verschieden, die Art der Informationsrepräsentation (als Zeichnung) jedoch identisch.

2 Das Problem

Dreidimensionale Computergrafiken verbreiten sich immer mehr. Im Küchenstudio erhält der Kunde heute ein fotorealistisches Bild seiner geplanten Küche; Häuser, Fahrzeuge und Gebrauchsgegenstände werden mit 3D-Modellen geplant, gebaut, beworben und in Büchern (Lexika, Gebrauchsanweisungen) dargestellt. In der Welt der Kommunikation (auch zwischen Menschen) spielen virtuelle Welten nicht nur zur Unterhaltung sondern auch als Benutzungsschnittstelle für Anwendungen (Tele-Shopping, Tele-Lernen, ...) eine immer bedeutendere Rolle.

Für Sehende ergeben sich hieraus keine Probleme, weil für sie entsprechende Visualisierungssysteme bestehen. Blinde können diese jedoch nicht nutzen. Um auch ihnen einen vergleichbaren Zugang zu ermöglichen, ist es wichtig, nicht die Grafik (das Abbild) zu beschreiben oder tastbar zu machen, sondern das *Modell* hinter der Grafik (das Abgebildete). Ähnliches wird von Begleitpersonen Blinder schon getan, wenn sie Bilder beschreiben: sie beschreiben nicht das flächige Bild ("Wanderer *unter* Bergen"), sondern das räumliche Abgebildete ("Wanderer *vor* Bergen"). Da eine verbale Beschreibung nicht leicht zu generieren ist und vor allem ein anderes Medium darstellt als eine Zeichnung, muß ein Weg gefunden werden, das gleiche Modell mit dem gleichen Medium (Zeichnung) darzustellen. Dabei müssen die unterschiedlichen Modalitäten der Erfassung berücksichtigt werden. Haptische Zeichnungen sind anders aufgebaut als visuelle [2]. Das Problem läßt sich also wie folgt darstellen:

> *Gesucht ist eine Methode, aus den gleichen Modellen, die auch für visuelle Zeichnungen realer Gegenstände verwendet werden, angemessene haptische Zeichnungen und für diese geeignete Zugangsmechanismen zu entwerfen.*

Da es für den Zugang zu existierenden haptischen Grafiken schon gewisse Methoden gibt (siehe hierzu auch Abschnitt 4), konzentriert sich dieser Beitrag auf den Entwurfsprozeß.

2.1 Andere nichtvisuelle Zugangsformen zu Grafiken für Blinde

In diesem Beitrag wird nicht vertieft auf das allgemeinere Problem des Zugangs Blinder zu Grafiken bzw. graphisch repräsentierter Information eingegangen. Eine ausführlichere Auseinandersetzung mit diesem Thema findet sich beispielsweise in [6]. Wie dort wird auch hier vor allem auf zwei Aspekte der nichtvisuellen Präsentation solcher Information Wert gelegt:

- Nicht die Grafik selbst, sondern die ihr zugrunde liegende Information (das *Modell*) muß präsentiert werden.

- Den *Zweck*, den die visuelle Grafik für Sehende erfüllen soll, muß auch die nichtvisuelle Darreichungsform erfüllen.

Das wichtigste Medium für die nichtvisuelle Mensch-Computer-Interaktion ist *Text*, der entweder per Sprachsynthese präsentiert oder auf einem Braille-Display ausgegeben werden kann. Die Umsetzung gegenständlicher Bilder (bzw. der auf ihnen dargestellten Informationen) in eine textuelle Form würde einen Wechsel des Mediums (linearer Text statt flächiger Grafik) bedeuten und ist zudem zur Zeit noch nicht automatisch durchführbar. Auf die Rolle der Sprache bei der Interaktion mit haptischen Grafiken wird im Abschnitt 4.2 eingegangen. *Nichtverbale* akustische Informationspräsentation bietet sich nur bei wenigen (aber bedeutenden) Formen von Grafiken an (Geschäftsgrafiken, numerische Diagramme, siehe [5]).

3 Haptisches Abbilden von 3D nach 2D

Methoden zum Herstellen von Abbildungen orientieren sich seit der Renaissance am menschlichen Sehsinn. Mit der Erfindung der fotografischen Kamera schien die Frage nach der optimalen Abbildungsmethode endgültig geklärt zu sein. Informierende Bilder [14] werden seit dem nach den Gesetzen der Optik erstellt, ob nun eine Kamera verwendet, ob gemäß ihrer Methode gezeichnet, oder ob "fotorealistisch" gerendert wird, immer spielt das Kameramodell die entscheidende Rolle.

Da sich gezeigt hat, daß herkömmliche Bilder, auch wenn sie tastbar gemacht wurden, Blinden, insbesondere Geburtsblinden, keine

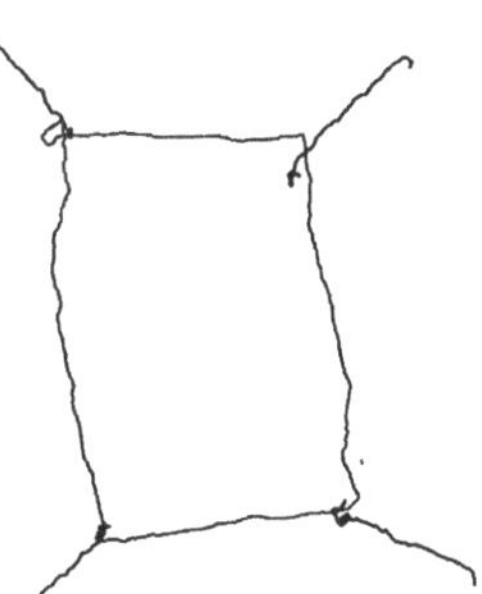

Abb 1: Von einer Blinden gezeichneter Tisch

Vorstellung des Abgebildeten vermitteln, wurden vereinzelt Versuche unternommen, Blinde selbst räumliche Gegenstände zeichnen zu lassen [4, 7]. Die so erhaltenen Zeichnungen sehen, wie z.B. Abb 1, wesentlich anders aus als solche, die von Sehenden oder von herkömmlichen Renderern erzeugt wurden. Aus diesen Zeichnungen und den beim Zeichnen erhobenen Daten lassen sich nun Methoden für das Rendern zur haptischen Wahrnehmung entwickeln, die im Abschnitt 3.2 beschrieben werden sollen.

3.1 Kognitionspsychologische Überlegungen

Bevor auf Implementierungsdetails eingegangen wird, seien hier noch einige grundlegende Überlegungen eingefügt, die bei der Implementierung berücksichtigt wurden.

Der Tastsinn unterscheidet sich in verschiedener Hinsicht vom Sehsinn:

- Der Tastsinn ist ein Nahsinn. Er ist prinzipiell auf die Reichweite der Arme begrenzt. Seine Reichweite kann z.B. durch einen Langstock erhöht werden; durch die Möglichkeit, sich fortzubewegen, kann die effektive Reichweite weiter vergrößert werden. Die Beschränkung auf die unmittelbare Umgebung bleibt allerdings erhalten.

- Der Tastsinn ist "objektiver" als der Sehsinn. Rechte Winkel realer Objekte werden auch als solche ertastet, parallele Linien und unterschiedlich große Gegenstände werden auch so wahrgenommen.

- Tasten erfolgt meist sequentiell. Das heißt, eine Linie muß mit dem Finger nachgezogen werden, bevor sie ganz wahrgenommen wird. Flächige Wahrnehmungen (auf der Handfläche oder auch der Rückenhaut [15]) sind erheblich ungenauer.

- Die haptische Auflösung ist wesentlich geringer als die visuelle. Linien, die dichter als ca. 1 mm beieinander liegen, werden oft nicht als separate Linien erkannt. Andererseits können feine Oberflächenstrukturen besser ertastet als gesehen werden.

Neben den Besonderheiten des Tastsinns müssen auch allgemeine Eigenschaften der Abbildungen und des Abgebildeten betrachtet werden:

- Man kann einen Gegenstand entsprechender Größe von allen Seiten ertasten, ihn eventuell sogar anheben und sein Gewicht abschätzen. Dadurch gewinnt man ein vollständigeres Bild seiner Struktur als durch die bloße Erfassung aus einer Richtung.

- Tastend werden immer Kräfte wahrgenommen (vor allem Reibungs- und Deformationskräfte), nicht reflektierte Wellen wie beim Sehen oder Hören. Eine Kraft, die dabei immer wirkt, ist die *Schwerkraft*, die den Dingen eine natürliche Richtung gibt (die nach unten), auch wenn sie ansonsten nicht ausgerichtet sind.

- Beim Ertasten eines Gegenstands und auch eines haptischen Bildes hat der Tastende immer einen *eigenen Standpunkt*, der den Dingen für ihn eine weitere Richtung gibt.

Gerade die letzten zwei Aspekte widersprechen sich in ihren Auswirkungen: Die natürliche Richtung in Abbildungen "nach unten", die vielen Blinden schon von taktilen Karten her bekannt ist, ist meist nicht identisch mit der "Erfassungsrichtung". Insbesondere bei der Kodierung dreidimensionaler Eigenschaften muß hierauf Rücksicht genommen werden.

Effekte, wie das Erkennen "auf einen Blick" treten bei haptischen Abbildungen nur auf, wenn der Erwartungsvorrat an Bildern beschränkt ist (Wiedererkennen eines bekannten Bildes). Im allgemeinen ist das Ertasten und Identifizieren von Bildern ein recht zeitaufwendiger Prozeß, der eine hohe kognitive Belastung mit sich bringt. Um diese Belastung gering zu halten, kann man einerseits den kognitiven Leseaufwand schon bei der Gestaltung simulieren und entsprechend berücksichtigen, andererseits kann man weitere Sinne, z.B. das Gehör (Sprache), einsetzen.

Schließlich muß beim Entwurf haptischer Bilder berücksichtigt werden, ob der Erfassende geburtsblind ist oder späterblindet. Späterblindete haben oft noch eine Vorstellung von visuellen Bildern. Sie sind bisweilen in der Lage, einfache Bilder mit perspektivischen Effekten zunächst zu ertasten, sie sich anschließend visuell vorzustellen und dann das Dargestellte zu identifizieren. Allerdings gilt das nur für sehr einfache geometrische Formen (Würfel u.ä.).

3.2 Implementation des haptischen Renderers

Unter Berücksichtigung dieser kognitionspsychologischen Überlegungen und der Analyse der vorkommenden Zeichnungen und Modelle wurde ein haptischer Renderer entwickelt, der nach dem in Abb 2 dargestellten Schema arbeitet:

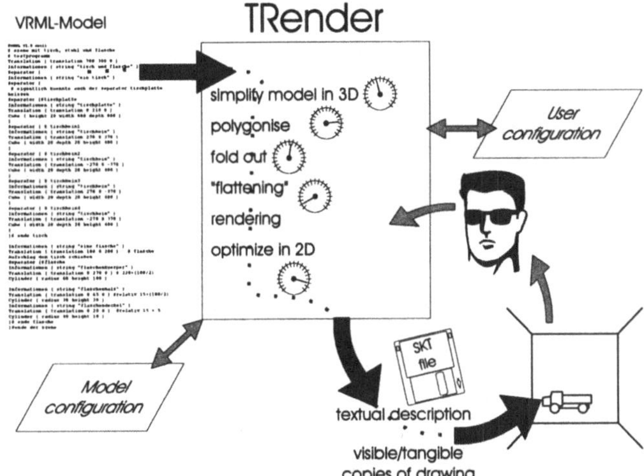

Abb 2: Schematische Darstellung der haptischen Rendering-Pipeline

Ausgangspunkt ist eine formale Beschreibung der darzustellenden Szene im gängigen VRML-Format (Virtual Reality Modelling Language, [13]). Dieses Format bietet sowohl den Vorteil einer weiten und wachsenden Verbreitung als auch den der semantischen Anreicherung der Modelle. Insbesondere die zweite Eigenschaft wird vom Renderer ausgenutzt.

Beim Rendern werden folgende Schritte durchgeführt:

- *Vereinfachung des 3D-Modells*: Die gesamte Szene wird untersucht, dabei werden zwei Manipulationen vorgenommen:
  - Sehr kleine Objekte werden ggf. entfernt
  - Objekte werden ggf. den ihnen nächsten 2D-Formen angenähert: Flache Quader werden zu Rechtecken, lange schmale zu Linien (Abb 3). Dabei stellt die Erhaltung der Objekttopologie ein besonderes Problem dar.

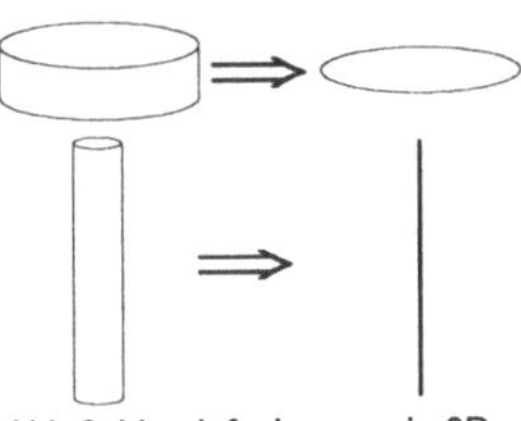

Abb 3: Vereinfachungen in 3D

226

- *Polygonalisierung*: Für die anschließende Manipulation der Einzelteile eignen sich Polygone besonders gut, daher werden alle Objekte in Polygone zerlegt. Hierbei wird die hierarchische Struktur des Modells beibehalten, ebenso bleibt die Information, aus welcher 3D-Form jedes Polygon entstand, erhalten.

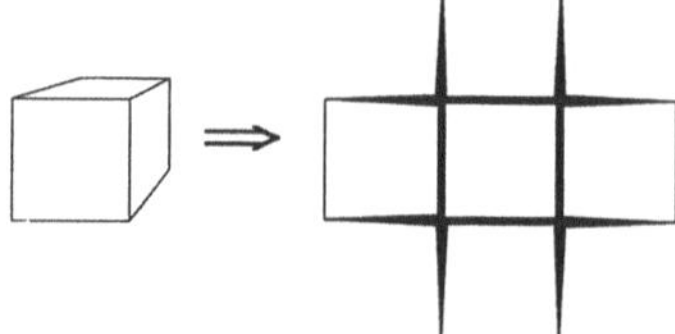

Abb 4: Auffalten eines Würfels

- *Auffalten* komplexer Objekte: Gemäß den gefundenen Methoden [7] werden nun Polygone um eine ihrer Kanten so rotiert, daß sie in einer Ebene mit anderen Polygonen des gleichen Objekts liegen. Die ursprüngliche Lage des Polygons wird gespeichert (Abb 4).

- *"Plätten"* komplexer Objekte: Im Gegensatz zum Auffalten werden hier Objektteile nicht um Polygonkanten rotiert, sondern um einen ihrer Endpunkte. Die Rotationsachse wird gemäß einer Energie-Minimierungsfunktion gewählt (Abb 5). Das Ergebnis sieht so aus, als hätte sich das Objekt unter großem Druck in eine Richtung verformt, bis es flach (platt) war.

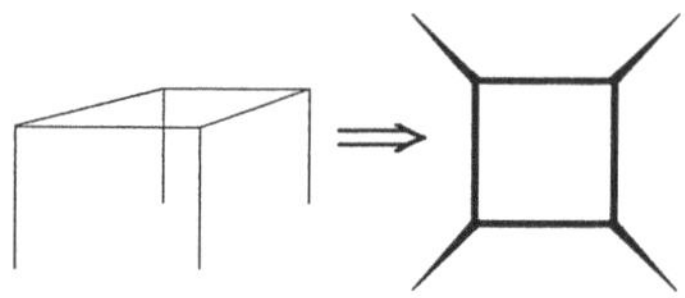

Abb 5: Ein "geplätteter" Tisch

- Generierung der *2D-Rohzeichnung*: Nachdem nun alle relevanten Polygone und Linien parallel zur geplanten Zeichenebene liegen, werden diese unter Verwendung des "painters"-Algorithmus gezeichnet: zunächst die hinten liegenden und dann die davor befindlichen, so daß ggf. Überlappungen entstehen.

- *Optimierung* der 2D-Zeichnung: Die so erhaltene Zeichnung enthält eventuell noch Linien, die zu eng beieinander liegen, oder verwirrende Überschneidungen. Bevor die Zeichnung wirklich zu Papier gebracht wird, werden die zu dichten Linien entflochten, und bei Überschneidungen wird die vordere deutlich von der hinteren Linie abgehoben (Abb 6).

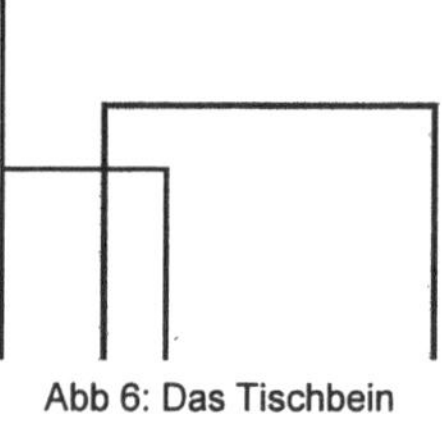

Abb 6: Das Tischbein steht *vor* dem Stuhl

- Ausgabe der *Zeichnung* und der *Beschreibung*: Das Endergebnis des Rendering-Prozesses ist eine Vektorgrafik-Datei mit den zu zeichnenden Linien und eine Beschreibungsdatei im SKT-Format [9], die den Namen des Objekts jeder gezeichneten Linie enthält. Über den Namen kann dann auch auf weitere Hintergrundinformation zugegriffen werden.

Der hier skizzierte haptische Renderer liefert nicht unmittelbar ein tastbares Bild. Er stellt eine Vektorgrafik zur Verfügung, die dann auf dem für den jeweiligen Zweck angemessenen haptischen Medium ausgegeben werden kann. Im einfachsten Fall wird die Grafik zunächst auf normales Papier gedruckt und anschließend auf Schwellpapier kopiert. Genauso gut wäre auch eine Ausgabe auf Tiefzieh-Folie möglich. Auch ein Online-Medium wie die Stiftplatte [12] kann verwendet werden, wenn auch mit den Nachteilen, die der relativ hohe Stiftabstand mit sich bringt. Schließlich könnte auch ein Ausdruck mittels Braille-Drucker in Frage kommen. Die hier erzielbaren Auflösungen sind jedoch in vielen Fällen unbefriedigend.

Beim Entwurf und der Implementierung des Renderers konnte (und mußte) auf die Modellierung einer Kamera verzichtet werden. Die Notation eines Blickwinkels existiert nicht. Auch Schatten und Texturen, die auf Oberflächen projiziert wurden und nicht tastbar sind, werden ignoriert. Dadurch wird die zu bearbeitende Datenmenge zunächst verkleinert. Da jedoch alle tastbaren Kanten und Flächen berücksichtigt werden müssen, müssen auch nicht sichtbare Flächen, wie Seiten- und Rückwände von Quadern und Innenseiten von Hohlkörpern (z.B. einer Tasse) berücksichtigt werden. Daher können Vorverarbeitungsalgorithmen der Computergrafik, die nicht sichtbare Flächen einfach aus dem Modell entfernen, hier nicht eingesetzt werden.

Tiefeninformation, die beim visuellen Rendern durch optische Verzerrungen repräsentiert wird, kann für die haptische Wahrnehmung mit mehreren Methoden dargestellt werden: Kanten und Flächen, die parallel zur Papieroberfläche liegen, werden durch einfache Linien gezeichnet. Dabei können weiter hinten liegende Objekte mit dünneren Linien gezeichnet werden. Konventionen folgend, die auch Sehende verstehen, können Linien, die "ins Bild hinein" laufen als schmale Keile gezeichnet werden (vorn dick, hinten dünn). Diese Methode wird auch von Blinden intuitiv verwendet und verstanden. Unter den weiteren Verfahren sei hier besonders die aus der Kartographie stammende Höhenlinien-Notation genannt.

Neben diesen in der Zeichnung selbst verwendeten Methoden kann auch erst beim Abtasten mit speziellen Hilfsmitteln wie dem Pantograph [11] die Tiefeninformation geliefert werden.

In beiden Fällen muß die grundsätzliche Entscheidung getroffen werden, welche Flächenrichtung parallel zur Papierfläche dargestellt werden soll, und welche senkrecht dazu: Einerseits können horizontale Flächen, andererseits solche, die senkrecht zur (im allgemeinen horizontalen) Erfassungsrichtung liegen (vertikal sind), verwendet werden. Beim Erfassen "von oben" fallen diese beiden Flächen zusammen, so daß sich hier eine elegante Lösung ergibt.

Eine wichtige Eigenschaft des Renderers ist seine Adaptierbarkeit. Weil sich die genannten Schritte teilweise widersprechen oder (alle gemeinsam angewendet) das Modell bis zur Unkenntlichkeit entstellen würden, sind Regler in das System integriert, die das Ausmaß für jede Manipulationsart einstellbar machen. Einerseits trägt diese Adaptierbarkeit der Tatsache Rechnung, daß es noch keine allgemeingültigen Regeln für das haptische Rendern gibt. Andererseits haben Versuche gezeigt, daß es große Unterschiede innerhalb der Gruppe der blinden Benutzer gibt. Manche bevorzugen die Ausfalt-Methode, während andere das "Plätten" bevorzugen. Auch Höhenlinien oder Schnittbilder werden beim Zeichnen und Ertasten verwendet.

Um den individuellen Präferenzen gerecht zu werden, kann der Renderer Benutzerprofile für jeden Benutzer anlegen und beim Rendern eines neuen Bildes auf diese zurückgreifen. Zusätzlich wird für jedes gerenderte Modell die Menge der Einstellungen gesichert; dadurch können unterschiedlich komplexe und verschiedenartige Modelle unterschiedlich behandelt werden.

4 Interaktionsmechanismen

Die Hardcopies der Grafiken können nun für sich allein oder mit Hilfe eines Tasttabletts und der zugehörigen SKT-Datei erkundet werden. Für die Erkundung existieren verschiedene Ansätze, die alle auf einer Korrespondenz von Position auf der Zeichnung und entsprechender Textinformation beruhen (siehe auch Nomad, [10]): Der blinde Benutzer tastet die Zeichnung ab und drückt an Stellen, die ihn interessieren oder die er nicht zuordnen kann, auf die Zeichnung, die auf dem Tasttablett liegt. Aus der Position auf dem Tablett und der hinterlegten Datenbasis (SKT-Datei) wird ein Text ermittelt, der durch Sprachsynthese ausgegeben wird.

In der Regel ertasten Blinde zunächst grob die ganze Zeichnung, um sich einen Überblick zu verschaffen, wo auf der Fläche Linien zu finden sind. Anschließend verfolgen sie mit wenigen Fingern einzelne Linien und erarbeiten sich so eine Vorstellung des Dargestellten, ein mentales Modell. Beim Ertasten der Zeichnung haben sie grundsätzlich die gleichen Freiheiten wie ein Sehender beim Anschauen. In welcher Reihenfolge Linien verfolgt werden, bleibt dem Erfassenden überlassen.

4.1 Das Bild im Kontext - der Kontext im Bild

Bei der Untersuchung der Interaktion wirklicher Benutzer mit der Zeichnung spielt die Situation, in der die Zeichnung auftaucht, eine entscheidende Rolle. Jede Zeichnung steht in einem Kontext. Falls die Zeichnung Teil eines Buches oder eines Online-Texts ist, bildet der Text den Kontext. Im Rahmen von VR-Anwendungen ist auch jeweils klar, in welchem Zusammenhang das Bild steht. Dieses Hintergrundwissen macht die Erkennung wesentlich einfacher. Ein Bild aus einem Küchenprospekt wird eher einen Herd mit vier Herdplatten enthalten als einen Spielwürfel mit vier Punkten auf der Oberseite ... Dies vorausgesetzt, kann die Darstellung einzelner Objekte unter Umständen etwas weniger genau sein, ohne daß die Erkennbarkeit des Gesamtbildes leidet.

Ebenso ist die Bedeutung eines Kreises am oberen Ende des Flaschenhalses klar, weil sich dort in der Regel deren (kreisrunde) Öffnung antreffen läßt. Die Zeichnung kann sich also auf Aspekte konzentrieren, die aus dem Zusammenhang nicht abzuleiten sind, jedoch graphisch gut darstellbar sind. Die grundlegenden Informationen über den räumlichen Aufbau der Szene und der einzelnen Objekte werden aus der Zeichnung selbst gezogen.

An dieser Stelle taucht die Frage auf, wie "ikonisch", d.h. symbolisch, haptische Grafiken sind und welche Rolle die Sprache bei der Interaktion mit haptischen Grafiken spielt. Natürlich sind Zeichnungen (auch haptische) vereinfachte Abbildungen. Da die hier behandelten Zeichnungen jedoch auf Modellen beruhen, können sie alle Informationen aus dem Modell enthalten, also maximale Genauigkeit haben. Alle dargestellten Eigenschaften (Winkel, Größenverhältnisse, Entfernungen, ...) entsprechen den Daten des Modells. Die Ikonizität taktiler Zeichnungen ist vergleichbar mit derjenigen visueller Zeichnungen. Die vereinfachte Darstellung eines Tisches (siehe Abb 5) hat zunächst auch einen gewissen ikonischen Charakter, d.h. die Grundinformation "Hier steht ein Tisch" könnte auch symbolisch, d.h. über Sprache zur Verfügung gestellt werden (siehe Abschnitt 4.2). Jede Zeichnung enthält aber auch zusätzliche Informationen wie Art und Größe des Tisches, Anzahl der Tischbeine, Gegenstände auf dem Tisch, usw. Diese sind nicht ikonisch repräsentiert, sondern abgebildet.

4.2 Rolle der Sprache

Da Zeichnungen immer auch in einem Zusammenhang erfaßt werden und da dieser Zusammenhang oft in Form eines Texts vorliegt, kann man davon ausgehen, daß ein System zur Sprachverarbeitung in vielen Fällen zur Verfügung steht. Wie auch bei visuellen Bildern ergänzen sich Bild und Sprache im Idealfall gegenseitig. Falls die Information im Bild redundant zur Textinformation ist, kann eine Umsetzung des Bildes in taktile Form ggf. entfallen. Dies gilt dann nicht, wenn die Information zwar doppelt dargestellt, aber grafisch leichter als sprachlich zu erfassen ist.

Im einzelnen kommt der Sprache, hier vor allem der gesprochenen Sprache, in folgenden Fällen Bedeutung zu:

- Ausgabe des Kontexts: Der Text, in dem Bezug auf ein Bild genommen wird, wird oft per Sprachsynthese oder Braille-Display ausgegeben.

- Ausgabe von Objektnamen: Um den Erkennensprozeß beim Erfassen von Bildern zu beschleunigen und zu vereinfachen, können Objektnamen auf Verlangen gesprochen werden [9, 10].

- Ausgabe von nicht tastbaren Informationen: Nicht tastbare (und ggf. nicht sichtbare) Informationen können sprachlich zur Verfügung gestellt werden; dazu gehören Farbe und Material von Objekten, deren Eigentümer oder Geschichte.

- Ausgabe von Navigationshinweisen: Benutzer können durch sprachliche Hinweise zu einem gegebenen Ziel hingeführt werden.

- Eingabe von Kommandos: Beim Ertasten eines Bildes ist es sehr hinderlich, die Finger von der Zeichnung weg und hin zur Tastatur (und wieder zurück) zu bewegen. Dabei kann der fließende Vorgang ins Stocken geraten. Daher bietet Spracheingabe über ein Spracherkennungssystem eine ideale Ergänzung.

- Suchbefehle bilden einen Sonderfall der möglichen Kommandos: Sie erfordern ggf. eine rudimentäre natürlich-sprachliche Schnittstelle und einen entsprechend großen Wortschatz.

- Eingabe von Objektnamen beim Zeichnen: Wie Objektnamen beim Ertasten ausgegeben werden und so die menschliche Erkennung erleichtern, kann automatische Spracherkennung das Kommentieren beim Erstellen von Zeichnungen sehr erleichtern [7].

5 Diskussion und Ausblick

Das vorgestellte Verfahren zur Herstellung von und zur Interaktion mit haptischen Bildern räumlicher Szenen stellt einen ersten Schritt in Richtung eines vollständigeren Zugangs Blinder zu Informationsquellen dar. Ein den Bedürfnissen Blinder angemessener Entwurf dieser Bilder ohne die Verkomplizierung durch die Gewohnheiten Sehender (perspektivische Verzerrungen usw.) bildet eine Grundlage dieses Vorgehens. Weitere Grundlagen sind die verfügbaren Medien, die bisher für diesen Zweck zwar ausreichen, aber kein komfortables Arbeiten ermöglichen. Ein Tasttablett mit einer in "Echtzeit" änderbaren Oberfläche in der Art von Schwellpapier wäre wünschenswert. Entsprechende Geräte befinden sich aber noch im Entwicklungsstadium.

Schon heute stehen allerdings Geräte zur haptischen Ein-/Ausgabe zur Verfügung, die eine echte Kraftrückkopplung ermöglichen [11]. Kombiniert mit Schwellpapierzeichnungen ließen sich insbesondere Eigenschaften wie die Schwerkraft o.ä. gut darstellen, ohne die Komplexität des Bildes zu erhöhen. Auch eine Navigationsunterstützung, die den suchenden Finger in die richtige Richtung "zieht", kann so realisiert werden.

Schließlich bietet das hier vorgestellte Verfahren Ansatzpunkte für die Forschung auf dem Gebiet der haptischen MCK insgesamt. Auch Sehende werden durch die immer mehr zur Verfügung stehenden Ausgabegeräte in die Lage versetzt, den haptischen Kanal zusätzlich zu den schon stark belasteten visuellen und akustischen Kanälen zu nutzen, so daß eine Beschäftigung mit meiner Thematik auch hier an Bedeutung gewinnt.

Danksagungen

Der Dank des Autors gebührt vor allem Immo Eitel, der im Rahmen seiner Studienarbeit erfolgreich an der Implementierung des haptischen Renderers arbeitete. Den blinden Versuchspersonen, die mit frühen Prototypen des Systems arbeiteten und wertvolle Hinweise zu dessen Optimierung gaben, sei ebenfalls gedankt. Schließlich sollte auch erwähnt werden, daß viele Anregungen zu dieser Arbeit auf Konferenz-Reisen an den Autor herangetragen wurden, die aus Mitteln des EU-Projekts HCM-IRS bezahlt wurden.

Literaturangaben

[1] Computer Zeitung 1 (11.01.1996), S. 14.

[2] P.K. Edman: Tactile Graphics. New York:American Foundation for the Blind 1992.

[3] G. Gordon (Ed.): Active Touch. (The Mechanism of Reception of Objects by Manipulation) Oxford, New York:Pergamon Press 1978.

[4] J.M. Kennedy: Drawing & the Blind. (pictures by touch) New Haven, London:Yale University Press 1993.

[5] M. Kurze: Giving Blind People Access to Graphics. (Example: Business Graphics) In: Proc. Software-Ergonomie '95 Workshop "Nicht-visuelle graphische Benutzungsoberflächen", Darmstadt, 22.02.1995.

[6] M. Kurze: Zugang zu Grafiken für Blinde. (Notwendigkeit - Probleme - Perspektiven). In: Ergonomie & Informatik 26, Stuttgart, 1995, 5-12.

[7] M. Kurze: TDraw - A computer-based tactile drawing tool for blind people. In: Proc. Assets '96. Vancouver, British Columbia, Canada, 1996: ACM, 131-138.

[8] M. Kurze; E. Holmes: 3D Concepts by the Sighted, the Blind and from the Computer. In: Proc. of the ICCHP '96 ; Linz 17-19 July 1996; München:Oldenbourg 1996, 551-556.

[9] J. Lötzsch: Computer-Aided Access to Tactile Graphics for the Blind. In: Proc. 4th International Conference on Computers for Handicapped Persons (ICCHP) ; Vienna 1994; Berlin, Heidelberg:Springer-Verlag (Lecture Notes in Computer Science 860) 1994, 575-581.

[10] D. Parkes: Nomad, an audio-tactile tool for the aquisition, use and management of spatially distributed information by visually impaired people. In: Proc. of the Second International Symposium on Maps and Graphics for Visually Handicapped People. London:A.F. & Dodds 1988

[11] C. Ramstein: MUIS: Multimodal User Interface System with Force Feedback and Physical Models. In: Proc. of the IFIP International Conference on Human-Computer Interaction Interact '95, Lillehammer, Norway 1995.

[12] W. Schweikhardt: Interaktives Erkunden tastbarer Grafiken durch Blinde. In: Software-Ergonomie '85; Stuttgart:Teubner 1985, 366-375.

[13] VRML: The Virtual Reality Modeling Language; Version 1.0 Specification; Jan '96, URL: http://vag.vrml.org/vrml10c.html

[14] B. Weidenmann (Hrsg.): Wissenserwerb mit Bildern. Bern:Verlag Hans Huber 1994.

[15] W. Zagler; P. Mayer: Technical Possibilities for the Improvement of Orientation and Mobility for Disabled Persons. In: Proc. of the 1st TIDE Conference ; Brussels April 1993; Amsterdam:IOS Press 1993, 35-40

Adresse des Autors

Martin Kurze
Freie Universität Berlin
Institut für Informatik; Arbeitsgruppe Interaktive Systeme
Takustr. 9
14195 Berlin
E-Mail: kurze@inf.fu-berlin.de;
URL: http://www.inf.fu-berlin.de/~kurze

Erfahrungsbasierte Fuzzy-Regelung von Lernwegempfehlungen als Beitrag zur Änderungsfreundlichkeit von Lernsystemen

Martin G. Möhrle und Andreas Henke

Brandenburgische Technische Universität Cottbus

Zusammenfassung

Zum Zwecke der rechnergestützten Individualisierung von Softwaresystemen und insbesondere von Lernsystemen wird eine erfahrungsbasierte Fuzzy-Regelung konzipiert, und die Ergebnisse ihrer praktischen Erprobung werden vorgestellt. Eine erfahrungsbasierte Fuzzy-Regelung gibt Lernwegempfehlungen ab und berücksichtigt dabei die Erfolge bzw. Mißerfolge früherer Empfehlungen. Sie besteht aus zwei miteinander verflochtenen unscharfen Regelkreisen. Im Rahmen der praktischen Erprobung der erfahrungsbasierten Fuzzy-Regelung wurden mit einem Experiment empirische Daten gesammelt, mit denen eine Prototypimplementierung der erfahrungsbasierten Fuzzy-Regelung getestet wurde. Die Erprobung belegt die Tragfähigkeit des Konzepts, das sich sowohl in Lernsystemen als auch in anderer Software umsetzen läßt.

1 Der Nutzen einer erfahrungsbasierten Regelung von Lernwegempfehlungen

In vielen Softwaresystemen ist eine *Individualisierung*, d. h. eine Anpassung des Systems *durch die Benutzer* an ihre Wünsche und Präferenzen, erwünscht und notwendig [1, S.175 f.]. Eine solche Individualisierung kann allerdings auch zu Problemen führen, etwa wenn die Benutzer nicht über das notwendige Wissen über die Gestaltungsdimensionen des Softwaresystems verfügen [1, S.181]. In diesem Fall scheint die Individualisierung *durch das System selbst* wünschenswert, die vor allem auf der Verwendung von Benutzermodellen und der gezielten Zuordnung eines Benutzers zu solch einem Benutzermodell beruht [2, S.124 ff.]. Hieran knüpft der vorliegende Aufsatz an, bei dem am Beispiel eines Lernsystems die Zuordnung zwischen Benutzer und Benutzermodell mit Hilfe einer erfahrungsbasierten Fuzzy-Regelung vorgenommen wird.

Die Individualisierung eines Softwaresystems und - im folgenden herausgestellt - eines Lernsystems kann an *unterschiedlichen Vorabeigenschaften* der Benutzer anknüpfen. Vorabeigenschaften können beispielsweise Vorbildung, bisherige praktische Erfahrung sowie individuelle Problemerkennungs- und -lösungsfähigkeiten sein [3, S.25]. Um auf Basis der Vorabeigenschaften ein Lernsystem zu individualisieren, sollten *unterschiedliche Lernwege* (ausgerichtet auf unterschiedliche Benutzertypen) bereitgestellt werden, und es sollte der für die jeweiligen Vorabeigenschaften eines Benutzers *geeignete* Lernweg empfohlen werden. Dem Autoren eines Lernsystems stellen sich dabei *zwei miteinander verflochtene Probleme*:

- Bei erstmaliger Benutzung eines Lernsystems bestehen noch *keine empirischen Erfahrungen hinsichtlich der Güte* der Lernwegempfehlung. Die Lernwegempfehlung beruht üblicherweise auf Vermutungen des Autors des Lernsystems.

- Im Laufe der Benutzung können sich die *ursprünglich angenommenen Beziehungen* zwischen Vorabeigenschaften und der angestrebten Lernproduktivität *ändern*. Erstens können bisher noch nicht berücksichtigte Vorabeigenschaften hinzutreten, zweitens können sich Vorabeigenschaften inhaltlich, aber nicht formal ändern.

Zur Überwindung der beiden Probleme und damit als Beitrag zur Änderungsfreundlichkeit von Lernsystemen wird in diesem Beitrag eine *erfahrungsbasierte Fuzzy-Regelung der Lernwegempfehlung* vorgeschlagen. Bei ihr fließen Erfahrungen, die man aus früheren Durchläufen durch ein Lernsystem sammelt, in die aktuelle Lernwegempfehlung ein. Die erfahrungsbasierte Fuzzy-Regelung wird im folgenden in *drei Schritten* spezifiziert:

- Die klassische *Regelungstechnik* verschmilzt mit den neueren Theorien der *unscharfen Logik* (*Fuzzy-Logik*) zu dem leistungsfähigen Instrument der *Fuzzy-Regelung*. Wir setzen in unserem theoretischen Konzept eine besondere Form davon ein, die als *erfahrungsbasierte* Fuzzy-Regelung bezeichnet sei und mit *zwei vernetzten Fuzzy-Regelkreisen* umgesetzt wird (Abschnitt 2).

- Das theoretische Konzept ist *tragfähig*, was wir durch eine *Erprobung* in drei Schritten nachweisen können: In einem umfangreichen *Experiment* wurden Meßdaten gewonnen. In einem *Prototyp* wurde der regelungstechnische Ansatz implementiert. Anhand der Meßdaten wurde sodann der Prototyp *getestet* (Abschnitt 3).

- Das vorgeschlagene Konzept kann in *vielfältiger Weise erweitert* werden (Abschnitt 4).

2 Das theoretische Konzept der erfahrungsbasierten Fuzzy-Regelung

Die erfahrungsbasierte Fuzzy-Regelung von Lernwegempfehlungen soll zunächst theoretisch konzipiert werden. Zum Verständnis des theoretischen Konzepts hilft die analoge Betrachtung dessen, wie sich ein *menschlicher Trainer* in einer gleichartigen Situation verhalten würde [4, S.49]. *Zwei Rückkopplungen* treten dabei hervor, die auf die Gestaltung eines Lernsystems in Form zweier vernetzter Fuzzy-Regelkreise übertragbar sind.

Zunächst zur Analogie: Wie würde sich ein menschlicher Trainer in einer gleichartigen Situation verhalten, wenn er also einen oder mehrere Probanden durch einen Kurs mit mehreren, nach Schwierigkeit des Lernstoffes gestuften Lernwegen leiten soll? Was würde er beim *ersten Probanden* tun? Wie würde sich sein Verhalten ändern, wenn er *mehrere Probanden* durch den Kurs geleitet hätte? In welcher Form würde sich sein (des Trainers) *Lernerfolg* offenbaren?

Plausibel scheint folgendes Vorgehen: Bevor der Trainer den ersten Probanden durch einen Kurs zu leiten hat, wird er *Kriterien* festlegen bzw. *Schwellenwerte* festsetzen, die auf den Vorabeigenschaften der Probanden aufbauen und der Lernwegempfehlung dienen. Diese Kriterien bzw. Schwellenwerte wird der Trainer aus seiner *Erfahrung*, seiner *Wertung der Schwierigkeiten* etc. gewinnen. Der Trainer wird sodann den ersten Probanden durch den Kurs geleiten. Anschließend, nach der Auswertung eines Abschlußtests und einer Abschlußbefragung des Probanden, wird er sich fragen, inwieweit seine Eingangseinschätzung in dem Sinne erfolgreich war, daß der Proband durch den für ihn geeignetsten Lernweg durch den Kurs gelaufen ist. Hier lassen sich - vereinfacht - *drei Fälle* unterscheiden:

- Der Proband hatte einen *akzeptablen* Lernerfolg bei *akzeptabler* Lernzufriedenheit.
- Der Proband hatte einen *akzeptablen* Lernerfolg, fühlte sich aber von Teilen des Kurses *unterfordert*.
- Der Proband hatte *keinen* akzeptablen Lernerfolg.

Im ersten Fall wird sich der Trainer in seiner Einschätzung *bestätigt* fühlen. Im zweiten und dritten Fall wird der Trainer seine Eingangseinschätzung *anzweifeln*: Wahrscheinlich hat er im zweiten Fall den Probanden zu niedrig eingestuft, so daß dieser Lernstoff bearbeiten mußte, der ihm schon vorher geläufig war. Im dritten Fall hat er hingegen den Probanden wahrscheinlich zu hoch eingestuft, so daß dieser ohne die notwendigen Grundlagen durch den Kurs laufen mußte.

Aus den Ergebnissen seiner Analyse wird der Trainer sein *zukünftiges Verhalten* ableiten: Im ersten Fall wird er ohne Änderung an seiner Eingangseinschätzung festhalten wollen. Im zweiten Fall wird er seine Eingangseinschätzung nach oben, im dritten Fall nach unten korrigieren wollen.

Die überprüfte und ggf. revidierte Eingangseinschätzung ist nun der *Maßstab für den nächsten Probanden*. An dessen Abschlußergebnissen wird der Trainer erneut seine Eingangseinschätzung messen, und dieser Kreislauf gilt für alle weiteren Probanden. Die über alle vorangegangenen Probanden gewonnene Erfahrung zeigt sich *im aktuellen Wert für die Eingangseinschätzung*. Bei einem pädagogisch geschickten Trainer wird sich die Eingangseinschätzung mit zunehmender Probandenzahl auf einen *stabilen Wert* einstellen.

Das Verhalten des menschlichen Trainers läßt sich auch in einem *technischen System* nachbilden, siehe grundsätzliche Ansätze hierzu bei [5, S.1] und schon sehr früh bei [6, S.337]. Hierzu liefert die *Fuzzy-Logik* in ihrer regelungstechnischen Anwendung die nötige Grundlage [7, S.9] [8, S.178] [9, S.3]. Das Verhalten des Trainers wird durch *zwei, miteinander vernetzte Regelkreise* simuliert. Der erste Regelkreis generiert eine *Empfehlung zur Einstufung eines Benutzers*. Der zweite Regelkreis korrigiert bei einer (nachträglich ermittelten) schlechten Einstufung den ersten Regelkreis. Die durch einen bestimmten Benutzer gewonnene *Erfahrung* kann hierdurch den folgenden Benutzern zugute kommen.

Im einzelnen sei folgende *Anordnung* vorgeschlagen: Ein Proband bearbeitet einen Anfangstest. Die Ergebnisse laufen in den Regler 1 als Istgröße. Dieser Regler erhält im einfachsten Fall als Führungsgröße einen *Schwellenwert*, anhand dessen die Lernwege getrennt werden. Regler 1 vergleicht den Istwert und den Schwellenwert, er liefert eine *Lernwegempfehlung* und gleichzeitig einen Wert für die Stärke der Lernwegempfehlung (Bild 1).

Der Proband läuft nun durch den empfohlenen Lernweg und bearbeitet einen Abschlußtest sowie eine Abschlußbefragung. Deren Ergebnisse gehen als *Istgröße* in den Regler 2 ein. Als Führungsgröße für diesen Regler dient der Wert für die *Stärke der Lernwegempfehlung*, den Regler 1 generiert hat. Regler 2 generiert aus beiden seine *Stellgröße*, nämlich einen Wert, um den der *Schwellenwert als die Führungsgröße von Regler 1 korrigiert* wird (Bild 2) [10, S.25-29).

Im Zuge mehrerer Durchläufe wird durch die skizzierte Anordnung der *Schwellenwert optimiert*, der über die Zuordnung der Benutzer auf einen der beiden möglichen Lernwege entscheidet. *Erfahrungen* mit den vorhergegangenen Benutzern bilden dafür die Basis.

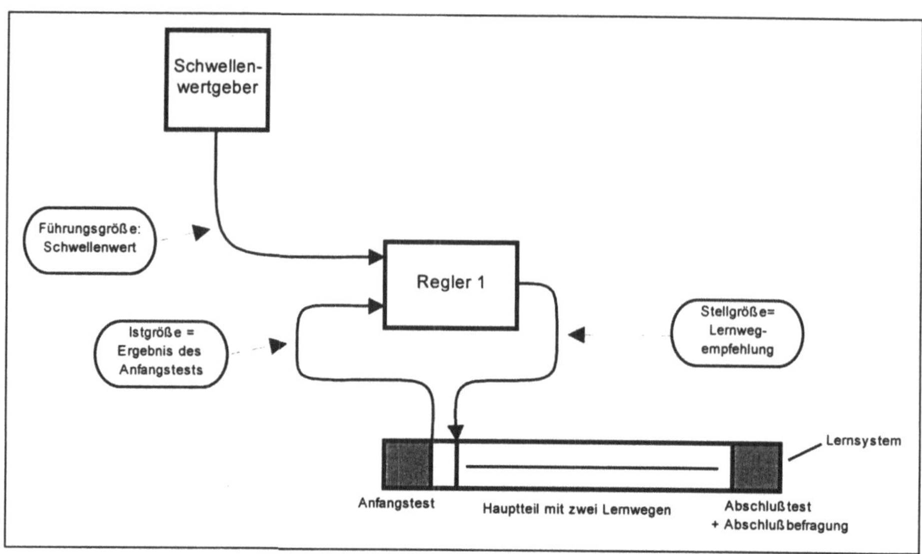

Bild 1: Der erste Regelkreis der erfahrungsbasierten Fuzzy-Regelung von Lernwegen

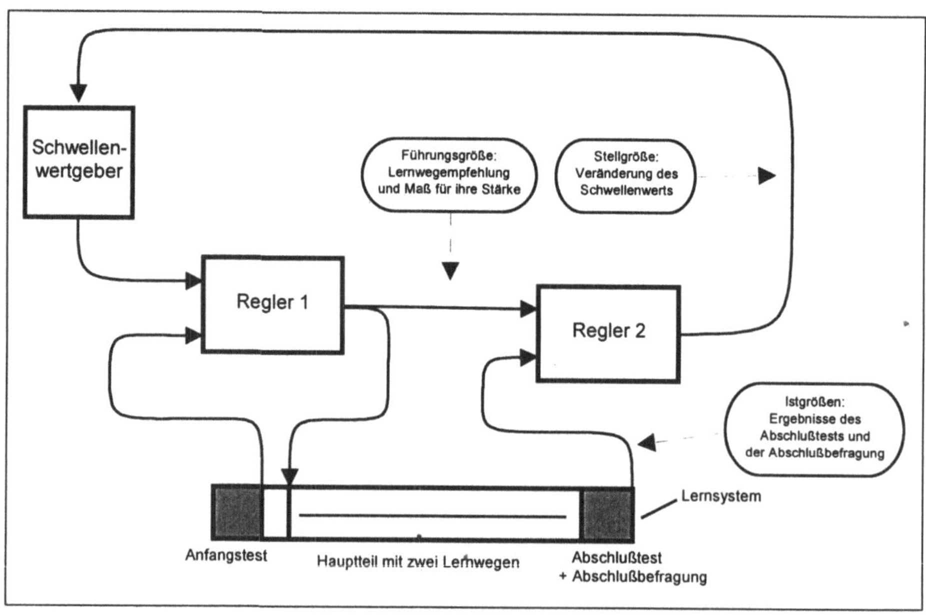

Bild 2: Ergänzung des ersten Regelkreises um einen zweiten, der über einen
variablen Schwellenwertgeber auf den ersten zurückwirkt

Eine besondere Rolle für das Funktionieren der Anordnung spielen die *beiden Regler*. Zu ihrer Gestaltung eignet sich die *Fuzzy-Logik*, mit der man die Regeln, die das Reglerverhalten bestimmen sollen, an die *natürliche Sprache* angelehnt formulieren kann. Am Regler 1 sei das Wirken eines Fuzzy-Reglers verdeutlicht:

(1) Verschiedene Istgrößen eines Benutzer am Eingang des Reglers werden in *unscharfe Werte* (Zugehörigkeiten zu unscharfen Mengen) übersetzt. Hierzu dienen *linguistische Variablen* [8, S.179], die aus jeweils einem oder mehreren Fuzzy-Sets bestehen. Beispielsweise umfaßt die linguistische Variable "Erfahrung" die Fuzzy-Sets "niedrig", "mittel" und "hoch".

(2) Die unscharfen Eingangswerte fließen in ein *Regelwerk* ein, das eine Anzahl verhältnismäßig einfacher Regeln enthält. Hier wird die *Inferenz* gebildet, die logische Verknüpfung der unscharfen Eingangswerte zu einem unscharfen Ausgangswert.

(3) Die unscharfen Ausgangswerte gehen in eine *erste linguistische Ausgangsvariable* ein, die ebenfalls aus mehreren Fuzzy-Sets besteht, und hier wird aus ihnen ein einzelner, scharfer Wert erzeugt. Diesen Schritt bezeichnet man auch als *Defuzzifizierung*.

(4) Der resultierende scharfe Wert wird in eine *zweite linguistische Ausgangsvariable* geleitet. Die *Führungsgröße* bestimmt die Form der dieser Variablen zugehörigen Fuzzy-Sets, was die Besonderheit des Ansatzes bildet. Hier entscheidet sich, ob dem Benutzer der *Lernweg für Anfänger oder für Fortgeschrittene* empfohlen werden soll. Auch liefert die linguistische Ausgangsvariable einen Wert für die *Stärke* der Empfehlung.

3 Erprobung des theoretischen Konzepts

Die zwei vernetzten Fuzzy-Regelkreise aus Abschnitt 2 bilden zunächst ein *theoretisches* Konzept. Es stellt sich die Frage, inwieweit sich dieses Konzept mit vertretbarem Aufwand umsetzen läßt?

- In einem *Experiment* mit 131 Studenten wurden Daten gesammelt (Abschnitt 3.1). In diesem Experiment wurde ein Lernprogramm *ohne* erfahrungsbasierte Fuzzy-Regelung verwendet.

- Parallel dazu wurde ein *Prototyp der erfahrungsbasierten Fuzzy-Regelung* entwickelt und in Turbo Pascal 6.0 implementiert (Abschnitt 3.2).

- Anhand der im Experiment gewonnenen Daten wurde der Prototyp *getestet* und *eingestellt* (Abschnitt 3.3).

3.1 Ein Experiment zur Sammlung von Daten

Die erfahrungsbasierte Fuzzy-Regelung sollte anhand *empirisch gewonnener Daten* erprobt werden. Hierfür wurde ein *Lernsystem mit zwei Lernwegen* entwickelt, das verschiedene Abfragen enthält. Durch *zufällige* Zuordnung von freiwilligen Probanden zu einem der beiden Lernwege konnte ein *breites Spektrum* von Verknüpfungen zwischen Vorabeigenschaften und Abschlußbewertungen festgehalten werden.

Als Grundlage des Experiments wurde ein Lernsystem zum Betriebssystem MS-DOS 4.0 und der darauf aufbauenden graphischen Benutzeroberfläche DOS-Shell erstellt. Das Lernsystem entstand mit *zwei Lernwegen*, jeweils einen für Anfänger und einen für Fortgeschrittene. Beide Lernwege wurden für die Zielgruppe "Studierende der ersten Semester" konzipiert [11, S.33-70). Bei der gewählten Zielgruppe hat eine EDV-Ausbildung, wenn überhaupt, entweder in der Schule stattgefunden, oder Kenntnisse im Umgang mit Computern wurden aufgrund eigenen Interesses selbst erworben.

Die zwei Lernwege enthielten eine *identische Anfangsabfrage* vor dem Einstieg in die Lektionen und eine ebenfalls *identische Abschlußabfrage* nach dem Durchlaufen der Lektionen. In der *Anfangsabfrage* wurden Fragen zur schulischen EDV-Ausbildung, zur privaten EDV-Nutzung und zur persönlichen Einstellung zur EDV gestellt. Die *Abschlußabfrage* überprüfte das Verständnis des gelernten Stoffes (Funktionen eines Betriebssystems), die Kenntnis und sichere Anwendung der Betriebssystembefehle sowie Fertigkeiten im Umgang mit der graphischen Benutzeroberfläche. Ergänzend konnten die Probanden ihre Einschätzung der *Qualität* des Lernsystems nach unterschiedlichen Kriterien und ihre *Lernmotivation* angeben.

Die zwei Lernwege wurden von Studenten des Wirtschaftingenieurwesens an der Universität Kaiserslautern bearbeitet. Hierbei durchliefen *131 Probanden* einmalig jeweils einen der beiden Lernwege, 71 Probanden den Lernweg für Fortgeschrittene und 60 Probanden den Lernweg für Anfänger. Die Zuteilung der Probanden zu den Lernwegen geschah *zufällig*.

Während der Bearbeitung wurden die *Eingaben jedes Benutzers* sowie die benötigten *individuellen Lernzeiten* in gesonderten Dateien *gespeichert*. Nach der Aufbereitung dieser Dateien entstanden hieraus *empirisch gewonnene Datensätze* für die weitere Erprobung der erfahrungsbasierten Fuzzy-Regelung.

3.2 Entwicklung eines Prototyps zur erfahrungsbasierten Fuzzy-Regelung

Parallel zum skizzierten Experiment wurde ein *Prototyp zur erfahrungsbasierten Fuzzy-Regelung* entwickelt. Hierfür wählten wir die *Programmiersprache Turbo Pascal 6.0*: Sie bietet die Möglichkeit der *objektorientierten Programmierung*, in der sich ein Fuzzy-Regler und die dazugehörigen unscharfen Mengen in besonders natürlicher Weise darstellen lassen [12, S.405 ff.] [13, S.17]. Zusätzlich stand eine in Turbo Pascal 6.0 *objektorientiert konzipierte Programmbibliothek* von [13] zur Verfügung, mit deren Hilfe die meisten Aspekte der Implementation eines Fuzzy-Reglers abgedeckt werden konnten.

Die Entwicklung des Prototyps umfaßte in Anlehnung an ein Ablaufschema von [14, S.60] *vier wesentliche Schritte*:

- Definition der *linguistischen Variablen* für die beiden Regler,
- Definition der *Fuzzy-Sets* für diese linguistischen Variablen,
- Definition der *Regeln* für die beiden Regler sowie
- *Stabilisierung* und *Dämpfung* des Schwellenwertgebers.

Im *ersten Schritt* wurden die *linguistischen Variablen* für die beiden Regler definiert. In Regler 1 fließen die Ergebnisse der Anfangsabfrage ein. Hierfür wurden entsprechend der drei Abfrageteile im Lernprogramm drei linguistische Eingangsvariablen festgelegt, nämlich die linguistischen Variablen "Schulische EDV-Ausbildung", "Private EDV-Nutzung" sowie "Persönliche Einstellung zur EDV". Deren Ausprägungen gehen in die zwei linguistischen Ausgangsvariablen "Vorkenntnisse" und "Lernwegempfehlung" ein. Regler 2 verarbeitet die Ergebnisse der Abschlußabfragen mit den drei linguistischen Variablen "Lernzeit", "Lernerfolg" und "Lernspaß". Deren Ausprägungen, genauso wie der aus Regler 1 stammende unscharfe Wert für die Lernwegempfehlung, gehen in die linguistische Ausgangsvariable "Veränderung des Schwellenwerts" ein.

Für die einzelnen linguistischen Variablen wurden im *zweiten Schritt* Fuzzy-Sets definiert, die den *Wertebereich* der linguistischen Variablen genau beschreiben. Bei der Definition der Fuzzy-Sets für Eingang und Ausgang der beiden Fuzzy-Regler flossen *persönliche Wertungen* ein; beispielsweise wurde für die linguistische Variable "schulische EDV-Ausbildung" ein Wertebereich von 0 bis 80 Punkten definiert, in dem drei Fuzzy-Sets angesiedelt wurden. Die Lage dieser Fuzzy-Sets "gut", "mittel" und "schlecht" spiegelt einerseits einen Teil des *Erfahrungswissens des Autors* wider, und dient andererseits als *Parameter zur Einstellung der Regler* [15, S.28-31].

Linguistische Eingangs- und Ausgangsvariablen sind über *Regeln* miteinander verknüpft. Diese Regeln liegen als *Produktionsregeln* in der Form "Wenn A, dann B" vor; sie werden im *dritten Schritt* festgelegt. Die Regeln der Fuzzy-Regler haben neben der *Lage* der Fuzzy-Sets den größten Einfluß auf das Regelverhalten der Fuzzy-Regler und enthalten einen *weiteren Teil des Erfahrungswissens* des Autors [16, S.239].

Zusätzlich zu den beschriebenen Elementen von Fuzzy-Reglern wurden im *vierten Schritt* Mechanismen zur *Stabilisierung* und *Dämpfung* eingeführt. Eine *stabilisierende* Wirkung wird durch die Verwendung von *"Wirkungsradien"* erzielt: Der Schwellenwert wird nur verändert, wenn der Parameter zur Lernwegempfehlung und der Schwellenwert innerhalb eines gewissen Abstandes zueinander liegen. Eine *dämpfende* Wirkung wird herbeigeführt, in dem frühere Schwellenwerte in *exponentiell geglätteter* Form in die Berechnung des aktuellen Schwellenwerts einfließen.

3.3   Test des Prototyps

Der erstellte Prototyp einer erfahrungsbasierten Fuzzy-Regelung konnte anhand der experimentell gewonnenen Daten *getestet* werden. Dabei bedurfte es eines *speziellen Testvorgehens*, bei dem die Verwendbarkeitsprüfung der Datensätze im Mittelpunkt stand. Die Ergebnisse des Tests belegen die *Tragfähigkeit des theoretischen Konzepts*: In mehreren Testläufen, bei denen sich die anfangs eingestellten Schwellenwerte sowie die Parameter zur Dämpfung und Stabilisierung unterschieden, erreichten die Schwellenwerte nach einer gewissen Anzahl an Datensätzen *ein ähnliches Niveau*.

Für den Test war ein *spezielles Vorgehen* nötig. Dies folgte aus der Konzeption des Experiments, bei dem die Probanden *zufällig* jeweils einem der beiden Lernwege des Lernsystems zugeordnet wurden (siehe Abschnitt 3.1). Daher konnten bei einer bestimmten Stellung des Schwellenwertgebers *nicht alle Datensätze* verwendet werden. Es wurden nur solche Datensätze verwendet, bei denen der vom Probanden durchlaufene Lernweg mit der bei dem aktuellen Schwellenwert ausgesprochenen Lernwegempfehlung übereinstimmten.

Das detaillierte Vorgehen beim Test orientiert sich an einem *Ablaufschema* (Bild 3): Zunächst wird ein *Anfangsschwellenwert* festgelegt. Sodann wird zufällig ein Datensatz eines Probanden gewählt.

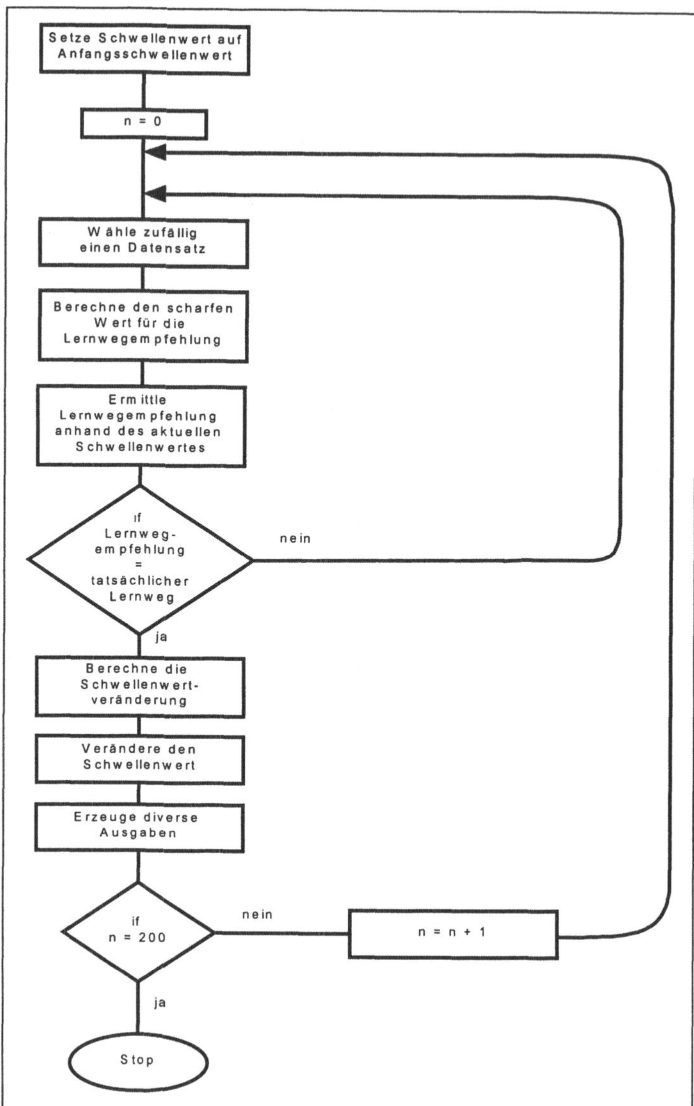

Bild 3: Ablaufschema für den Test des Prototyps

Jeder Datensatz enthält die Information, welcher Lernweg vom Probanden durchlaufen wurde. Anhand der Ergebnisse der Anfangsabfrage und des aktuellen Schwellenwerts wird die *Lernwegempfehlung* ausgesprochen. Stimmt der Lernweg, den der Proband durchlaufen hat, mit der ermittelten Lernwegempfehlung überein, wird der Datensatz weiterverwendet, ist dies nicht der Fall, wird ein neuer Datensatz ausgewählt.

Für jeden ausgewählten Datensatz wird die *Schwellenwertveränderung* anhand der ausgesprochenen Lernwegempfehlung und den vom Probanden tatsächlich erreichten Ergebnissen aus der Abschlußabfrage berechnet. Der *Schwellenwert* wird entsprechend *verändert* und für den folgenden Datensatz *übernommen*.

Das gesamte Vorgehen wird solange wiederholt, bis eine Grenze (z.B. 200 Datensätze) erreicht ist.

Bei den verschiedenen Testläufen zeigte sich ein ähnliches Verhalten (Bilder 4 und 5):

- Zunächst bewegt sich der Schwellenwert von seinem Anfangswert auf ein bestimmtes Niveau zu. Dieses Niveau hat er nach etwa 150 Datensätzen erreicht.

- Sodann pendelt der Schwellenwert um dieses Niveau. Die Parameter zur Stabilisierung und Dämpfung bestimmen dabei die Schwankungsbreite.

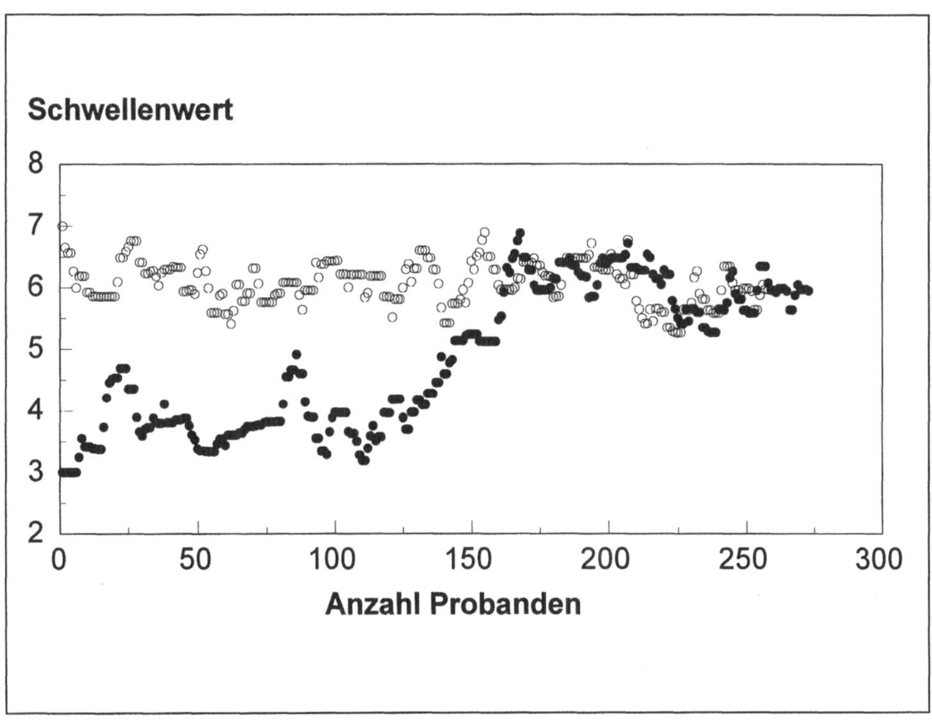

Bild 4: Konvergenzverhalten der vernetzten Fuzzy-Regler mit Parametereinstellung I.
Der Schwellenwert kann zwischen 0 und 10 schwanken. In der Graphik sind zwei Testläufe mit deutlich unterschiedlichen Anfangseinstellungen für den Schwellenwert dargestellt. Im ersten Testlauf startete der Schwellenwert bei dem Wert 3, im zweiten beim Wert 7.

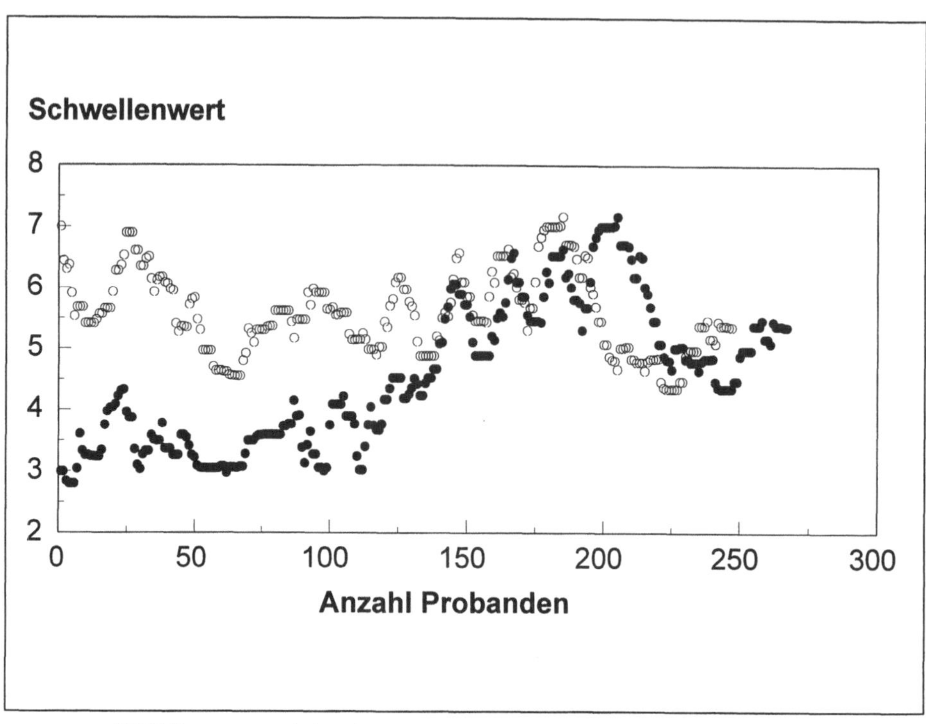

Bild 5: Konvergenzverhalten der vernetzten Fuzzy-Regler mit Parametereinstellung II.
Der Schwellenwert kann zwischen 0 und 10 schwanken. In der Graphik sind zwei Testläufe mit deutlich
unterschiedlichen Anfangseinstellungen für den Schwellenwert dargestellt. Im ersten Testlauf startete der
Schwellenwert bei dem Wert 3, im zweiten beim Wert 7.

4   Ausblick

Mit einem Konzept zur erfahrungsbasierten Fuzzy-Regelung lassen sich Lernwegempfehlungen aussprechen und mit ansteigender Probandenzahl an geänderte Rahmenbedingungen anpassen. Zwei miteinander vernetzte Fuzzy-Regler verändern hierfür einen zentralen Entscheidungsparameter. Wir haben dieses Konzept in einen Prototyp umgesetzt und mit Daten erprobt, die wir empirisch in einem Experiment gewonnen haben. Die Erprobung belegt die Tragfähigkeit des Konzepts.

Sowohl das theoretische Konzept als auch der Prototyp wurden in dieser Untersuchung *bewußt einfach* gehalten. Es bestehen verschiedene Variationsmöglichkeiten, mit denen sich der Nutzen des theoretischen Konzepts steigern läßt. Sie zielen auf die *technische Gestaltung*, die *Anwendungsbreite* und *-granularität*:

- *Technische Gestaltung:* Der Autor könnte Regeln und linguistische Variablen variieren und verfeinern. Dadurch könnte sich das Regelverhalten verbessern [9, S.122-123].

- *Anwendungsbreite:* Neben Lernsystemen eignen sich auch andere Software-Produkte für das vorgestellte Konzept. Es könnte beispielsweise auch in Hilfesystemen Einsatz finden.

- *Anwendungsgranularität:* Anstelle einer einzelnen Lernwegempfehlung in einem Lernsystem könnte man sich auch *mehrere* Lernwegempfehlungen jeweils für einzelne Lektionen vorstellen. Die Ausgangsvariablen für eine dieser Regelungen können als Eingangsvariablen in eine darauf folgende Regelung eingehen. Insbesondere diese Erweiterung bietet nach unserem Ermessen ein beträchtliches Potential zur Steigerung der Änderungsfreundlichkeit von Lernsystemen.

Literaturverzeichnis

[1]   Herczeg, Michael: Software-Ergonomie. Grundlagen der Mensch-Computer-Kommunikation. Bonn et al., 1994: Addison-Wesley.

[2]   Zeidler, Alfred; Zellner, Rudolf: Software-Ergonomie: Techniken der Dialoggestaltung. München, Wien, 1992: Oldenbourg.

[3]   Keller, Rolf; Müller, Konstantin: Computerunterstützter Unterricht. Leitfaden zur Planung und Evaluation von CUU-Projekten. Bern, 1992: Bundesamt für Industrie, Gewerbe und Arbeit.

[4]   Steppi, Hubert: CBT - Computer Based Training: Planung, Design und Entwicklung interaktiver Lernprogramme. Stuttgart, 1989: Klett.

[5]   Leonhard, Werner: Einführung in die Regelungstechnik: lineare und nichtlineare Regelvorgänge; für Elektrotechniker, Physiker und Maschinenbauer, 4. Auflage. Braunschweig, Wiesbaden, 1987: Vieweg.

[6]   Kaspers, Walter; Küfner, Hans-Jürgen: Messen, Steuern und Regeln für Maschinenbauer. Braunschweig, 1977: Vieweg.

[7]   Zimmermann, Hans-Jürgen; Altrock, Constantin von: Prinzipien und Anwendungspotential der Fuzzy-Mengentheorie, in: Künstliche Intelligenz, 5(1991), 4, S.6-12.

[8]   Preuß, Hans-Peter: Fuzzy-Control - heuristische Regelung mittels unscharfer Logik. Teil 1, in: atp - Automatisierungstechnische Praxis, 34(1992), 4, S.176-184.

[9]   Traeger, Dirk: Einführung in die Fuzzy-Logik, 2. Auflage. Stuttgart, 1994: Teubner.

[10] Henke, Andreas: Adaptive Individualisierung der Lernwegempfehlung: Entwurf eines Fuzzy-Regelkreises als Bauelement eines tutoriellen Systems. Universität Kaiserslautern 1994.

[11] Henke, Andreas: Entwicklung und Evaluation eines Lernprogrammes mit zwei Lernwegen, Entwurf eines Experimentes zur Individualisierung der Lernwegsteuerung. Universität Kaiserslautern 1992.

[12] Barth, Gerhard; Welsch, Christoph: Objektorientierte Programmierung, in: Informationstechnik, 30(1988), 6. S.404-421.

[13] Tilli, Thomas: Fuzzy-Logik: Grundlagen, Anwendungen, Hard- und Software, 2. Auflage. München, 1992: Franzis.

[14] Cox, Earl: Fuzzy Fundamentals, in: IEEE Spectrum, 29(1992), 10, pp.58-61.

[15] Cox, Earl: Fuzzy Fundamentals, in: IEEE Spectrum, 30(1993), 2, pp.27-31.

[16] Preuß, Hans-Peter: Fuzzy-Control - heuristische Regelung mittels unscharfer Logik. Teil 2, in: atp - Automatisierungstechnische Praxis, 34(1992), 5, S.239-246.

Adressen der Autoren

Prof. Dr. Martin G. Möhrle
Brandenburgische Technische Universität Cottbus
Postfach 10 13 44
D-03013 Cottbus
Email: moehrle@rz.tu-cottbus.de

Dipl.-Wirtsch.-Ing. Andreas J. Henke
Lange Rötterstraße 65
68167 Mannheim
Email: 0621378824-0001@t-online.de

Berufliche Rehabilitation durch computergestützte Telekooperation

Gebrauchstauglichkeit der internetbasierten Telearbeitsumgebung BSCW und multi-modale Adaptionsoptionen für Behinderte

Michael Pieper, Henrike Gappa und Dirk Hermsdorf

GMD - Forschungszentrum Informationstechnik GmbH
Institut für Angewandte Informationstechnik
Forschungsgruppe: Mensch-Maschine Kommunikation

Zusammenfassung

Technischer Fortschritt an sich bedeutet nicht zwingend eine Erweiterung der Anwendungspotentiale von Informationstechnik. Normalerweise werden die speziellen Nutzungsanforderungen behinderter Endbenutzer bei der Systementwicklung nicht gleichberechtigt mitbebedacht. Diese Gruppe von Endbenutzern wird zumeist nicht in den Designprozeß miteinbezogen. Die informationstechnischen Endprodukte sind daher oft wegen kleiner aber für die Behinderten wesentlicher Mängel nicht einsetzbar. Deshalb werden bei der Implementation des TEDIS Telearbeitsplatzes Verfahren des partizipativen Systemdesigns angewandt. Dies betrifft auch die technischen Aspekte der Telearbeitsplatzumgebung. Auf der Grundlage strukturierter Interviews zu Fragen der Gebrauchstauglichkeit (Usability) des Telearbeitsplatzes werden behindertengerechte Anpassungen der WWW-basierten Telekooperationsumgebung BSCW vorgenommen. Zukünftiges Ziel der Forschungs- und Entwicklungsarbeiten ist die Implementation einer generischen Mensch-Maschine-Schnittstelle für den Internet-Zugang, die universell an unterschiedlichste Beeinträchtigungen schwerbehinderter und älterer Menschen angepaßt werden kann

1. Problemstellung

In Europa gibt es derzeit etwa 100 Mio. ältere und 50 Mio. behinderte Menschen. Nach neueren amtlichen Schätzungen sind allein in Deutschland 6,4 Mio. Menschen schwerbehindert, Tendenz steigend. Mit Blick auf die unter dem Stichwort Informationsgesellschaft diskutierten gesellschaftlichen Umbrüche und der demographischen Tatsache, daß bei einer sich ständig erhöhenden Lebenserwartung verstärkt von altersbedingten Beeinträchtigungen auszugehen ist, die den Zugang zu neuen Medien trotz sozialisationsbedingt zunehmender Akzeptanz („ageing computer professionals") verstellen, erhalten daher Forschungs- und Entwicklungsarbeiten zur Bereit-stellung von angepaßten Zugangstechnologien zu computergestützten Informations- und Kommunikationssystemen für Endbenutzer mit Spezialanforderungen zunehmend eine auch gesellschaftspolitische Bedeutung.

Die Anpaßbarkeit moderner Telekommunikationstechnologien eröffnet prinzipiell vielfältige Möglichkeiten, behinderte und ältere Menschen trotz körperlicher Beeinträchtigungen in Stand zu setzen sowohl weiterhin ein selbstbestimmtes Leben zu führen als auch gleichberechtigt an den gesellschaftlichen und wirtschaftlichen Tätigkeiten der Gemeinschaft, in der sie leben, teilzunehmen. Dennoch formuliert die Fachzeitschrift *Arbeitsrecht im Betrieb* in einer ihrer zurückliegenden Ausgaben, daß trotz einer technischen Entwicklung, bei der körperliche Anstrengungen immer stärker in den Hintergrund gedrängt werden, die Zahl der arbeitslosen Schwerbehinderten trotz der Beschäftigungspflicht nach dem Schwerbehindertengesetz immer weiter zunimmt [2]. Der Zugang zu modernen Telekommunikationssystemen kann allerdings Behinderten durchaus Möglichkeiten eröffnen, wieder am Arbeitsleben teilzunehmen. Vor allem die häufig eingeschränkte räumliche Mobilität Behinderter, die sich besonders nachteilig auf ihre berufliche Integration auswirkt, wird durch Telearbeit in ihren negativen Folgen gemindert

Ziel des GMD-Projektes TEDIS (Teleworking for Disbled People) ist daher die Entwicklung einen beispielhaften Telearbeitsplatz für schwerbehinderte Endbenutzer mit Spezialanforderungen. Er soll industriell einfach und kostengünstig nachgebaut werden können und einer großen Gruppe schwerbehinderter Menschen zugute kommen. Das Bundesministerium für Bildung, Wissenschaft, Forschung und Technologie (BMBF) fördert TEDIS im Rahmen des Förderschwerpunktes „Telekooperation". Grundsätzlich, so die Vorgaben des Projektträgers soll keine Neuentwicklung vorgenommen, sondern auf verfügbare Kooperationssysteme zurückgegriffen werden, die als Referenzsysteme behinderungsgerecht auszugestalten sind.

2. Der TEDIS Feldtest

Im Rahmen eines ersten Feldversuchs wird derzeit diesen Anforderungen entsprechend die internetbasierte Telekooperationsumgebung BSCW (Basic Support for Cooperative Work) zur Ausgestaltung der Telearbeitsplätze von zwei schwerst körperbehinderten Telearbeitern eingesetzt.

2.1 BSCW (Basic Support for Cooperative Work)

BSCW wurde von der GMD-Forschungsgruppe für Kooperationssysteme entwickelt und ist im World Wide Web (WWW), der Multimedia-Oberfläche des Internet unter der Netzadresse „http://bscw.gmd.de/" frei verfügbar. Das System bietet Basisfunktionen zur Unterstützung des Informationsaustausches und des Nachvollzuges von Arbeitstätigkeiten in räumlich dezentralisierten Arbeitsgruppen, die - zumeist zeitlich asynchron - im Internet zusammenarbeiten. Gängige Standardapplikationen zur Unterstützung herkömmlicher Büroarbeiten aber auch weitreichendere Anwendungssoftware zur multimedialen Dokumentenbearbeitung können in diese Arbeitumgebung integriert werden. BSCW-Server werden derzeit von UNIX-Plattformen unter Solaris 2.x und SunOS 4.1.x unterstützt. Systemversionen für Windows und Macintosh Umgebungen sind in Arbeit.

2.2 Das Feldtestszenario

Der TEDIS Feldtest wird in Zusammenarbeit mit dem auch international renommierten "Forschungsinstitut Technologie-Behindertenhilfe" (FTB) durchgeführt. Träger dieses Forschungsinstituts, das Teil eines umfassenden, sich auch auf soziale, medizinische und berufliche Rehabilitation erstreckenden Behindertenzentrums ist, ist die Behindertenhilfe der Evangelischen Stiftung Volmarstein.

Derzeit arbeiten zwei schwerst körper- und mobilitätsbehinderte Erwachsene, die gemeinsam in einer Wohnung bei Dortmund leben, mit Hilfe des Telearbeitssystems BSCW der Verwaltung des 50 km entfernten FTB in Volmarstein bei Hagen zu. Andreas (*Name geändert*) ist 37 Jahre alt und von einer Muskeldystrophie betroffen, die sich derzeit vor allem durch eingeschränkte Einsetzbarkeit der oberen und unteren Extremitäten äußert. Boris (*Name geändert*) ist 32 Jahre alt. Aufgrund einer cerebral bedingten Bewegungsstörung ist seine Motorik durch vornehmlich spastische Bewegungen gekennzeichnet. Diese Beeinträchtigung wirkt sich im Sinne einer Tetraplegie aus. Beide Behinderte sind deswegen auf einen elektrischen Rollstuhl angewiesen, den sie selbst mit Hilfe eines Joysticks steuern. Beide sind arbeitslos. Computer sind ihr Hobby. Profundes Fachwissen haben sie sich autodidaktisch an eigenen PCs beigebracht.

Zur Zeit beschäftigen sich die beiden Telearbeiter unter Inanspruchnahme des relationalen Datenbanksystems „Microsoft Access" mit dem Aufbau der Multimedia-Datenbank „Technische Hilfen am FTB". Diese Datenbank soll im Endausbau eine Auflistung und Beschreibung aller am Forschungsinstitut Technologie-Behindertenhilfe entwickelten technischen Hilfen in Form eines auch über CD-ROM oder Online Dienste abfragbaren „FTB-Katalogs" enthalten. Die Arbeitsaufgabe besteht im wesentlichen darin, Kurzbeschreibungen für derzeit ca. 200 Einzelprodukte zu verfassen und diese mit entsprechendem Bildmaterial aufzubereiten. Die Bildverarbeitung erfolgt durch Scannen verschiedener Bilder aus bereits vorhandenen Ausstellungskatalogen und durch Verwertung und Aufbereitung von Bildmaterial aus dem Videobestand des FTB. Ziel ist es, Interessenten und Besuchern des FTB eine Zusammenstellung und Übersicht über verschiedene technische Hilfen zu geben, die in den Räumlichkeiten des FTB ausgestellt und von Betroffenen auf ihre persönliche Eignung erprobt werden können.

Andreas, der bereits einschlägige Aufträge für das FTB im Rahmen von Werkverträgen abgewickelt hat, konzentriert sich im Rahmen dieser Aufgabenstellung im wesentlichen auf die Bild-, Video- und Graphikbearbeitung. Im letzten Jahr hat er für das FTB Prospekte für die jährliche Fachmesse REHA gestaltet. Bei seiner Arbeit am PC verwendet er bislang keine behinderungsspezifischen Anpassungen.

Boris befaßt sich mit der Textverarbeitung. Er kann beim Tippen nur seine rechte Hand einsetzen und daher Eingaben eher langsam durchführen. Aus diesem Grund gibt er Texte bislang überwiegend nicht selbst vollständig ein, sondern verarbeitet eingescannte Texte sowohl inhaltlich als auch gestalterisch weiter. Als behinderungsspezifische Anpassung verwendet er keine Mouse sondern einen Trackball. Weiterhin hat Boris die Verzögerungszeit beim Mouse-Doppelklick angepaßt.

Bevor einzelne, von Andreas und Boris aufbereitete Datensätze in den Hilfsmittelkatalog aufgenommen werden, werden sie zur gestalterischen Abstimmung über das BSCW-System mit dem FTB in Volmarstein ausgetauscht. Der Zugang zum Internet erfolgt über ISDN und den Internet-Provider T-Online.

3. Usability

Vor Beginn der eigentlichen, vorstehend beschriebenen Telearbeit wurde ein auf die Handhabung von BSCW bezogener "Usability-Test" durchgeführt. Zum einen sollte dadurch ermittelt werden, inwieweit die beiden behinderten Pilotanwender nach einer ursprünglichen Systemeinweisung und einer etwa einmonatigen Übungs- und Eingewöhnungszeit die BSCW-Handhabung beherrschten. Zum anderen sollten erste Hinweise auf die Gebrauchstauglichkeit des ursprünglich nicht unter Berücksichtigung behinderungsspezifischer Spezialanforderungen ausgestalteten BSCW-Systems für behinderte Endbenutzer ermittelt werden.

3. 1 Methodisches Vorgehen

Dazu mußte von den Pilotanwendern eine Standardaufgabe abgearbeitet werden, zu deren Durchführung sie anschließend befragt wurden. Die Standardaufgabe unterteilte sich in drei Teile, die aufeinander aufbauend den von BSCW angebotenen Funktionsumfang allumfassend abdeckten. Inhaltlich bezog sich die Abwicklung der Standardaufgabe auf eine Validierung der vom TEDIS-Projektteam vorgenommenen „Ist-Analyse" des Feldtestrahmenbedingungen vor Beginn der eigentlichen Pilotanwendung. Insofern trug die Standardaufgabe auch methodisch zur Einlösung des in TEDIS verfolgten partizipativen Forschungs- und Entwicklungsansatzes bei, der technischorganisatorische Systemanpassungen in den umfassenden Dialog mit betroffenen Endbenutzern einbindet. Der erste Teil der Standardaufgabe umfaßte folgende Arbeitsschritte:

I.

- Einrichtung einer neuen BSCW-Arbeitsumgebung TEDIS-DO (Für "TEDIS-Dortmund").

- Einrichtung von Nutzungsberechtigungen für die neueingerichtete Arbeitsumgebung für drei Mitglieder des TEDIS-Projektteams in Sankt Augustin und für die beiden in Dortmund ansässigen Pilotanwender selbst.

- Anlegen von zwei Ordnern (files) mit den Bezeichnungen "Befragungen" und "Mitteilungen" in der neueingerichteten Arbeitsumgebung TEDIS-DO.

Nach Abwicklung des ersten Teils der Standardaufgabe legte die Feldtestbetreuerin des TEDIS Projektteams die von ihr dokumentierten „Interviewergebnisse" der erwähnten „Ist-Analyse" in den Ordner „Befragungen". Dieses Dokument wurde durch Abarbeitung der nachfolgend aufgelisteten Arbeitsschritte im zweiten Teil der Standardaufgabe von den Pilotanwendern auf Korrektheit überprüft:

II.

- „Upload" des zwischenzeitlich in den Ordner "Befragungen" eingelegten Dokuments "Interviewergebnisse" zur Vornahme von Korrekturen auf der lokalen Festplatte.

- „Download" des geänderten Dokuments "Interviewergebnisse" in den Ordner Befragungen auf dem Internet/BSCW-Server.

- Erstellung einer kurzen Mitteilung mit der Bezeichnung "Begründung" in dem die vorgenommenen Korrekturen kurz begründet werden.

- Ablage dieses Dokuments "Begründung" in den anderen Ordner "Mitteilungen".

Nach Abarbeitung dieses Teils der Standardaufgabe wurde seitens des TEDIS-Projektteams eine "Stellungnahme" zu der „Korrektur**begründundung**" der Pilotanwender in den Ordner „Mitteilungen" gelegt. Anschließend wurde der nachfolgend umschriebene dritte und letzte Teil der Standardaufgabe von den Pilotanwendern abgearbeitet:

III.

- Einlagerung des Ordners "Mitteilungen", dessen Dokumente sich ja alle auf Mitteilungen zu der durchgeführten Befragung beziehen, als Unterverzeichnis in den Ordner "Befragungen".

- Löschung des Dokuments „Begründung" im Ordner „Mitteilungen", weil der gesamte Vorgang der Textkorrektur und Korrekturbegründung auch in der abschließenden Stellungnahme zur Korrekturbegründung enthalten ist.

Nach erfolgreicher Abarbeitung der vorstehend beschriebenen Standardaufgabe durch die Pilotanwender wurde ein strukturiertes Interview mit dem Ziel durchgeführt, eine Einschätzung der beiden behinderten Endbenutzer über die Gebrauchstauglichkeit der Telekooperationsumgebung BSCW zu erhalten. Der Leitfaden des Interviews orientierte sich dabei an den im softwareergonomischen Evaluationsverfahren EVADIS II dokumentierten Gütekriterien [5].

Das EVADIS II-Verfahren ist auf Grundlage eines zweidimensionalen Orientierungsrahmen entwickelt worden. Die erste Dimension besteht aus dem IFIP-Modell für Benutzerschnittstellen und läßt sich wie folgt skizzieren: Der Anwender interagiert mit einem rechnerbasierten System auf zumindest vier unterschiedlichen Schnittstellenebenen der sogenannten Terminal- (Input/Output), Dialog-, Werkzeug- und Organisationsschnittstelle. Die Terminalschnittstelle regelt die Art der Eingabe von Daten, Parametern und Kommandos sowie andererseits die Art der Darstellung der Rechnerausgabe gegenüber dem Benutzer. Die Dialogschnittstelle bezieht sich auf die Art der Steuerung des Dialogs mit dem System (z.B. kommando-, menue-, icon-gesteuert) und die Art der vom System im Dialogzusammenhang angebotenen Hilfefunktionen, Systemzustands- und Fehlermeldungen. Die Werkzeugschnittstelle bezieht sich auf die Handhabung von Anwendungsprogrammen. Die Organisationsschnittstelle beschreibt Beziehungen zwischen Anwendern innerhalb desselben organisatorischen oder soziotechnischen Kontextes, der durch die Grenzen eines Kooperationssystemen unterliegenden Kommunikationsnetzes gegeben ist.

Die zweite Dimension des Orientierungsrahmens besteht aus einer Reihe von aufgaben-, organisations-, und softwareergonomischen Kriterien aus den Arbeitswissenschaften. Dazu gehören auch im engeren oder weiteren Sinne benutzerbezogene Kriterien wie Anforderungs- vielfalt, Ganzheitlichkeit, Durchschaubarkeit, Rückmeldung, Autonomie, Kooperations- und Kommunikationsförderlichkeit, Lern- und Entwicklungsmöglichkeiten, Datenschutz/Daten- sicherheit, Belastung und Beanspruchung. Eine nähere Erläuterung dieser Kriterien soll an dieser Stelle nicht vorgenommen werden, da dies zu weit führen würde und für die dargestellte Untersuchung nicht wesentlich ist.

3.2 Ergebnisse

Die folgende Darstellung der Ergebnisse des Usability-Interviews wird übersichtlichkeits- halber unter den Aspekten Ein-/Ausgabeschnittstelle, Dialogschnittstelle und Werkzeug- schnittstelle zusammengefaßt.

3.2.1 Ein-/Ausgabeschnittstelle

3.2.1.1 Informationsdarstellung

Der Bildschirm präsentiert sich deutlich strukturiert und unterteilt die Dialogbereiche je nach Aufgabenstellung. Für Endbenutzer mit motorischen Einschränkungen in den oberen Extremitäten ist bei der Bearbeitung der Dialogbereiche die Optimierung der erforderlichen Mausbewegungen von großer Bedeutung.

Hervorhebungen werden in BSCW entsprechend der durchgeführten Operation dargestellt, so daß beispielsweise bereits angewählte Links farblich verändert dargestellt werden. Beim Einsatz von Farben wurden wahrnehmungspsychologische Erkenntnisse berücksichtigt und nicht mehr als 4 Farben eingesetzt, so daß sie auch als Strukturierungshilfe wahrgenommen werden können. Auf Farbblindheit wurde jedoch - beispielsweise durch doppelte Codierung - keine Rücksicht genommen.

Fehlermeldungen werden in einem Fenster verbunden mit einem akustischen Signal angezeigt. Dies wurde als hilfreicher Prompt wahrgenommen.

Die Dialogbereiche sind gut strukturiert auf dem Bildschirm angebracht, so daß Übersicht- lichkeit gewährleistet bleibt. Von den Pilotanwendern wurde jedoch eine Zoom- bzw. Verkleinerungsfunktion als wünschenswert erachtet, um z.B. umfassende Übersicht über alle in einem Workspace vorhandenen Ordner oder ausschließlich in einunddemselben Ordner enthaltenen Dokumente erlangen zu können. Eine Sortierfunktion, z.B. nach dem Dokument- erstellungsdatum, wäre ebenfalls ein weiterer sinnvoller Beitrag zur Übersichtlichkeit.

Die Darstellung von Texten/Zahlen und hier insbesondere die Schriftgröße und der -typ konnten von den Benutzern ohne Schwierigkeiten wahrgenommen werden. Auch die Länge der Texte wurde als angebracht empfunden. Dies ist ein wichtiges Kriterium für Endbenutzer mit Leseschwierigkeiten oder Perzeptionsstörungen, z.B. als Folge einer Hirnschädigung. Zu den hier wesentlichen Perzeptionsstörungen gehören beispielsweise mangelnde Figur-Grund- Differenzierung, die Erfassung von Ganzheiten und Formanalyse und -synthese.

3.2.1.2 Eingabe

Inbezug auf Fehlerrobustheit wurde die gleichlautende Bezeichnung "Add" in divergierenden Icons kritisiert, da das Symbol nicht als das für die Differenzierung entscheidende Merkmal sofort wahrgenommen wurde. Es kam hier bei den Pilotanwendern zu Verwirrungen und als Folge wurden ungewünschte Aktionen ausgelöst. Symbole sollten generell mit deutlichen Kontrasten und prägnantem Design dargestellt werden, um Endbenutzern mit Perzeptionsstörungen die Handhabung zu erleichtern.

Die Cursorpositionierung dient in BSCW als Eingabeaufforderung. Die Eingabemarke befindet sich jedoch nicht immer an der im Arbeitskontext erforderlichen Stelle. Darüberhinaus wurde eine Möglichkeit vermißt, Eingabefelder für die Abfolge von mehreren gleichartigen Eingaben geöffnet zu lassen (z.B. durch eine "Weiter"-Abfrage). Unter dem Aspekt, daß insbesondere Endbenutzer mit Beeinträchtigungen in den oberen Extremitäten zur Optimierung ihres Arbeitseinsatzes darauf angewiesen sind, möglichst wenig Maus- bzw. Cursoransteuerungen vornehmen zu müssen sowie die Anzahl einzugebender Zeichen so gering wie möglich zu halten, müssen hier Softwareanpassungen vorgenommen werden.

3.2.1.3 Sprache

Die in den Icons benutzten Abkürzungen implizierten für die Benutzer nicht immer die damit verbundene Aktion. Aus manchen Abkürzungen, z.B. "des" für „description", konnte keine Wortbedeutung entwickelt werden. Dies weist auf ein weiteres grundsätzlicheres Problem hin, nämlich die mangelnde Berücksichtigung der Muttersprache des Endbenutzers. Die meisten im Internet bzw. World-Wide-Web vorhandenen Websites sind in englisch gehalten, wodurch interessierte Anwender mit geringen Englischkenntnissen - und dies betrifft ältere und behinderte Endbenutzer in besonderem Maße - von der Nutzung ausgeschlossen sind. Dies ist ein informationspolitisches Problem, welches einer grundsätzlichen Lösung bedarf, indem man entweder in mehrere Sprachen übersetzte Versionen oder zumindest Übersetzungshilfen anbietet.

3.2.2 Dialogschnittstelle

Die häufig auftretende Anzeige von Nummerncodes bei der Ausgabe von Fehlermeldungen im WWW ist wenig hilfreich für eine Lösung des aufgetretenen Problems. In vielen Fällen gibt BSCW selbst aber nicht nur einen Nummerncode wieder, sondern nennt Gründe für den Fehler mit Hinweisen zu deren Behebung. Diese Angaben wurde jedoch von den Pilotanwendern als unverständlich eingestuft und führten demzufolge auch nicht zu Problemlösungsversuchen. Die Diskrepanz zwischen Systemexperten und der eigentlichen Zielgruppe, nämlich anwendungsorientierten Endbenutzern, wird hier deutlich. In diesem Zusammenhang sind bessere Abstimmungen notwendig.

3.2.3 Werkzeugschnittstelle

Die Funktionalität des Anwendungssystems erschien als ausreichend für die Abwicklung der oben beschriebenen Arbeitsaufgabe. Genauere Aussagen, z.B. bezüglich der Einbindung von Multimedia-Dokumenten, lassen sich erst nach einem umfassenden Einsatz des Systems für derartige Aufgabenstellungen treffen.

Als Funktionalitätserweiterung wurde vorgeschlagen, die BSCW-Benutzerschnittstelle um eine Möglichkeit, z.B. E-mail Adressen über einen Zwischenspeicher einlesen zu können, zu erweitern. Durch diese und ähnliche Maßnahmen könnten Online-Zeiten verkürzt und die für motorisch behinderte Endbenutzer zeitraubende Eingabe effizienter gestaltet werden.

4. Multi-modale Adaptionsnotwendigkeiten für motorisch Behinderte

Insbesondere mit Blick auf neuartige Multimediatechnologien ergeben sich derzeit eine Menge neuer Möglichkeiten, die zu einem ganzen Bündel äußerst facettenreicher, behinderungsangepaßter Interaktionsmodi zwischen behindertem Endbenutzer und Computersystem zusammengefaßt werden können [6]. Computergestützte Rehabilitationstechnologien („Assistive Technologies") bieten mittlerweile eine Vielfalt unterschiedlichster akustischer, visueller und taktiler Anpassungsmöglichkeiten für Benutzerschnittstellen, darunter sog. „Stereophone Earcons" (z.B. der Klang eines Mülleimers, der die Löschprozedur eines MAcintosh-Betriebssystems für Blinde symbolisiert)[4], oder sog. „Virtual Lipreaders" (z.B. das animierte menschliche Gesicht auf einem graphischen Display, das seine Lippen im Einklang mit Spracheingabe bewegt und tauben Menschen so die Spracherkennung ermöglicht) [1]. Während diese beiden Beispiele zwei Möglichkeiten der Anpassung von Benutzerschnittstellen an *rezeptive* Behinderungen illustrieren, werden in TEDIS wenngleich weniger spektakuläre, so doch aus der Sicht behinderter Endbenutzer überaus bedeutsame Anpassungen an *expressive* Behinderungsmodalitäten vorgenommen.

Grundsätzlich kann für körperbehinderte Endbenutzer die Steuerbarkeit von BSCW erheblich erleichtert werden, wenn als Ersatz für die feinmotorisch äußerst anspruchsvolle Cursorführung per Maus ein vorprogrammierter Tastatur-Cursor zur Verfügung stünde. Da dies die browser- und nicht die BSCW-spezifische Handhabungsfunktionalität betrifft wird in TEDIS derzeit die außschließliche Bedienbarkeit eines Internet-Browsers über Computertatstatur realisiert, darunter auch behindertengerechte Spezialtastaturen.

Diese Adaptionen werden unter Verwendung des „MS Internet Control Pack" implementiert, einem Baukasten von derzeit 8 sogenannten „Active X TM Controls", die u.a. in Visual Basic eingebunden werden können. Alternativ werden auch Sourcecodemodifikationen des als Public Domain Software zur Verfügung stehenden HotJava-Browsers in Betracht gezogen.

Im einzelnen sollen folgende Anpassungen für motorisch behinderte Anwender realisiert werden, die durch vertieftes Nachfragen im Rahmen sog. narrativer Usability-Interviews ermittelt wurden:

- tabben durch Links, Formulare und Checkboxen von HTML Dokumenten
- Textmarkierung durch Tastaturkommandos
- Auswahl von Hypertextlinks durch Eingabe des ersten Buchstaben des entsprechenden Labels
- tabben durch Hypertextlinks in vorgebbarer Schrittgröße
- Annäherungstoleranzen für die Ansteuerung von Schaltflächen und Links in HTML-Dokumenten

- Scanningverfahren, d.h. sequentielles Markieren von Schaltflächen und Links und
  Auswahl und Auslösung aktuell markierter Stellen über Tastatur
- Erkennung und Ergänzung wiederkehrender Routineeingaben nach Eingabe von nur
  wenigen Buchstaben

5. Ausblick: Der generische Ansatz

Grundsätzlich wird in TEDIS versucht, bereits existierende Telekooperationssysteme mit
einer generischen und somit möglichst universell einsetzbaren graphischen Benutzeroberfläche auszustatten, deren Objekte (Fenster, Kommandomenüs, Schaltflächen etc.) sich
auf multimediale Weise, etwa visuell, akustisch oder taktil, manifestieren und so an
spezifische körperliche Beeinträchtigungen des jeweiligen Endbenutzers anpassen.

Zwischen Systemplattform und Benutzeroberfläche, so die Grundidee, ist eine Software-
schicht zu implementieren, die durch behinderungsabhängige Spezifikationen von Ein-/Aus-
gabetreibern eine schnelle und unkomplizierte Anpassung von Applikationen an die Spezialbedürfnisse behinderter Endbenutzer ermöglicht.

In diesem Zusammenhang beteiligt sich TEDIS maßgeblich an der internationalen Diskussion
um Architekturprinzipien für die Ausgestaltung generischer Benutzeroberflächen im WWW
[3]. Ein umfassendes behinderungsspezifisch anpassbares Design von HTML-Dokumenten
basiert auf der einfach nachvollziehbaren Grundidee, daß Teile eines Multimediadokumentes,
die in einem Format präsentiert werden, das Endbenutzern mit bestimmten, zumeist
rezeptiven Behinderungen nicht oder nur schwer zugänglich ist, in Darstellungsformen
überführbar sein muß, die sich diesen Endbenutzern leichter erschließen. Im Extrem führt
diese Idee zur Ableitung eines „Design for All", das den uneingeschränkten Zugang zu
WWW-basierten Multimediadokumenten für jedermann sicherstellen soll. Die Funktionalität
und die Ästhetik der Multimediaapplikation oder des Multimediadokumentes soll dabei
möglichst uneingeschränkt erhalten bleiben.

Weniger trivial ist in diesem Zusammenhang die Frage, in welcher Form Informationen am
besten von einer Darstellungsart in die andere transformiert werden können. Die derzeit vorherrschende pragmatische Lösung besteht in der Forderung, alle möglichen Darstellungsformen, also Graphiken, Bilder, Videos, Sounds und gestalteten Text (Schriftarten, -größen,
Überschriften, Absätze...) textuell zu umschreiben. Text in Form von ASCII- oder besser
noch SGML-Code kann multimedial jederzeit zumindest an die Spezialbedürfnisse von
Behinderten mit rezeptiven Handicaps angepasst werden. Insbesondere das webbezogene
SGML-Derivat HTML bietet viel mehr textbezogene Strukturinformationen als ASCII, bindet
auch multimediale Darstellungsinhalte wie Graphiken und Bilder in textuelle Darstellungs-
entsprechnungen ein und ist daher von zunehmender Bedeutung für die Anbindung von
Treiber- und Konvertierungsprogrammen für Braillezeilen, synthetische Sprachausgaben und
ähnliches mehr.

Um den Implementationsanspruch einer generischen, also möglichst universell an unterschiedlichste Beeinträchtigungen behinderter Menschen anpaßbaren Mensch-Maschine-
Schnittstelle für den Internet-Zugang zu unterstreichen, sollen zusätzlich die nachfolgenden

nicht browser- sondern HTML-bezogenen Adaptionen für WWW-Dokumente und WWW-Applikationen wie BSCW für sehbehinderte und blinde Endbenutzer vorgenommen werden.

- Einfügen von Labels (Größe der Liste, Anzahl der Unterlisten) in Listen von HTML-Dokumenten
- Umwandlung von Tabellen in eine sequentielle Darstellung
- Auflistung aller im Dokument vorhandenen Hypertextlinks
- Auflistung aller im HTML Code vorkommenden „Header x" Überschriften
- Schlüsselwörter wie 'Link', 'Radiobutton' etc. vor die entsprechenden Stellen im HTML Dokument

Literaturverzeichnis

[1] H.-H. Bothe: Artificial Visual Speech - Generated by Fuzzy Inference Methods. In: I. Placencia Porrero and R. Puig de la Bellacasa (Hg.): The European Context for Assistive Technology. Proc. 2nd TIDE Congress, Amsterdam 1995, 1995, 302 -305

[2] Bund-Verlag GmbH (Hg.): Arbeitsrecht im Betrieb, 2/95, S. 86ff.

[3] P. Fontaine: Writing Accessible HTML Documents, Center for Information Technology Accommodation, General Services Administration, Washinghton, DC. USA, 1995

[4] E. Leimann; H.-H. Schulte: Earcons and Icons - An Experimental Study. In: k. Nordby at al. (Hg.): Human-Computer Interaction - Interact '95, London, 1995,49 - 54

[5] R. Oppermann, Reinhard; B. Murchner; H. Reiterer; M. Koch: Softwareergonomische Evaluation. Der Leitfaden EVADIS II. (Mensch Computer Kommunikation 5/2), de Gruyter, Berlin: 1992

[6] M. Pieper: "Calling the Blind" is "Watched by the Deaf" - Directions for Multimodal CSCW-Adaptations to Receptive Disabilities. In: C. Stephanidis (Hg.): Towards User Interfaces for All: Current Efforts and Future Trends, Heraklion: ICS-FORTH, 1995

Adressen der Autoren

Dr. Michael Pieper u. M. Ed. Henrike Gappa
GMD - Forschungszentrum Informationstechnik GmbH
Institut für Angewandte Informationstechnik
Forschungsgruppe Mensch-Maschine Kommunikation
Schloß Birlinghoven
53754 Sankt Augustin
e-mail: michael.pieper@gmd.de
         henrike.gappa@gmd.de

Cand. Dipl. Inf. Dirk Hermsdorf
Fountain Head Apartments
1221 Bob Pettit Boulevard #214
Baton Rouge, LA 70820
U.S.A.
e-mail: dirkherm@bit.csc.lsu.edu

Der Benutzungsfragebogen ISONORM 9241/10: Ergebnisse zur Reliabilität und Validität

Jochen Prümper

Fachhochschule für Technik und Wirtschaft Berlin

Zusammenfassung

In diesem Beitrag wird die testtheoretische Güte des Benutzungsfragebogens ISONORM 9241/10 [17] dokumentiert. Grundlage hierzu sind Beurteilungen von 1265 Benutzern von insgesamt 178 Softwareprogrammen. Zur Bestimmung der testtheoretische Güte werden Untersuchungen zur Reliabilität, Validität und Normierung durchgeführt. Zur Reliabilitätsbestimmung wird die *Wiederholungsmethode* und die *Konsistenzanalyse* herangezogen. Die *Wiederholungsmethode* zeigt, daß sich der ISONORM 9241/10 durch eine hohe Merkmalskonstanz auszeichnet. Die *Konsistenzanalyse* zeigt die Leistungsfähigkeit des ISONORM 9241/10 als Meßinstrument, da die einzelnen Skalen des ISONORM 9241/10 gute innere Konsistenzen aufweisen. Im Rahmen der Validitätsbestimmung werden Untersuchungen zur *kriterienbezogenen Validität* und zur *Konstruktvalidität* angestellt. Zur Überprüfung der *inneren kriterienbezogenen Validität* werden die ISONORM 9241/10-Bewertungen mit den Bewertungen von zwei Benutzungsfragebögen QUIS [21] und BBD [22] korreliert. Es zeigt sich, daß der ISONORM 9241/10 zu vergleichbaren Ergebnissen gelangt, und gleichzeitig den Vorteil in sich birgt, eine ausdrückliche Operationalisierung von ISO 9241 Teil 10 darzustellen. Zur Überprüfung der *äußeren kriterienbezogenen Validität* wird unter Verwendung von EVADIS II [9] überprüft, inwiefern die Befragung durch Benutzer mit der Evaluation durch Experten korrespondiert. Es zeigt sich, daß die ISONORM 9241/10-Benutzungsanalysen einen signifikanten Bezug zu einer systematischen, leitfadenorientierten Expertenevaluation aufweisen. Zur Überprüfung der *Konstruktvalidität* werden zwei Versionen eines Softwareprogramms mit und zwei Versionen eines Softwareprogramms ohne graphische Benutzungsschnittstelle herangezogen. Softwareergonomische Veränderungen von einer Version zur nächsten sind mittels ISONORM 9241/10 meßbar; zusätzlich lassen sich differenzierte Aussagen über diese Verbesserungen anstellen. Die *Normierung* erlaubt die Einordnung einer einzelnen Software in Relation zu den Beurteilungen anderer Programme.

1 Einleitung

Der Benutzungsfragebogen ISONORM 9241/10 wurde erstmals auf der Software-Ergonomie Fachtagung '93 vorgestellt [17]. Er stellt eine Operationalisierung der ISO 9241 Teil 10 „Grundsätze der Dialoggestaltung" dar, welchem eine zentrale Rolle im Rahmen der „Ergonomischen Anforderungen für Bürotätigkeiten mit Bildschirmgeräten" [4] zukommt. Die dort formulierten Forderungen sind unabhängig von einer bestimmten Dialogtechnik und sollen sowohl bei der Leistungsbeschreibung, der Gestaltung als auch der Bewertung von Dialogsystemen angewandt werden. Erklärtes Ziel der Anwendung dieser Norm ist es, Benutzungsschnittstellen zu gewährleisten, „die gebrauchstauglicher und konsistenter sind und eine höhere Produktivität ermöglichen" [4, S. 4]. Zu diesem Zweck werden in ihr sieben allgemeine ergonomische Grundsätze definiert, die mit Empfehlungen und Beispielen konkretisiert und erläutert werden. Bei diesen Grundsätzen handelt es sich um: Aufgabenangemessenheit, Selbstbeschreibungsfähigkeit, Steuerbarkeit, Erwartungskonformität, Fehlertoleranz, Individualisierbarkeit und Lernförderlichkeit.

Bei der Entwicklung des ISONORM 9241/10 stand im Vordergrund, ein effizient einsetzbares, robustes Verfahren zu entwickeln, das zur Überprüfung der Normkonformität von ISO

9241/10 herangezogen werden kann. Ein weiters Anwendungsgebiet ist die Beurteilung von Prototypen im Rahmen eines iterativen Systemdesigns.

Erste Untersuchungen wurden hierzu bereits vorgelegt [12,14,15,16,17]. Es wurde dargestellt, daß der ISONORM 9241/10 ein nützliches formatives Evaluationsverfahren ist, wenn er im Rahmen eines partizipativen Designkonzeptes dazu genutzt wird, in Kombination mit einer visualisierten Gruppendiskussion erste konkrete Schwachstellen einer Software zu identifizieren [17].

In einer summativen Evaluationsstudie über die ergonomische Qualität von Softwareprodukten mit und ohne graphische Benutzungsschnittstellen (GUI) konnte - im Sinne einer ersten Validierung - erwartungsgemäß gezeigt werden, daß Software mit GUI auf allen sieben Skalen des ISONORM 9241/10 signifikant bessere Werte erreichte als solche ohne GUI [15].

Dennoch weisen die Autoren des ISONORM 9241/10 selbst darauf hin, daß zur Überprüfung der testtheoretischen Qualität des Verfahrens - insbesondere hinsichtlich der Reliabilität und Validität - noch weitere Untersuchungen angestellt werden müssen [15,17]. Dieses Defizit zu beheben ist Thema des vorliegenden Beitrags. Hierzu erfolgt zunächst eine kurze Beschreibung des ISONORM 9241/10 sowie der Benutzer, die an den Evaluationen teilgenommen haben. Anschließend wird der Nachweis der Reliabilität anhand der beiden Verfahren *Wiederholungsmethode* und *Konsistenzanalyse* geleistet. Zur Validitätsbestimmung werden Untersuchungen zur *kriterienbezogenen Validität* und zur *Konstruktvalidität* angestellt.

1.1 Beschreibung des Verfahrens

Um dem Anspruch eines ökonomisch einsetzbaren Verfahrens gerecht zu werden, wurde der Umfang des ISONORM 9241/10-Fragebogens so zugeschnitten, daß jeder einzelne der sieben Grundsätze von ISO 9241/10 über fünf Items operationalisiert wurde. Insgesamt umfaßt das Verfahren damit 35 Items.

Der „Granulationsgrad" [19] der Beschreibungssprache wurde so gewählt, daß die spezifischen Eigenschaften unterschiedlicher Softwaresysteme hinreichend genau differenziert werden können und er gleichzeitig auf möglichst viele unterschiedliche Programme anwendbar ist. Die Durchführungszeit für den ISONORM 9241/10 beträgt ca. 10 Minuten.

Jeder der sieben Grundsätze von ISO 9241/10 wurde über fünf bipolare Aussagen konkretisiert. Beantwortet werden diese Aussagen auf einer sieben-stufigen Skala von „---" (1) bis „+++" (7). Abbildung 1 liefert ein Beispielitem des ISONORM 9241/10 für den Grundsatz „Steuerbarkeit".

Die Software ...	---	--	-	-/+	+	++	+++	*Die Software ...*
erzwingt eine unnötig starre Einhaltung von Bearbeitungsschritten.	□	□	□	□	□	□	□	erzwingt keine unnötig starre Einhaltung von Bearbeitungsschritten.

Abb. 1: Beispielitem des ISONORM 9241/10 für den Grundsatz „Steuerbarkeit"

Den Benutzungsfragebogen ISONORM 9241/10 gibt es in deutscher, englischer, französischer und niederländischer Sprache.

1.2   Beschreibung der Gesamtstichprobe

Bislang liegen von 1265 Benutzern (51.8% Frauen, 48.2% Männer) Beurteilungen von insgesamt 178 Softwareprogrammen vor. Das Durchschnittsalter der Befragten lag bei rund 34 Jahren, sie arbeiteten im Durchschnitt seit 77 Monaten mit Computern und seit 25 Monaten mit der von ihnen beurteilten Software. Die Frage „Wie gut beherrschen Sie die beurteilte Software?" beantworteten die befragten Personen auf einer Skala von 1 (sehr schlecht) bis 7 (sehr gut) im Durchschnitt mit 5.1. Mit der beurteilten Software arbeiteten die Personen im Mittel 14 Stunden pro Woche, mit Computern überhaupt 24 Stunden pro Woche. Insgesamt arbeiteten die Befragten durchschnittlich mit 5.8 verschiedenen Softwareprogrammen.

2   Reliabilität

Unter der *Reliabilität* oder Zuverlässigkeit eines Meßverfahrens versteht man den Grad der Genauigkeit, mit dem es ein bestimmtes Merkmal mißt, gleichgültig, ob es dieses Merkmal auch zu messen beansprucht (letzteres ist eine Frage der Validität) [7]. Zwei Vorgehensweisen der Reliabilitätsbestimmung sollen im folgenden beleuchtet werden: die *Wiederholungsmethode* und die *Konsistenzanalyse*.

2.1   Wiederholungsmethode

Die *Wiederholungsmethode* - oder auch Bestimmung der *Retest-Reliabilität* - schlägt sich in einem Stabilitätsmaß nieder und gibt an, inwiefern eine Person bei Befragung zu einem späteren Zeitpunkt zu ähnlichen Beurteilungen desselben Sachverhaltes gelangt.

Zur Bestimmung der Retest-Reliabilität wurde eine Substichprobe von 49 Benutzern gebeten, ihre ISONORM 9241/10-Beurteilung zu einem zweiten Zeitpunkt für dieselbe Software zu wiederholen. Zwischen den Meßzeitpunkten t_1 und t_2 lagen im Mittel 6.7 Monate.

Tabelle 1 gibt die Retest-Reliabilitäten des ISONORM 9241/10 wieder. Die niedrigste Retest-Reliabilität beträgt bei dem Grundsatz Lernförderlichkeit $r_{t1/t2} = .59$, die höchste Retest-Reliabilität bei dem Grundsatz Fehlertoleranz $r_{t1/t2} = .68$. Die durchweg hochsignifikanten Stabilitäts-Koeffizienten zeigen, daß Benutzer auch nach einem längeren Zeitraum zu ähnlichen Beurteilungen desselben Softwaresystems gelangen. Der ISONORM 9241/10 legt damit einen zufriedenstellenden Grad an Stabilität an den Tag.

2.2   Konsistenzanalyse

Die *Konsistenzanalyse* untersucht als Methode zur Bestimmung der Skalenzuverlässigkeit den Grad der Homogenität der einzelnen Items im Hinblick auf das zu erfassende Merkmal. Diese innere Konsistenz kennzeichnet als instrumentelle Reliabilität gewissermaßen die Leistungsfähigkeit des Meßinstrumentes und ist ein von den Bedingungen des Meßdurchführung unabhängiger Aspekt der Reliabilität [7].

Zur Bestimmung der *inneren Konsistenz* konnte wiederum die Gesamtstichprobe analysiert werden. Die Berechnung der inneren Konsistenz erfolgt über Cronbachs alpha.

Es zeigt sich (vgl. Tab. 1), daß die einzelnen Skalen des ISONORM 9241/10 gute innere Konsistenzen aufweisen.

Grundsatz	$r_{t1/t2}$*	alpha[⊚]
Aufgabenangemessenheit	.67	.81
Selbstbeschreibungsfähigkeit	.62	.86
Steuerbarkeit	.64	.84
Erwartungskonformität	.60	.84
Fehlertoleranz	.68	.87
Individualisierbarkeit	.63	.89
Lernförderlichkeit	.59	.83

Anmerkung: * $p < .001$; $N = 49$ Benutzer
[⊚] N zwischen 1208 und 1251 Benutzern; Bestimmung der inneren Konsistenz über *Cronbachs alpha*

Tab. 1: Retest-Reliabilität und innere Konsistenz des ISONORM 9241/10

3 Validität

Die *Validität* oder Gültigkeit eines Tests gibt den Grad der Genauigkeit an, mit dem ein Meßverfahren das, was es zu messen beansprucht, tatsächlich mißt [7]. Bei der Validität sollen verschiedene Aspekte unterschieden werden: zum einen die *kriterienbezogene Validität* und zum anderen die *Konstruktvalidität*.

3.1 Kriterienbezogene Validität

Bei der *kriterienbezogenen Validität* werden die Meßergebnisse mit einem sog. Außenkriterium in Verbindung gesetzt. Bezüglich der kriterienbezogenen Validität läßt sich eine Unterscheidung zwischen innerer und äußerer Validität fällen. Von *innerer Validität* spricht man, wenn der Skalenwert mit den Skalenwerten anderer schon als valide anerkannter Meßverfahren korreliert wird. Unter *äußerer Validität* versteht man das Korrelieren der Skalenwerte mit einem Schätzurteil oder mit einer objektiv bewerteten Kriteriumsleistung [7].

3.1.1 Innere kriterienbezogene Validität

Zur Überprüfung der inneren kriterienbezogenen Validität soll der Frage nachgegangen werden, ob der ISONORM 9241/10 zu vergleichbaren Ergebnissen gelangt wie andere Benutzungsfragebögen.

Für diese Validierung wurden zwei Fragebögen herangezogen, die ebenfalls Benutzungsfreundlichkeit von Software erheben. Es handelt sich dabei um die deutsche Übersetzung [6] der Langform des QUIS [21] und um den BBD [22]. Da beide Verfahren keinen Bezug zu ISO 9241/10 aufweisen, muß sich die Validitätsuntersuchung auf den Vergleich der jeweiligen Gesamtwerte beschränken.

Für diese Studie wurden 31 Benutzer gebeten, eine ihnen bekannte Software sowohl anhand des ISONORM 9241/10 als auch mittels des QUIS und des BBD zu beurteilen. Diese Benutzer arbeiteten durchschnittlich seit rund 18 Monaten mit der von ihnen evaluierten Software und seit rund 58 Monaten überhaupt mit Computern. Die Frage „Wie gut beherrschen Sie die beurteilte Software?" beantworteten sie auf einer Skala von 1 (sehr schlecht) bis 7 (sehr gut) im Durchschnitt mit 5.6.

Die Korrelation zwischen ISONORM 9241/10 und QUIS beträgt $r = .73$ ($p < .001$; $N = 31$ Benutzer), die zwischen ISO 9241/10 und BBD $r = .71$ ($p < .001$; $N = 31$ Benutzer).

Diese Korrelationen zeigen, daß der ISONORM 9241/10 in der Gesamtbetrachtung zu ähnlichen Ergebnissen gelangt wie andere Benutzungsfragebögen; jedoch den Vorteil in sich birgt, eine explizite Operationalisierung von ISO 9241 Teil 10 zu sein.

3.1.2 Äußere kriterienbezogene Validität

Zur Bestimmung der äußeren kriterienbezogenen Validität soll überprüft werden, ob die Befragung durch Benutzer mit der Evaluation durch Experten korrespondiert.

EVADIS II [9] ist ein solches softwareunterstütztes, leitfadenorientiertes Evaluationsverfahren für Ergonomieexperten.

Für diese Validitätsstudie wurden 13 verschiedene Softwareprogramme von einem Software-Ergonomen in Zusammenarbeit mit einem erfahrenen Benutzer der jeweiligen Software anhand von EVADIS II evaluiert und mit den entsprechenden ISONORM 9241/10-Beurteilungen durch Benutzer verglichen. Für die ISONORM 9241/10-Analysen wurde eine Substichprobe von 383 Benutzern herangezogen. Jedes der 13 Softwareprogramme wurde dabei durchschnittlich von 29.5 (N_{min} = 11, N_{max} = 79) Benutzern beurteilt.

Die Korrelation zwischen dem EVADIS II-Gesamtwert und dem ISONORM 9241/10-Gesamtwert beträgt r = .59 (p < .01; N = 13 Softwareprogramme).

Es zeigt sich also, daß die ISONORM 9241/10-Benutzeranalysen einen signifikanten Bezug zu einer systematischen, leitfadenorientierten Expertenevaluation aufweisen.

3.2 Konstruktvalidität

Bei der *Konstruktvalidität* leitet man Hypothesen aus dem zugrundeliegenden Konstrukt ab und überprüft diese empirisch mittels des zu validierenden Meßinstrumentes [7].

Im vorliegenden Fall liegt der Bestimmung der Konstruktvalidität die Hypothese zugrunde, daß bei der neuen Version einer Software ergonomische Qualitätsverbesserungen zu verzeichnen sind. Im Sinne einer Validierung sollten diese Qualitätsverbesserungen mittels ISONORM 9241/10 meßbar sein. Dabei soll insbesondere ein Augenmerk darauf gelegt werden, ob sich für verschiedene Softwareprogramme differenzierte Aussagen hinsichtlich der Qualitätsverbesserung bei einzelnen Grundsätzen anstellen lassen.

Für diese Validitätsstudie sollten zwei Versionen eines Softwareprogramms mit und zwei Versionen eines Softwareprogramms ohne graphische Benutzungsschnittstelle herangezogen werden. Exemplarisch für diese Fragestellung soll MS-Word für DOS (mit den Versionen 4.0 und 5.0) sowie MS-Excel für Windows (mit den Versionen 4.0 und 5.0) einer näheren Betrachtung unterzogen werden.

Eine Substichprobe von 118 Benutzern wurde für diese ISONORM 9241/10-Analysen herangezogen. MS-Word 4.0 wurde dabei von 22 Benutzern, MS-Word 5.0 von 52 Benutzern, MS-Excel 4.0 von 28 Benutzern und MS-Excel 5.0 von 16 Benutzern evaluiert. Zwischen den entsprechenden Gruppen gab es keine signifikanten Unterschiede hinsichtlich der erhobenen Expertise- und Demographie-Daten.

Abbildung 2 verdeutlicht die Ergebnisse für den Versionsvergleich von MS-Word 4.0 zu 5.0 und Abbildung 3 den Versionsvergleich von MS-Excel 4.0 zu 5.0 für die einzelnen Skalen des ISONORM 9241/10.

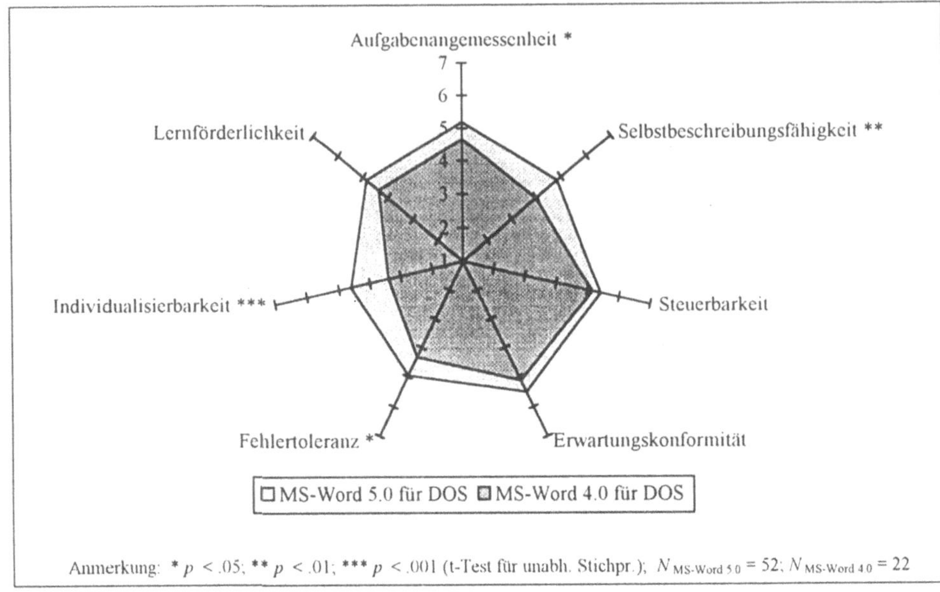

Anmerkung: * $p < .05$; ** $p < .01$; *** $p < .001$ (t-Test für unabh. Stichpr.); $N_{\text{MS-Word 5.0}} = 52$; $N_{\text{MS-Word 4.0}} = 22$

Abb. 2: ISONORM 9241/10 - Versionsvergleich von MS-Word 4.0 und MS-Word 5.0

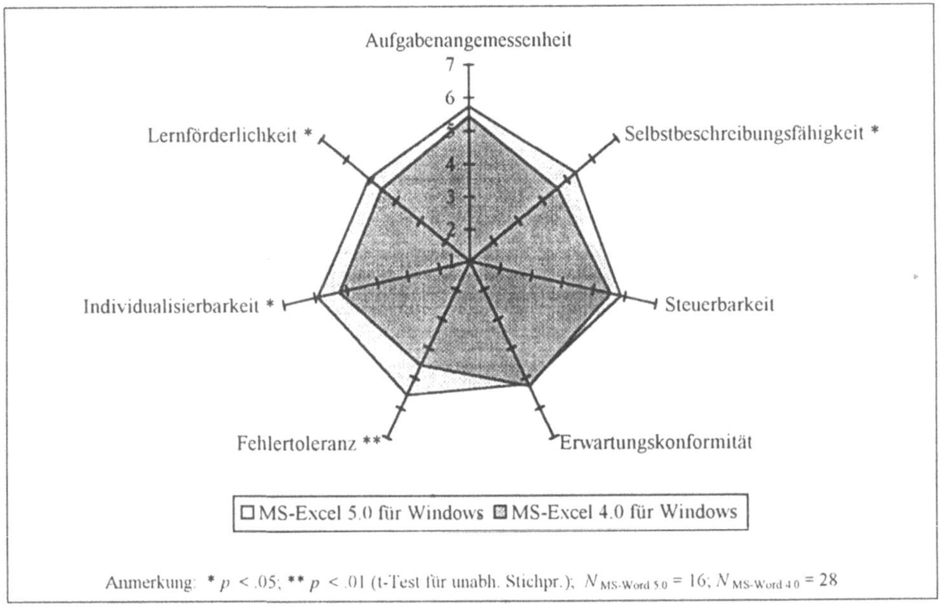

Anmerkung: * $p < .05$; ** $p < .01$ (t-Test für unabh. Stichpr.); $N_{\text{MS-Word 5.0}} = 16$; $N_{\text{MS-Word 4.0}} = 28$

Abb. 3: ISONORM 9241/10 - Versionsvergleich von MS-Excel 4.0 und MS-Excel 5.0

Es zeigt sich, daß sich MS-Word 5.0 gegenüber MS-Word 4.0 hinsichtlich der vier Grundsätze Aufgabenangemessenheit, Selbstbeschreibungsfähigkeit, Fehlertoleranz und Individualisierbarkeit signifikant verbessert hat. Dabei unterscheidet sich der ISONORM 9241/10-Gesamtwert signifikant auf dem 5%-Niveau mit folgenden Mittelwerten: $\bar{x}_{\text{MS-Word 5.0}} = 5.06$ und $\bar{x}_{\text{MS-Word 4.0}} = 4.45$.

Bei MS-Excel sieht das Bild anders aus: MS-Excel 5.0 verbesserte sich gegenüber MS-Excel 4.0 signifikant in puncto Selbstbeschreibungsfähigkeit, Fehlertoleranz, Individualisierbarkeit und Lernförderlichkeit. Der ISONORM 9241/10-Gesamtwert unterscheidet sich dabei signifikant auf dem 5%-Niveau wie folgt: $\bar{x}_{\text{MS-Excel 5.0}} = 5.52$ und $\bar{x}_{\text{MS-Excel 4.0}} = 5.03$.

Damit zeigt sich, daß Veränderungen von einer Version zur nächsten mittels ISONORM 9241/10 meßbar sind, und daß sich zusätzlich differenzierte Aussagen über diese Verbesserungen anstellen lassen.

4 Normierung des ISONORM 9241/10

Unter der Normierung eines Meßinstrumentes versteht man das Erstellen eines Bezugssystems, das die Einordnung eines individuellen Meßergebnisses ermöglicht. Ziel ist es, die Vergleichbarkeit einzelner Messungen zu gewährleisten [7].

Im vorliegenden Fall soll diese Normierung aus der Ermittlung der Verteilung der Meßwerte und bestimmter Verteilungsparameter bestehen. Dadurch besteht die Möglichkeit, den ISONORM 9241/10-Wert einer einzelnen Software in Relation zu den Beurteilungen anderer Programme zu setzen.

Grundlage dieser Normierung sollten nur die Softwareprogramme sein, die von mindestens sieben Benutzern beurteilt wurden. Dieses Kriterium erfüllten 41 Softwareprogramme. Jedes dieser 41 Softwareprogramme wurde dabei durchschnittlich von 24.6 ($N_{\text{min}} = 7$, $N_{\text{max}} = 106$) Benutzern beurteilt.

Tabelle 2 gibt die Skalenmittelwerte, das jeweilige Minimum und Maximum sowie die Standardabweichungen an. Alle sieben Skalen sind normalverteilt.

Grundsatz	$\bar{x}$	min.	max.	s
Aufgabenangemessenheit	4.87	2.60	5.76	.70
Selbstbeschreibungsfähigkeit	4.55	2.50	5.70	.65
Steuerbarkeit	5.01	3.20	6.13	.75
Erwartungskonformität	4.94	2.97	6.09	.72
Fehlertoleranz	4.29	2.59	5.75	.79
Individualisierbarkeit	4.28	2.41	5.92	.92
Lernförderlichkeit	4.52	2.56	5.83	.70
Anmerkung: $N = 41$ Programme				

Tab. 2: Skalenwerte des ISONORM 9241/10

Der niedrigste ISONORM 9241/10-Wert eines Softwaresystems findet sich bei dem Grundsatz Individualisierbarkeit ($\bar{x} = 2.41$); der höchste Wert bei einer anderen Software bei dem Grundsatz Steuerbarkeit ($\bar{x} = 6.13$). Der ISONORM 9241/10-Gesamtmittelwert beträgt $\bar{x} = 4.64$ ($\bar{x}_{\text{min}} = 2.97$; $\bar{x}_{\text{max}} = 5.75$; $s = .65$; $N = 41$ Programme).

Abbildung 3 zeigt Softwareprogramme geordnet nach ihrem ISONORM 9241/10-Gesamtwert. Eine nähere Betrachtung verdeutlicht, daß 34 der dort aufgeführten 41 Softwareprogramme einen ISONORM 9241/10-Gesamtwert von $\bar{x} \geq 4.0$ erreichen und damit entsprechend der zugrundeliegenden Fragebogenskala im positiven Bereich angesiedelt sind. Allerdings erreichen lediglich 14 Programme einen ISONORM 9241/10-Gesamtwert von $\bar{x} \geq 5.0$. Etwa zwei Drittel (65.8 %) der untersuchten Softwareprogramme erreichen diesen Wert, der auf der Fragebogenskala einer Konnotation von „+" (Plus Eins) entspricht, nicht.

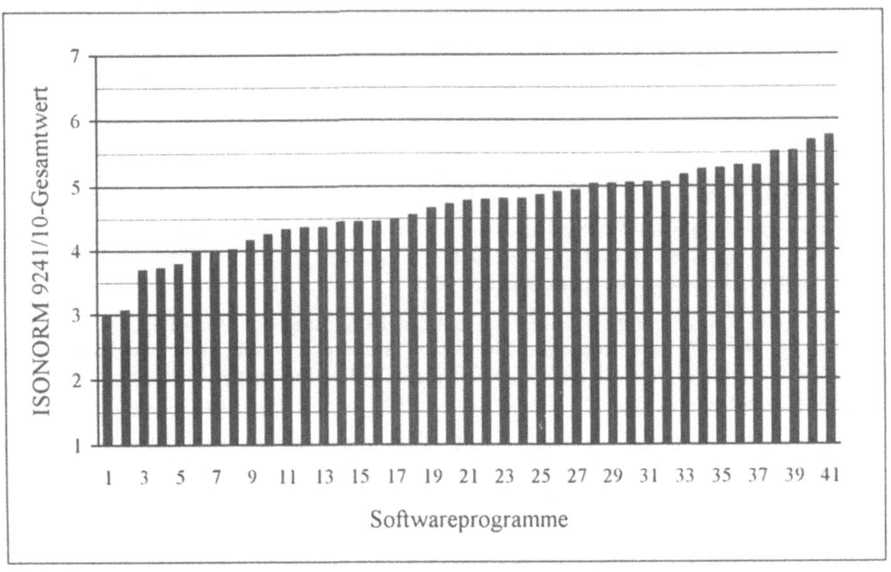

Abb. 3: Softwareprogramme geordnet nach ISONORM 9241/10-Gesamtwert

5 Zusammenfassung

Ziel dieses Beitrags war die Dokumentation der testtheoretische Güte des Benutzungsfragebogens ISONORM 9241/10. Zu diesem Zweck wurden verschiedene Untersuchungen zur Reliabilität, Validität und Normierung durchgeführt.

Zur Reliabilitätsbestimmung wurden die *Wiederholungsmethode* und die *Konsistenzanalyse* herangezogen. Aufgrund der signifikanten Stabilitäts-Koeffizienten konnte mittels der *Wiederholungsmethode* gezeigt werden, daß sich der ISONORM 9241/10 durch eine hohe Merkmalskonstanz auszeichnet. Anhand der *Konsistenzanalyse* zeigte sich die Leistungsfähigkeit des ISONORM 9241/10 als Meßinstrument, da die einzelnen Skalen des ISONORM 9241/10 gute innere Konsistenzen aufwiesen.

Im Rahmen der Validitätsbestimmung wurden Untersuchungen zur *kriterienbezogenen Validität* und zur *Konstruktvalidität* angestellt. Zur Überprüfung der *inneren kriterienbezogenen Validität* wurden die ISONORM 9241/10-Bewertungen mit den Bewertungen der beiden Benutzungsfragebögen QUIS [21] und BBD [22] korreliert. Es zeigte sich, daß der ISONORM 9241/10 zu vergleichbaren Ergebnissen gelangt, und gleichzeitig den Vorteil in sich birgt, eine ausdrückliche Operationalisierung von ISO 9241 Teil 10 darzustellen. Zur Überprüfung der

261

äußeren kriterienbezogenen Validität wurde mittels EVADIS II [9] überprüft, ob die Befragung durch Benutzer mit der Evaluation durch Experten korrespondiert. Es konnte gezeigt werden, daß die ISONORM 9241/10-Benutzungsanalysen einen signifikanten Bezug zu einer systematischen, leitfadenorientierten Expertenevaluation aufweisen. Zur Überprüfung der *Konstruktvalidität* wurden zwei Versionen eines Softwareprogramms mit und zwei Versionen eines Softwareprogramms ohne graphische Benutzungsschnittstelle herangezogen. Es zeigte sich, daß Veränderungen von einer Version zur nächsten mittels ISONORM 9241/10 meßbar sind, und daß sich zusätzlich differenzierte Aussagen über diese Verbesserungen anstellen lassen.

Die *Normierung* erlaubt für den ISONORM 9241/10 ein Bezugssystem herzustellen, das die Einordnung einer einzelnen Software in Relation zu den Beurteilungen anderer Programme ermöglicht.

Neben diesen Ergebnissen sollten auch Überlegungen zur *Ökonomie* und *Nützlichkeit* eines Meßverfahrens angestellt werden. Ein Meßverfahrens ist dann *ökonomisch*, wenn es „eine kurze Durchführungszeit beansprucht, wenig Material verbraucht, einfach zu handhaben, als Gruppentest durchführbar, schnell und bequem auszuführen ist" [7, S. 12]. Diese Kriterien werden durch den ISONORM 9241/10 ausnahmslos erfüllt. *Nützlich* ist ein Meßverfahren dann, wenn es ein Merkmal mißt, für dessen Untersuchung ein praktisches Bedürfnis besteht. Als direkte Operationalisierung der ISO 9241/10 erfüllt der ISONORM 9241/10 auch dieses Kriterium.

Die schriftliche Befragung von Benutzern kann und darf nicht die einzige Methode der Wahl sein, wenn eine Software umfangreich evaluiert werden soll. Aufgrund der Analyse einer Vielzahl von Evaluationsverfahren kann deswegen auch der Schluß gezogen werden, daß es die „beste" Evaluationsmethode nicht gibt, sondern daß die jeweilige Eignung vom Gegenstand und Einsatzzweck der Evaluation bestimmt wird [20]. Jede Evaluationsmethode hat ihre spezifischen Vorteile und Schwerpunkte. Um ganzheitliche Aussagen bezüglich der software-ergonomischen Qualität von Benutzungsschnittstellen machen zu können, ist deshalb der Einsatz einer Kombination unterschiedlicher Evaluationsmethoden empfehlenswert [10]. Hier reicht das Spektrum von der heuristischen Evaluation [8], über die software-unterstützte, leitfadenorientierte Evaluation [11], bis hin zur handlungsorientierten Fehleranalyse [2,3,5,13,18], um nur einige Beispiele zu nennen.

Auf eine systematische und normbasierte Befragung durch Benutzer sollte jedoch auf keinen Fall verzichtet werden, da es z.B. vorkommen kann, daß Experten eine Software für effektiv und effizient halten, daß aber Benutzer mit der Software nicht zufrieden sind [1]. Mit dem ISONORM 9241/10 wurde ein reliables und valides Instrument entwickelt, das für diesen Zweck zum Einsatz kommen kann.

Literaturverzeichnis

[1]   W. Dzida: Qualitätssicherung durch software-ergonomische Normen. In: E. Eberleh, H. Oberquelle & R. Oppermann (Hg.): Einführung in die Software-Ergonomie. DeGruyter, Berlin, 1994, 374-406.

[2]   M. Frese & D. Zapf (Hg.): Fehler bei der Arbeit mit dem Computer - Ergebnisse von Beobachtungen und Befragungen im Bürobereich. Huber, Bern, 1991.

[3]   M. Hassenzahl, J. Prümper & U. Sailer: Die Priorisierung von Problemhinweisen in der software-ergonomischen Qualitätssicherung. In diesem Band.

[4]   ISO 9241/10: Ergonomische Anforderungen für Bürotätigkeiten mit Bildschirmgeräten - Teil 10: Grundsätze der Dialoggestaltung. CEN, Brüssel, 1995.

[5]   A. Kensik, J. Prümper & M. Frese: Ergonomische Gestaltung von Software auf Grundlage handlungsorientierter Fehleranalysen. In: H.D. Böcker (Hg.): Software-Ergonomie '95 - Anwendungsbereiche lernen voneinander. Teubner, Stuttgart, 1995, 217-232.

[6]   A. Kinder: Testen und Bewerten von Software durch Benutzer. Justus-Liebig-Universität, Gießen, 1991.

[7]   G.A. Lienert & U. Ratz: Testaufbau und Testanalyse. Psychologie Verlags Union, München, 1994.

[8]   Nielsen, J. Usability Engineering. Cambridge: AP Professional, 1993.

[9]   R. Oppermann, B. Murchner, H. Reiterer & M. Koch: Software-ergonomische Evaluation. Der Leitfaden EVADIS II. De Gruyter, Berlin, 1992.

[10] R. Oppermann & H. Reiterer. Software-ergonomische Evaluation. In: E. Eberleh, H. Oberquelle & R. Oppermann (Hg.): Einführung in die Software-Ergonomie. DeGruyter, Berlin, 1994, 335-371.

[11] R. Oppermann, C. Wick, T. Geis, M. Koch, P. Lutz, J. Prümper, H. Reiterer & W.W. Strapetz: Der ISO 9241 Evaluator. In: Ergonomie & Informatik 27 (1996), 13-17.

[12] J. Prümper: Benutzerorientierte, iterative Software-Entwicklung in der Praxis. In: W. Coy, P. Gorny, I. Kopp & C. Skarpelis (Hg.): Menschengerechte Software als Wettbewerbsfaktor. Teubner, Stuttgart, 1993, 630-647.

[13] J. Prümper: Fehlerbeurteilungen in der Mensch-Computer Interaktion. Reliabilitätsanalysen und Training einer handlungstheoretischen Fehlertaxonomie. Waxmann, Münster, 1994.

[14] J. Prümper: GUIs sind noch weit entfernt von einer optimalen Bewertung. ISO 9241 Teil 10 als Leitlinie für SW-Entwicklung und -Bewertung. In: Computerwoche 25 (1993), 29 & 46.

[15] J. Prümper: Software-Evaluation based upon ISO 9241 Part 10. In: T. Grechenig & M. Tscheligi (Hg.): Human Computer Interaction. Springer, Berlin, 1993, 255-265.

[16] J. Prümper: Viele Programme erfüllen die ISO-Norm 9241/10 ungenügend. Anwenderbeurteilung zeugt von Nachholbedarf. In: Computerwoche 21 (1994), 47-48.

[17] J. Prümper & M. Anft: Die Evaluation von Software auf Grundlage des Entwurfs zur internationalen Ergonomie-Norm ISO 9241 Teil 10 als Beitrag zur partizipativen Systemgestaltung - ein Fallbeispiel. In: K.H. Rödiger (Hg.): Software-Ergonomie '93 - Von der Benutzungsoberfläche zur Arbeitsgestaltung. Teubner, Stuttgart, 1993, 145-156.

[18] J. Prümper, T. Heinbokel & H.J. Küting: Virtuelle Prototypen als Werkzeuge zur benutzerzentrierten Produktentwicklung. Anwendung einer handlungstheoretischen Fehlertaxonomie auf reale und simulierte Oberflächen von Waschmaschinen. In: Zeitschrift für Arbeitswissenschaft 4 (1993), 160-167.

[19] M. Rauterberg: Läßt sich die Gebrauchstauglichkeit interaktiver Software messen? Und wenn ja, wie? In: Ergonomie & Informatik 16 (1992), 3-18.

[20] H. Reiterer: Ergonomische Kriterien für die menschengerechte Gestaltung von Bürosystemen, Anwendung und Bewertung. Universität Wien, Dissertation, 1990.

[21] B. Shneiderman: Designing the user interface. Addison-Wesley, Reading MA, 1987.

[22] P. Spinas: Arbeitspsychologische Aspekte der Benutzerfreundlichkeit von Bildschirmsystemen. ADAG, Zürich, 1987.

Adresse des Autors

Prof. Dr. Jochen Prümper
FHTW-Berlin
Fachgebiet Wirtschaftspsychologie
Treskowallee 8
10313 Berlin
Email: pruemper@aol.com

Beurteilung multimedialer Anwendungen und Systeme

Martin Riegel

TÜV Informationstechnik GmbH

Zusammenfassung

Abb. 1: Multimedia-Symbol

Multimediale Anwendungen und Systeme, wie Firmenpräsentationen auf CD-ROM (Point Of Information - POI), Katalog- und Verkaufssysteme in Kiosk-Terminals (Point Of Sale - POS) oder Weiterbildungs- und Lernprogramme, die beispielsweise Online über das Internet bereitgestellt werden (Computer Based Training - CBT), entbehren zur Zeit anerkannte ergonomische Qualitätsstandards. Der vorliegende Beitrag befaßt sich daher mit Bewertungskriterien, -maßstäben und -verfahren zur ganzheitlichen Beurteilung der Anwenderschnittstelle.

Ziel ist es nicht, eine wissenschaftliche Diskussion über Detailfragen anzufachen, sondern vielmehr einen praxisorientierten Ansatz zu präsentieren. Das beim RWTÜV bzw. der TÜV Informationstechnik (TÜViT) entwickelte und eingesetzte Ergonomie-Prüfverfahren, hat sich als offen und flexibel bewährt. Bewertungskriterien und -maßstäbe müssen jedoch immer dynamisch auf die zu beurteilende Anwendung, das Gesamtsystem und die Zielgruppe angepaßt werden. Eine Aufgabe, die in der Umsetzung teilweise große Probleme aufwirft. Fast täglich stehen neue technische und gestalterische Möglichkeiten gerade im Multimedia-Umfeld zur Verfügung, für die zunächst keine validierten Erkenntnisse vorliegen. Blickt man z.B. auf einige WWW-Applikationen (World Wide Web), so wird schnell klar, daß die Frage nach der Definition und Umsetzung ergonomischer Anforderungen leider oftmals unbeantwortet bleibt.

1 Bedeutung ergonomischer Beurteilungen

Verbindliche Qualitätsstandards zur Gestaltung von ergonomischen Multimedia-Applikationen (MM-A) fehlen weitestgehend. Vorhandene Ansätze zur Softwareergonomie, wie die Anforderungen aus der Norm ISO 9241 [1] und hier insbesondere der Teile 10 bis 17, könhen der rasanten technischen Entwicklung gerade in Bezug auf immer neue Möglichkeiten in Film und Ton kaum folgen.

Die Festschreibung jeweils geeigneter ergonomischer Bewertungskriterien sowie das Aufstellen von Bewertungsmaßstäben ist anderseits Grundlage für die Entwicklung und natürlich die Beurteilung von Applikationen bei denen die Bedienungsfreundlichkeit eine übergeordnete Bedeutung hat [2]. Allzu offensichtlich ist die Problematik in der sich nicht nur die Ersteller befinden, denn welche ergonomischen Vorgaben sollen beachtet werden wenn keine definierten und meßbaren Anforderungen bestehen?

Einige Aspekte aus der Sicht des Erstellungsprozesses, der Anwendersituation, der Normungsarbeit und der Evaluationspraxis präzisieren dies:

1.1 Erstellungsprozeß

Bereits in der Konzeptionsphase einer geplanten MM-A wird die Bedeutung von Kriterien zur ergonomischen Beurteilung deutlich. Alle Anforderungen an die zu entwickelnde Applikation werden im Pflichtenheft dokumentiert. Gegen diese Anforderungen wird während und am Ende des Projektes geprüft, wobei Bewertungkriteren und -maßstäbe zur Anwendung kommen. Die Praxis zeigt, daß Forderungen an die Bedienerfreundlichkeit überwiegend nur am Rande Erwähnung finden und überdies oft allzu allgemein formuliert sind: „Die Eingangs-animation ist so zu gestalten, daß sie die Zielgruppe anspricht und zum Fortfahren anregt." Es bedarf bei diesem Exempel einiger wesentlicher weiterer Vorgaben, z.B. zu den Eigenschaften, Neigungen, Fähigkeiten und Fertigkeiten der Zielgruppe sowie der Definition nebst Bewertungsmaßstab von *ansprechend* und *anregend*, um Interpretationsspielräume von vornherein zu minimieren.

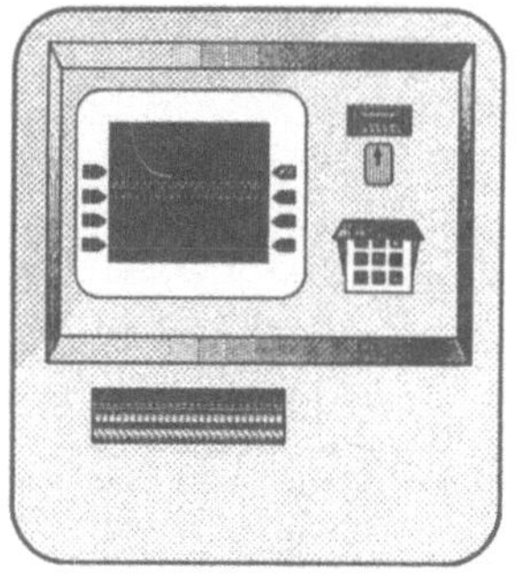

Abb. 2: Kiosk-Terminal

In der Umsetzungsphase erstellen dann Konzeptionisten, Screen-Designer, Programmierer etc. meist unter extremen Zeit- und Kostenrestriktionen die MM-A, wie z.B. für einen Kiosk-Terminal (Abb. 2). Für Anforderungen, die nicht eindeutig festgelegt wurden, bleibt dementsprechend wenig oder gar keine Zeit. „Nice to have - but not necessary" hört man selbst vom Projektmanagement immer häufiger. Daneben findet man gerade bei kreativen Entwicklungsprozessen eine stark ablehnende Haltung gegenüber Vorgaben jedweder Art. Schon das differenzierte Aufstellen von Anforderungen an das Erscheinungsbild bzw. Design, wird als massiver Eingriff in die Gestaltungsfreiheit angesehen. Somit definiert jeder Entwickler in jedem Projekt aufs neue eigene ergonomische Anforderun-gen. Gegen diese Vorgehensweise spricht der vermeidbare Zeitaufwand ebenso wie die subjektive Ausrichtung derartiger Anforderungen. Eine Überprüfung auf Erfüllung der Forderungen könnte folglich nur vom Entwickler, im Rahmen einer damit nicht sehr aussagekräftigen Selbstprüfung, erfolgen.

Ferner werden die meisten MM-A aufgrund fehlender Ressourcen und Fachkompetenz nicht im Unternehmen selbst hergestellt. Ganz im Outsourcing-Trend liegt die Beauftragung von externen Dienstleistern, die sich in den letzten Jahren auf die Gestaltung von jeweils bestimmten MM-A spezialisiert haben. Der Auftraggeber muß nicht nur die ergonomischen Vorgaben im Pflichtenheft genau formulieren, sondern vielmehr auch die fertiggestellte Applikation in Bezug auf die Erfüllung der Vorgaben kontrollieren. Diese Aufgabe fällt oft fachfremden Unternehmensbereichen zu, wie dem Einkauf oder der Qualitätssicherung. Für die betroffenen Mitarbeiter sind daher Bewertungskriterien und ein leicht handhabbares Beurteilungsverfahren unerläßlich. Wobei das allein sicherlich auch nicht ausreichend ist. Denn ohne weitergehendes Know-how können die Bewertungskriterien kaum interpretiert werden und eine Beurteilung avanciert zum Vabanquespiel. Schulungen zur Vermittlung von Grund-kenntnissen, zur Einweisung in den Gebrauch der Beurteilungsverfahren sowie zur *Eichung* der betroffenen Mitarbeiter auf einen einheitlichen Qualitätslevel bleiben unentbehrlich.

1.2 Anwendersituation

Multimedia - die integrierte Nutzung unterschiedlicher Medien mit dem Ziel der Informations-vermittlung bzw. Kommunikation. Texte, Grafiken, fotorealistische Bilder, Videofilme, Animationen oder auch auditive Sequenzen werden zusammengefügt und über immer leistungsfähigere Trägermedien On- und Offline verbreitet. Fast unbegrenzte Gestaltungs- und Einsatzmöglichkeiten zeigen sich in kommerziellen, wie auch in privaten Bereichen. Entscheidend bei MM-A ist jedoch die Interaktivität. Der Nutzer ist nicht länger ein machtloser Konsument, der alles akzeptieren muß was ihm dargeboten wird. Er kann die

Applikation selber gezielt steuern und damit zu jeder Zeit in den Informationsablauf eingreifen. Endlich stehen die Nutzer im Mittelpunkt und können eigene Entscheidungen treffen. Ein neuer, weitgehend unbekannter Zustand, der die Ersteller derartiger Anwendungen zum Umdenken zwingt. Mehrwert, Attraktivität und intuitive Bedienung bei optimaler Anpassung an die sensorischen, motorischen und kognitiven Leistungen der Nutzer, sind daher nun die vehementen Forderungen an MM-A. Forderungen, die infolge einer geänderten Erwartungshaltung auf neuen weitgefächerten Bewertungsmaßstäben der Nutzer beruhen. Damit werden die konventionellen Bewertungsmaßstäbe aus der Softwareergonomie nicht ungültig. Jedoch ist durch die Verschiebung in der Bedeutung einzelner Faktoren und durch die Ergänzung um neue technische Möglichkeiten, eine spezielle Ausrichtung auf multimediale Applikationen gefordert, welche auch die Möglichkeiten und Besonderheiten der Trägermedien einschließt.

Anwendertests, bereits in der Entwicklungsphase, müssen fernerhin zu festen Bestandteilen von MM-Projekten werden, um der fortwährenden Neuerung des Erfahrungshintergrundes sowie der Erwartungshaltung der Zielgruppe jederzeit Rechnung tragen zu können.

1.3   Normungsarbeit

Trotz beharrlichen Bemühungen nationaler und internationaler Normungsinstitutionen (hier vorwiegend: DIN [u.a. 3 u. 4], IEEE [5], ISO [z.B. 1], VDI [z.B. 6]) nimmt die Umsetzung von Standards bekanntermaßen viel Zeit in Anspruch. Dies führt dazu, daß zwischen verabschiedeten Normen bzw. dem Stand der Technik auf der einen Seite und aktuellen Produkten auf der anderen Seite eine gewisse Lücke klafft. Gerade in dem dynamischen und sich schnell weiterentwickelnden Multimedia-Markt scheint diese Lücke zur Zeit besonders groß zu sein. Gründe hierfür sind neben der Entwicklungsgeschwindigkeit sicherlich in der Interdisziplinarität von MM-A und in dem Fehlen validierter Erkenntnisse zu suchen.

Um ein einheitlichen, gleichbleibend hohen Qualitätsstandard in Bezug auf die Ergonomie von MM-A gewährleisten zu können, wird ein gemeinsames Entwickeln von handhabaren Bewertungskriterien und -maßstäben aus der Praxis heraus notwendig.

1.4   Evaluationspraxis

Verschiedene Ansätze aus der Wissenschaft [7, 8], spezielle Forschungsprogramme, wie ESPRIT, Projekt MUSiC (Metrics for Usability Standards in Computing) oder vom BMBF, Projekt SANUS (Sicherheit und Gesundheitsschutz bei der Arbeit an Bildschirmen auf der Basis internationaler Normen Und Standards) [9], aber auch aus der Industrie (beispielsweise individuelle Anforderungskataloge von Herstellern oder Prüfmethoden von TÜV's [z.B. 10]) zeigen die Spannbreite gegenwärtiger Beurteilungsverfahren. In der Softwareergonomie haben sich einige Verfahren für bestimmte Voraussetzungen bzw. Anlässe etabliert. Andere sind noch wenig bekannt und fanden bisher kaum Anwendung. Ausschlaggebend für die Durchsetzung in der Praxis war und ist durchweg die Möglichkeit des flexiblen, unkomplizierten und schnellen Einsatzes.

Übergeordnet kann als gängige Vorgehensmethode zur Beurteilung der ergonomischen Qualität, die gesonderte Definition von Anforderungen und die Nutzung eines speziellen Verfahrens aufgrund jeweils eines bestimmten Anlasses gesehen werden (siehe Abb. 3). Ideal wäre hingegen die allgemeingültige Definition von Anforderungen respektive von Bewertungskriterien, nebst universeller Nutzung eines einzigen Beurteilungsverfahrens.

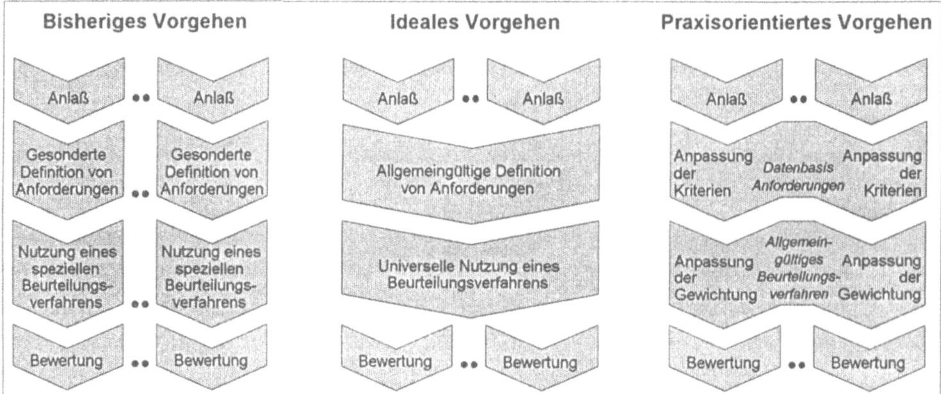

Abb. 3: Vorgehensmethoden zur Beurteilung der ergonomischen Qualität

Mit dem bei der TÜViT eingesetzten Prüf- und Beurteilungsverfahren „Ergonomie-Testat" ist ein erster Schritt in Richtung eines *idealen* Vorgehens unternommen worden. Das praxisorientierte, systematische Vorgehen verknüpft eine standardisierte Datenbasis von Bewertungskriterien und ein strukturell allgemeingültiges Beurteilungsverfahren, mit auf den Anlaß anpassbaren Freiheitsgraden. Ein Freiheitsgrad besteht in der autonomen Zusammenstellung der Bewertungskriterien, die jeweils in Bezug auf das Produkt, die Benutzerinformationen, die Zielgruppe und die Randbedingungen aus der Datenbasis ausgewählt und ggf. angepaßt werden. Daneben ist immer eine individuelle Gewichtung der Bedeutsamkeiten der einzelnen Bewertungskriterien notwendig. Dazu später mehr.

Der einfache Aufbau und die leichte Anpassbarkeit auf unterschiedliche Voraussetzungen haben sich in der Vergangenheit bewährt. Nachfolgend soll daher dieses Prüfverfahren näher erläutert werden.

2 Das TÜViT- Prüfverfahren

Mit dem TÜViT-Prüfverfahren werden u.a. multimediale Applikationen, wie Firmenpräsentationen oder Katalog- und Verkaufssysteme ebenso wie Weiterbildungs- und Lernprogramme beurteilt. Letztendlich mit dem Ziel, die Akzeptanz der Zielgruppe in Bezug auf die Gestaltung der MM-A herauszufinden. Nicht die Applikation, die noch schriller, noch verrückter oder noch umfangreicher und komplexer ist, begeistert die Anwender bei ausgedehnter Nutzung. Aus der Erfahrung heraus, sind es vielmehr kreativ und interessant gestaltete Applikationen die weitestgehend selbsterklärungsfähig sowie leicht und sicher bedienbar sind.

2.1 Aufbau der Prüfung

1.		Grundlagen
2.		Gewichtung
3.		Anwendertest
4.		Bewertung
5.		Ergebnisse

Abb. 4: Fünfstufiges Prüfverfahren

Die Prüfung ist, unter Einbeziehung von Anwendern, als Expertenrating konzipiert. Anwendererfahrungen werden somit zwar berücksichtigt, fließen aber erst nach Bewertung durch die Ergonomen mehr oder weniger stark in das Ergebnis ein.

Das erweiterte Prüfverfahren basiert auf einer fünfstufigen Vorgehensweise. Abbildung 4 zeigt hierzu die Stufen in ihrer zeitlichen Rangfolge.

2.1.1 Grundlagen

Zu Beginn werden die Prüfungsgrundlagen der betrachteten Applikation bzw. des Gesamtproduktes, der Benutzerinformationen und der Zielgruppe analysiert und zusammengestellt.

Für die MM-A sind u.a.

- die vorgesehenen Einsatzbedingungen,
- die bestimmungsgemäße Verwendung,
- die Handlungsabläufe zur sachgerechten Nutzung sowie
- die möglichen Gefahrenpotentiale zu erfassen.

Analog ist die Zielgruppe zu definieren. Hierbei ist dem Leitgedanken, immer vom „schwächsten Glied in der Kette" auszugehen, zu folgen. Also beispielsweise den Anwendern von Konsumerprodukten ein Recht auf „Unwissenheit, Dummheit und unter Umständen sogar Analphabetentum" zuzuschreiben. Da sich natürlich das Wissen und Können der Zielgruppe in Bezug auf eine Anwendung im Wandel der Zeit ändern kann, ist eine entsprechende Definitionen zu Beginn jeder Prüfung neu erforderlich.

Aus der Datenbasis werden dann alle relevanten Prüfkriterien, welche die ergonomischen Anforderungen implizieren, ausgewählt und zusammengestellt. Soweit notwendig, sind im Einzelfall Prüfkriterien noch auf Besonderheiten der MM-A hin anzupassen. Diese Aufgabe obliegt den Ergonomen, die dabei in der Regel nationale und internationale Normen, Standards, Richtlinien, Studien etc. heranziehen und interpretieren müssen.

In einem sogenannten Ergonomie-Prüfkatalog werden die Prüfkriterien gebündelt und hierarchisch angeordnet, wie Abbildung 5 am Beispiel eines Kiosk-Terminals (vgl. Abb. 2) verdeutlicht.

Die höchste Ebene teilt das Produkt in Komponenten. Eine Komponente wäre zum Beispiel das Software-Interface. Jede Komponente wird entsprechend ihrer Elemente oder Funktionen sowie den dazu notwendigen Handlungsabläufen in Merkmale und weiterhin in Prüfkriterien gegliedert. Ein Merkmal der oben genannten Komponente ist beispielsweise die „Grafische Gestaltung", ein weiteres wären die „Tonsequenzen". Sind alle relevanten Merkmale zusammengestellt, werden den Merkmalen entsprechende Prüfkriterien mit dazugehörigen Prüfmethoden zugeordnet. In dem Merkmal 1.2 „Grafische Gestaltung" werden unter anderem die Prüfkriterien Bildaufteilung, Farbgestaltung und Typographie geführt.

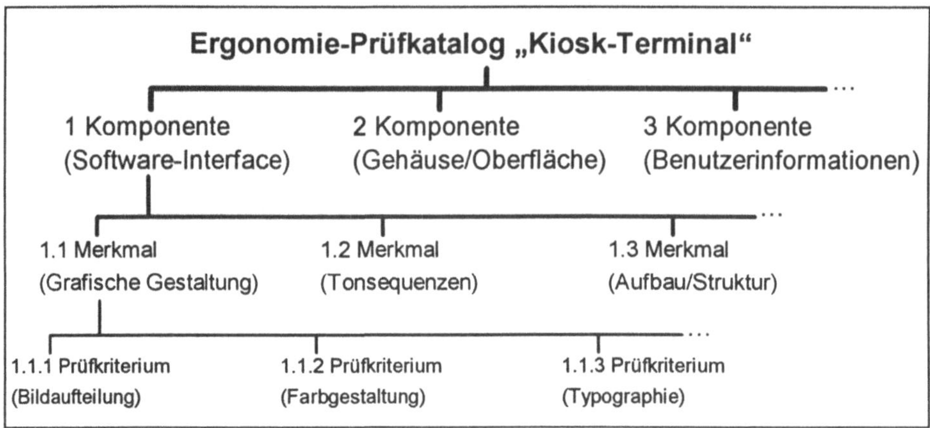

Abb. 5: Hierarchischer Aufbau des Ergonomie-Prüfkataloges

Jedes Prüfkriterium wird in dem produktorientierten Ergonomie-Prüfkatalog explizit ausformuliert, wie anhand der Funktionstasten eines Informations-Terminals, Komponente Gehäuse/ Bedienoberfläche, beispielhaft dargestellt. Neben der *Realisierung* und der *Prüfmethode* wird unter dem Punkt *Beurteilung* eine bindende Meßlatte definiert. Diese zeigt für 0, 1, 2 und 3 Bewertungspunkte, die jeweils zu erfüllenden Anforderungen an (Abb. 6). So ergeben sich z.B. für diese Komponente 60 bis 100 Prüfkriterien.

Prüfkriterium: 2.1.5 Selbstbeschreibungsfähigkeit der Funktionstasten

Realisierung:

Auf die Nutzergruppe (hier: computerbegeisterte Kunden in der Altersgruppe von 16 bis 35 Jahren) abgestimmtes, schnelles und bedienungssicheres Erlernen der Funktionen durch das Produkt selbst. Keine zusätzlichen (schriftlichen o.ä.) Informationen nötig.

Prüfmethode:

Test der Steuerbarkeit des Systems über die Funktionstasten bei der Erstnutzung. Überprüfung der Erwartungskonformität und der intuitiven Erlernbarkeit anhand vorgegebener Aufgaben.

Beurteilung:

3 Alle Funktionstasten erklären sich bei der Nutzung vollständig selbst.

2 Die Funktionstasten sind weitestgehend selbstbeschreibungsfähig. Durch wenige einmalige Informationen sind sie schnell und bedienungssicher zu nutzen.

1 Nur in Kenntnis zusätzlicher Informationen, die durch das Gesamtsystem erwartungskonform bereitgestellt werden, sowie nach mehrmaliger Nutzung sind alle Funktionstasten in ihren Wirkweisen verständlich und nachvollziehbar.

0 Die Nutzung der Funktionstasten erklärt sich nicht von selbst. In Kenntnis aller zusätzlichen Informationen und nach mehrmaliger Nutzung sind einige Funktionstasten bzw. ihre Wirkweisen immer noch nicht nachvollziehbar.

Abb. 6: Auszug aus einem Ergonomie-Prüfkatalog

2.1.2 Gewichtung

Alle Komponenten, Merkmale und Prüfkriterien werden nun einer Gewichtung unterzogen. Das heißt, der Stellenwert (die Bedeutsamkeit) untereinander wird aus der Perspektive der Anwender prozentual gewichtet. Dabei ergibt die Summe der Gewichtungen aller

- Komponenten des Produktes 100%,
- Merkmale einer Komponente 100%,
- Prüfpositionen eines Merkmals 100%.

Die Bedeutung der Gewichtungen zeigt sich einleuchtend bei der Betrachtung von zwei Schaltern mit unterschiedlichen Funktionen. Obwohl sich zum Beispiel der Schalter zur Regulierung der Farbkontraste und der Ein-/Aus-Schalter an ein und demselben Bildschirm befinden, kommt ihnen eine unterschiedliche Bedeutung zu. Der häufig oder täglich genutzte Ein-/Aus-Schalter hat eine Schlüsselfunktion für dieses Produkt. Hingegen wird der Schalter zur Einstellung des Farbkontrasts relativ selten gebraucht und würde daher mit nur 25% gegenüber dem Ein-/Aus-Schalter mit 75% gewichtet.

2.1.3 Anwendertest

Usabilty Tests mit Kleingruppen begleiten die Beurteilungen, um z.B. grundlegende Designfehler aufzudecken. Hierzu werden unter möglichst realen Rahmenbedingungen typische Aufgaben bearbeitet. Nach Festlegung des Testdesigns in dem Untersuchungsziele, Arbeitsaufgaben, -abläufe etc. definiert werden, erfolgt die Auswahl und Einweisung der Probanden. Ergonomen verfolgen die Abarbeitung der gestellten Aufgaben, die durch *lautes Denken* und eine abschließende Befragung zumeist sehr aufschlußreiches Analysematerial liefern.

Wie bereits angedeutet, fließen die Ergebnisse dieser einfachen Anwendertests nicht direkt in die Bewertung mit ein. Vorangestellt ist die Filterung der Aussagen, mit der u.a. eine wenig repräsentative Zusammensetzung der Testgruppe und die geringe Anzahl der Probanden kompensiert werden kann.

Neben der Ergebnissen aus dem Anwendertest sind weitere Daten zu erheben. Die Aufnahme der physikalischen Meßgrößen erstreckt sich bei technischen Produkten von Abmessungen bis hin zu Formgebungen. Nicht meßbare Größen (Wie hautsympatisch ist das Material einer Bedienoberfläche von einem TouchScreen-Monitor? etc.) werden in ihren Qualitäten beschrieben.

2.1.4 Bewertung

Die ergonomische Bewertung der aufgenommenen Ergebnisse erfolgt jeweils in einem interdisziplinären Expertenteam. Im Rating werden innerhalb der Bewertungsskala, welche die Punkte von 0,0 bis 3,0 (Forderungen nicht erfüllt bis erfüllt) aufweist, die Bewertungsergebnisse der einzelnen Prüfkriterien ermittelt. Grundlage der Bewertung ist die zu Beginn der Prüfung schriftlich fixierte Meßlatte für die Beurteilungen der Prüfkriterien.

Die Punktebewertungen der einzelnen Prüfkriterien werden mit den zugehörigen Gewichtungsprozenten multipliziert und anschließend aufsummiert. Normiert ergeben sich so die realen Bewertungsergebnisse jedes Merkmals. Ebenso werden danach die Bewertungsergebnisse der Komponenten und schließlich des gesamten Produktes ermittelt (Abb. 7).

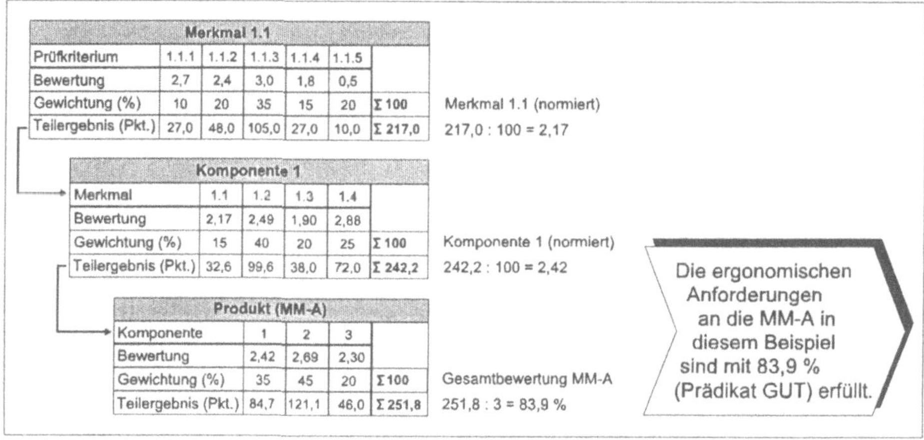

Prüfkriterium	1.1.1	1.1.2	1.1.3	1.1.4	1.1.5	
Bewertung	2,7	2,4	3,0	1,8	0,5	
Gewichtung (%)	10	20	35	15	20	Σ 100
Teilergebnis (Pkt.)	27,0	48,0	105,0	27,0	10,0	Σ 217,0

Merkmal 1.1 (normiert) 217,0 : 100 = 2,17

Komponente 1					
Merkmal	1.1	1.2	1.3	1.4	
Bewertung	2,17	2,49	1,90	2,88	
Gewichtung (%)	15	40	20	25	Σ 100
Teilergebnis (Pkt.)	32,6	99,6	38,0	72,0	Σ 242,2

Komponente 1 (normiert) 242,2 : 100 = 2,42

Produkt (MM-A)				
Komponente	1	2	3	
Bewertung	2,42	2,69	2,30	
Gewichtung (%)	35	45	20	Σ 100
Teilergebnis (Pkt.)	84,7	121,1	46,0	Σ 251,8

Gesamtbewertung MM-A 251,8 : 3 = 83,9 %

Abb. 7: Beispiel zum Bewertungsverfahren

Die endgültige Bewertung der Erfüllung ergonomischer Anforderungen durch ein Produkt, erfolgt durch ein Vergleich des optimalen Bewertungsergebnisses (100% der ergonomischen Anforderungen erfüllt; jedes Prüfkriterium 3,0 Punkte) mit dem realen Bewertungsergebnis.

2.1.5 Ergebnisse

Nach Abschluß der Bewertung auf der Ebene der Prüfkriterien werden die Teilergebnisse jeder Komponente und das Gesamtergebnis berechnet. In Abhängigkeit des Gesamtergebnisses wird das Siegel 'Ergonomie TÜV geprüft' in drei Stufen verliehen:

Stufe	Bewertung	Bezeichnung und Prüfprädikat	TÜV-Siegel
①	70% - 80%	„Ergonomie TÜV geprüft" „Ergonomics TÜV tested"	
②	80% - 90%	„Ergonomie TÜV geprüft gut" „Ergonomics TÜV tested good"	
③	über 90%	„Ergonomie TÜV geprüft sehr gut" „Ergonomics TÜV tested very good"	

Abb. 8: Erfüllungsgrad der Anforderungen und Bewertungsstufen

Das Gesamtergebnis zeigt prozentual, auf 100% normiert, den Erfüllungsgrad der ergonomischen Anforderungen für die überprüften MM-A. Zum Bestehen müssen im Gesamtergebnis mindestens 70% aller Anforderungen erfüllt sein.

Es bleibt anzumerken, daß, wenn eine Komponente weniger als 50% der ergonomischen Anforderungen erfüllt (im Mittel also 1,5 Punkte), dieses ein Nichtbestehen der gesamten MM-A zur Folge hat. Gleiches gilt, wenn außerordentlich wichtige Anforderungen *KO-Kriterien* nicht erfüllt werden. Notwendige Warnhinweise die fehlen, oder falsche zum Mißbrauch anregende Informationen in der Gebrauchsanleitung bzw. der Online-Hilfe, könnten hier exemplarisch genannt werden.

Wurde die MM-A aus ergonomischer Sicht als nicht bestanden gewertet, ist eine Nachprüfung vorgesehen. Anhand des Prüfberichtes können in einem solchen Fall sofort gezielt Verbesserungsmaßnahmen eingeleitet werden. Die überarbeitete Version wird dann erneut eingereicht. Auch bei Weiterentwicklungen der MM-A bzw. der Benutzerinformationen, wäre ebenfalls eine Nachprüfung erforderlich, wenn das vorhandene TÜV-Prüfsiegel weiterhin genutzt werden soll.

Für den europaweiten und internationalen Einsatz, ist die Prüfplakette „Ergonomics TÜV tested" entwickelt worden. MM-A die das Prüfverfahren positiv abgeschlossen haben, können nach einer zusätzlichen Konformitätsprüfung der englischen/amerikanischen Variante zusätzlich das adäquate internationale Prüfsiegel nutzen. Mit diesem Qualitätssiegel ist es möglich, weltweit die besondere ergonomische Qualität der Produkte, und damit deren Zusatznutzen, zu kommunizieren und sich von den Mitbewerbern abzuheben.

2.2 Prüfungstiefe

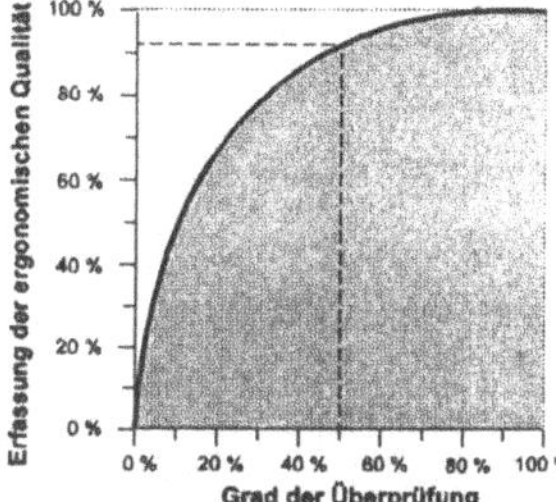

Abb. 9: Diagramm Prüfungstiefe

Der Umfang von Softwareprogrammen steigt stetig. Das hat zu Folge, daß für komplexe MM-A in einem wirtschaftlichen Rahmen keine hundertprozent-Prüfung möglich ist. Unter der Voraussetzung, daß eine streng homogene Programmentwicklung mit einheitlichen Vorgaben und Randbedingungen vorlag, hat in guter Näherung das Diagramm in Abbildung 9 Gültigkeit. Bei dem etwa exponentiellen Kurvenverlauf ist mit einem Überpüfungsvolumen von 50% eine Einschätzung der ergonomischen Qualität von über neunzig Prozent möglich. - Eine meist hinreichende Genauigkeit, zumal subjektive Einflüsse bei der Beurteilung ergonomischer Kriterien in jedem Fall die Berücksichtigung einer gewissen Streubreite von min. 3% bedingen.

2.3 Erfahrungen

Das vorgestellte Ergonomie-Prüfverfahren ist seit über fünf Jahren im Einsatz. Während dieser Zeit wurde eine Vielzahl informationstechnischer Produkte und Anwendungen, vornehmlich aus den Bereichen Unterhaltungselektronik, Kommunikationstechnik sowie Computer und Zubehör, beurteilt. Das Verfahren mußte im Laufe der Zeit nur geringfügig modifiziert werden. Im Wesentlichen handelte es sich dabei um den Punkt Anwendertests, der neu ausgerichtet und optimiert wurde.

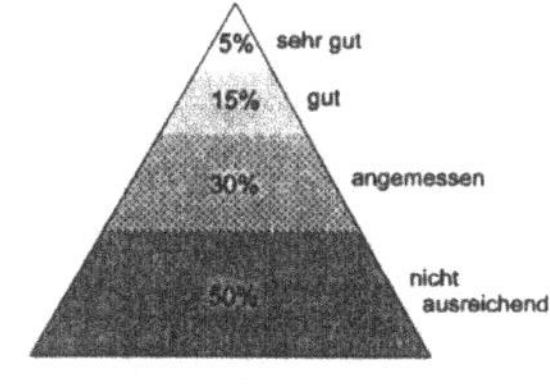

Abb. 10: Verteilung der
Beurteilungsergebnisse

Rückblickend betrachtet kann eine Verteilung der Beurteilungsergebnisse, wie in Abbildung 10 aufgezeigt, verallgemeinert werden: Circa 50% der Erstprüfungen von Produkten schließen mit gewissen Mängeln ab. Verbesserungen, aufgrund der differenzierten Prüfungsergebnisse, führen dann in der Regel zu einem positiven Abschneiden. Sehr selten (5%), und bei einigen Produktgruppen noch immer unerreicht, wird eine sehr gute ergonomische Qualität bescheinigt. Für eine diesbezügliche Aussage, die sich ausschließlich auf MM-A bezieht, liegt bislang noch keine ausreichende Datenbasis vor. Es bleibt aber zu vermuten, daß sich eine ähnliche Verteilung einstellen wird.

3 Resümee und Ausblick

Die Hauptproblematik liegt damit weniger bei den Prüf- bzw. Bewertungsverfahren, von denen einige praxiserprobte vorliegen, als vielmehr in der Definition von Bewertungskriterien und angemessen -maßstäben.

Neben klassischen Bewertungkriterien aus der Softwareergonomie oder z.B. auch der Dokumentationsprüfung, müssen neue, eigens auf multimediale Produkte abgestimmte, zur Anwendung kommen. Diese sind bislang weitgehend noch rudimentär (vgl. Abb. 11). Hier liegt eine gewaltige Aufgabe für die kommenden Jahre, in der vor allem die Praxis gefordert ist ihre gesammelten Erkenntnisse und Erfahrungen einzubringen. Ersteller, Prüfinstitutionen sowie Anwender sind aufgerufen gemeinsam aktiv werden, um alle Belange gleichermaßen zu berücksichtigen.

Kriterien für die Qualität von Multimedia-Produkten* **Auszug aus dem Katalog des Deutschen MultiMedia Verbands [11]**		
Kommunikationsziel • Gibt es eine genaue Zieldefinition? • Ist sie zu erkennen? • Frage nach der Zielgruppe und der Kernaussage /Kerninformation **Online** • Welche Dateigröße hat die Grafik („Ladezeit")? • Wird der Nutzer über die Dateigröße informiert? • Wird auf andere Informationsangebote („Links") verwiesen ? **Innovative Idee** • Ist die Idee neu und attraktiv? • Ist die Idee außergewöhnlich? **Ablauf** • Kann ein linearer Teil eines Programms abgebrochen / unterbrochen werden? • Was passiert nach der Interaktion (dem „Klick")? • Wie erkennt der Nutzer, daß seine Interaktion registriert wurde? **Reaktionszeit** • Wie schnell reagiert das System auf die Interaktion („Klick")? **Navigation / Orientierung** • Gibt es eine „One-Step-Back"-Funktion? • Wie erfährt der Nutzer, „wo" er sich befindet? • Wie wird ihm gezeigt, wie weit ein linearer Teil vorgeschritten ist? • Gibt es „Kennzeichnungsmöglichkeiten" („Bookmarks")? **Screen-Design** • Wie ist die Übersicht? • Wie ist die Bildaufteilung? **Funktionalität der Bildelemente** • Typographie? • Farbgestaltung? • Ist die Oberfläche bewegt? **Buttons** • Sind die Buttons eindeutig?	**Inhalte / Erzählstruktur** • Umfang des Inhalts? • Wie sind die Inhalte gegliedert? • Wie sind die Inhalte miteinander verknüpft? • Wie viele Ebenen müssen „durchgeklickt" werden, bis man zu der relevanten Information kommt? **Sound-Qualität** • Kann die Lautstärke verändert werden? • Kann der Sound ausgeschaltet werden? • Gibt es eine „Atmo" (Sound-Teppich)? Stereo? • Sind Sound-Ideen vorhanden (Klickgeräusche, Blättergeräusche)? **Sprache** • Ist der Text verständlich? • Wurde der Text von einem Profi gesprochen? • Wird der geschriebene Text gelesen? • Kann man dieses Feature ausschalten? **Bilder** • Wurden die Bilder speziell für das Programm bearbeitet? **Video** • Erfüllt das Video den Kommunikationszweck (Authentizität)? • Passen die Videos zum Thema? **Bildqualität** • Wiederholungsfrequenz / Bilder pro Sek.? • Bildqualität - Farben? • Bildqualität - Rauschen? • Erfüllt ein Präsentator seinen Zweck, oder wäre besser auf ihn verzichtet worden? **Technik** • Ist eine Installation notwendig? • Ist eine De-Installation vorgesehen? • Wie hoch ist der Speicherbedarf? • Wie absturzsicher ist das Produkt, wurde ausreichend getestet?	**USP zum Buch / Film** • Was unterscheidet das Produkt von einem gedruckten Produkt („Mehrwert")? **Simulation versus Selektion** • Können Objekte bewegt, verändert werden? • Können Daten eingegeben und bearbeitet werden? • Werden Schieberegler genutzt? **Additional Value** • Ist eine Weiterverwendung in anderen Applikationen möglich (Texte, Bilder, Tonsequenzen, ...)? • Sind Querverweise („Links") vorhanden? **Vernetzung** • Besteht eine Anbindungsmöglichkeit (zum Beispiel zu einem Online-Dienst, oder zu einer Datenbank)? • Besteht eine Anbindungsmöglichkeit (zum Beispiel zu einer CD-ROM)? **Ausdrucksmöglichkeit** • Können Informationen / Inhalte / Berechnungen ausgedruckt werden? **Hilfefunktion** • Gibt es eine direkte konkrete Hilfe oder nur eine allgemeine Hilfe? • Ist die Hilfefunktion verständlich? **Thema- und Titelfaszination** • Ist das Produkt zeitgerecht (zu früh / zu spät)? • Ist das Produkt einzigartig? • Ist das Produkt phantasiefördernd? *** Die Kriterien beziehen sich auf „Packed media" oder „Standalones" wie zum Beispiel Kiosk-Terminals.**

Abb. 11: Qualitätskriterien für die Beurteilung von Multimedia-Produkten

In diesem Zusammenhang erscheint eine zentrale Koordinierung der Aktivitäten besonders wesentlich, da zur Zeit Vielerorts parallel analoge Projekte laufen. Unbedingt vermieden werden sollte es, unterschiedliche Interpretationen von Bewertungskriterien und darüber hinaus abweichende Bewertungsmaßstäbe zu generieren und verbreiten. Die dadurch entstehende Verwirrung und das aufwendige nachträgliche Zusammenfügen kann wohl kaum im Sinne der Beteiligten sein.

Literaturverzeichnis

[1] ISO 9241 (z.T. draft): Ergonomic requirements for office work with visual display terminals (VDTs). International Organization for Standardization, Part 1-17 (unterschiedliche Erscheinungsdaten)

[2] J. Wandmacher: Software-Ergonomie. de Gruyter, Berlin; New York, 1993

[3] DIN 66 285 (identisch mit RAL-GZ 901): Anwendungssoftware - Gütebedingungen und Prüfbestimmungen. Beuth Verlag GmbH, Berlin, 1990

[4] DIN 66 234, Teil 8: Bildschirmarbeitsplätze - Grundsätze ergonomischer Dialoggestaltung. Beuth Verlag GmbH, Berlin, 1988

[5] IEEE Standards Collection Software Engineering. The Institute of Electrical and Electronics Engineers, Inc., New York, 1994

[6] VDI 5005: Software-Ergonomie in der Bürokommunikation. VDI-Gesellschaft Entwicklung Konstruktion Vertrieb, Düsseldorf, 1990

[7] J. Englisch: Ergonomie von Softwareprodukten: methodische Entwicklung von Evaluationsverfahren. BI-Wiss.-Verlag, Mannheim; Leipzig; Wien; Zürich, 1993

[8] B. Shneidermann: Designing the User Interface: Strategies for Effective Human-Computer Interaction. Addison-Wesley, 1987

[9] Das SANUS Handbuch. Arbeit, Technik und Bildung GmbH, Chemnitz, 1996

[10] M. Riegel: Prüfung - Zertifizierung - Marketing - Gebrauchsanleitungen auf dem RWTÜV-Prüfstand. in: Technische Dokumentation optimieren, Dr. Josef Raabe Verlag, Berlin; ..., 1995

[11] dmmv: Multimedia Annual. Metropolitan Verlag, Köln, 1996

Adresse des Autors

Dipl.-Ing. Eur. Erg. Martin Riegel
TÜV Informationstechnik GmbH
Ein Unternehmen der CUBIS-Gruppe
Bereich Benutzerinformationen, Produktergonomie
Im Teelbruch 122
D-45219 Essen
E-Mail: M.Riegel@tuvit.cubis.de

Bildschirmarbeit gemäß EU-Richtlinie

Christian Stary*, Thomas Riesenecker-Caba, Michael Kalkhofer

*Universität Linz, Institut für Wirtschaftsinformatik, FORBA - Forschungs- und
Beratungsstelle Arbeitswelt Wien

Zusammenfassung

Zur Verbesserung der Gesundheit und Sicherheit am Bildschirmarbeitsplatz wurde bereits 1990 seitens der EU
eine Richtlinie definiert. Um die in der Bildschirmrichtlinie angeführten grundlegenden Software-ergonomischen Anforderungen in die Praxis umzusetzen, ist es für alle EU-Mitgliedsstaaten notwendig, nationale Gesetze
zu erlassen und für die betriebliche Umsetzung der in der Richtlinie vorgesehenen Anforderungen an die
Software Rechnung zu tragen. Da es keine direkt einsetzbaren Verfahren zur Bewertung von Bildschirmarbeit
gemäß den Aussagen der Richtlinie gibt, wurden in Österreich zwei Projekte durchgeführt. In dem ersten der
beiden Projekte wurde die Richtlinie ob ihres Bezugs zu Software-ergonomischen Kriterien analysiert sowie bestehende Bewertungsverfahren der Software-Ergonomie erhoben und ob ihres Einsatzes bewertet. In einem
zweiten Projekt wurden die betrieblichen Rahmenbedingungen erhoben, unter denen ein Software-ergonomi-
sches Meßinstrument zum Einsatz kommen kann. Daran anschließend wurde EU-CON, ein EU-konformes Be-
wertungs- und Gestaltungsverfahren entwickelt. Der vorliegende Beitrag faßt die wesentlichsten Erkenntnisse
zusammen und stellt EU-CON vor.

Schlüsselbegriffe: Bildschirmarbeit, Software-Ergonomie, Mensch-Computer Interaktion, Bildschirmrichtlinie,
Evaluierung, Gestaltung.

1 Gesetzliche Notwendigkeiten

Mit dem zunehmend verbreiteten Einsatz von Computersystemen am Arbeitsplatz wurde auch
die Regulierung von Arbeitsbedingungen vorangetrieben. Mit der Einführung interaktiver
Software am Arbeitsplatz stiegen damit nicht nur die Anteile an Tätigkeiten, welche mit Hilfe
interaktiver Software bewältigt werden, sondern auch die Belastungen von Beschäftigten,
welche durch technologische oder organisatorische Faktoren bedingt sind.

Die Vermeidung von Gesundheitsgefährdungen sowie die Unterstützung der individuellen
Aufgabenbewältigung stehen im Mittelpunkt der Diskussion über einen EU-weiten Arbeitnehmerschutz am Bildschirmplatz. Dementsprechend wurde bereits 1990 eine Richtlinie des
Rates der EU zur Mensch-Maschine-Kommunikation erlassen [2]. Nach der EU-Bildschirmrichtlinie 90/270/EWG vom 29.5.1990 haben betriebliche Bildschirmarbeitsplätze

- den auszuführenden Aufgaben und
- den Benutzereigenschaften und der menschlichen Informationsverarbeitung angepaßt,
- bezüglich des Erfahrungsschatzes der Software-Benutzer flexibel, sowie
- bezüglich der Abläufe einsichtig und
- gegen unwissentliche Kontrolle geschützt zu sein.

Nach EU-Recht sind sämtliche Richtlinien in nationales Recht der Mitgliederstaaten überzu-
führen. Dies ist in diesem Bereich bereits mehrheitlich geschehen [6]. Jedes Unternehmen ist
demnach verpflichtet, bei der Planung, Anschaffung und dem Betrieb von Software die Aussagen der Richtlinie nachweislich einzuhalten. Da jedoch die Bestimmungen der Richtlinie,

und damit in der Folge die nationalen Gesetze zum Arbeitnehmerschutz kaum Hinweise enthalten, wie die einzelnen Bestimmungen in die Praxis umzusetzen sind, etwa durch Vorgabe von Instrumenten, sind umfangreiche Aktivitäten zur praktischen (betrieblichen) Umsetzung erforderlich:

1. die Erhebung der wissenschaftlichen Grundlagen (Kriterien, Prinzipien, Verfahren und Methoden) zur Bewertung und Gestaltung interaktiver Arbeitsplätze mit Bezug zur Richtlinie,
2. die Erfassung der Rahmenbedingungen für den betrieblichen Einsatz eines Bewertungsinstruments und
3. die Operationalisierung der Richtlinie entsprechend den Ergebnissen aus 1. und 2.

Diese Aktivitäten wurden im Rahmen zweier österreichischer Projekte gesetzt (Bundesministerium für Wissenschaft und Forschung [GZ 190.134/2-II/8/94] und Bundesministerium für Arbeit und Soziales [GZ 120/19-GrA/95]), deren Ergebnisse in der Folge zusammengefaßt werden.

2 Methodische Möglichkeiten

Im Rahmen des ersten Projekts wurden folgende Aktivitäten gesetzt:

- *Analyse der EU-Richtlinie*: Ableitung von Kriterien der Software-Ergonomie zur Bewertung von Software mit Bezug zur Richtlinie.
- *Analyse bestehender Bewertungsverfahren*: Auswertung bestehender Verfahren der Software-Ergonomie bezüglich ihrer Ziele, Kriterien, Methoden, Hilfsmittel, empirischer Absicherung, Bewertungstiefe, Aufwand und Handhabbarkeit.
- *Abgleich der EU-Richtlinieninterpretation mit der Verfahrensanalyse*: Abdeckung der EU-Anforderungen durch bestehende Verfahren, Erstellung eines Defizitkatalogs zur Erarbeitung eines richtliniengetreuen Instruments zur Bewertung.

Die Analyse der Richtlinie zeigte, daß eine Aussage den aufgabenbezogenen Aspekt von Bildschirmarbeitsplätzen direkt anspricht, 2 Aussagen sich vornehmlich an den technischen Eigenschaften von Benutzerschnittstellen orientieren sowie mehrheitlich menschliche und soziale Aspekte angesprochen werden. Eine in diesem Zusammenhang durchgeführte Clusteranalyse zeigt starken Bezug traditioneller Software-ergonomischer Kriterien zu den Aussagen der Richtlinie [7, S. 101-132]:

1. Mit Ausnahme der Kooperationsförderlichkeit haben alle Kriteriencluster direkten Bezug zu mehreren Aussagen der Richtlinie.
2. Zentrale Bedeutung haben Aspekte der Steuerbarkeit und Adaptivität (Flexibilität), gefolgt von der Aufgabenangemessenheit, Erwartungskonformität, Erlernbarkeit, des Datenschutzes und der Datensicherheit.
3. Die Aussagen zur Benutzerfreundlichkeit, Adaptivität, Kontrolle sowie Grundsätze der Ergonomie haben aufgrund ihrer Abstraktion von konkret meßbaren Kriterien sieben von acht Kriteriencluster involviert, gefolgt von Aussagen zur Anzeige von Abläufen, zur Aufgabenangemessenheit und zur benutzergerechten Anzeige von Information.

Bei der ebenfalls im ersten Projekt durchgeführten Analyse der vor allem in Europa entwickelten Bewertungs- und Gestaltungsverfahren zeigte sich, daß die Kriterien der DIN-Norm 66 234 Teil 8 [1] Aufgabenangemessenheit, Selbstbeschreibungsfähigkeit, Steuerbarkeit,

Erwartungskonformität und Fehlerrobustheit, sowie die zusätzlichen Kriterien des ISO-Standards 9241 Part 10 [4] Individualisierbarkeit und Erlernbarkeit am häufigsten eingesetzt werden. Im Rahmen des Projekts wurden 18 Verfahren nach einem eigens entwickelten Analysekatalog aufbauend auf den Arbeiten von Hampe-Neteler und Rödiger [3] untersucht. Neben arbeitspsychologischen Ansätzen wurden vor allem holistische Ansätze wie EVADIS II [5] oder der in Schweden eingesetzte Software-Checker [10] untersucht.

Neben den Normen wurden vor allem die Richtlinie als auch arbeitspsychologische Verfahren als Ausgangspunkt der Entwicklungen genannt. Arbeitspsychologische Verfahren beziehen sich jedoch kaum direkt auf die Aussagen der Richtlinie. Sie enthalten den meisten Bezug zur ersten Aussage der Richtlinie, nämlich der Aufgabenangemessenheit, und zwar über ihren Bezug zur Erfassung von Aufgaben und Tätigkeiten am Arbeitsplatz. Die am häufigsten verwendeten Hilfsmittel zur Bewertung von Bildschirmarbeit sind Fragebögen, Prüffragen, Interviews bzw. Antwortblätter. Zur Unterstützung der Auswertungen kann sehr oft Software eingesetzt werden. Zur Erstellung der Aufgabenstellung liefern diverse Handbücher die notwendige Vorinformationen. Viele der Verfahren stellen hohe Anforderungen an die Evaluateure, insbesondere Verfahrenskenntnisse (und damit inkludiert eine umfassende Schulung) sowie Softwarekenntnisse der zu untersuchenden Arbeitsaufgaben. Einige Verfahren benötigen nur bei schwierigen Aufgabenstellungen die Mithilfe von Experten.

Aus den Ergebnissen der Verfahrensanalyse wird klar, daß keines der erhobenen Verfahren direkt zur Operationalisierung der Aussagen der EU-Richtlinie herangezogen werden kann, obwohl einige Verfahren engen Bezug zu mehreren Aussagen und Betrachtungsebenen der Richtlinie aufweisen. Der verfahrensbezogene Grund dafür liegt zum einen in der Unvollständigkeit der Verfahren bezüglich der Aussagen der Richtlinie (bislang gab es auch keine wissenschaftliche Analyse der Aussagen), und zum anderen in der fehlenden empirischen Absicherung vor allem jener Verfahren, welche direkten Bezug zur Richtlinie besitzen.

3 Betriebliche Gegebenheiten

In dem zweiten Schritt zur Umsetzung der EU-Richtlinie, d.h. bevor mit der eigentlichen Entwicklung eines Bewertungsinstruments begonnen werden konnte, wurden zunächst die betrieblichen Rahmenbedingungen des Einsatzes eines Bewertungsinstruments erhoben [8].

Die zentralen Fragestellungen der Erhebung waren folgende:
1. Welcher Stellenwert wird der Software-Ergonomie im Rahmen des Arbeitnehmerschutzes beigemessen?
2. In welcher Art und Weise wird die Bewertung der Bildschirmarbeitsplätzen nach Software-ergonomischen Kriterien durchgeführt?
3. Wer zeichnet für die Evaluierung verantwortlich und welche (außer-)betrieblichen Stellen sind mit der Durchführung betraut?
4. Was geschieht mit den Ergebnissen der Evaluierung?

Es wurden bis zu zehn Vertreter aus vier Gruppen mittels strukturierter Interviews befragt, um sämtliche Sichtweisen der betrieblichen 'Wirklichkeit' zu erfassen: Interessenvertretungen der Arbeitgeber und Arbeitnehmer, Ministerien, Betriebe und Beratungs- und Wissenstransfereinrichtungen im Bereich Gesundheitsschutz am Arbeitsplatz.

Die Erhebungen zeigen zunächst, daß die Forderungen der Richtlinie zwar von den betrieblichen Verantwortlichen und den inhaltlich Betroffenen (Betriebsärzte, Sicherheitstechniker, etc.) grundsätzlich als wichtig erachtet werden, jedoch die Umsetzung des Gesetzes, d.s. Vorschriften im Rahmen der gesetzlichen Verordnung, und die betriebliche Erfahrung im Umgang mit Software-ergonomischen Kenngrößen, vor allem psychischen Belastungen fehlen. Ein Problem stellt in diesem Zusammenhang die reale Meßbarkeit von psychischen Belastungen, die als Folgewirkung unzureichend ausgestalteter Software entstehen können, dar. Die zur Zeit gebräuchlichste Gestaltungsinformation an Büroarbeitsplätzen bezieht sich auf Anforderungen der Hardware-Ergonomie (z.B. Lichteinfall, Lärmbelästigung, Eigenschaften des Arbeitsplatzes - Beschaffenheit von Tisch und Stuhl), die quantitativ überprüfbar sind. Hier kann auf bekannte Meßmethoden zurückgegriffen werden. Die Ergebnisse Hardware-ergonomischer Gestaltung von Arbeitsplätzen sind damit eindeutig belegbar, da quantifizierbar. Dem gegenüber ist eine 'objektive' Messung Software-ergonomischer Kriterien nicht möglich.

Ein weiteres Ergebnis der betrieblichen Befragungen betrifft die Einbeziehung von Standardsoftware bei der Arbeitsplatzanalyse. Standardsoftware wird bei der Bewertung als 'gesondert zu behandeln' eingestuft und als kaum bzw. umständlich änderbar eingeschätzt. Dies widerspricht jedoch der Konzeption dieser Art von Software. Sie enthält im Regelfall sehr viele Möglichkeiten, individuelle Anpassungen zugunsten von Tätigkeiten und Benutzervorstellungen vorzunehmen. Software-technisch betrachtet ist es das Ziel der Entwicklung derartiger Softwareprodukte, eine Menge von Funktionen und Datenstrukturen zu finden, welche in möglichst vielen Einsatzbereichen eingesetzt werden können. Dieses Ziel bedeutet nicht, daß auch die Benutzung dieser Funktionen standardisiert zu erfolgen hat. Die Softwareentwickler von Standardprodukten haben dies erkannt und erlauben den Benutzern möglichst viele Adaptierungen, etwa die Sequenz von Aktivitäten betreffend (Makros). Damit soll es weitgehend möglich sein, Standardsoftware an Prozesse und Benutzer anzupassen (Customizing). Hersteller von gesamtbetrieblichen Informationssystemen (z.B. SAP) investieren in die Entwicklung eigener Phasenpläne, um ihre Kunden möglichst an den optimalen Einsatz ihrer Software heranzuführen. Denn spätestens bei Zertifizierungsbestrebungen oder organisatorischen Entwicklungen, wie etwa der Einführung von Total Quality Management, kommt diesem Anpassungsprozeß entscheidende Bedeutung zu.

Die vorherrschende betriebliche Meinung ist, daß es den Produzenten obliegt, die Erfüllung der Software-ergonomischen Forderungen der Richtlinie zu garantieren. Daher wird der Wunsch nach einem Gütesiegel (für die Erfüllung von Software-ergonomischen Kriterien) zur leichteren Nachvollziehbarkeit von mehreren Seiten geäußert. Dabei wird aber meist übersehen, daß ein Prüfsiegel nur Aussagen über ein Produkt liefern kann. Der Einsatz an einem Arbeitsplatztyp oder einem einzelnen Arbeitsplatz wird jedoch mit einem Siegel nicht abgedeckt. Geprüfte Software kann daher den Arbeitgeber nicht von der Pflicht entbinden, beim Ankauf die Anpaßbarkeit und Benutzungsfreundlichkeit von Software zu berücksichtigen.

Die österreichische Behörde möchte weder durch das Gesetz noch die in Ausarbeitung befindliche Verordnung Betrieben bestimmte Vorgehensweisen auferlegen. In der Erreichung der Ziele des Arbeitnehmerschutzes sollen die Betriebe die freie Wahl der Meßmethoden besitzen. Aus diesem Grund wird die Bewertung von Bildschirmarbeitsplätzen in Zusammenhang mit Software-ergonomischen Kriterien in den Betrieben unterschiedlich gehandhabt.

Resultierend aus dem betriebswirtschaftlichen Druck soll in den meisten Fällen die Evaluierung so rasch wie möglich und mit geringem Aufwand durchgeführt werden. Dies führt dazu, daß in den Betrieben ähnliche Tätigkeiten (z.B. im Bereich des Rechnungswesens oder bei Sekretariatsarbeitsplätzen) zusammengefaßt und ein für diesen Tätigkeitsbereich 'repäsentativer' Arbeitsplatz bewertet wird.

Bei der Bewertung lassen sich zwei mögliche betriebliche Herangehensweisen differenzieren: Jene, welche von den meisten Betrieben bevorzugt wird und auf einer repräsentativen Bewertung von Arbeitsplatztypen (z.B. Arbeitsplatz einer Sekretärin) in Verbindung mit Produktevaluationen (z.B. Analyse eines Textverarbeitungssystems) basiert; jene, welche eine personenabhängige Herangehensweise vorsieht, bei der Arbeitsplätze in bezug zu einzelnen Beschäftigten untersucht werden, um deren betriebliche Einbindung und deren Belastungen zu analysieren und gegebenenfalls zu beheben.

Aufgrund der Einführung der Richtlinie in die Bestimmungen des österreichischen ArbeitnehmerInnenschutzgesetzes 1995 sahen sich die Interessenvertretungen aber auch Beratungsstellen veranlaßt, Instrumente zur Bewertung von Bildschirmarbeitsplätzen auszuarbeiten bzw. für die Bewertung zu empfehlen. Obwohl hier in den letzten Monaten verstärkte Anstrengungen unternommen und mehrere Instrumente vorgestellt wurden (ähnlich den Anstrengungen in der BRD von Technologieberatungsstellen, wie etwa Oberhausen, oder im SANUS-Projekt), finden diese in den Betrieben keine oder nur am Rande Anwendung. Die Instrumente liefern dabei Anregungen und Wissen für die verantwortlichen Stellen bzw. dienen als Basis für die Softwareentwicklung und die damit verbundenen Anforderungen (z.B. Pflichtenhefterstellung). Dabei muß jedoch berücksichtigt werden, daß die Untersuchung in Betrieben mit mehr als 250 Beschäftigten stattfand, die in der Regel für diese Aufgabenstellung über eine eigene Technikabteilung verfügen.

In den Interviews mit den verantwortlichen betrieblichen Stellen wurde eine weitgehende Unklarheit über Inhalt und Effizienz der veröffentlichten Verfahren festgestellt. Mitunter wurden diese Verfahren nur am Rande wahrgenommen, da betriebsintern mit eigenen Entwicklungen gearbeitet wird. Ein betrieblich handhabbares Instrument sollte dabei mehrere Eigenschaften erfüllen. Es sollte
- eine rasche Evaluierung ermöglichen,
- möglichst automatische Auswertungen liefern,
- dem Gesetz Genüge tun,
- übersichtlich und durchgängig sein,
- ein leichtes Erkennen von Mängeln bei der Bildschirmarbeit unterstützen und
- unmittelbares Feedback über die Umsetzung der notwendigen Maßnahmen bei der Mängelbehebung liefern.

Es werden neben dem Aufzeigen der Mängel vor allem Anleitungen gewünscht, um die festgestellten Mängel in kurzer Zeit und mit wenig Aufwand beheben zu können. Die Dauer der Bewertung - so die herrschende Meinung in den Unternehmen - soll sich im Rahmen von einer Stunde bewegen und mit möglichst geringen Kosten für den Betrieb verbunden sein.

Da in den Betrieben vornehmlich eine produktbezogene Bewertung bzw. eine Evaluierung von Klassen von Bildschimarbeitsplätzen oder Software stattfindet, bleiben Fragen der Arbeitsorganisation im Rahmen der Bewertung von Bildschirmarbeit weitgehend unberührt. Es existieren von seiten der Beschäftigten und Unternehmen Vorbehalte gegen eine Untersuchung arbeitsorganisatorischer Abläufe und Inhalte. Diese Bedenken beziehen sich auf die Offenlegung von innerbetrieblichen Vorgängen und die Transparenz von Kompetenzen. Nicht zuletzt ist die Furcht um den Arbeitsplatz für die Beschäftigten ein Hemmnis, da Evaluierungen für Rationalisierungsuntersuchungen gehalten werden.

Im Zusammenhang mit der Mängelbehebung wird in den Betrieben immer wieder die Notwendigkeit einer umfassenden Schulung der Betroffenen angeführt, da viele Belastungen auf den Aspekt einer unzureichenden Schulung zurückgeführt werden. Da Beschäftigte sich mit einzelnen Softwareprodukten oder -funktionen zuwenig oder gar nicht auskennen, entstehen Unsicherheiten und in weiterer Folge das Gefühl bei den Beschäftigten, überfordert zu sein.

Umfassende (Nach-)Schulung kann nach betrieblicher Meinung zu unmittelbaren Verbesserungen führen. Erst wenn auch diese Aktivitäten zu keinem befriedigenden Ergebnis führen, sollten Mängelbehebungen im Bereich der Software erfolgen. Diese Herangehensweise ist jedoch nur bedingt wirtschaftlich. Schulungen können bei Adaptierungsproblemen, welche durch bestehende Softwarefunktionen gelöst werden könnten, kostenintensiver als technische Mängelbehebungen sein.

Zusammenfassend läßt sich erkennen, daß die in der EU-Bildschirmrichtlinie enthaltenen (und fast 1:1 in das österreichische ArbeitnehmerInnenschutzgesetz übernommenen) Forderungen des Arbeitnehmerschutzes nicht in der intendierten Tragweite wahrgenommen werden. Vielmehr führt eine Interpretation durch betriebliche Praktiker zu Mißverständnissen und zu einer unzureichenden Auseinandersetzung mit der Richtlinie.

4 EU-CONform Evaluation and Engineering of VDU-Work

Aufbauend auf den konzeptionellen und praktischen Erkenntnissen sowie den Anforderungen zur Operationalierung, welche in Kapitel 2 und 3 beschrieben sind, wurde nun die Entwicklung eines EU-richtliniengetreuen Instruments vorangetrieben. Das Ergebnis ist EU-CON - Acronym für *EU-CONform Evaluation and Engineering of VDU Work.* Das Verfahren richtet sich zum einen an die von Bildschirmarbeit betroffenen Mitarbeiter eines Unternehmens (d.s. die Benutzer interaktiver Softwaresysteme) sowie die für die in einem bzw. für ein Unternehmen inhaltlich Verantwortlichen für die Umsetzung der Richtlinie (Sicherheitsvertrauensleute, -techniker, Betriebsarzt, etc.). EU-CON setzt sich dementsprechend aus drei Teilen zusammen:
- *Leitfaden*: Dieser wird in der Vorbereitungs- und Durchführungsphase benutzt, um den Kontext des Bildschirmarbeitsplatzes zu erfassen und die Beschäftigten zur Bewertung zu befähigen.
- *Handbuch für Evaluateure und Gestalter:* Dieses Hilfsmittel dient der Auswertung der Erhebungsergebnisse durch den Leitfaden. Im Mangelfall werden Handlungsanleitungen zur Verfügung gestellt, um den Ursachen nachzugehen bzw. die Mängel zu beheben.
- *Vorgehensmodell*: Damit wird der strukturierte Einsatz des Leitfadens und des Handbuchs unterstützt. Es unterstützt die Aktivitäten in den Phasen: Vorbereitung, Durchführung, Auswertung, und im Falle von Mängeln Ursachenaufdeckung und Verbesserungen.

Damit setzt das Verfahren einige Anforderungen mehrfach um: Er unterstützt nicht nur die verschiedenen Benutzergruppen (Evaluateure, Beschäftigte), sondern unterstützt sie phasengerecht. Bei der Vorbereitung wird nur jene Information erhoben, welche dem Evaluateur Hilfe bei der Mängelbehebung bietet. Damit verkürzt sich die Zeit der Durchführung. Werden Mängel indiziert, hilft das Handbuch, die notwendigen Schritte zu setzen.

Der *Leitfaden* setzt sich dabei aus mehreren Komponenten zusammen:
- <u>Merkblatt</u>: Das Merkblatt gibt anhand einer kurzen Darstellung den befragten Personen einen Überblick über die derzeitigen rechtlichen Bestimmungen und den Hintergrund für die auszufüllenden Fragen, da während der Testphase offensichtlich wurde, daß diese Bestimmungen und die damit verbundenen Ziele des Arbeitnehmerschutzes nur den wenigsten bekannt sind. Darüberhinaus werden die Inhalte der Bewertung erklärt, um Vorbehalte gegenüber den Zielen der Erhebung auszuräumen, und den Beschäftigten die mit der Bewertung verbundene Möglichkeit der Verbesserung der individuellen Arbeitssituation klar darzustellen.
- <u>Fragebogen</u>: Der Fragebogen setzt die Aussagen der Richtlinie zur Bewertung der Eigenschaften von Bildschirmarbeitsplätzen um. Der Fragebogen ist der Arbeitsgegenstand der Bewertung aus der Sicht der Beschäftigten. Die erste wesentliche Aufgabe bei der Bewertung stellt die Identifikation individueller Aufgaben und Arbeitsschritte dar, auf denen die Messung und danach die Bewertung der Aufgabenangemessenheit basiert. Durch die Erfassung der subjektiven Wahrnehmung von Aufgaben und des subjektiven Zugangs zur Aufgabenbewältigung fällt die Diskussion der Gesamtorganisation von Arbeitsschritten, welche eine objektive, zusätzliche Erhebung, etwa im Sinne einer Workflow-Modellierung, erfordert, weg.
  Die zweite wesentliche Aufgabe bei der Bewertung stellt die Beantwortung der aufgabenunabhängigen Fragen dar, welche im Anschluß an die Fragen zur Aufgabenangemessenheit zu bearbeiten sind. Dabei werden vor allem kognitive Faktoren und technische Aspekte der Adaptierbarkeit angesprochen.
- <u>Informationsblatt</u>: Um nicht mit beispielhaften Antworten die Befragten zur Beantwortung der Fragen in eine bestimmte Richtung zu lenken, aber trotzdem Anhaltspunkte zur Beantwortung der Fragen zur Verfügung zu stellen, wurden sämtliche Beispiele und Erläuterungen in einem sogenannten Informationsblatt zur Unterstützung zusammengefaßt. Das Informationsblatt enthält folglich Musterantworten und erklärende Beispiele, damit Beschäftigte etwa vor Ausfüllen des Fragebogens die Bearbeitung nachvollziehen können und die Bedeutung der Fragen für sich klarstellen.

Die ausgefüllten Fragebögen dienen schließlich gemeinsam mit dem Handbuch für Evaluateure und Gestalter der Auswertung der Antworten und Bestimmung von Verbesserungsmaßnahmen im Mangelfall.

Das *Handbuch* besteht aus mehreren Teilen:
- Teil 1 vermittelt <u>Hintergrundinformation</u> zur Bewertung von Bildschirmarbeitsplätzen gemäß der EU-Richtlinie.
- Teil 2 gibt einen <u>Überblick</u> über die Anwendung des Leitfadens und den Einsatz des Fragebogens.

- Teil 3 führt die grundlegenden <u>Software-ergonomischen Kenngrößen</u> ein, welche der Entwicklung des Fragebogens zugrunde gelegt wurden.
- Teil 4 schließlich enthält den Inhalt des <u>Fragebogens</u>, welcher mit Hintergrundinformation und Handlungsanleitungen zur Mängelbehebung <u>erweitert</u> wurde.

Somit wird nicht nur die retrospektive, sondern auch die prospektive Gestaltung von Benutzungsschnittstellen unterstützt.

Das *EU-CON-Vorgehensmodell* unterstützt nicht nur die traditionellen Phasen der Bewertung, sondern auch die daran anschließende mögliche Umgestaltung des Bildschirmarbeitsplatzes.

1. <u>Vorbereitung.</u> Die Vorbereitungsphase der Bewertung nach EU-CON sollte die folgenden beiden betrieblichen Aktivitäten umfassen:
   1.1 *Briefing der Beschäftigten* durch den Evaluateur (gegebenenfalls nach Durchsicht des Handbuchs zur Evaluierung und Gestaltung). Die Inhalte sollten jenen entsprechen, welche die Beschäftigten auch auf dem Merk- und Informationsblatt finden können. Wichtig ist dabei, daß die Beschäftigten die Chance durch die Erhebung erkennen, an *ihrem* Arbeitsplatz organisatorische, technische, soziale und individuelle Verbesserungen durchsetzen zu können.
   1.2 *Ausgabe der Fragebögen* an die Beschäftigten mit dem Merk- sowie dem Informationsblatt, das Beispiele zu den einzelnen Fragen enthält.
   1.3 *Erstellen der Aufgabenliste*: Diese Aktivität ist die wesentlichste in dieser Phase, da das Verständnis der eigenen Aufgabenbereichs, vor allem die Granularität der Aktivitäten, entscheidend für die Erarbeitung von Vorschlägen und die Ableitung von Verbesserungsmaßnahmen ist.
2. <u>Durchführung der Erhebung.</u> Nach erfolgter Vorbereitung der Inhalte und des weiteren Vorgehens wird zunächst die Datenerhebung durchgeführt:
   2.1 *Beantwortung der Fragen* mit Hilfe des Merk- und Informationsblatts im Fragebogen durch *jeden* Beschäftigten. Bei Bedarf sollte der Evaluateur für Fragen als Auskunftsperson zur Verfügung stehen.
   2.2 *Abgabe der Fragebögen* an den Evaluateur zur weiteren Auswertung.
3. <u>Auswertung der Ergebnisse.</u> In diesem Schritt wird sowohl einzeln als auch kumulativ bewertet:
   3.1 *Einzelauswertung der Fragen jedes Fragebogens* durch den Evaluateur (gegebenenfalls unter Zuhilfenahme des Handbuchs zur Evaluierung und Gestaltung).
   3.2 *Sammelauswertung und Interpretation jedes Fragebogens* durch den Evaluateur, gegebenenfalls unter Zuhilfenahme des Handbuchs zur Evaluierung und Gestaltung.
   3.3 *Im Mangelfall Erarbeitung von Lösungsvorschlägen* durch den Evaluateur, gemeinsam mit dem Betroffenen, gegebenenfalls unter Zuhilfenahme des Handbuchs zur Evaluierung und Gestaltung.
4. <u>Um/Neugestaltung des Bildschirmarbeitsplatzes.</u> In diesem Schritt werden Verbesserungsmaßnahmen gesetzt und bewertet.
   4.1 *Umsetzung von Verbesserungsmaßnahmen* nach Ursachenidentifikation (gegebenenfalls unter Anleitung des Handbuchs) durch einschlägige Experten (Management, Techniker, Ergonomen, etc.).

4.2 *Überprüfung der gesetzten Maßnahmen* durch den Evaluateur, gemeinsam mit dem Betroffenen, gegebenenfalls unter Zuhilfenahme des Handbuchs zur Evaluierung und Gestaltung - im Mangelfall erneutes Durchlaufen der Bewertungsaktivitäten.

Vor allem durch das zuletzt erwähnte Bündel an Maßnahmen unterscheidet sich EU-CON von traditionellen Bewertungs- oder Gestaltungsverfahren der Software-Ergonomie. Es unterstützt nämlich nicht nur die Ursachenidentifikation problematischer Situationen der Bildschirmarbeit, sondern auch die Behebung von Mängeln.

Erste Tests mit 66 Personen in unterschiedlichen Branchen (Handel, Gewerbe, Produktion) und unterschiedlich granularen Aufgabenbeschreibungen führten zur Verbesserung der Struktur der Elemente des Leitfadens sowie zur Erhöhung der Verständlichkeit der Fragen.

5 Schlußfolgerungen

Der Beitrag befaßte sich mit der Entwicklungsgeschichte sowie den Ergebnissen mehrerer österreichischer Projekte zur Umsetzung der EU-Richtlinie zur Mensch-Maschine-Kommunikation. Neben den wissenschaftlichen, politischen und betrieblichen Rahmenbedingungen wurden auch neuartige Konzepte in dem Bewertungs- und Gestaltungsverfahren EU-CON umgesetzt:

- Die *ganzheitliche Messung und ursachenbekämpfende Umgestaltung* durch die Erweiterung des Phasenmodells der Bewertung um eine Gestaltungsphase, welche Verbesserungen in strukturierter und ursachenspezifischer Form ermöglichen.
- *Adäquate Hilfsmittel und Werkzeuge für jede Phase der Anwendung.*
- Der *explizite Bezug zu individuell wahrgenommenen Aufgaben* im Unterschied zur traditionellen Messung der Aufgabenangemessenheit, welche in den meisten Verfahren repräsentativ erfolgt. Durch die differenzierte Betrachtung der (Um-)Gestaltung der Bildschirmarbeit nach organisatorischen, kognitiven und technischen Merkmalen kommen den einzelnen Dimensionen der Aufgabenangemessenheit [9] entscheidende Bedeutung zu. Diese Dimensionen wurden allerdings bisher kaum untersucht und erfordern zumindest eine inhaltslogische Analyse der Aufgabenangemessenheit [siehe Beitrag von Ch. Stary und A. Totter in diesem Band].
- *Durchgehende Anwendbarkeit* des Verfahrens, d.h. unabhängig von der Ausprägung der Software.
- *Einfache Handhabbarkeit*, da die Anwendbarkeit des Leitfadens für alle Benutzer(gruppen) gegeben ist und sich zwischen 30 und 60 Minuten bewegt.

In der Folge gilt es nun das Verfahren bezüglich Objektivität, Validität und Reliabilität abzusichern. Sind diese Merkmale erfüllt, dann kann von einem wissenschaftlich abgesicherten Bewertungsverfahren gesprochen werden, welches nicht nur empirisch abgesichert ist, sondern auch der wirtschaftlichen und gesetzgebenden Realität Rechnung trägt.

Literatur

[1]   DIN 66 234 Teil 8: Bildschirmarbeitsplätze, Grundsätze der Dialoggestaltung. Berlin, 1988: Beuth.

[2]   EG-Richtlinie 90/270/EWG: Mensch-Maschine Schnittstelle, in: Richtlinie des Rates vom 29. Mai 1990 über die Mindestvorschriften bezüglich der Sicherheit und des Gesundheitsschutzes bei der Arbeit an Bildschirmgeräten (Fünfte Einzelrichtlinie im Sinne von Artikel 16 Absatz 1 der Richtlinie 89/391/EWG), in: Amtsblatt der Europäischen Gemeinschaften, Vol. 33, L 156, Mindestvorschriften (Artikel 4 und 5), Absatz 3, S. 18, 21.6.1990.

[3]   Hampe-Neteler, W.; Rödiger, K.H.: Software-Ergonomie. Verfahren der Evaluierung und Standards zur Entwicklung von Benutzungsoberflächen, Bericht, Universität Bremen, 1992.

[4]   ISO 9241 Part 10: Dialogue Principles, ISO/DIS, 1993.

[5]   Oppermann, R. et al.: Software-ergonomische Evaluation. Der Leitfaden EVADIS II. Berlin, 1992: deGruyter.

[6]   Rauterberg, M.; Vossen, P.; Krueger, M.(eds): Proceedings Workshop 'The EU-Directive 90/270 on VDU Work', Int. Conference on Occupational Ergonomics and Safety. Zürich, 1996.

[7]   Stary, Ch.; Riesenecker-Caba, Th.; Flecker, J.: EU-konforme Bewertung von Bildschirmarbeit - Schritte zur Operationalisierung: Wissenschaftliche Grundlagen und Vorarbeiten zur Entwicklung eines EU-richtliniengetreuen Bewertungsinstruments für Bildschirmarbeit. Zürich, 1995: vdf.

[8]   Stary, Ch. et al: Leitfaden zur Software-ergonomischen Bewertung von Bildschirmarbeit, Endbericht, Bundesministerium für Arbeit und Soziales. Wien, September 1996.

[9]   Stary, Ch.; Totter, A.: Cognitive and Organizational Dimensions of Task Appropriateness, in: Proceedings ECCE'8, pp. 127-131, European Association of Cognitive Ergonomics, 1996.

[10] TCO; Swedish Confederation of Professional Employees: SOFTWARE CHECKER - An Aid to the Critical Examination of the Ergonomic Properties of Software, Handbook and Checklist. Schweden, 1992.

Adressen der Autoren

Christian Stary
Universität Linz, Institut für Wirtschaftsinformatik
Communications Engineering
Freistädterstraße 315, A - 4040 Linz
stary@ce.uni-linz.ac.at

Thomas Riesenecker-Caba, Michael Kalkhofer
FORBA - Forschungs- und Beratungsstelle
Arbeitswelt
Aspernbrückengasse 4/5, A - 1020 Wien
forba@xpoint.at

Objektorientierte Dialogspezifikation mit ODSN

Gerd Szwillus

Fachbereich Mathematik/Informatik, Universität - GH Paderborn

Zusammenfassung

Wir diskutieren die zentrale Rolle eines - in der Spezifikationssprache ODSN notierten - objektorientierten Kontrollmodells im Entwicklungsprozeß graphischer interaktiver Benutzungsschnittstellen. In diesem Papier geht es vorwiegend um die Strukturierung des Entwurfsprozesses in sinnvolle und effizient miteinander kooperierende Entwurfseinheiten, die eine separate Behandlung der relevanten Entwurfsbereiche erlaubt, ohne den Zusammenhang dazwischen außer acht zu lassen.

1 Einführung

Das Design graphischer Benutzungsschnittstellen impliziert die simultane Behandlung verschiedener Aspekte; zu den wichtigsten dabei zählen:

- **visuelles Design** (z.B. Metaphern, Aussehen, Gestaltung, Layout),
- **strukturelles Design** (z.B. Abfolge von Bildschirmen, Struktur von Menüs, Informationsverteilung auf verschiedene Fenster oder Dialogboxen),
- **Interaktionstechniken** (z.B. Einsatz von Tastatur, Maus, Touch-Screen, Spracheingabe, textuelle Ausgabe, Graphikausgabe mit den jeweils typischen Ein/Ausgabetechniken),
- **Softwarestrukturierung** (z.B. Objektstrukturen, aktive und passive Komponenten).

Heutzutage werden graphische Benutzungsschnittstellen typischerweise mit sogenannten *Interface Buildern* [5], einer unterliegenden Graphik-Basissoftware, wie *X-Windows* und einer prozeduralen oder objektorientierten Programmiersprache, wie etwa *C* oder *C++*, programmiert. Der Grundgedanke dieser Art von Entwicklung:

- Anordnen und Gestalten der *Widgets* mit dem Interface Builder,
- Anbinden von Verhalten an die Ereignisse, die „auf" den Widgets ausgelöst werden mit der Programmiersprache unter Verwendung von Routinen des Basissystems für graphische Operationen.

Die Konsequenz dieser Arbeitsweise ist es, daß die Betonung auf der visuellen Gestaltung der Standard-Widgets liegt, die einheitlich im Interface Builder durchgeführt wird. Alle anderen Aspekte aber werden teilweise im Interface Builder behandelt und teilweise, verteilt auf viele „kleine" *Callback*-Funktionen, in Code der Programmiersprache versteckt. So werden zwar einzelne, ggf. komplexe Dialogboxen im Interface Builder interaktiv gestaltet; die Abfolge von Dialogboxen, oder das Auftauchen von Fenstern, oder gar welche graphischen Elemente in den Fenstern erscheinen wird in Code definiert - untereinander und mit semantischen Aktionen vermischt. Somit leistet diese Arbeitsweise eine sehr gute Spezifikation der Widgethierarchie, aber macht eine separate und kohärente Behandlung und Darstellung der anderen Aspekte (bis auf die Softwarestrukturierung) nahezu unmöglich.

Man erkennt an verschiedenen modernen Ansätzen zur Spezifikation von Benutzungsschnittstellen, daß ein Bedarf für die konsequente Behandlung der einzelnen Aspekte vorhanden ist:

- Janssen's Dialognetze [6] oder auch der THESEUS-Ansatz [11] befassen sich explizit mit dem strukturellen Design von Benutzungsschnittstellen,

- OBJECTION [13] und andere Ansätze versuchen, den Entwurf und Einsatz anwendungsspezifischer Graphiken ähnlich einfach und effizient zu gestalten wie die Verwendung von Standardwidgets,

- Petri-Netz-Ansätze [1,2] modellieren Interaktionstechniken und stellen ihren Bezug zum strukturellen Design her.

In diesem Papier soll eine Modellierungstechnik unter dem Begriff *objektorientiertes Kontrollmodell* präsentiert werden, die - wurzelnd in den Ideen der klassischen UIMS-Ansätze [4,9] - eine Strukturierung von Benutzungsschnittstellen vornimmt, die es erlaubt, in separaten, sich gegenseitig ergänzenden Spezifikationsschritten, die angesprochenen Aspekte explizit zu behandeln und miteinander in Bezug zu setzen. Es entsteht dabei ein formales Modell, welches das strukturelle Design als zentralen Gestaltungsspielraum aufspannt. Dieses kann dann schrittweise angereichert werden um das visuelle Design der beteiligten Objekte - entweder als Standardelemente oder als applikationsspezifische Graphiken -, und um die Anbindung spezieller Interaktionstechniken. Die im Kontrollmodell entstehende Strukturierung in Objekte mit Attributen und Methoden kann unmittelbare Vorgabe für die Softwarestruktur der gesamten Applikation sein.

Dadurch daß das Kernmodell formal spezifiziert wird, profitiert man von den bekannten Vorteilen eines expliziten Modells: das Modell kann **analysiert** werden, es kann gegen eine formale Spezifikation **verifiziert** werden, man kann daraus **Prototypen** ableiten und es **separat** von der Programmentwicklung und anderen Designschritten entwickeln. Das objektorientierte Kontrollmodell-System ist z. Zt. in Implementation als Sprache *ODSN (Objektorientierte Dialogspezifikations-Notation)*. Der Vorläufer von ODSN, die Sprache DSN/2 [7], ist nicht objektorientiert, verfügt aber über das lauffähige Werkzeug KAP [8] zur Definition und Ausführung von Prototypen von DSN/2-Kontrollmodellen. Die Arbeit mit KAP zeigt, daß Modularisierung und Strukturierung in Objekte unbedingt erforderlich ist, um die Komplexität realistisch großer Modelle zu beherrschen. Die (positiven) Erfahrungen der Modellierung von Geräte-Schnittstellen mit KAP motivierten uns, DSN/2 zu ODSN zu erweitern und eine entsprechende - zur Zeit noch in Arbeit befindliche - Werkzeugumgebung zu entwickeln.

2 Zustände: Die zentrale Bedeutung des Kontrollmodells

Die Beschreibung der möglichen Zustände, die eine Benutzungsschnittstelle annehmen kann, und die zwischen diesen Zuständen gegebenen Übergänge, beinhalten zwangsläufig wesentliche Designentscheidungen. Der „aktuelle" Zustand bestimmt, welche Aktionen der Benutzer (oder auch das System) durchführen kann und welche nicht - eine ganz entscheidende Information für den Benutzer: „Wo bin ich ? Wie kam ich hierher ? Warum kann ich etwas nicht tun ?". In modernen Benutzungsschnittstellen findet sich dieses Phänomen beispielsweise

- beim „Ausgrauen" von Menüpunkten,

- in kontextabhängig sich verändernden Menüs,

- in objektabhängigen *Drop-Down*-Menüs,

- oder beim impliziten Moduswechsel durch Bewegen der Maus.

Für gute *Benutzbarkeit („Usability")* ist das Design eines gut verständlichen oder auch möglichst unauffälligen Kontrollverhaltens daher unumgänglich. ODSN erlaubt es, dieses Verhalten in einer objektorientiert strukturierten Sprache explizit zu spezifizieren. Das resultierende *objektorientierte Kontrollmodell* wird damit zur zentralen Instanz der Benutzungsschnittstelle, aber auch des Entwicklungsprozesses. Durch Anreicherung, Verfeinerung und Modifikation entsteht schrittweise ein Modell, das

- es erlaubt, früh im Entwicklungsprozeß **Prototypen** zu bauen,
- damit das Kontrollmodell und seine Auswirkung auf die **Usability** zu testen,
- es **formal** zu **analysieren** auf mögliche Designfehler,
- und es als Ausgangspunkt für eine **Generierung** oder auch (nur) eine **formale Spezifikation** für die Programmiertätigkeit zu verwenden.

Die schon erwähnten klassischen Ansätze zur Spezifikation von Dialogen verwendeten ebenfalls den Zustandsbegriff; dabei wurden die Zustände einer Benutzungsschnittstelle explizit dargestellt. In modernen Oberflächen würde dabei eine unbehandelbar hohe Zahl von Zuständen entstehen - daher verloren diese Ansätze an Bedeutung. Mit PPS [10] hat Dan Olsen die Idee eingebracht, statt dessen Objekte mit „lokalen" Zuständen zu definieren, deren Gesamtheit dann den „großen" Zustandsraum bestimmt, ohne diesen jemals explizit komplett aufzuzählen (oder zu zeichnen). Diesen Grundgedanken, der in die Dialogbeschreibungssprache DSN [3] einfloß, wird jetzt in ODSN ausgebaut zu einem komplexen, objektorientierten Kontrollmodell. Wir zeigen, daß dies eine relevante Abstraktionsebene ist, die separat spezifizierbar, definierbar, ausführbar und analysierbar ist. Außerdem definiert sie klare Schnittstellen zur Präsentation (Ausgabe und Eingabe) und zur Applikation.

3 Das objektorientierte Kontrollmodell

3.1 Beschreibung einzelner Objekte

3.1.1 Zustandsraum

Eine ODSN-Spezifikation beschreibt Objekte der Benutzungsschnittstelle, die zu jedem Zeitpunkt jeweils einen wohldefinierten Zustand besitzen und diesen Zustand durch das Eintreten von Ereignissen wechseln können. Dabei wird für jedes Objekt der Zustandsraum durch *Felder* spezifiziert; dies sind Listen von sogenannten *Flags*. Zu jedem Zeitpunkt ist jeweils **genau eines** der Flags eines Feldes „gültig" - es können niemals zwei Flags eines Feldes gleichzeitig angenommen werden. So ist etwa durch die Felder HighlightState, BorderState und Colour, notiert als

```
HighlightState    (#NotHigh #High)
BorderState       (#Normal #Thick)
Colour            (#Red #Blue #Yellow #Green)
```

festgelegt, daß ein Objekt mit diesen Feldern als Zustandsraum-Definition unter anderem im Zustand (#High #Thick #Green) sein und insgesamt 16(=2*2*4) verschiedene Zustände annehmen kann.

3.1.2 Ereignisse

Der Zustand eines Objektes ändert sich ausschließlich durch das Eintreten von Ereignissen. Dies kommen typischerweise von **außerhalb** des Objektes und werden entweder vom Benutzer oder von kooperierenden Objekten ausgelöst bzw. erzeugt. Diese Ereignisse werden im Objekt - als INPUT EVENTS - definiert. So legt

```
INPUT EVENTS
    Highlighting  (iHighlight   iDeHighlight)
    Border        (iNormalize   iThicken)
    Picking       (iHit)
```

fest, daß das Objekt auf die angegebenen Eingabeereignisse reagieren kann.

3.1.3 Zustandsänderungen

Die Zustandsänderungen, die ein Ereignis hervorruft, werden durch die *Übergangsregeln* eines Objekts definiert. Regeln enthalten auf der „linken" Seite eine Aufzählung der Flags, die gesetzt sein müssen, sowie ein Ereignis, das die Regel schalten läßt; auf der „rechten" Seite steht die Liste der Flags, die durch das Schalten der Regeln gesetzt werden. Außerdem kann eine Regel rechts auch Ereignisse (eigene oder solche anderer Objekte) enthalten, die dann durch das Schalten der Regel ausgelöst werden. So besagt die Regel

#Green #NotHigh iHighlight → #High #Blue,

daß bei gesetzter Farbe #Green und falls das Objekt nicht-„highlighted" (#NotHigh) ist, das Ereignis iHighlight einerseits die Farbe auf #Blue ändert und außerdem der Zustand „highlighted" (#High) gesetzt wird. Für jeden aktuellen Zustand kann man dem Kontrollmodell durch Analyse entnehmen, welche Ereignisse in diesem Zustand überhaupt zum Schalten einer Regel führen können. Wäre etwa die obige Regel die einzige Regel mit dem Ereignis iHighlight, so würde dieses Ereignis zum Beispiel bei Farbe #Yellow oder im Highlight-Zustand #High nicht zum Schalten irgendeiner Regel führen und wäre somit ein nicht erlaubter Übergang. Insgesamt entsteht durch die Definition von Zustandsraum, verarbeiteten Ereignisse und Regeln eine Beschreibung von Objektverhalten, die festlegt,

- welche Zustände ein Objekt überhaupt annehmen kann (Zustandsraum),
- welche Ereignisse prinzipiell von diesem Objekt entgegengenommen werden (Aufzählung der INPUT EVENTS),
- welche Ereignisse genau in jedem Zustand als Eingaben „erlaubt" sind (Analyse der linken Seiten der Regeln),
- wie der dann resultierende neue Zustand aussieht (Analyse der rechten Seiten der Regeln).

3.2 Objektstrukturen

Derartige Beschreibungen können als Typbeschreibungen (Klassen) oder als individuelle Objektdefinitionen in ODSN spezifiziert werden. Klassenbeschreibungen können dann beliebig oft instanziiert werden. Die Notation für *statische Instanziierung* sieht wie folgt aus:

CLASS Rectangle   ... ENDCLASS
OBJECT Canvas;   OBJECT r1, r2: Rectangle; ... ENDOBJECT

Dem Objekt Canvas stehen unter den Namen r1 und r2 zwei Objekte vom Typ Rectangle zur Verfügung, die in der darüber angedeuteten Klassenbeschreibung definiert sind. Um auch dynamische Effekte - wie Objekterzeugung und -zerstörung - modellieren zu können, gibt es auch die Möglichkeit der dynamischen Instanziierung über Regeln: Nach Definition eines Objekttyps kann ein Objekt durch Schaltenlassen einer Regel erzeugt, zerstört, aktiviert und deaktiviert werden. Die folgende Regel etwa erzeugt ein neues Objekt vom Typ Rectangle, wenn sie schaltet: #CreatingR iPos → NEW Rectangle.iCreate.

NEW ist ein Schlüsselwort; iCreate ist ein vordefiniertes Ereignis, das implizit für jede Klasse definiert ist. Analog kann man einem derart erzeugten Objekt u.a. die Ereignisse iDestroy, iActivate und iDeActivate schicken. Dieses Konzept beinhaltet auch das Ansprechen dynamisch erzeugter Elemente, soll hier aber nicht genauer behandelt werden (s. jedoch [12] und Abschnitt 3.4 Datenfluß).

3.3 Objektkooperation

Damit die Objekte einer solchen Spezifikation kooperieren können, müssen sie Informationen austauschen. Wir hatten bereits gesagt, daß jedes Objekt ausschließlich durch Auslösen „seiner" Eingabeereignisse agieren kann. In ODSN können, aufgrund der Objektstruktur, Ereignisse sowohl **vom Benutzer** als auch **von anderen Objekten** kommen; zur Definition einer Benutzungsschnittstelle ist es unumgänglich, eindeutig festzulegen, welche Ereignisse der Benutzer auslösen können soll, und welche intern erzeugt werden. Zu diesem Zweck markieren wir die Eingabeereignisse als „extern" (vom Benutzer kommend) und „intern" (von anderen Objekten kommend). Extern ist die Voreinstellung, daher fügt man i.w. das Schlüsselwort intern zu den internen Eingabeereignissen zu. So beschreibt die folgende Modifikation der Eingabeereignisse der Objekte vom Typ Rectangle, daß vom Benutzer lediglich das Ereignis iHit empfangen wird, während die anderen Ereignisse ausschließlich von anderen Objekten kommen können:

```
class Rectangle
      INPUT EVENTS
            intern Highlighting  (iHighlight, iDeHighlight)
            intern Border        (iNormalize, iThicken)
            Picking              (iHit)
      ...
      ENDCLASS
```

Das Objekt Canvas kann nun durch Verwenden von r1.iThicken auf der rechten Seite einer Regel dem Rectangle-Objekt r1 den „Befehl" geben, seinen Rand fett darzustellen.

Andererseits bedingt Objektkooperation auch, daß ein Objekt (etwa r1) einem anderen Objekt (etwa Canvas) Informationen über seinen Zustand geben können muß. Dies geschieht durch **Bekanntmachen** einiger (oder aller) Flags eines Objekts - syntaktisch wiederum mit dem Schlüsselwort intern. Ist etwa der HighlightState als intern vereinbart,

```
      intern HighlightState  (#NotHigh #High),
```

dann kann man in Canvas etwa das Flag r1.#High auf der linken Seite einer Regel abfragen. Gemäß dem Axiom, daß ein Objekt nur durch Ereignisse aktiv beeinflußt werden kann, darf dieses Flag jedoch nicht von außen gesetzt werden - also auf der rechten Seite einer Regel auftauchen.

Durch die Festlegung der Schnittstellen von Objekten über das interne Bekanntmachen einiger Eingabeereignisse und einiger Flags ist ein klares Konzept der Kooperation von Objekten definiert. Diese folgt der Idee, daß die „Bedienung" eines Objektes durch den menschlichen Benutzer intern nachgebildet wird: Der Benutzer kann Ereignisse „auf" einem Objekt auslösen und einige der Zustandsgrößen (Flags) in ihrer Veränderung beobachten. Genauso können nun andere Objekte Ereignisse auslösen und einige der Zustandsflags in ihren Regeln abfragen. Dadurch entsteht in der Beschreibung eine strukturgebende Symmetrie zwischen interner und externer Kooperation. Gleichgültig, ob ein Objekt letztlich „direkt" vom Benutzer bedient wird oder über ein Mittlerobjekt erreicht wird, bleibt die Denkweise beim Entwickeln eines Objektes gleich.

3.4 Datenfluß

Vielfach wird bei Ereignissen, die der Benutzer auslöst, nicht nur das Ereignis selbst, sondern auch „kleine" Zusatzinformation transportiert. Ein Mausklick etwa trägt typischerweise die Koordinaten des getroffenen Punktes mit sich. Diese Beobachtung machen wir zum generellen

Prinzip für die Kooperation von Objekten untereinander und mit dem Benutzer. So mag es etwa von Bedeutung für das Objekt Rectangle sein, bei welchen x- und y-Koordinaten das Ereignis iHit auftritt. Wir erweitern daher das „reine" Eingabeereignis iHit um die Angabe, daß auch ganzzahlige Werte x und y mitgeliefert werden: Picking (iPos {int x, y}).

An diesem Punkt kümmern wir uns nicht darum, **wie** diese Werte entstehen - wir legen auch nicht fest, daß iHit durch eine Mausoperation realisiert wird. Wir sagen nur **daß** das Ereignis iHit unter den Namen x und y ganze Zahlen „mitliefert". Als Datentypen stehen hier Standardtypen, aber auch der Typ oid zur Verfügung, was für *Objektidentifikator* steht.

Um Datenflußwerte festzuhalten, geben wir auch den Feldern - wie den Ereignissen - die Möglichkeit, Werte zu speichern. Damit ergibt es sich nun, daß die Ereignisse und Flags, die „links" in einer Regel auftreten, Daten „mitbringen" können, die dort in der Regel zur Verfügung stehen. Sie können in mehrfacher Weise verwendet werden:

1. Als Test, ob eine Regel schalten kann durch einen einfachen Vergleich der Werte etwa mit einer Konstanten.
2. Zum Setzen der Zusatzinformation eines Flags oder erzeugten Ereignisses auf der rechten Seite der Regel.
3. Zum zielgerechten Ansprechen von Objekten, deren Objektidentifikation bekannt ist, um diesen Ereignisse zu senden.

Bis zu diesem Punkt der Definition des Kontrollmodells ist eine sinnvolle, in sich geschlossene Einheit entstanden, die einen wesentlichen Gestaltungsspielraum aufspannt, der die Definition der Benutzungsschnittstelle und ihre gute Benutzbarkeit direkt beeinflußt. Nicht behandelt wurden bislang die Aspekte:

- **Visuelles Design** (wie sehen die Objekte aus?)
- Mit welchen **Interaktionstechniken** agiert der Benutzer ?
- Wie ist die **Applikation** angeschlossen ?
- Wie sieht die **Softwarestruktur** der Benutzungsschnittstelle mitsamt Anwendung aus ?

Für die Anreicherung des Gesamtmodells um diese Informationen schafft jedoch gerade das Kontrollmodell ideale Voraussetzungen.

4 Anbindung der Präsentation

4.1 Der Gestaltungsspielraum

Nach der Definition des Kontrollmodells liegen einige Informationen über die visuelle Präsentation und die Eingabetechniken bereits fest:

- Welche Objekttypen mit welchem Zustandsraum müssen überhaupt dargestellt werden?
- Welche Benutzerereignisse diesen Objekten vom Benutzer „geschickt" werden; mit welchen Zusatzinformationen müssen diese behaftet sein?

Nicht festgelegt ist im Kontrollmodell,

- Wie sehen die Objekttypen genau aus (s. Abschnitt 4.2 Objektgestaltung)?
- Wann genau sind welche Objekte sichtbar (s. Abschnitt 4.3 Objektsichtbarkeit)?
- Wie werden die Benutzereignisse „auf" den Objekten ausgelöst (s. Abschnitt 4.4 Benutzerereignisse)?

4.2 Objektgestaltung

Um die Gestalt eines Objekttyps festzulegen, steht dem Entwickler der im Kontrollmodell definierte Zustandsraum des Objekttyps zur Verfügung. Wir unterscheiden explizit bei der Zuordnung eines „Objektaussehens" die *Gestalt* und die *Projektion* dieser Gestalt auf ein Darstellungsmedium - etwa eine zweidimensionale Graphikfläche. Den Objekttypen des Kontrollmodells selbst ordnen wir eine - mit den Daten des Zustandsraums parametrisierte - Gestalt zu. Dabei muß die Gestaltform eines Objekttyps so gewählt, daß die Darstellungsfunktion, mit der diese Gestalt später dem Benutzer gezeigt werden soll, diese auch darstellen kann. Mögliche Gestaltformen bei Einsatz **konventioneller zweidimensionaler Graphik** sind die dort typischen Elemente wie Kreise, Linien, Rechtecke, Bitmaps, Kombinationen hiervon, aber auch flächige Widgets aus Interaktionsbibliotheken; ist die Darstellung die perspektivische Projektion einer **dreidimensionalen (virtuellen) Welt**, dann sind räumliche Objekte, wie Kugel, Kegel, Quader, Flächen und wiederum Kombinationen hiervon Kandidaten für Objektgestalten.

Die Gestaltdefinition des Objekttyps kann beliebige Komponenten des Zustandsraums - aber auch nur diese Information ! - benutzen, um die Gestalt zu beeinflussen. Bei dieser Festlegung wird also entschieden, ob eine intern gespeicherte Zustandsinformation dem Benutzer nach außen sichtbar gemacht wird, oder nicht. Unser Beispiel-Objekttyp mit dem Zustandsraum

```
HighlightState   (#NotHigh #High)
BorderState      (#Normal #Thick)
Colour           (#Red #Blue #Yellow #Green)
```

etwa könnte als 2D-Gestalt ein Rechteck besitzen, dessen Farbe vom Feld Colour gesetzt wird, dessen Rand, je nach gesetztem BorderState-Flag normal oder fett ist, und dessen HighlightState als Information zur inversen Darstellung ausgewertet wird. Wir könnten aber auch einen 3D-Quader mit ähnlicher Auswertung der Felder für eine 3D-Welt festlegen.

4.3 Objektsichtbarkeit

4.3.1 Darstellende Objekte

Die oben angesprochene Projektionen oder Darstellungsfunktionen, die Objektgestalten erst auf einem entsprechenden Medium sichtbar machen, werden ebenfalls Objekten zugeordnet. Darstellende Objekte werden entsprechend ausgezeichnet; neben ihrer Gestalt, die wie üblich definiert wird, haben sie die Eigenschaft, daß sie ganz oder teilweise dem Darstellen anderer Objekte dienen. Dieses Phänomen findet sich vielfach in graphischen Benutzungsschnittstellen, Paradebeispiel sind natürlich alle denkbaren Varianten von **Fenstern**, die typischerweise auf einer rechteckigen Zeichenfläche die Gestalt anderer Objekte zeigen.

In Ergänzung zum bislang definierten Kontrollmodell fügen wir die Information, welches Objekt welche anderen Objekte zeigt, in Form sogenannter shown-by-Klauseln hinzu. So könnten wir etwa mit

```
OBJECT Canvas;
    OBJECT r1: Rectangle SHOWN BY Canvas;
    OBJECT r2: Rectangle SHOWN BY r1;
    ...
ENDOBJECT
```

spezifizieren, daß das Rechteck r1 von Canvas gezeigt wird, während das Rechteck r2 von Rechteck r1 dargestellt wird. Auch hier enthält das Kontrollmodell lediglich die Information **daß** diese Darstellungsbeziehungen existieren - nicht wie sie genau aussehen.

Durch die SHOWN-BY-Klauseln eröffnen wir uns die Möglichkeit, schon bei der Definition des Kontrollmodells festzulegen, welche Abhängigkeiten in der graphischen Darstellungshierarchie bestehen. Dadurch können wir diese auf Benutzungsprobleme hin abzuprüfen, etwa ob vom Kontrollmodell her erlaubte Benutzerereignisse sich auch nur auf „sichtbare" Objekte beziehen, oder ähnliches.

4.3.2 Darstellungsfunktionen

Haben wir ein Objekt als „darstellend" im Kontrollmodell festgelegt, müssen wir eine Darstellungsfunktion dafür auswählen. Die gewählten zu zeigenden Gestalten müssen dabei zu dieser Darstellungsfunktion „passen". So können wir zum Beispiel festlegen,

- daß eine Teilfläche des Objektes Canvas sich wie ein Fenster verhält (also eine zweidimensionale Zeichenfläche aufspannt, die man z.B. scrollen und pannen kann), oder
- eine solche Fläche als Kamera auf eine 3D-Szene agiert, die man dann im Objekt Canvas einstellen, fokussieren, zoomen etc. kann.

Die *Darstellungsfunktion* beinhaltet die Information,

- wie **Koordinaten** der Objektgestalten in solche der Präsentation „umgerechnet" werden,
- wie mit **Farben** und Oberflächen etc. umgegangen wird, und
- wie gleichzeitige Darstellungen **mehrerer** Objektgestalten behandelt werden.

Die Koordinatentransformation beinhaltet Effekte wie Zooming und Panning im 2D-Fall, aber auch Blickpunkt oder Blickrichtung bei dreidimensionalen Szenerien. Auch Verzerrungen, wie Fish-Eye-View, oder verschiedene semantische Sichten können hierauf abgebildet werden. Die Abbildung mehrerer Objektgestalten besitzt Eigenschaften, die nicht aus der Definition der einzelnen Gestalten heraus abgeleitet werden können. Ein Beispiel aus dem Bereich 2D-Fenster wäre etwa der Umgang mit überlappenden Teilobjekten, die durch „z-Werte" unterschieden werden, wobei das Fenster - also das darstellende Objekt - diese z-Staffelung verwalten muß.

Die Darstellungsfunktion kann nun wiederum - genau wie die Gestalt der Objekte - aus dem Zustandsraum des darstellenden Objektes heraus parametrisiert werden. So könnte der Zoomfaktor, die Koordinate der linken oberen Ecke einer Zeichenfläche, oder die Position von Lichtquellen und/oder Kamera einer dreidimensionalen Szenerie dadurch beeinflußt werden. Zusätzlich kann der Zustandsraum auch die eigentliche - nicht-darstellende - Gestalt eines solchen Objekttyps beeinflussen. Dazu können im Fall von Fenstern etwa die Farbe und Dicke der Umrandung, oder auch der Text im Fenstertitel zählen. Für die Präsentation gilt, daß das Kontrollmodell den Rahmen aufspannt, innerhalb dessen sich die visuelle Gestaltung abspielt; dabei handelt es sich um strukturelle Eigenschaften, die sinnvoll abgespalten werden und somit separat bereits Entwurfsgegenstand sein können.

4.4 Benutzerereignisse

Nachdem definiert ist, wie die Objekte gestaltet sind und durch welche anderen Objekt sie wie dargestellt werden, kann man definieren, welche Benutzerereignisse in welcher Weise „auf" diesen Objekten ausgelöst werden können. Auch hier ist das Kontrollmodell eine präzise Vorgabe dafür, **welche** als extern angegebenen Ereignisse mit **welchen** zugehörigen Datenflüssen realisiert werden müssen, ohne zu präjudizieren wie diese Zuordnung aussieht. Die technische Umsetzung der Benutzungsschnittstelle - etwa als 2-D-Fensteroberfläche - bedingt das Vorhandensein bestimmter Ereignistypen, die auf den Basiselementen der Umgebung basieren - etwa Mausklick, Doppelklick, Entry-Events, Exit-Events etc. Diese „technischen" Ereignisse müssen nun den „logischen" Ereignissen des Kontrollmodells zugeordnet werden. Entsprechend der zweigeteilten Darstellungskonzeption (Gestalt, Darstellungsfunktion) muß hierbei ebenfalls zweistufig vorgegangen werden: Die auftretenden technischen Ereignisse beziehen sich zunächst nur auf die darstellenden Objekte. Dieses Objekt muß analysieren,

- ob es „selbst" gemeint ist (etwa wenn ein Fensterrand mit der Maus „verzogen" wird),
- oder ob es das Ereignis an eines (und wenn ja, an welches) der dargestellten Objekte weiterreichen soll (etwa wenn eine Rechteck in der Zeichenfläche des Fensters angeklickt wird). In diesem Fall muß das technische Ereignis, das auf der Darstellung eingetreten war, „umgerechnet" werden in ein Ereignis auf der Objektgestalt des getroffenen Objektes.

Ähnliche Vorgänge werden in modernen Fenstersystemen ständig durchgeführt: Das Anklicken eines Bildschirmpixels (mit absoluten Bildschirmkoordinaten), welches aber in der Zeichenfläche eines Fensters liegt, wird umgerechnet auf ein Anklicken einer Zeichenflächenposition in **deren** Koordinatensystem. Dieser Transformationsprozeß ist Eigenschaft der Darstellungsfunktion für die darstellenden Objekte. Zusätzlich zur Gestaltdefinition muß dann noch vereinbart werden, welche Ereignisse auf den Objektgestalten definiert sind, und welchen „logischen" Ereignissen aus dem Kontrollmodell sie zugeordnet werden.

5 Anbindung der Applikation

Die Kopplung zwischen Applikation und Benutzungsschnittstelle muß bei modernen direktmanipulativen Schnittstellen sehr eng sein, um effizientes *semantisches Feedback* leisten zu können. Die Applikation wirkt in das Kontrollmodell hinein; umgekehrt wird die Applikation vom Kontrollmodell angestoßen, um semantische Berechnungen durchzuführen. Wir bilden diese Mechanismen ab auf ein ereignisorientiertes Konzept, wie es im Kontrollmodell ohnehin schon vorhanden ist.

5.1 Die Schnittstelle zur Applikation: Applikationsereignisse

Typischerweise erfolgt durch ein Benutzerereignis der Anstoß zum Aufruf einer Applikationsroutine. Wir gehen davon aus, daß die Applikation sich als Sammlung von Routinen präsentiert - im Sinne der üblichen *Callback*-Strukturen. Der Aufruf einer Applikationsroutine korrespondiert zu **zwei** Ereignissen im Kontrollmodell: dem **Aufruf** der Routine (und Versorgung mit Parametern) und der **Fertigmeldung** der Routine (mit Rücklieferung von Ergebnissen). Um dies im Kontrollmodell einzufangen, führen wir den Ereignistyp „Applikationsereignis" ein. Ein Applikationsereignis definiert einerseits einen Namen für das Ereignis und legt fest, welche Routine mit welchen Daten aufgerufen wird und welche Daten diese Routine zurückliefert. Der Aufruf einer Routine wird dann durch Erzeugen des Aufruferereignisses (auf der rechten Seite von Regeln) im Kontrollmodell spezifiziert; die Rückmeldung der Applikationsroutine mit den Ergebnissen kann auf der linken Seite von Regeln als Ereignis abgefangen werden. Durch diese Konzeption ist grundsätzlich Nebenläufigkeit zwischen Benutzungsschnittstelle und Applikation modellierbar. Es ist eine Eigenschaft des Kontrollmodells, ob es nach Absetzen einer Berechnungsfunktion der Applikation auf das Ergebnis „wartet" oder nicht.

Es zeigt sich, daß das Kontrollmodell gut geeignet ist, um die Schnittstelle zwischen Oberfläche und Applikation zu definieren: Hat man sich auf die zu verwendenden Applikationsereignisse geeinigt, kann separat an Oberfläche und Applikation gearbeitet werden. Dem Applikationsentwickler liegen dann Funktionsprototypen als Vorgaben für seine Arbeit vor; für den

Entwickler der Benutzungsschnittstelle ist eindeutig definiert, **wie** diese Applikationsfunktionen angebunden sind.

5.2 Objektorientierte Strukturierung der Applikation

Die Angabe von Applikationsereignissen in Objekten bzw. Klassen legt es nahe, die Applikation in Analogie zum Kontrollmodell zu strukturieren. Verwendet man ein Applikationsereignis in der Definition einer Klasse oder eines Objektes, kann man dies als Vorgabe dafür betrachten, daß es in der eigentlichen Applikation eine entsprechende Klasse bzw. ein entsprechendes Objekt gibt, das eine derartige Funktion als Methode besitzt. Das schließt nicht aus, das in der Applikation zusätzliche, für die Benutzungsschnittstelle unsichtbare Klassen und Objekte existieren. Die Objektstruktur des Kontrollmodells liefert aber eine sinnvolle Vorgabe für die objektorientierte Strukturierung der Applikation.

6 Werkzeuglandschaft

In Anlehnung an das bereits existierende Prototypwerkzeug KAP [8] ist eine Werkzeuglandschaft in Arbeit, die mit verschiedenen dedizierten Editoren und angeschlossenen Generierungskomponenten die Erzeugung von Prototypen, aber auch Teile der endgültigen gesamten Applikation erlauben wird. Zur Zeit ist die Definition der textuellen Version von ODSN abgeschlossen, und wir arbeiten an einem Simulator. Bis auf eine Generatorkomponente ist eine vergleichbare Werkzeuglandschaft im Tool KAP implementiert; allerdings für das nicht objekt-orientierte und auch in anderer Hinsicht „einfachere" DSN/2.

Literatur

[1] B. d'Ausbourg. G. Durrieu. P. Roche: Deriving a formal model of an interactive system from ist UIL description in order to verify and test its behaviour. In: Proceedings of the 3rd Eurographics Workshop on Design. Specification and Verification of Interactive Systems (DSV-IS'96). Namur. 1996.

[2] J. Acott et al: A Formal Description of Low Level Interaction and its Application to Multimodal Interactive Systems. In: Proceedings of the 3rd Eurographics Workshop on Design. Specification and Verification of Interactive Systems (DSV-IS'96). Namur. 1996.

[3] M.B.Curry. A.Monk: Dialogue modelling of graphical user interfaces. Technical Report. Department of Psychology. University of York. 1991.

[4] M. Green: A Survey of Three Dialogue Models. In: ACM Transactions on Graphics 5 (1986). 3. 245-275

[5] P. Griebel. M.Pöpping: Motifation - Ein Benutzungsschnittstellen-Entwicklungssystem. In: Informatik Fachberichte 293. Springer Verlag. 1991. 445-454.

[6] C. Janssen: Dialognetze zur Beschreibung von Dialogabläufen in graphisch-interaktiven Systemen. In: K.-H. Rödiger (Hrsg.): Software-Ergonomie '93 - von der Benutzungsoberfläche zur Arbeitsgestaltung. Stuttgart. Teubner. 1993. 67-76.

[7] K. Kespohl. G. Szwillus: Prototyping Device Interfaces with DSN/2. In: Proceedings of the 3rd Eurographics Workshop on Design. Specification and Verification of Interactive Systems (DSV-IS'96). Namur. 1996.

[8] K. Kespohl. G. Szwillus: KAP - A Prototyper for Technical Device Interfaces. In: CHI'96 Conference Companion. Formal Demonstration. 1996

[9] B. Myers: User-Interface Tools: Introduction and Survey. In: IEEE Software (1989). 15-23.

[10] D.R. Olsen: Propositional Production Systems for Dialogue Description. In: Human Factors in Computer Systems. CHI'90 Conference Proceedings. 1990. 57-63

[11] E. Schlungbaum. T. Elwert: Modellierung von graphischen Benutzungsoberflächen im Rahmen des TA-DEUS-Ansatzes. In: H.-D. Böcker (Hrsg.): Software-Ergonomie '95 - Mensch-Computer-Interaktion. Anwendungsbereiche lernen voneinander. Stuttgart. Teubner. 1995. 331-348

[12] G. Szwillus: Ein objektorientiertes Kontrollmodell für graphische Benutzungsschnittstellen. In: Softwaretechnik '96. 1996.

[13] H. Uhr et al: OBJECTION: Entwicklung anwendungsspezifischer interaktiver Graphik. Software-Ergonomie '97 - Usability Engineering: Integration von Mensch-Computer Interaktion und Software-Entwicklung. Dresden. 1997

Adresse des Autors:

Gerd Szwillus
Universität - GH Paderborn. Fachbereich Mathematik/Informatik (17)
Fürstenallee 11. 33102 Paderborn
szwillus@uni-paderborn.de

Die Gebrauchstauglichkeitprüfung von Software im Dienste einer gesundheitlichen Prävention[1]

Johannes K. Triebe & Marion Wittstock

Zusammenfassung

Im Hinblick auf die Umsetzung der Europäischen Bildschirmrichtlinie in nationale Regelungen durch Inkrafttreten des Arbeitsschutzgesetzes und der entsprechenden Unfallverhütungsvorschrift wurde ein Verfahren zur Erfassung von Belastungs- und Beanspruchungsfaktoren an Computerarbeitsplätzen entwickelt. Der "Fragebogen für computergestützte Tätigkeiten in Büro und Verwaltung" (FCT-BV) wurde an einer breit gestreuten Stichprobe (N=188) erprobt, wobei sich zeigte, daß software-ergonomische Qualitätsmerkmale einen bedeutsamen Beitrag zum gesundheitlichem Wohlbefinden leisten.

1. Ausgangssituation

Mit der Verabschiedung des neuen Arbeitsschutzgesetzes wird nunmehr auch in Deutschland die Umsetzung europäischer Richtlinien in nationales Recht in die Wege geleitet. In der EU-Rahmenrichtlinie Arbeitsschutz wird den Arbeitgebern als allgemeine Verpflichtung "die Planung der Gefahrenverhütung mit dem Ziel einer kohärenten Verknüpfung von Technik, Arbeitsorganisation, Arbeitsbedingungen, sozialen Beziehungen und Einflüssen der Umwelt auf den Arbeitsplatz" [6, Artikel 6(2)g] aufgetragen und in der EU-Bildschirmrichtlinie steht ausdrücklich: "Der Arbeitgeber ist verpflichtet, eine Analyse der Arbeitsplätze durchzuführen, um die Sicherheits- und Gesundheitsbedingungen zu beurteilen, die dort für die beschäftigten Arbeitnehmer vorliegen" [8, Artikel 3(1)], um anschließend geeignete Maßnahmen zur Ausschaltung festgestellter Gefahren zu ergreifen. Darüberhinaus enthält die Bildschirmrichtlinie Mindestanforderungen an die Hardware-Ergonomie, die Umgebungseinflüsse und erstmalig in einer solchen Richtlinie zum Arbeitsschutz auch *Mindestanforderungen an die eingesetzte Software*, deren zentrale Rolle bei der Gestaltung von Tätigkeiten betont wird.

Zwar wurde die Aufforderung zur systematischen Arbeitsanalyse auch in die nationalen Bestimmungen [11] übernommen, aber wegen der Breite und Vielfalt ihres Geltungsbereichs wurde bewußt darauf verzichtet, ein bestimmtes Analyseverfahren vorzuschreiben. Die Arbeitgeber werden stattdessen ausdrücklich aufgefordert, "sich über den neuesten Stand der Technik und der wissenschaftlichen Erkenntnisse auf dem Gebiet der Gestaltung der Arbeitsplätze zu informieren" [8], was der Berücksichtigung gesicherter arbeitswissenschaftlicher Erkenntnisse entspricht, wie sie in der Bundesrepublik Deutschland üblich ist und auch in der UVV explizit gefordert wird.

Als eine gültige Zusammenstellung der gesicherten (arbeits-)wissenschaftlichen Erkenntnisse zum Thema Bildschirm- und Computerarbeit können die einschlägigen nationalen, europäischen und internationalen Normensammlungen (DIN/CEN/ISO, vgl. insbesondere ISO 9241)

---

[1] Grundlage der vorliegenden Darstellung bildet das von den Autoren für die GBA mbH. Berlin. im Auftrag der Bundesanstalt für Arbeitsschutz. Dortmund. bearbeitete Projekt F1394 [9].

gelten. Und tatsächlich zeigen sich dort, wo in der EU-Bildschirmrichtlinie und in der UVV "Arbeit an Bildschirmgeräten" konkretere Forderungen zur Mensch-Maschine-Schnittstelle aufgestellt werden, enge Bezüge zur ISO 9241, vor allem Part 10, wobei sich die UVV direkter an den Formulierungen der Norm orientiert, was weiter nicht verwunderlich ist, fußt dieser Teil der ISO doch auf dem seit langem in Deutschland anerkannten Teil 8 der DIN 66234.

Alle hier genannten Regelwerke betrachten den Bildschirmarbeitsplatz als System, wie es z.B. in den Arbeitswissenschaften und der Arbeitspsychologie üblich ist [vgl. auch 12]. Am konsequentesten wird dieser Ansatz in der jüngsten Version der ISO 9241 Part 11 "Guidance on usability" verfolgt. In ihr wird beschrieben, daß die Überprüfung der Gebrauchstauglichkeit eines Softwareprodukts als dem bestimmenden Faktor für die Arbeit an Bildschirmgeräten sinnvoll nur *im Rahmen seines Nutzungskontextes* stattfinden kann, welcher sich aus dem Benutzer, seiner Aufgabe, seinen Arbeitsmitteln und der Arbeitsumgebung zusammensetzt. Ferner sind das zu prüfende Softwareprodukt und die Ziele seines Einsatzes zu beschreiben. In den Erläuterungen zur weiteren Untergliederung der einzelnen Komponenten wird dabei ausdrücklich vor der Gefahr gewarnt, in eine technikzentrierte Begrifflichkeit zu verfallen, also Aufgaben des Benutzers und Ziele des Softwareeinsatzes nur als Funktionen des Produkts wiederzugeben statt sie als Schritte zur Zielerreichung darzustellen. Als Kriterien für die Gebrauchstauglichkeit eines Softwareprodukts und die mit ihm zu erreichende Arbeitsqualität werden in ISO 9241 Part 11 Effektivität, Effizienz und Zufriedenheit vorgeschlagen [siehe im einzelnen 3, Teil 11, 3.1-3.5].

Einige der in den Beispielen von ISO 9241-11 genannten Maße (wie Anzahl der Versuche bis zur fehlerfreien Ausführung oder Zeitmessungen zur Aufwandsbestimmung) scheinen vielfach zwar eher für kurzzeitige Laboruntersuchungen an einfachen Aufgaben geeignet zu sein, denn als Korrelat der Beanspruchung des Benutzers, die über einen längeren Zeitraum des regelmäßigen Arbeitens mit einem bestimmten Softwareprodukt entsteht. Da aber auch auf andere Maße (z.B. Fehlbeanspruchungen und ihre Folgen) und Erfassungsmöglichkeiten ausdrücklich hingewiesen wird, steht es dem Untersucher frei, auch weitergehende Vorstellung auf der Grundlage von ISO 9241-11 zu entwickeln.

Nach der vorgelegten Konzeption hängt es von der jeweils konkreten Konstellation der Nutzungsbedingungen ab, ob eine Software die an sie gestellten Anforderungen erfüllt, denn die allgemeinen Bewährungskriterien müssen in Abhängigkeit von den aktuellen Nutzungsbedingungen spezifiziert werden. Ein solches Vorgehen erscheint u.U. zwar aufwendig, aber es ist methodisch begründet, durchaus durchführbar und aussichtsreich. Obwohl im Unterschied zu technischen (auch z.B. hardwarebezogenen) Normen ein bestimmtes Prüfverfahren nicht vorgeschrieben wird (und sinnvollerweise auch nicht werden sollte), sind in ISO 9241 - 11 hinreichend klare Anforderungen formuliert; außerdem wird ausdrücklich auf die Möglichkeit hingewiesen, subjektive Urteile des Benutzers als Indikatoren heranzuziehen [vgl. 3, Teil 11, 5.3.1].

Im ganzen gesehen kann die Umsetzung der software-ergonomischen Normen in Prüf- und Meßvorschriften auf Modellvorstellungen zurückgreifen, die sowohl in der sozialwissenschaftlichen Evaluationsforschung [vgl. z.B. 7] als auch in der psychologischen Eignungsdiagnostik [vgl. 13] seit langem bekannt sind. Generell läßt sich u.E. die These vertreten:

Soweit Eignungsdiagnostik möglich und praktizierbar erscheint, kann nicht ernsthaft behauptet werden, die ergonomische 'Eignung' von Software - die sich in dieser Analogie um die 'Mitarbeit' in einem Unternehmen 'bewirbt' - sei letzten Endes einer auf empirische Methoden gestützten Überprüfung nicht zugänglich.

Das gesamte, zur ergonomischen Eignungsprüfung bzw. Evaluation von Software anhand der ISO- bzw. CE-Normen erforderliche Vorgehen läßt sich zusammenfassend in einem Modell darstellen, das sich eng an einen von DZIDA [2] in die Normenentwicklung eingebrachten Entwurf anlehnt, und in modifizierter Form bei Triebe & Wittstock [9, S.81ff.] dargestellt wurde.

2. Ergebnisse einer Pilotstudie zum Einfluß software-ergonomischer Qualitätsmerkmale auf das gesundheitliche Befinden im Kontext der Bedingungen von Bildschirm-Arbeit

Es ist offensichtlich, daß jede Untersuchung, die sich mit den Beanspruchungsfolgen der Gestaltung von Benutzungsschnittstellen befaßt, stets den Gesamtzusammmenhang von Arbeitsaufgaben, Benutzern und Arbeitssystem einbeziehen muß. Dies schließt einfache Kurzverfahren als Erhebungsinstrumentarium in vielen Fällen aus. Gleichwohl besteht für die Praxis im Hinblick auf die aus der europäischen Bildschirmrichtlinie und ihrer Umsetzung in nationales Recht sich ergebenden Verpflichtungen ein erheblicher Bedarf an Untersuchungsmethoden, die mit vertretbarem Aufwand zu brauchbaren und zuverlässigen Aussagen führen.

Wir gehen davon aus, daß sich auf längere Sicht ein mehrstufiges Vorgehen als die Methode der Wahl erweisen dürfte. Dabei könnte mit Hilfe eines Fragebogens, der die unter Feldbedingungen zu berücksichtigenden Gegebenheiten hinreichend differenziert erfaßt, zunächst ein Überblick über die an den Bildschirm-Arbeitsplätzen eines Unternehmens vorliegenden Beanspruchungen gewonnen werden. Hierauf im Sinne eines Screeningverfahrens aufbauend, wären dann solche Arbeitsplätze oder Einflußfaktoren zu identifizieren, die sich für die Beschäftigten als belastungs-kritisch erweisen und einer genaueren Untersuchung mit Hilfe aufwendigerer objektiver Verfahren bedürfen. Erst in diesem Zusammenhang könnte es z.T. erforderlich werden, auch labormäßig konzipierte Untersuchungen - eventuell unter Einschluß psychophysiologischer Meßverfahren - in das gesamte Beurteilungskonzept mit einzubeziehen.

Aus dem Gesagten folgt, daß es zur Überwindung der derzeitigen Probleme bei der software-ergonomischen Evaluation darauf ankommt, bisherige Ansätze so zu integrieren, daß ein in der Praxis einsetzbares Instrument entsteht, welches dann auf eine langfristig zeit- und aufwandsparende Weise seinen Beitrag zur Verbesserung des Softwarelebenszyklus leisten könnte. Deshalb entschieden wir uns für die Entwicklung eines ökonomisch einsetzbaren Instrumentariums, den *Fragebogen für computergestützte Tätigkeiten in Büro und Verwaltung (FCT-BV)*, der dazu dienen sollte, auf empirischem Wege zunächst an einer möglichst breit gestreuten Stichprobe verschiedenartiger Bürotätigkeiten zu überprüfen, inwieweit systematische Zusammenhänge zwischen ergonomischer Software-Gestaltung und dem gesund heitlichen Wohlbefinden professioneller Benutzer bestehen. Das Verfahren sollte ohne besondere Betreuung und zusätzliche Erläuterungen prinzipiell bei allen erfahrenen Inhabern von Computerarbeitsplätzen einsetzbar sein und umfaßt folgende Themenbereiche:

- Benutzerurteile zur Bewertung der ergonomischen Qualität von Software in Anlehnung an Gestaltungsgrundsätze der ISO/DIN-Normen;

- bürotypische Beeinträchtigungen des psychophysischen Wohlbefindens und ihre mehr als nur kurzfristigen Manifestationen in Form gesundheitlicher Beschwerden;

- arbeits- und kompetenzbezogene Benutzermerkmale;

- aus Sicht der psychologischen Handlungstheorie relevante Tätigkeitsmerkmale;

- Fragen zu Computerbenutzung, Hard- und Softwareausstattung, Schulung und betrieblichen / organisatorischen Rahmenbedigungen des Computereinsatzes.

Von den ausgefüllten Fragebögen konnten letztlich $N = 188$ in die Auswertung einbezogen werden. Entsprechend unserer auf eine breite Streuung ausgerichteten Verteilungsstrategie stammen diese Fragebögen aus 63 verschiedenen Betrieben (wobei auch Familienunternehmen und Freiberufler als einzelner Betrieb gelten). Die größte Gruppe aus einem einzelnen Betrieb umfaßt 14 Freiwillige aus mehreren Abteilungen. Die prozentuale Verteilung einer Reihe wichtiger Stichprobenmerkmale zeigt einen eher hohen Bildungsstand und das Überwiegen von Untersuchungsteilnehmern mit bereits längerer Computererfahrung. Kleine und mittlere Unternehmen sind, wie beabsichtigt, gut repräsentiert (fast 38% stammen aus Unternehmen mit weniger als 100 Beschäftigten), und Nicht-Vollzeitbeschäftigte sind in hinreichender Zahl vertreten. Nach Tätigkeitsarten bildet die bürotypisch wichtige Kategorie der Sachbearbeiter mit fast 24% die größte Einzelgruppe unserer Stichprobe; Sekretariatstätigkeiten sind zusammengefaßt mit knapp 13% vertreten. Spezialistentätigkeiten mit hohem Anteil an Bildschirmarbeit sind durch die Kategorie Technisches Zeichnen / CAD mit gut 11% repräsentiert. Eine Gruppe eher sporadischer Computerbenutzer ist mit 9% durch die Vorgesetztentätigkeiten in unserer Stichprobe vertreten. EDV-Spezialisten, die einerseits dem Bereich Technik / Service / Organisation angehören (13,3%), oder gar andererseits als Software-Entwickler (8%) unmittelbar mit den Problemen einer benutzergerechten, gebrauchstauglichen Programmgestaltung befaßt sein sollten, sind in unserer Pilotstudie stärker vertreten, als dies in weiterführenden Untersuchungen angestrebt werden sollte (sie waren für uns als Sondergruppe von Interesse).

2.1 Software-ergonomische Qualitätsmerkmale

Einen zentralen Bestandteil des FCT-BV bildete die Beurteilung einer längeren Liste aus den Forderungen der ISO-Norm 9241-10 bzw. der VBG 104 abgeleiteter software-ergonomischer Qualitätsmerkmale (SQ-Merkmale), anhand deren der Befragte das bei seiner Arbeit am häufigsten benutzte Programm einstufen sollte. Die Erfassung erfolgte über einen Fragenkomplex von 44 Unterfragen wie im folgenden Beispiel:

Von dieser Software läßt sich sagen...

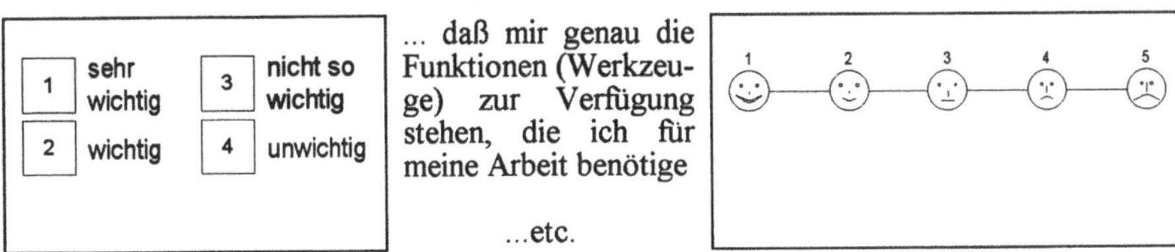

Jedes dieser Statements war somit zum einen hinsichtlich seiner Wichtigkeit (4-stufige Skala) einzustufen, zum andern sollte der Befragte angeben, wie zufrieden er diesbezüglich mit dem von ihm am häufigsten benutzten Programm ist, indem er den entsprechenden Gesichtsausdruck ankreuzte (5-stufige Skala, die zuvor bereits inklusive verbaler Bezeichnungen von 'sehr zufrieden' bis 'sehr unzufrieden' eingeführt worden war).

Zu den bezüglich der Wichtigkeit an erster Stelle rangierenden SQ-Merkmalen gehörten: Die Möglichkeit, Fehlbedienungen rückgängig machen zu können; Verlustsicherheit bezüglich bereits geleisteter Arbeit; die Verfügbarkeit genau der Funktionen (Werkzeuge), die für die Arbeit benötigt werden; daß das Programm fatale Fehler gar nicht erst zuläßt; und daß nichts zu umständlich funktioniert. Die hier aus räumlichen Gründen nicht mitgeteilten Häufigkeitsverteilungen zeigten, daß bei ca. drei Viertel aller SQ-Merkmale mehr als 70% der Einstufungen auf die Antwortkategorien 'sehr wichtig' bzw. 'wichtig' entfielen. Damit läßt sich zunächst schon festhalten, daß die Zielrichtung der software-ergonomischen Qualitätsnormen - aus denen die Statements abgeleitet sind - bei den von uns befragten Benutzern auf eine breite Zustimmung stößt: Die Wichtigkeit der in der ISO-Norm 9241-10 bzw. der VBG 104 enthaltenen Anforderungen steht außer Frage.

Auch die Zufriedenheitseinstufungen zeigten insgesamt ein im positiven Bereich liegendes Gesamtbild, das für einen durchaus schon erkennbaren Erfolg bisheriger software-ergonomischer Bemühungen seitens der Hersteller spricht. Am unzufriedensten waren die Benutzer unter den als wichtig eingeschätzten Merkmalen besonders mit der Qualität der im Handbuch gelieferten Informationen, der Verständlichkeit von Fehlermeldungen, den Möglichkeiten des Datenaustauschs zwischen Programmen und der tatsächlichen Kontextspezifik von Hilfeinformationen.

Nunmehr ist eine Besonderheit der im FCT-BV realisierten Zufriedenheitsmessung näher zu erläutern, mit deren Hilfe wir den Versuch machen, die Urteile der Benutzer im Sinne belastungs- / beanspruchungstheoretischer Überlegungen zu interpretieren und auf diese Weise mit Fragen der Gesundheitsprävention in Verbindung zu bringen, wie dies den Intentionen der europäischen Bildschirmrichtlinie entspricht. Diese Besonderheit ergibt sich aus der Weiterverarbeitung des *kombinierten* Urteils (Doppelbewertung) bezüglich Wichtigkeit *und* Zufriedenheit.

Die folgende Matrix zeigt, wie nach unseren Überlegungen die verschiedenen möglichen Antwortkombinationen zur Wichtigkeit und Zufriedenheit interpretiert werden sollten:

EINSTUFUNG	sehr wichtig	wichtig	nicht so wichtig	unwichtig
sehr zufrieden	Entlastung	Entlastung	Überraschung	Überraschung
zufrieden	Entlastung	Entlastung	Überraschung	Überraschung
teils - teils	Belastung	Belastung	Entlastung	Entlastung
unzufrieden	Belastung	Belastung	Abwertung	Abwertung
sehr unzufrieden	Belastung	Belastung	Abwertung	Abwertung

Nach unserer (Arbeits-)Hypothese, die auch schon im alltagspsychologischen Sinne plausibel erscheint, lassen sich einige Antwortkombinationen ohne weiteres als Faktoren potentieller Belastung oder Entlastung des Benutzers interpretieren.[2] *Belastung* entsteht, wenn dem Benutzer bestimmte software-ergonomische Qualitätsmerkmale zwar 'sehr wichtig' oder 'wichtig' sind, er aber mit ihrer Realisierung bei der von ihm benutzten Software (d.h. seinem hauptsächlichen Arbeitsmittel) 'unzufrieden' oder 'sehr unzufrieden' ist. Im Grenzfall ist dies auch noch für solche Fälle anzunehmen, bei denen er nur 'teils-teils' zufrieden ist.

Entlastung entsteht aus einer weitgehenden Übereinstimmung zwischen Wichtigkeit und Zufriedenheit, da anzunehmen ist, daß in diesen Fällen die Erwartungen bzw. Anforderungen bezüglich des Arbeitsmittels und die bei der täglichen Arbeit festgestellten Benutzungsqualitäten keine bedeutsamen Diskrepanzen aufweisen. Auch hier sehen wir den Grenzfall von Entlastung dort, wo der Benutzer bei Merkmalen, die ihm 'nicht so wichtig' oder 'unwichtig' erscheinen, wenigstens 'teils-teils' zufrieden ist.

Daneben gibt es psychologisch komplexere Antwortkombinationen, die sich nicht so ohne weiteres dieser Interpretation zuordnen lassen. *Abwertung* im Sinne der sprichwörtlichen 'Saure-Trauben-Reaktion' dürfte dort mit im Spiel sein, wo der Benutzer Merkmalen, mit deren Realisierung er 'unzufrieden' oder 'sehr unzufrieden' ist, auch keine weitere Wichtigkeit zumißt (z.B. weil er glaubt, daß sie technisch nicht realisierbar seien). Diese Reaktion vermindert das Diskrepanzerleben und erleichtert es, bestimmte Mängel des Arbeitsmittels hinzunehmen, ohne sich ständig darüber ärgern zu müssen. Als *Überraschung* bezeichnen wir schließlich eine weitere komplexe Reaktion, wenn der Benutzer feststellt, daß ausgerechnet diejenigen Merkmale seines Arbeitsmittels zufriedenstellend realisiert sind, die ihm weniger wichtig erscheinen. Wir nehmen an, daß es sowohl bei der Abwertungs- als auch bei der Überraschungsreaktion sehr stark vom konkreten Fall (z.B. der Konstellation von Tätigkeit, Software und organisationalem Kontext) abhängen dürfte, ob man dahinter eher eine Belastungs- oder eine Entlastungsindikation vermuten kann.

Im Sinne der erläuterten Interpretation berechneten wir für jeden Teilnehmer unserer Untersuchung als zusammenfassenden (quantitativen) Ausdruck seiner Urteile zu den software-ergonomischen Qualitätsmerkmalen eine Variable *"Software-Qualitätsbilanz"* (SQ-Bilanz), indem wir die Differenz aus den als 'entlastend' und den als 'belastend' einzuschätzenden prozentualen Anteilen der kombinierten Wichtigkeits-/Zufriedenheitsbeurteilung bildeten. Wir gehen also - als Arbeitshypothese - davon aus, daß sich 'Belastungen' und 'Entlastungen' gegeneinander aufrechnen lassen, und daß von einer Entlastung des Benutzers durch ergonomisch gestaltete Software umso eher gesprochen werden kann, je positiver eine solche Bilanzierung ausfällt. Die SQ-Bilanz umfaßt einen Wertebereich zwischen -100 und +100 (%). Ist sie negativ,

---

[2] Darüber hinaus läßt sich im Sinne von ISO 9241-11 auch argumentieren, daß gerade im Wichtigkeitsurteil die Anforderungsspezifik und der Nutzungskontext ihren Ausdruck finden. Besonders bei einer in Bezug auf Tätigkeiten und eingesetzte Software so heterogenen Stichprobe wie in der vorliegenden Untersuchung läßt sich ein zusammenfassender Zufriedenheitsindex nur sinnvoll interpretieren, wenn er unter gleichzeitiger Berücksichtigung der Wichtigkeitseinstufung gewonnen wurde.

so liegen mit großer Wahrscheinlichkeit schwerwiegende ergonomische Mängel der beurteilten Software vor, da ein Benutzer in solchen Fällen hauptsächlich die Erfahrung macht, daß für ihn wichtige Merkmale seines Arbeitsmittels nicht oder wenig zufriedenstellend realisiert sind. In unserer Stichprobe ergab sich für die SQ-Bilanz zwar eine deutlich in den positiven Bereich verschobene Verteilung, doch zeigte sich auch, daß für gut 21% der Fälle diese Bilanz negativ ausfiel und somit erhebliche Bedenken in Bezug auf die ergonomische Qualität (Gebrauchstauglichkeit) der dort eingesetzten Software angebracht sind.

2.2 Gesundheitliches Befinden

Neben dem Themenbereich der software-ergonomischen Qualitätsmerkmale bildeten gesundheitliche Beschwerden einen zweiten Schwerpunkt des FCT-BV. Die Erfassung erfolgte anhand 5-stufiger Häufigkeitseinstufungen (1=sehr oft,...,5=nie) über einen Fragenkomplex mit 32 Unterfragen, die im Rückgriff auf die einschlägige Literatur [vgl. besonders 1 und 5] zusammengestellt worden waren.

Die Ergebnisse zeigten, daß im ganzen gesehen gesundheitliche Beschwerden in der von uns erfaßten Stichprobe keine dominante Rolle spielen. Ins Auge fallende häufigere Verwendungen der Antwortkategorien 'sehr oft' und 'oft' fanden sich hauptsächlich bei Müdigkeit, Rückenschmerzen, Abschaltstörungen nach Feierabend, Bewegungsarmut, HWS-Beschwerden, Abwechslungsdrang, Kopfschmerzen und Augenproblemen. Charakteristisch ist zugleich die mit einer Ausnahme (Sodbrennen u.ä.) durchgängige Antworttendenz der beiden Geschlechter, bei denen die Männer stets ein selteneres Auftreten der einzelnen Beschwerden eingestehen. Insgesamt ist festzustellen, daß die in der vorliegenden Untersuchung an führender Stelle rangierenden gesundheitlichen Beeinträchtigungen weitgehend mit Befunden der einschlägigen Literatur übereinstimmen und dem für Büroarbeiten charakteristischen Bild von einerseits streßverwandten psychischen Symptomen, andererseits einer einseitigen körperlichen Belastung geschuldeten physischen Beschwerden entsprechen.

2.3 Multiple Regressionsanalyse zur Vorhersage des psychophysischen Befindens aufgrund ausgewählter Charakteristika der Bildschirmarbeit

Ein präventiver Gesundheitsschutz der an Bildschirmarbeitsplätzen Beschäftigten bildet die zentrale Intention sowohl der europäischen Bildschirmrichtlinie als auch der zu ihrer Umsetzung in deutsches Recht vorgesehenen VBG 104 [11]. Von daher interessiert besonders die Frage, inwieweit sich aufgrund von Daten, die mit Hilfe des FCT-BV erhoben werden können, gesundheitsrelevante Aussagen (Vorhersagen) gewinnen lassen.

Bei dem gewählten Verfahren der multiplen Regressionsanalyse, die hier nur zusammenfassend in ihrem Ergebnis dargestellt werden kann [zu Einzelheiten siehe 9], wird das gesundheitliche Gesamtbefinden (durchschnittliche Häufigkeit gesundheitlicher Beschwerden) als abhängige Größe betrachtet und als Kriteriumsvariable bezeichnet. Sie soll anhand der Werte einiger anderer, unabhängiger Variablen und der zwischen ihnen bestehenden Zusammenhänge (Kovarianzen) geschätzt bzw. vorhergesagt werden.

Für ein schrittweises Analyseverfahren wurden zunächst aus den Fragen des FCT-BV zwölf Variablen als aus theoretischen Gründen wahrscheinlich relevante Einflußgrößen auf das gesundheitliche Gesamtbefinden ausgewählt. Da sich für das Kriterium selbst signifikante Geschlechtsunterschiede fanden (F=9,01 mit p.< .01), war es überdies zwingend, das Geschlecht als 0/1-codierte sog. 'Dummy- Variable' zusätzlich in die Gruppe der Prädiktoren einzubeziehen [vgl. 4, S.145ff.].

In einem ersten Analyseschritt wurden zunächst *sämtliche* Prädiktoren in die Regressionsgleichung aufgenommen. Dabei ergab sich ein multiples R = .65, bzw. ein R^2 = .42 als Ausdruck der damit erklärten Kriteriumsvarianz. Das Ergebnis ist zwar mit einem F= 5,49 hochsignifikant, doch sind einige Prädiktoren in die Regressionsgleichung aufgenommen, deren Beitrag zur Vorhersage nur unwesentlich ist. In einem *zweiten Schritt* wurden daher nunmehr der Reihe nach die *Variablen mit geringem Beitrag entfernt*, bis ein vorzugebendes kritisches Signifikanzniveau (POUT, hier p=.10) unterschritten würde. Im vorliegenden Fall wurden so der Reihe nach die zunächst mit einbezogenen Prädiktoren Schulungs-Qualität, Kompetenzausnutzung, ergänzende Hilfen, Anforderungswechsel, Computererfahrung und Alter aus der Regressionsgleichung entfernt. Das damit erreichte Resultat zeigt zusammenfassend *Abbildung 1*, in der die zur Vorhersage verbliebenen Prädiktoren - geordnet nach abnehmender Größe ihres standardisierten Regressionskoeffizienten (Beta) - dargestellt und zusätzlich ihre Einzelkorrelationen mit dem Kriterium veranschaulicht werden.

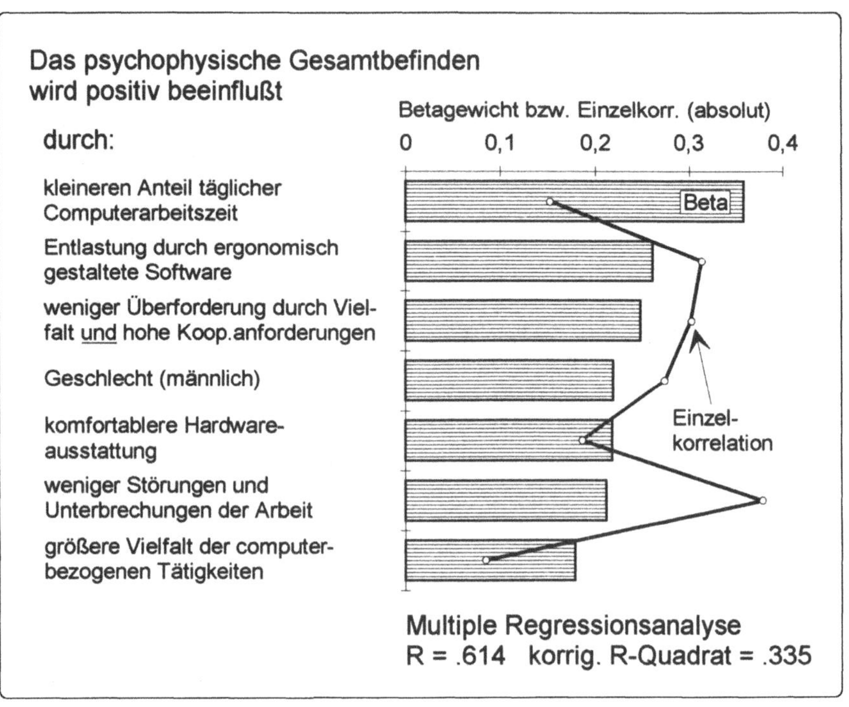

Abb. 1: Ergebnis der multiplen Regression

Im Endergebnis der vorliegenden Analyse läßt sich somit das gesundheitliche Befinden - in der in unserer Untersuchung operationalisierten Form - aus nur *sieben Prädiktoren* mit einem multiplen R = .61 vorhersagen (F = 9,15 mit p < .001). Der im Kriterium erklärte Varianzanteil liegt bei 38% (Stichprobe), und im Sinne eines möglicherweise für die Population zutreffenden Wertes bei 33,5%.

Mehrere Feststellungen sind angesichts dieses Resultats im vorliegenden Zusammenhang von besonderem Interesse:

(1) Software-ergonomische Gestaltungsansätze zur Verbesserung der Benutzungsschnittstelle stehen tatsächlich im Dienste einer gesundheitlichen Prävention. Die europäische Bildschirmrichtlinie, die auf Software bezogenen Normierungsansätze und die Vorschriften der VBG 104 weisen nach unseren Befunden in die richtige Richtung. Es gibt keinen vernünftigen Grund, auf ihre baldige Anwendung zu verzichten.

(2) Der Anteil der Computernutzung an der täglichen Arbeitszeit wird - für sich alleine betrachtet - in seiner Bedeutung für das gesundheitliche Befinden unterschätzt (Einzelkorrelation r =.15) und erhält erst im Gesamtzusammenhang mit den anderen Prädiktoren eine wesentliche Bedeutung (sog. Supressorvariable).

(3) Angesichts der weiter fortschreitenden Verbreitung von Bildschirmarbeit werden zur Gesundheitsprävention vorgeschlagene Konzepte von Mischarbeit, die auf eine Reduktion der am Bildschirm verbrachten Arbeitsanteile abzielen, in Zukunft wenig Chancen einzuräumen sein. Sinnvoller erscheint es, am jeweiligen Arbeitsplatz die Vielfalt der mit dem Computer bearbeiteten Aufgaben zu steigern, da dies nach den vorliegenden Befunden ebenfalls zur Verbesserung des gesundheitlichen Befindens beiträgt.[3]

(4) Während Tätigkeiten mit höherer Aufgabenvielfalt (unabhängig von ihrer Charakterisierung durch Bildschirmarbeit) nach zahlreichen Befunden in der Literatur positiv zu werten sind [vgl. 10], kann nach unseren Feststellungen eine große Aufgabenvielfalt dann zur Überforderung - mit negativen Folgen für die Gesundheit - führen, wenn sie zugleich mit hohen Kooperations- und Kommunikationserfordernissen verbunden ist.

Alles in allem erweist sich der FCT-BV schon in seiner jetzigen Form als ein nützliches Verfahren zur Erfolgskontrolle der mit den internationalen Normen und der europäischen Bildschirmrichtlinie angestrebten Maßnahmen zur gesundheitlichen Prävention. Es dürfte sich lohnen, ihn weiterzuentwickeln, mit den Ergebnissen objektiver Verfahren zu vergleichen, und zu standardisieren. Danach könnte er wahrscheinlich als ein relativ zeitökonomisches Screening-Verfahren eingesetzt werden, dessen Ergebnisse zeigen, ob und in welchen Bereichen eines Unternehmens und der von ihm eingesetzten Software Problemschwerpunkte liegen, die den Einsatz aufwendigerer Untersuchungen erforderlich machen.

Selbstverständlich sind - wie bei jeder an einer Stichprobe entwickelten multiplen Regression - die Ergebnisse weiterer Untersuchungen abzuwarten, wobei insbesondere eine sogenannte Kreuzvalidierung anzustreben wäre, d.h. die Erprobung der hier gewonnenen Regressionskoeffizienten an einer zweiten Stichprobe aus derselben Population. Die Besonderheiten unserer Stichprobe (z.B. mehrheitlich Benutzer mit längerer Computererfahrung) könnten auch dazu geführt haben, daß so unbestreitbar wichtige Gesichtspunkte wie die Qualität der Benutzerschulung oder die Verfügbarkeit arbeitsplatznaher kollegialer Unterstützung bei auftauchenden Softwareproblemen hier keinen relevanten Einfluß auf das gesundheitliche Befinden hatten.

---

[3] In der vorliegenden Untersuchung war eine größere Vielfalt der computerbezogenen Tätigkeiten auch mit einem höheren Anteil täglicher Computerarbeitszeit verbunden (Korrelation r = .30).

Literaturverzeichnis

[1] Becker.P. & Minsel.B: Psychologie der seelischen Gesundheit. Band 1. Göttingen: Hofgrefe. 1982: Band 2. Göttingen: Hofgrefe. 1986

[2] Dzida.W.: Evaluation model - Testing products for conformity with ergonomic standards. Paper submitted to ISO TC159 /SC4 /WG5. St.Augustin: GMD. 1990

[3] Europäische Norm. Entwurf. EN 29241. Brüssel: Europäisches Komitee für Normung. Oktober 1994 (identisch mit ISO Normenreihe 9241)

[4] Gaensslen.H. & Schubö.W.: Einfache und komplexe statistische Analyse. München: Ernst Reinhardt. 1973

[5] Hacker.W. & Richter.P.: Psychische Fehlbeanspruchung - Psychische Ermüdung. Monotonie. Sättigung und Streß. Berlin: DVW. 1980 (2.Aufl. 1984)

[6] Rahmenrichtlinie über die Durchführung von Maßnahmen zur Verbesserung der Sicherheit und des Gesundheitsschutzes der Arbeitsnehmer bei der Arbeit. 89/391/EWG

[7] Ravden.S.J. & Johnson.G.I.: Evaluating usability of human-computer interfaces: A practical method. Chichester: Ellis Horwood. 1989

[8] Richtlinie des Rates über die Mindestvorschriften bezüglich der Sicherheit und des Gesundheitsschutzes bei der Arbeit an Bildschirmgeräten. 90/270/EWG

[9] Triebe.J.K. & Wittstock.M.: Anforderungskatalog für Softwareentwicklung - Auswahl und Anwendung. Schriftenreihe der Bundesanstalt für Arbeitsschutz - Forschung - Fb743. Bremerhaven: Wirtschaftsverlag NW. 1996

[10] Ulich.E.: Arbeitspsychologie. Zürich / Stuttgart: Verlag d. Fachvereine / Poeschel. 1992 (2.Aufl.)

[11] Unfallverhütungsvorschrift (UVV) "Arbeit an Bildschirmgeräten". VBG 104. Grundentwurf März 1995

[12] VDI-Richtlinien Bürokommunikation (Entwürfe): VDI 5005: Software-Ergonomie in der Bürokommunikation. Düsseldorf. 1988. VDI 5010: Strategie und Vorgehen zur Einführung von Bürokommunikation. Düsseldorf. 1988

[13] Wiggins.J.S.: Personality and prediction: Principles of personality assessment. Reading. Mass.: Addison-Wesley. 1973

Adresse der Autoren

Dr. phil. Johannes K. Triebe & Marion Wittstock. Dipl.-Psych. - Kienitzer Str. 114. D-12049 Berlin. Tel. (030) 691 9180. Email: triwitt@berlin.snafu.de

OBJECTION: Entwicklung anwendungsspezifischer interaktiver Graphik

Holger Uhr

Universität - GH
Paderborn

Peer Griebel

Knoll System AG,
Saulgau

Manfred Pöpping

ProDV Software
GmbH, Dortmund

Gerd Szwillus

Universität - GH
Paderborn

Zusammenfassung

Das vorliegende Papier beschreibt die Arbeitsweise des Werkzeuges OBJECTION, einer Entwicklungsumgebung für graphische Interaktionselemente. OBJECTION verwendet einen objektorientierten Ansatz zur Strukturierung der Benutzungsschnittstelle und basiert ganz wesentlich auf graphischen Constraints, die zur eleganten Behandlung von Layoutregeln eingesetzt werden. OBJECTION ermöglicht es, die manuelle Programmierung applikationsspezifischer graphischer Interaktion auf den unumgänglich notwendigen Umfang zu beschränken.

1 Einführung

Bei der Erzeugung graphischer Benutzungsschnittstellen erhält der Programmierer für den Einsatz von Standard-Widgets, wie Fenstern, Buttons, Scrollbars usw. gute Entwicklungsunterstützung durch sogenannte *Interface-Builder [3,4,5]*. Diese Werkzeuge erlauben es ihm, Widgets auf der Oberfläche zu plazieren, ihre Attribute einzustellen und *Callbacks*, also Methoden, die beim Eintritt bestimmter Ereignisse, wie etwa dem Drücken eines Buttons, ausgeführt werden sollen, zu spezifizieren.

Eine derartige Unterstützung fehlt vollständig, sobald man in den Bereich der applikationsspezifischen, interaktiven Graphikelemente wechselt (zum Beispiel Entity-Relationship-Diagramme im Software-Engineering). Die dort eingesetzten Graphikelemente zeichnen sich typischerweise dadurch aus, daß sie nicht aus einem vorgefertigten Widget-"Katalog" stammen, sondern speziell für eine Applikation bzw. ein Anwendungssystem entwickelt werden bzw. in einem Fachgebiet in ihrem Aussehen standardisiert sind. Die dargestellten Graphikelemente sind zumeist direkt manipulierbar und repräsentieren Daten oder Objekte der Applikation; vielfach ist ihre räumliche Anordnung oder die graphische Verknüpfung mit Linien bedeutungstragend.

Bisher muß der Entwickler solche dedizierten interaktiven Graphiken (auch *applikationsspezifische Widgets* genannt) weitgehend ohne Werkzeugunterstützung implementieren. Die jeweilige Umgebung bietet ihm lediglich graphische Grundfunktionen zum Zeichnen von Linien, Kreisen u. ä. an, aber das Verhalten der Objekte zu programmieren und insbesondere die interaktive Manipulierbarkeit zu erreichen, bedeuten für den Programmierer hohen Entwicklungsaufwand.

Das Ziel von OBJECTION ist die Unterstützung des Benutzungsschnittstellen-Entwicklers in diesem Bereich der applikationsspezifischen Widgets. OBJECTION ist ein Entwicklungswerkzeug, das dem Programmierer hilft bei

- der Erzeugung des **graphischen Erscheinungsbildes** der Widgets,
- der Festlegung ihres **Verhaltens** inklusive der Reaktion auf Mausereignisse wie Selektion oder Verschieben,

- der **Layoutkontrolle** durch Anbindung graphischer Constraints und
- dem effizienten **Testen** der fertiggestellten Komponenten durch einen eingebauten Interpreter.

In Verbindung mit einem „klassischen" Interface-Builder, zum Abdecken von Anordnung und Parametrisierung der Standard-Widgets, wird somit der noch von Hand zu programmierende Anteil stark vermindert und auf den unumgänglich notwendigen Teil reduziert. Im folgenden werden wir die Leistungsfähigkeit von OBJECTION zunächst an einer Beispielentwicklung diskutieren und anschließend die Komponenten von OBJECTION skizzieren.

2 Was kann OBJECTION ? Ein Beispiel

2.1 Leistungsbeschreibung des Petrinetz-Editors PE

Die zu entwickelnde Applikation **Petrinetz-Editor (PE)** soll die Erzeugung von Elementen des Petrinetzes und die Verbindung der Elemente mit Linien ermöglichen, wobei nur korrekte Petrinetze erzeugt werden sollen. Der Benutzer soll die Elemente auf der Zeichenfläche frei anordnen können, wobei PE auf die Einhaltung bestimmter Layout-Regeln achtet. Außerdem soll das Petrinetz animiert werden können, so daß die Marken „zu wandern" beginnen („Markenspiel").

2.1.1 Bedingungs-Ereignis-Systeme

Es gibt verschiedene Varianten von Petrinetzen; wir betrachten hier sogenannte *Bedingungs-Ereignis-Systeme (B/E-Systeme)*, deren Eigenschaften wir im folgenden kurz skizzieren; einfache Beispielnetze sind in Abbildung 1 dargestellt[1]. Ein *B/E-System* ist ein gerichteter Graph, bei dem die Menge der Knoten sich aufteilt in *Bedingungen* und *Ereignisse*. Kanten verlaufen nur von Bedingungen zu Ereignissen und umgekehrt („Bipartitheit" des unterliegenden Graphen). Eine Bedingung nimmt immer genau einen von zwei Zuständen an: Sie ist entweder *markiert* (erfüllt) oder nicht *markiert* (nicht erfüllt). Die Bedingungen, von denen Kanten zu einem Ereignis verlaufen, nennt man

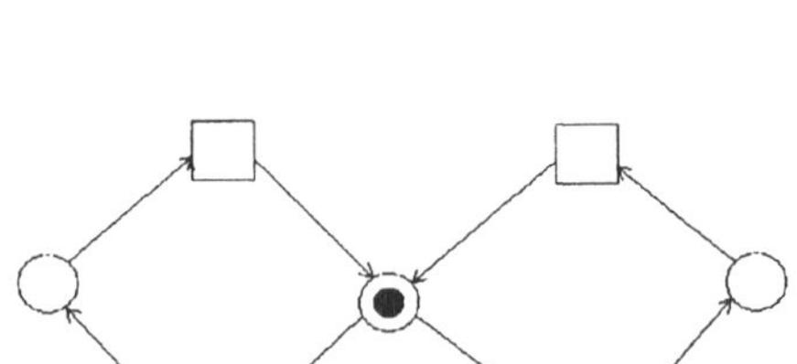

Abbildung 1: B/E-Systeme ohne und mit Konflikt

Vorbedingungen des Ereignisses, die Bedingungen, zu denen Kanten von einem Ereignis verlaufen, entsprechend *Nachbedingungen* des Ereignisses. Ein Ereignis, für das gilt, daß alle Vorbedingungen erfüllt und alle Nachbedingungen nicht erfüllt sind, kann *schalten*. Beim Schalten werden die Markierungen von allen Vorbedingungen entfernt und alle Nachbedingungen markiert. Gibt es mehrere sich gegenseitig ausschließende Möglichkeiten

---

[1] B/E-Netze sind Spezialfälle von Stellen-Transitionen-Netzen.

für Ereignisse bei einer bestimmten Belegung des Netzes zu schalten (ein sogenannter *Konflikt*), ist das Verhalten des Netzes *nichtdeterministisch*.

2.1.2 Das Schaltverhalten im Editor PE

In PE soll das Schaltverhalten des Netzes simuliert werden. Dazu gibt der Benutzer im Netz eine Belegung der Bedingungen mit Marken vor - durch Anklicken der zu erfüllenden Bedingungen. Auf Befehl des Benutzers beginnt das Netz, ausgehend von dieser Belegung das Markenspiel. Dabei sollen auch Konflikte zwischen mehreren schaltfähigen Ereignissen durch zufällige Auswahl einer der möglichen Schaltvorgänge aufgelöst werden. Zwischen zwei Schaltvorgängen soll eine Wartezeit von einigen Sekunden eingehalten werden, damit der Benutzer Zeit hat, die Markierungen in der neuen Belegung zu erkennen.

2.1.3 Layoutkontrolle im Editor PE

Der Benutzer soll die Knoten des Netzes beliebig verschieben können, wobei die Kanten zwischen den Knoten automatisch nachgezogen werden sollen. Um die Sichtbarkeit der Knoten und Kanten zu gewährleisten, soll der Editor PE automatisch einen gewissen Mindestabstand zwischen zwei Knoten einhalten. Außerdem soll verhindert werden, daß versehentlich ein Knoten aus dem Sichtbereich der Zeichenfläche geschoben wird.

2.2 Definition der Klassen in OBJECTION

Der **erste Definitionsschritt** zur Erzeugung von PE mit Hilfe von OBJECTION besteht in der „Anmeldung" und Beschreibung der beteiligten Objekttypen (Klassen), nämlich *Bedingungen, Ereignisse* und *Kanten* des Netzes.

Mit Hilfe des *Klasseneditors* von OBJECTION werden diese drei Klassen eingeführt (s. Abbildung 4) und zunächst das graphische Aussehen der Elemente im *Outfit-Editor* (ein einfaches Zeichenprogramm) definiert: Ereignisse werden graphisch als Quadrate, Bedingungen als Kreise dargestellt. Im Inneren von Bedingungen liegt außerdem noch ein ausgefüllter Kreis, der die Markierung darstellt. Ist eine Bedingung nicht erfüllt, hat der Kreis die Farbe des Hintergrundes und wird dadurch „unsichtbar". Kanten werden durch Linienobjekte dargestellt, die vom Start- zum Zielknoten verlaufen und an der Seite des Zielknotens eine Pfeilspitze aufweisen.

Im Klasseneditor selbst werden die Attribute der Klassen mit ihrem Datentyp angegeben. Jede Bedingung erhält ein logisches Attribut, das angibt, ob die Bedingung markiert ist oder nicht. Jede Bedingung und jedes Ereignis haben als Attribute jeweils eine Liste der eingehenden und der ausgehenden Kanten-Objekte. Jede Kante hat als Attribute den Start- und den Endknoten.

An diesem Punkt der Entwicklung sind die Klassen von PE in ihrem Aussehen („outfit") und ihrer inneren Struktur (Attribute, Standardmethoden) definiert. Der OBJECTION-Benutzer hat das ausschließlich durch interaktive Spezifikation im Werkzeug (Anklicken und Eintippen von Namen) durchgeführt.

2.3 Anbindung von Verhalten

Im folgenden muß den definierten Objekten spezifisches Verhalten zugeordnet werden. Dies geschieht in OBJECTION mit zwei verschiedenen Techniken: **deklarativ**, durch Anbindung von

Constraints und (konventionell) **prozedural** durch explizite Programmierung. Nach unserer Erfahrung eignen sich die deklarativen Angaben eher zur Beschreibung von Layoutbeziehungen zwischen graphischen Elementen, tendenziell also eher analoge, ungerichtete Abhängigkeiten, während die prozeduralen Anteile typischerweise für diskret „schaltende" Entscheidungen, Erzeugungsaktivitäten bzw. gerichtete Zuweisungen geeignet erscheinen. Da es aber durchaus Verhaltenseffekte gibt, die mit beiden Techniken beschrieben werden können, ist hier eine (nicht-triviale) Designentscheidung zu treffen.

2.3.1 Anbindung von deklarativ beschriebenem Verhalten

Deklarativ beschreibt man das Verhalten von Objekten in OBJECTION mit Hilfe von *graphischen Constraints* [1]. Ohne auf dies Konzept näher einzugehen, sei gesagt, daß Constraints zwischen (üblicherweise) numerischen Variablen definierte Bedingungen sind, die automatisch - von einem sogenannten *Constraint-Solver* - eingehalten werden, wenn die Wertebelegungen sich ändern. In OBJECTION werden Constraints zu Objekten hinzugefügt, indem entsprechende Anweisungen („Füge das Constraint X ein") an den unterliegenden Constraint-Solver PARCON [2] in die Initialisierungmethode geschrieben werden[2].

Im **zweiten Definitionsschritt** des Beispiels PE notieren wir für jedes Ereignis und jede Bedingung vier Constraints, nämlich jeweils eines dafür, daß das Objekt rechts von der linken, links von der rechten, unterhalb der oberen und oberhalb der unteren Begrenzungskante der Zeichenfläche bleibt, um dieses Objekt garantiert **innerhalb der Zeichenfläche** zu halten. Sind die Constraints einmal festgelegt worden, muß sich der Programmierer nicht weiter um die Position kümmern: Wird ein Objekt vom Benutzer des Editors PE über den Rand geschoben, sorgt der Constraint-Solver selbsttätig dafür, daß die Koordinaten des Objektes so geändert werden, daß es wieder im sichtbaren Bereich landet. Formal handelt es sich bei diesen Constraints um einfache Ungleichungen zwischen Koordinaten.

Um die **Kanten an** den **Knoten** zu **fixieren**, werden beim Erzeugen der Kanten Constraints eingetragen, die zwei Punkte auf der Zeichenfläche identifizieren, sogenannte *Punktgleichheitsconstraints*. Damit werden die Anfangs- und Endpunkte jeder Kante mit den Anschußpunkten der entsprechenden Ereignisse bzw. Bedingungen identifiziert. Auf diese Weise werden die Kanten beim Bewegen der Knoten automatisch mitgezogen; aufgrund der Ungerichtetheit sämtlicher Constraints in OBJECTION gilt aber auch die Umkehrung: Bei Verschieben von Kanten wandern die angebundenen Knoten mit.

---

[2] Hier hätte man bei der Entwicklung von OBJECTION aus rein technischer Sicht die Sicht auf die prozedurale Umsetzung des Constraint-Spezifizierungs-Mechanismus verzichten und stattdessen die Spezifikation von Constraints durch Dialogboxen einbinden können; dies ändert aber nichts an der deklarativen Natur dieser Spezifikation.

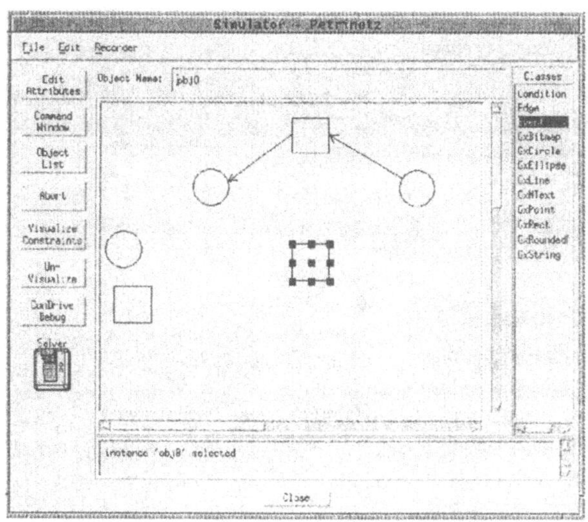

Abbildung 2: Der Petrinetz-Editor PE im OBJECTION-Simulator

Damit sich **Knoten nicht** gegenseitig **überdecken**, ist zwischen jeweils zwei Knoten ein Constraint notwendig, das den Abstand der Knoten nach unten begrenzt. Um dies zu erreichen, wird in der Initialisierungsmethode eines Knotens ein Mechanismus notiert, der zwischen jedem auf der Zeichenfläche existierenden Knotenobjekt und dem neuen Knoten ein sogenanntes *Abstandsconstraint* (mit einem definierten Mindestabstand) einsetzt. Bewegt der PE-Benutzer einen Knoten auf einen anderen zu, so weicht der andere Knoten aus, um den Mindestabstand einzuhalten. Gleichzeitig werden natürlich weiterhin die Kanten mitgezogen und ein Hinauslaufen über den Rand verhindert.

Die in diesem Beispiel verwendeten Constraints sind in OBJECTION bereits vordefiniert und können direkt über einen Namen eingebunden werden. Der Benutzer kann die Wirkung seiner Constraint-Definitionen unmittelbar während der Erstellung im Simulator (s. Abbildung 2) erproben und ggf. modifizieren. OBJECTION verfügt über eine große Zahl von Standard-Constraints, die im Laufe der Zeit durch die Realisierung vieler Beispiele entstanden sind. Zusätzlich bietet OBJECTION aber auch die Möglichkeit, neue Constraints aus bestehenden zu komponieren bzw. gänzlich neu zu schreiben (s. Abschnitt 3.3 Der Constrainteditor).

2.3.2 Anbindung von prozedural beschriebenem Verhalten

Einige Beziehungen zwischen Objektattributen können, wie eben beschrieben, durch Constraints formuliert werden. Andere Beziehungsarten eignen sich weniger gut für diesen Ansatz; zumeist handelt es sich dabei um konstruktive oder Entscheidungen treffende Abhängigkeiten. Für diese Art von Beziehungen sieht OBJECTION die Verhaltensbeschreibung mit Hilfe von prozeduralen Anweisungen, notiert in der Programmiersprache *Objective-C* vor. Hier wird deutlich, daß OBJECTION als Werkzeug für den Programmierer von Benutzungsschnittstellen vorgesehen ist - nicht als Spezifikationswerkzeug für Entwickler, die keine Programmiererfahrung haben.

Soweit möglich, kann man prozedurale Programmierung vermeiden, indem man auf die deklarative Beschreibungstechnik zurückgreift - es bleibt aber meistens ein Rest von Verhalten, der entweder gar nicht oder nur mühsam auf Constraints abbildbar ist. Trotz des Rückgriffs auf Programmierung stellt OBJECTION hier eine erhebliche Unterstüzung für den Programmierer dar: Die zu schreibenden Codeteile sind stets lokale, objektbezogene Methoden - ihr „Umfeld", also die Einbindung in das Gesamtsystem, wird von OBJECTION geleistet. Die

Erfahrung zeigt, daß die Codegröße sehr klein im Verhältnis zur Gesamtgröße der (erzeugten) Programms ist.

Im **dritten Definitionsschritt** von PE wird mit Hilfe solcher Programmierung die **Schwärzung** des **Markierungskreises**, die Überprüfung der **Bipartitheit des Graphen** beim Einfügen neuer Kanten und die Definition des **Schaltverhaltens** geleistet.

Aus Platzgründen wollen wir hier die Beschreibungen im einzelnen nicht angeben, erwähnen aber, des es sich bei den im Beispielfall PE notwendigen prozeduralen Verhaltensbeschreibungen lediglich um 70 Zeilen Objective-C-Code handelt.

2.4 Ergebnis der Arbeit mit OBJECTION

Betrachtet man das bis zu dieser Stelle erreichte, so kann im Simulator von OBJECTION das Petrinetz schon in seinem Kernverhalten getestet und benutzt werden. Die Elemente des Netzes verhalten sich, was das Markierungsverhalten betrifft, wie man es von einem korrekten Petrinetz erwartet. Der Benutzer kann beliebig Elemente hinzufügen und verbinden, sowie Bedingungen markieren[3]. Entspricht das Verhalten des Netzes nicht den Erwartungen des OBJECTION-Benutzers, können die Methoden sofort geändert und erneut getestet werden. Ebenso kann der Effekt der graphischen Constraints getestet und die Constraints können bei Nichtgefallen geändert werden.

Die gesamte Funktionalität des Prototypen der Anwendung PE beschränkt sich auf eine Zeichenfläche, die vom Simulator in OBJECTION zur Verfügung gestellt wird. Das Erzeugen neuer Objekte wird durch einen Doppelklick auf den Namen einer der Klassen, die rechts neben der Zeichenfläche angezeigt werden, angestoßen. Das Senden eines Methodenaufrufs an ein Objekt erfolgt über die automatisch generierte Attributbox des Objektes, die einen Eintrag für jedes Attribut und jede Methode des Objektes enthält und in die man auch die Parameter der Methoden eintragen kann.

Um das Ergebnis der Arbeit zu einer eigenständigen Applikation zu machen, kann man sich den Objective-C-Code jeder Klasse ausgeben lassen. Durch Einsatz eines konventionellen Interface-Builders konstruiert man die Widget-Umgebung für PE, definiert entsprechende Menüs, Buttons und Fenster mit Zeichenflächen, in denen man dann die - von OBJECTION erzeugten - Klassen darstellen läßt. Der eigenen Applikation wird dann natürlich auch der Constraint-Solver PARCON durch Linken hinzugefügt.

2.5 Andere Beispiele

Neben dem oben beschriebenen Beispiel wurden bereits eine Reihe anderer mit Hilfe von OBJECTION verwirklicht. Eines der komplexesten ist ein Syntaxdiagramm-Editor, der in starkem Maße von den graphischen Constraints Gebrauch macht, um ein ansprechendes und übersichtliches Layout zu erzeugen.

---

[3] Um das Beispiel kurz zu halten, wurde auf das Entfernen von Elementen verzichtet, doch wäre das Hinzufügen entsprechender Methoden problemlos möglich.

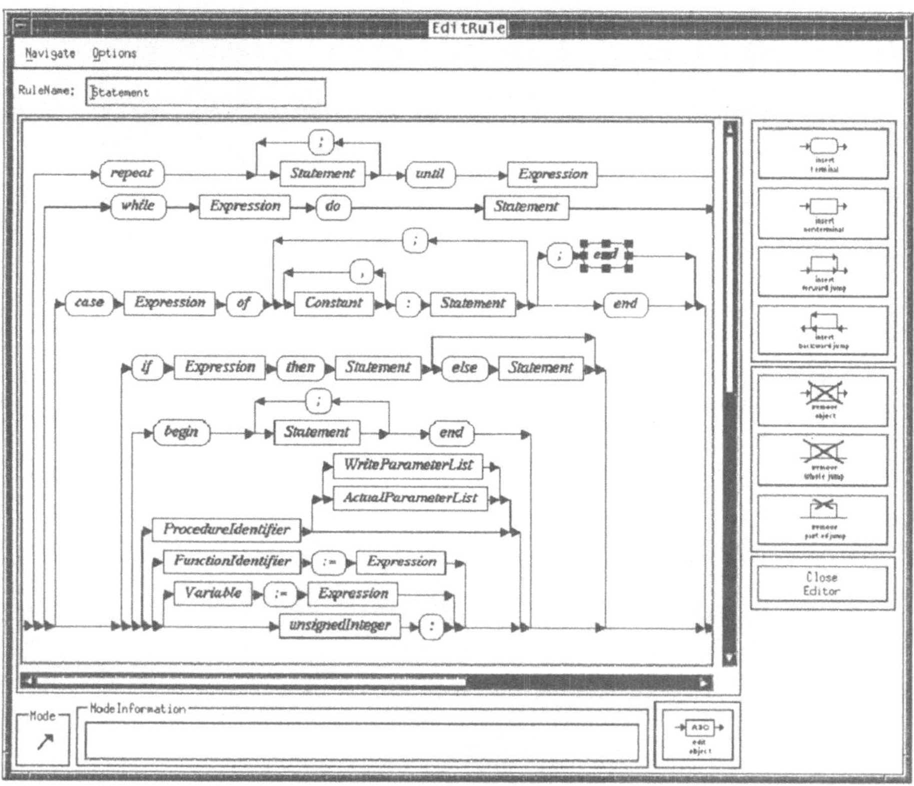

Abbildung 3: Der Syntax-Diagramm-Editor

Abbildung 3 zeigt die fertige Applikation, komplett mit den im Interface-Builder erzeugten Buttons. Obwohl bei dem in der Abbildung gezeigten Diagramm z. B. 2478 Constraints zum Einsatz kommen, ist der Constraintsolver schnell genug, um das Verschieben der Elemente in Echtzeit zu ermöglichen.

Andere Beispiele waren die Simulation eines Kanban-Fertigungssystems, bei dem sich Waren über Fließbänder bewegen und von Maschinen verarbeitet werden, sowie Editoren für elektronische Schaltungen, Structured-Chart-Diagramme und ein Spiel, bei dem eng aneinanderliegende Münzen in ein Loch geschoben werden müssen (wobei das Abstandsconstraint einen Großteil der Arbeit erledigt).

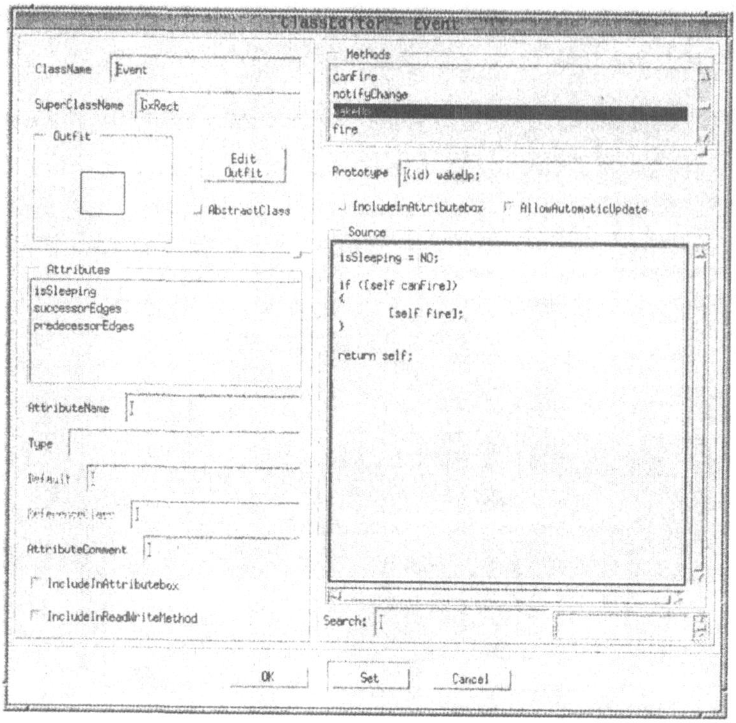

Abbildung 4: Der Klasseneditor

3 Das Werkzeug OBJECTION

Die Entwicklungsumgebung OBJECTION setzt sich zusammen aus einer Reihe von *Editoren*, dem *Constraintsolver* PARCON [2] und dem GxToolkit, einer *Bibliothek von Graphikobjekten*, die einerseits die Fähigkeit besitzen, mit der Maus selektiert, verschoben oder in der Größe verändert werden zu können und andererseits speziell für die Kommunikation mit einem Constraintsolver vorbereitet sind. Die Editoren benutzen als gemeinsame Datenbasis ein sogenanntes *Projekt*, in dem die vom Benutzer erstellten und die standardmäßig vorhandenen Klassen und Constraints beschrieben sind. Benutzt werden die Projektdaten außerdem vom Simulator, in dem die Klassen getestet werden können, und vom Codegenerator, der Objective-C-Code ausgibt.

3.1 Der Klasseneditor

In diesem Editor werden die Attribute und Methoden einer Klasse definiert (s. Abbildung 4). Für jede neu angelegte Klasse erzeugt der Klasseneditor automatisch eine Reihe von Standardmethoden, etwa zum Initialisieren und Löschen von Objekten oder zum Reagieren auf Maus-Aktionen. Die Methoden in OBJECTION werden in der Programmiersprache Objective-C, einer objektorientierten Erweiterung von C programmiert[4]. Dem Klasseneditor ist der Outfiteditor angeschlossen, eine Art einfaches Zeichenprogramm mit dem der Benutzer das graphische Erscheinungsbild eines Objektes definiert, indem er es aus Standardelementen wie Kreisen, Rechtecken, Linien oder Bitmaps zusammensetzt. Dabei können die Teile des Outfits auch einzeln von Methoden angesprochen werden.

3.2 Der Simulator

Da OBJECTION über einen eingebauten Objective-C-Interpreter verfügt, kann das Verhalten der Objekte im Simulator (s. Abbildung 2) bereits vollständig getestet werden. Der Benutzer kann sich hier auf einer Zeichenfläche Instanzen aller im Projekt definierten Klassen erzeugen. Für jedes Objekt kann in einem Fenster eine Attributbox geöffnet werden. Dort werden die Werte sämtlicher Attribute des Objektes angezeigt und können auch verändert werden. Außerdem werden die Methoden, die das Objekt „versteht", dort aufgelistet, und es können Methodenaufrufe an das Objekt geschickt werden.

Eine andere Möglichkeit, Methoden eines Objektes aufzurufen, ist Interaktion mit dem Objekt über die Maus. Wird ein Objekt mit der Maus selektiert, verschoben oder in der Größe verändert, werden jeweils entsprechende Methoden aufgerufen. Durch Modifikation dieser Methoden kann das Objekt spezifisch auf die Ereignisse reagieren.

3.3 Der Constrainteditor

In den (relativ seltenen) Fällen, in denen der Benutzer Constraints zwischen Objekten einsetzen möchte, die bisher noch nicht definiert wurden, kommt der *Constrainteditor* zum Einsatz. Zur Konstruktion von neuen Constraints gibt es zwei verschiedene Ansätze:

- Zum einen kann ein Constraint hierarchisch aus bereits bestehenden zusammengesetzt werden. So kann etwa aus mehreren Punktgleichheits-Constraints ein zusammengesetztes Constraint erzeugt werden, das zwei Rechtecke und eine Linie so verknüpft, daß die Endpunkte der Linie immer auf den Mittelpunkten der Rechtecke liegen (s. Abbildung 5). Auf Wunsch könnte bei diesem Beispiel die Linie beim Aufruf des Constraints automatisch neu erzeugt werden.

- Die andere Möglichkeit zur Constrainterzeugung sollte dank der großen Zahl bereits existierender Constraints relativ selten notwendig sein. Sie besteht darin, neue Constraints direkt in der Syntax des Constraintsolvers PARCON zu formulieren, wozu sich der Benutzer allerdings in die Interna von PARCON einarbeiten und insbesondere die oft überraschenden Wechselwirkungen von Constraints aufeinander verstehen muß.

---

[4] Objective-C ist als Teil des GNU-C-Compilers `gcc` frei verfügbar.

4 Ausblick

OBJECTION kann, wie bereits festgestellt wurde, momentan nicht als alleiniges Entwicklungswerkzeug für Applikationen eingesetzt werden, da die Möglichkeit fehlt, wie in einem Interface-Builder Standard-Widgets einzusetzen. Als zukünftige Möglichkeit bietet sich deshalb die Verknüpfung von OBJECTION mit einem Interface-Builder wie etwa MOTIFATION [3] zu einer kombinierten Entwicklungsumgebung an. Die Kombination der Eigenschaften beider Programme würde dazu führen, daß die Programmiertätigkeit bei der Entwicklung graphischer Benutzungsschnittstellen sich dann auf den unumgänglich notwendigen Rest beschränken könnte. Der Unterschied zwischen Standard-Widgets und „normalen" Graphik-Objekten würde verwischt und wesentlich flexibler agierende Benutzungsschnittstellen wären möglich. OBJECTION ist für den nicht-kommerziellen Einsatz frei verfügbar und kann über http://www.uni-paderborn.de/cs/Manfred.Poepping.html geladen werden. Die Umgebung ist unter LINUX, SUN-OS und Solaris lauffähig; das Paket enthält auch den Constraint-Solver PARCON, der auch unabhängig von OBJECTION als leistungsfähiger Constraint-Solver einsetzbar ist.

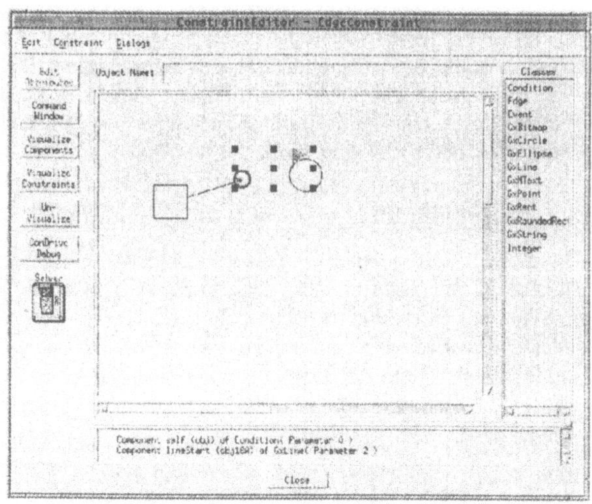

Abbildung 5: Der Constraint-Editor

Literatur

[1] J. Gosling: Algebraic Constraints. PhD Thesis, Carnegie-Mellon University, Pittsburgh, 1983.

[2] P. Griebel: Parcon - ein iterativer, paralleler Constraint-Solver. GI-Workshop „Neue Architekturkonzepte zur Gestaltung graphischer Systeme", Bonn, 1993.

[3] P. Griebel, M.Pöpping: Motifation - Ein Benutzungsschnittstellen-Entwicklungssystem. In: Informatik Fachberichte 293, Springer Verlag, 1991, 445-454.

[4] B. Myers: User-Interface Tools: Introduction and Survey. In: IEEE Software (1989), 15-23.

[5] TeleSoft AB (Hrsg).: TeleUSE User's Guide. TeleSoft AB, Release 1.1, 1990.

Adressen der Autoren

Holger Uhr
Uni - GH Paderborn,FB 17
Fürstenallee 11
33102 Paderborn
huhr@uni-paderborn.de

Peer Griebel
Knoll Informations-
systeme GmbH
Schwarzachstr. 20
88348 Saulgau
peer.griebel@t-online.de

Manfred Pöpping
ProDV GmbH
Martin-Schmeißer-Str.14
44227 Dortmund
poepping@prodv.de

Gerd Szwillus
Uni - GH Paderborn,FB 17
Fürstenallee 11
33102 Paderborn
szwillus@uni-paderborn.de

Auf dem Weg zur Blickmaus:
Die Beeinflussung der Fixationsdauer durch kognitive und kommunikative Aufgaben

Boris Velichkovsky[1], Andreas Sprenger[1] und Marc Pomplun[2]

[1]Technische Universität Dresden, [2]Universität Bielefeld

Zusammenfassung

Die Entwicklung eines subjektzentrierten Paradigmas der Mensch-Computer-Interaktion benötigt umfangreiche Erkenntnisse über die Mechanismen der Informationsverarbeitung des Menschen. Neue psychologische und technologische Entwicklungen ermöglichen es nicht nur, wie bis jetzt, die statistischen Charakteristiken von Menschen einzubeziehen, sondern auch Systeme zu bauen, die für die aktuellen Foci der Aufmerksamkeit und die Absichten miteinander interagierender Personen sensibel sind. Dazu können moderne Methoden der Augenbewegungsregistrierung einen wertvollen Beitrag leisten. Die wichtigsten Schritte in diese Richtung sind mit der Benutzung von Augenbewegungsdaten als ein Ausgabemedium, für blickabhängige Bildbearbeitung und für die Disambiguierung verbaler wie auch nichtverbaler Informationen verbunden. In diesem Beitrag berichten wir über experimentelle Untersuchungen, die die Beeinflussung der Fixationsdauer durch eine Reihe von kognitiven und kommunikativen Aufgaben genau analysieren können. Speziell geht es um Suchaufgaben, perzeptive, semantische, metakognitive (selbst-referentielle) und deiktische kommunikative Aufgaben. Die Ergebnisse sprechen für eine selektive Beeinflussung von verschiedenen Segmenten der Fixationsdauer durch diese Aufgaben, wobei explizite kommunikative Anforderungen die Anzahl von besonders langen Fixationen (>500 ms) in einer signifikanten Weise erhöhen. Das öffnet den Weg zu einer gezielten Benutzung verschiedener Gruppen der Fixationen bei der Interaktion mit Computern. Die Frage der Trainierbarkeit der menschlichen Okulomotorik im Hinblick auf die kommunikativen Leistungen wird zusätzlich behandelt.

Einleitung

Wichtige technologische Veränderungen reifen oft unauffällig in den interdisziplinären Nebenfeldern. Neue Forschungsmethoden aus den Neuro- und Kognitionswissenschaften können in der nächsten Zukunft solche Veränderungen im Bereich der Mensch-Computer-Interaktion auslösen. Diese Veränderungen ermöglichen die Entwicklung technischer Systeme, die für momentane Leistungsfähigkeiten und Intentionen des Nutzers sensibel sind. Zum erstenmal zeigen die Methoden der Augenbewegungsmessung und bildgebende Verfahren aus der Neurophysiologie (Neuroimaging) menschliche kognitive Prozesse anschaulich auf. Insbesondere die Blickregistrierung findet zunehmend ihre Anwendung für alternative Varianten der Interaktion mit Computern [8, 12, 25] und wird in den Fachliteratur als eine der wichtigsten mittelfristigen Perspektiven der Entwicklung auf diesem Feld geschätzt [13]. In unserem Beitrag wollen wir einige für die Benutzung der Augenbewegungsdaten notwendige Fragen stellen und experimentell beantworten.

Augenbewegungsmessung und Mensch-Computer-Interaktion

Moderne Eyetracking Systeme bieten vielfältige Anwendungsmöglichkeiten bei relativ geringer Beanspruchung des Nutzers – sie arbeiten zunehmend berührungslos, schnell und mit hoher Genauigkeit. Frühere Methoden waren ausschließlich für Laboruntersuchungen konzipiert und mit starken Einschränkungen für die Probanden verbunden (wie die

Kopffixierung mit Beißbrett oder eine Rekalibrierung nach jedem Lidschlag - siehe [29]). Die sprunghafte Entwicklung der Computertechnologie hat dazu geführt, daß videobasierte Systeme vermehrt eingesetzt werden. Diese Systeme benutzen Infrarot-Kameras und bestimmen aus dem Bild der Pupille und teilweise dem Cornealreflex einer Lichtquelle den Blickort. Je nach Anwendungszweck werden derzeit verschiedene Systeme eingesetzt, um die Vorteile entsprechend zu nutzen. So verlangen berührungslos arbeitende Systeme vom Nutzer außer einer kurzen Kalibrierphase nur eine relativ konstante Sitzposition, wobei Kopfbewegungen meist zugelassen sind. Die Genauigkeit der Systeme liegt unter 0,5 Winkelgrad, der Blickort bleibt über die gesamte Meßperiode stabil. Nachteilig wirkt sich aber die relativ geringe zeitliche Auflösung von 50 bzw. 60 Hz auf die Erkennung von Sakkaden und kurzen Fixationen aus [21, 26].

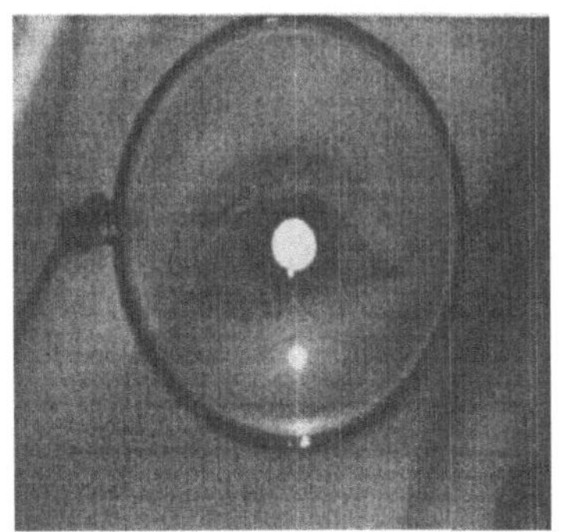
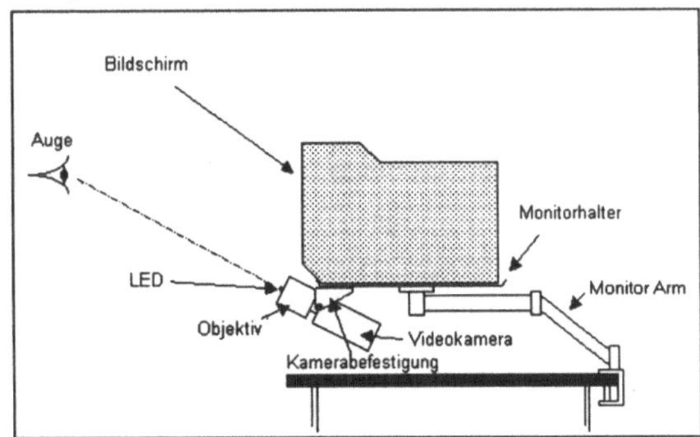

Abb. 1: Menschliches Auge unter Infrarotbeleuchtung einer Registrierungskamera. Für die Blickorterfassung wird die Pupillenfläche und der darunter liegende Cornealreflex benutzt. Störreflexe, wie durch das Brillenglas, werden kompensiert. Rechts: Aufbau eines berührungslos arbeitenden Systems am Arbeitsplatz.

Eine weitere Gruppe von Eyetracking-Systemen verwendet Infrarot-Miniaturkameras, die an einem Kopfgestell befestigt sind. Solche Systeme erlauben deutlich höhere Abtastraten bis zu 600 Hz bei 0,25 Winkelgrad Meßgenauigkeit [20]. Auch hier sind Kopfbewegungen – teilweise auch Körperbewegungen im Raum – zulässig. Probanden bemerken lediglich das sehr leichte Kopfgestell, das Gesichtsfeld ist überwiegend nicht eingeschränkt. Die Verwendung der Blickortinformation für kommunikative Aufgaben erfordert Systeme, die eine leistungsfähige Datenübertragung aufweisen. Neueste Systeme kommunizieren per TCP/IP, so daß die Blickortdaten auch über große Distanzen per Internet transferiert werden können [25].

Der räumlichen Auflösungsmöglichkeit eines Meßsystems kommt eine besonders große Bedeutung zu. Das foveale Sehfeld ist zwar im Mittel 2 Winkelgrad groß, die Menschen können jedoch in der Regel visuell zwei Punkte im Abstand von nur einer Bogenminute unterscheiden. Bei der Anwendung von Augenbewegungen in der Mensch-Computer-Interaktion hat sich eine Meßgenauigkeit von mindestens 0,5 Winkelgrad als sinnvoll herausgestellt. Neuronale Netze können dabei den Meßfehler der Eyetracking-Systeme um bis zu 70% reduzieren [1, 16]. Durch eine Blickortbestimmung bekommen wir auch Hinweise über die Verteilung der Aufmerksamkeit. In der Tat ist die Blickrichtung der einzige relativ zuverlässige (obwohl, nicht ideale - [9, 11]) Hinweis auf den Fokusort der visuellen Aufmerksamkeit. Einzelne Parameter der Augenbewegungen bieten weiterhin Aufschluß über kognitive Verarbeitungsprozesse und emotionale Zustände. Je nach Meßsystem liegen verschiedene Parameter zur Auswertung vor. Beispielhaft seien hier die Lidschlußaktivität, Pupillendurchmesser sowie die besonders interessanten Parameter der Blickrichtung und der Fixationsdauer erwähnt.

Es ergibt sich eine Reihe möglicher Anwendungen. Die erste und naheliegendste Anwendung ist die der „Blickmaus". Das Hauptanwendungsgebiet ist in dem Bereich, wo die feingesteuerte Benutzung von Händen unmöglich ist. Beispielsweise können körperbehinderte oder ältere Menschen, die weder die Extremitäten noch die Sprache benutzen können, mithilfe dieses Systems mit der Umwelt kommunizieren. Augenbewegungsdaten können zur Steuerung verschiedener elektronischer Systeme eingesetzt werden, so z.B. e-mail, Fax, Telefon, Videophon und Sprachgeneratoren.

Ein weiterer Ansatz zur blickortgesteuerten Interaktion verfolgt weniger die Idee, Augenbewegungen als Ersatz für die Computermaus zu benutzen. Vielmehr werden die Rohdaten auf einer semantisch höheren Ebene interpretiert und in einer neuen Form von Multimedia-Anwendung ohne explizite Befehlseingabe seitens des Menschen benutzt. Dabei wird kontinuierlich die Zuwendung der Aufmerksamkeit zu einzelnen Objekten gemessen (z.B. [12, 21]). Als Bezeichnung für dieses „noncommand" Interaktionsprinzip wurde der Begriff „Interest and Emotion Sensitive Media" [6] vorgeschlagen. Mit anderen Worten, die Möglichkeit der Augenbewegungsmessung ermöglicht eine quasi-passive Beeinflussung von elektronischen Medien. Dies läßt sich erreichen durch Messung

- des Interessensschwerpunktes durch Identifizierung von Aufmerksamkeitsbereichen in komplexen Bildern,
- von emotionalen Reaktionen, die sich durch die Auswertung von Lidschlägen und Pupillendurchmesser zeigt (siehe [8, 21, 23]).

Einige frühere Versuche in diese Richtung waren aber wenig erfolgreich, da sie den Unterschied zwischen Augenbewegungen und verschieden Aufmerksamkeitsformen nicht beachteten. Jacobs [12] nannte treffend die spezifische Schwierigkeit das "Midas touch problem": Wenn alle Augenbewegungen unmittelbare Konsequenzen für die Arbeit des Computers hätten, wäre jede Interaktion schlicht unmöglich. Offensichtlich sollen Fixationen für diese Anwendungen mit zeitlichen Filtern in expliziten Bereichen der Fixationsdauer und/oder Fixationsgruppen eingeordnet werden. Moderne Systeme dieser Art sind mit Einführung zeitlicher Filtern im Bereich von 500 ms verbunden, was als eine empirisch gefundene Startvariable eingeschätzt werden kann (siehe z.B.[20]). Wie begründet ist diese Wahl und was bedeuten verschiedene Segmente der Fixationsdauer bei menschlichen Augenbewegungen?

Das vorläufige Modell der visuellen Informationsverarbeitung

Die Beantwortung dieser Fragen setzt voraus, daß wir eine Klassifikation möglicher Aufgaben für das menschliche visuelle System und die Okulomotorik aufstellen und den Einfluß dieser Aufgaben auf Parameter der Augenbewegungen überprüfen. In der letzten Zeit hat die Entdeckung von sogenannten *Expreß-Sakkaden* [7] das allgemeine Interesse an Augenbewegungen geweckt. Diese Bewegungen haben eine außergewöhnlich kurze Latenz von etwa 100 ms, die mit Abstand kürzeste Latenz unter allen menschlichen Bewegungen überhaupt. Die Expreß-Sakkaden sind als okulomotorische Komponenten der Orientierungsreaktion zu interpretieren: Sie sind gewissermaßen immer von der Außenwelt reflexartig erzwungen und widersetzen sich Versuchen der bewußten Kontrolle. Ihre potentielle Benutzung für die Interaktion mit Computern ist deshalb mehr als fraglich.

Eine Auflistung der mehr zentral-gesteuerten Aufgaben ist zur Zeit hauptsächlich auf Basis des gesunden Menschenverstands möglich (siehe aber [3, 5, 22]). Jede relevante Portion der Information soll zuerst gefunden werden. Die bedeutenste Funktion der Augenbewegung ist die Realisierung von solchen *Suchaktivitäten*, die Sakaden aber auch mit anderen motorischen Aktivitäten (Kopf- und Torsodrehung, Lokomotionen) vereinigen. Wenn ein potentiell relevantes Objekt oder Merkmal gefunden ist, soll es natürlich perzeptiv beschrieben werden. Solche perzeptive *Identifikation* ohne Augenbewegungen ist kaum möglich (obwohl diese Bewegungen höchstwahrscheinlich von einem anderen Typ als Suchsakkaden sind). Nach der perzeptiven Identifikation oder vielleicht parallel mit deren letzter Phase kann die Information *semantisch kategorisiert* werden. Danach (oder wiederum parallel damit) findet für einige Fragmente der Information auch meta-kognitive, selbst-referentielle Bearbeitung statt. Diese Einschätzung kann auch aus der Perspektive von anderen Menschen passieren, da wir soetwas wie die „Theorie des Geistes" (Vorstellungen über Vorstellungen von anderen Menschen über dieselben Sachverhalte) haben [23]. Nur dann ist das Wissen für eine etwaige *Kommunikation* vorbereitet. Daß die Augenbewegungen auch bei der Kommunikation eine wichtige Rolle spielen, wurde schon vor mehr als 100 Jahren von Ewald Hering [10] betont, blieb aber bis heute wenig erforscht.

In einer Reihe zum Teil noch laufender Experimente wurde an der Universität Bielefeld und an der TU Dresden ein Versuch unternommen, die Parameter der Augenbewegungen bei den oben erwähnten Aufgaben zu untersuchen.

Augenbewegungen in Suchaufgaben

Ein typisches Muster der Suchsakkaden ist in der Abbildung 2 dargestellt [24, 17]. Die Aufgabe besteht in dem Auffinden eines Unterschiedes zwischen dem linken und rechten Teil des Bildes der in Form und in Farbe variierten Objekte. Man kann sehr deutlich hochamplitudige Augenbewegungen sehen, die skandierend große Flächen mehr oder weniger systematisch abtasten und zwar in der Art und Weise, daß eine unmittelbare Wiederholung des „Besuches" von denselben Objekten bzw. Objektgruppen extrem unwahrscheinlich ist. Der neurophysiologische Mechanismus im Hintergrund dieser Variante des „travelling salesman problem" ist der sogenannte IOR-Effekt (Inhibition of return), wobei unmittelbar nach einer Fixation (anscheinend nur im Suchmodus) alle Sakkaden zur selben Lokation gehemmt werden [11].

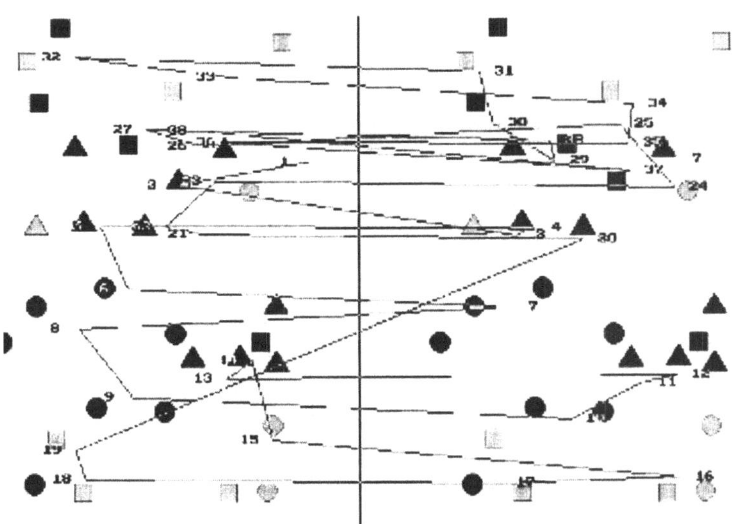

Abb.2: Ein typisches Muster der sakkadischen Augenbewegungen bei Suchaufgaben

Die Fixationsdauer von Suchsakkaden in solchen Experimenten liegt etwa zwischen 120 und 250 ms, signifikant abhängig von der Komplexität des betrachteten Bildausschnitts. Die Zeiten können sich in der Umgebung der Zielobjekte, die den gesuchten Unterschied aufweisen, vergrößern, was gewöhnlich kurz vor der Antwort geschieht. Hier sind aber nicht mehr Suchsakkaden, sondern okulomotorische Komponenten von relativ höheren Prozessen aktiv. Auf dieser letzten Stufe des Suchvoganges treten nun durchaus auch wiederholte Fixationen in derselben Region auf. Für die mögliche Anwendung der Fixationsortdaten bedeuten diese Erkenntnisse, daß auch das Segment der Fixationsdauer unmittelbar über der Spannweite der „Expreß-Fixationen", bis mindestens 250 ms, im regelmäßigen Betrieb für die Blickmausinitiation gesperrt bleiben sollte.

Verarbeitungstiefe und Fixationsdauer

Wenn eine potentiell wichtige Information im visuellen Umfeld gefunden ist, kann sie in verschiedenen „Tiefen" bearbeitet werden: Man bleibt sozusagen entweder auf der Oberfläche, wobei nur physikalische Merkmale der Objekte kodiert sind, oder man versucht auch semantische Beziehungen zu rekonstruieren. Die etwaige Verarbeitungstiefe ist eine Funktion der Situation, der Aufgabe sowie der Fertigkeiten (und Fähigkeiten) des Menschen. In der Kognitionspsychologie bilden die Untersuchungen der Verarbeitungstiefe ein selbständiges Ressort, wobei die meisten Daten mindestens über perzeptive, semantische und selbst-referenzielle Stufen der Verarbeitung sprechen [3, 24]. Mit weitgehend ähnlichen Ergebnissen sind aus dieser Perspektive die Prozesse der Bearbeitung der verbalen und nichtverbalen Informationen untersucht worden. Einige Beweise sprechen dafür, daß auch Parameter der Augenbewegungen eine deutliche Korrelation mit der Verarbeitungstiefe aufzeigen. Abbildung 3 zeigt ein Beispiel des Materials aus einem von unserer letzten Experimente [26], wo durch die variable Reduzierung der sensorischen Vorlage (gezeigt sind nur 3 Stufen) die Schwerigkeit der Bearbeitung kontrolliert wurde. Die Probanden sollten solche Bilder auf dem perzeptiven, semantischen oder metakognitiven (selbst-referentiellen) Niveau bearbeiten.

Abb. 3: Versuchsmaterial aus den Experimenten zur Verarbeitungstiefe

Bei der Betrachtung der Augenbewegungsdaten begrenzen wir uns auf eine Analyse der Fixationsdauer. In der Abbildung 4 sind die Verteilungen der Fixationsdauer der mehr als 27.000 Sakkaden von insgesamt fast 50 Probanden für jeweilige Verarbeitungstiefe dargestellt.

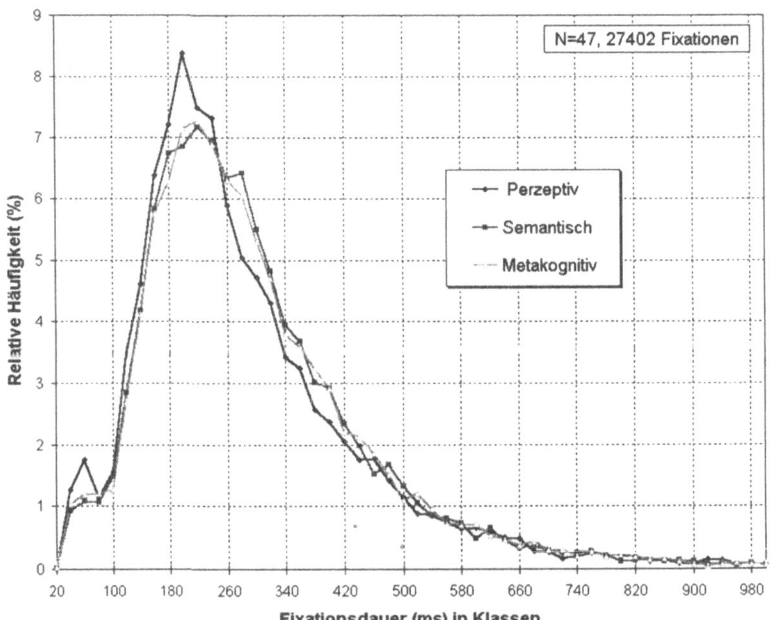

Abb. 4: Verteilung der Fixationsdauer in Abhängigkeit von der Verarbeitungtiefe

Man kann leicht sehen, daß die Kurven für die semantische und metakognitive Enkodierung weitgehend identisch sind. Beide unterscheiden sich aber von der der Fixationen bei perzeptiver Enkodierung. Diese Unterschiede liegen im Intervall zwischen etwa 120 und 450 ms – mit Übergang von einem großen Anteil der „perzeptiven" Fixationen zu dem großen Anteil der Fixationen bei der „tieferen" Verarbeitung bei einer Fixationsdauer von etwa 250 ms. Was sagen uns diese Daten nun? Offensichtlich läuft auch nach der Fixationsdauer von 250 ms eine interne Informationsverarbeitung weiter, diesmal mit Akzent auf Semantik und persönlicher Bedeutung. Bevor diese Information bearbeitet wird, wäre die Benutzung der Blickortdaten zwecks Kommunikation einfach verfrüht. In welchem Bereich der Fixationsdauer kann man Spuren kommunikativer Intentionen der Menschen entdecken?

Gibt es spezifische „kommunikative Fixationen"?

Die Blickrichtung für Ziele der Kommunikation zu benutzen ist eine natürliche menschliche Fähigkeit. In der Face-to-Face Interaktion mit anderen Menschen verwenden wir diese Fähigkeit zur Disambiguierung von verbalen Anweisungen, so daß – und zwar sehr früh in der Ontogenese – ein Zustand der „gemeinsamen Aufmerksamkeit" entstehen kann: Kooperierende Partner fixieren zur gleichen Zeit dieselben Objekte, wodurch das Problem der referentiellen Mehrdeutigkeit wegfällt (siehe [18]). Diese Besonderheit der natürlichen Kommunikation kann einen großen Teil ihrer Vorteile gegenüber technologisch vermittelten Formen der Kommunikation erklären, besonders bei der Übergabe von praktischem Wissen und konstruktiven Anweisungen. In der Abbildung 5 ist eine der Versuchsanordnungen unserer Experimente gezeigt, die es erlaubten, die relevanten Aspekte der natürlichen Kommunikation, in erster Linie die Berücksichtigung der Blickrichtung, auf elektronisch vermittelte Kommunikation zu übertragen [23, 27].

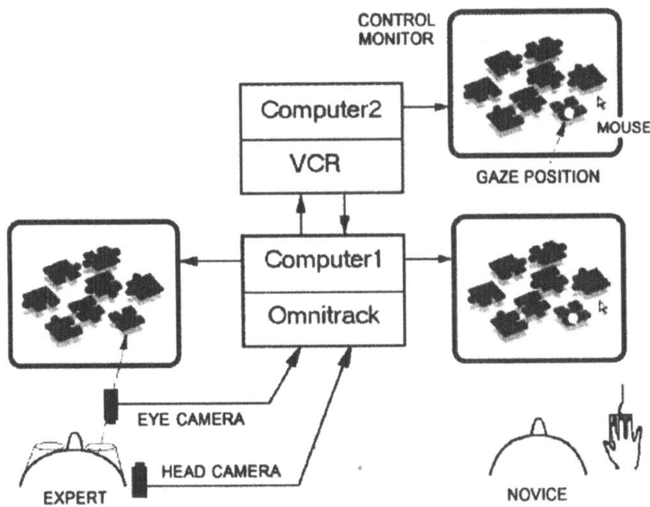

Abb. 5: Versuchsanordnung in Experimenten zum Transfer der Blickrichtungsdaten

Die Aufgabe bestand darin, ein computergestütztes Puzzlespiel zu lösen, dessen Teile mit einer Computermaus bewegt werden konnten. An den Experimenten nahmen jeweils Paare von Probanden teil, nämlich ein Experte, der die Puzzles trainiert hatte und sie mühelos und schnell lösen konnte, und ein Novize, der die Puzzlebilder nicht kannte. Nur der Novize konnte die Puzzleteile bewegen, und der Experte, der auf seinem Monitor dieselbe Puzzleszene wie der Novize betrachtete, versuchte, ihm so gut wie möglich dabei zu helfen. Die sprachliche Kommunikation der Teilnehmer sowie deren Augenbewegungen wurden dabei registriert. Es galt generell eine der folgenden drei Kommunikationsbedingungen:

- „Sprache": Experte und Novize konnten sich nur verbal verständigen.
- „Sprache und Maus": zusätzlich zur verbalen Kommunikation konnte der Experte einen Mauszeiger auf dem Bildschirm bewegen, der für beide Teilnehmer sichtbar war.
- „Sprache und Blick": zusätzlich zur verbalen Kommunikation wurde immer die aktuelle Blickposition des Experten oder des Novizen auf dem Monitor des Partners angezeigt.

Wir verzichten an dieser Stelle auf die Darstellung der umfangreichen Ergebnisse dieser Experimente. Im Kontext dieses Beitrages sind die Daten über Besonderheiten der Augenbewegungen von speziellem Interesse. Diese Daten zeigen, daß explizite Intention mit dem eigenem Blick, die die verbale Anweisung unterstützt, zu einem signifikanten Zuwachs an der Sakkaden mit Fixationsdauer länger als 500 ms führt. Es zeigt sich auch eine negative Korrelation zwischen der Anzahl solcher Fixationen bei dem Experten und der Zeit des Problemlösen von Novizen: Offensichtlich können solche lange Fixationen als eine wertvolle Quelle der Information zusätzlich zu sprachlichen Anweisungen bei der Interaktion mit dem Computer benutzt werden.

Schlußfolgerungen und Aufgaben für die Zukunft

Im Hinblick auf die am Anfang gestellten Fragen über die funktionelle Bedeutung der verschiedenen Segmente der Fixationsdauer in der menschlichen Informationsverarbeitung liefern diese Experimente erstaunlich deutliche Ergebnisse. In der Reihenfolge der Aufgaben von der Informationssuche in der visuellen Umgebung, über perzeptive, semantische und selbst-referentielle Verarbeitung zur Kommunikation des Wissens an Partner bei kooperativem Problemlösen scheinen scheinen mit einer Verlängerung der Fixationsdauer verknüpft zu sein. Eine zusätzliche Validierung durch neurophysiologische Daten zeigte, daß die Veränderungen der Verarbeitungstiefe eine Veränderung der Ebene der entsprechenden Hirnmechanismen bedeuten [2, 28]. Besonders bei kommunikativen Intentionen findet man außergewöhnlich lange Fixationen, die genau dem empirisch gefundenen Schwellenwert für Blickmausanwendungen von 500 ms entsprechen. Eine anschauliche Zusammenfassung dieser Ergebnisse ist in Abbildung 6 zu finden.

Unsere praktische Empfehlung zur Überwindung des „Midas touch problem" ist deshalb die Einstellung des zeitlichen Filters auf Werte über 450-500 ms. Diese Schwelle scheint auch aus einem anderem Grund von Bedeutung zu sein. Bei der sprachlichen Beschreibung einer Szene (wie auch beim lauten Lesen) eilen Augen und visuelle Aufmerksamkeit der Sprache immer voran, so daß eine zeitliche Spanne zwischen Augen und Stimme (Eye-Voice Span) entsteht. Diese Spanne ist sehr variabel, beträgt aber in der Regel etwa 400-500 ms [27]. Mit anderen Worten, nur wenn der Blickort mit einer Latenz von etwa 500 ms expliziert wird, werden der sprachliche Fokus und die visuelle Aufmerksamkeit auf demselben Objekt konvergieren. Solche Redundanz verstärkt die Zuverlässigkeit der Angaben und schafft Möglichkeiten zum Zustand der „gemeinsamen Aufmerksamkeit"

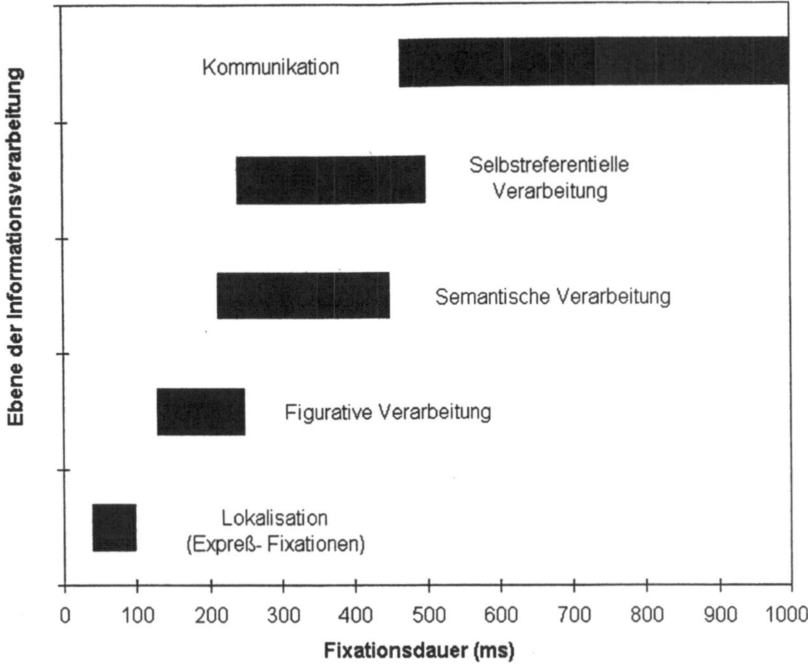

Abb. 6: Fixationsdauer als Indikator der Ebenen der Informationsverarbeitung

Es gibt weitere technische Möglichkeiten, die Blickmaus in ein völlig sicheres Kommunikationsmittel zu verwandeln. Man könnte anstelle individueller Fixationen die kommulativen Zeiten nahliegender Gruppen von Fixationen als die Auslösezeit für Schaltknöpfe verwenden, besonders wenn dazu die Technik des Aufbaus von sogenannten „Aufmerksamkeitslandschaften" verwendet wird [15, 27]. Im Gegensatz zur manuell gesteuerten Computer-Maus braucht eine Blickmaus nicht unbedingt ein visuelles Feedback, ein solches z.B. in Form von „Eyecons" [8, 25] fügt der Interaktion jedoch zusätzlich ein hohes Maß an Zuverlässigkeit bei. In einer weiteren Entwicklung könnte eine eklektische Kombination von Blickort, Blickdauer, Pupillendurchmesser und Blinzelhäufigkeit als Indizes von Interesse oder verschiedener Interpretationen einer Szene von einem neuronalen Netzwerk erkannt werden, das durch Lernabschnitte individuell eingestellt wird [14, 19, 25].

Natürlich bedeuten solche Sicherungsmaßnahmen weitere Verzögerungen bei der Initiation der Blickmaus, die noch weiter über dem kritischen Wert 500 ms liegen können. Das scheint aber nicht ein permanenter Nachteil dieser Technologie zu sein. Mehrere Beobachtungen zeigen, daß infolge des Übungseffekts der kritische Werte unter 500 ms verlagert werden kann, ohne daß die Zuverlässigkeit der Antworten gefährdet wird. Nach einem Bericht [4] können geübte Nutzer mit einem Schwellenwert von unter 250 ms arbeiten. Nach unserer Klassifikation entspricht das einer Verarbeitung im perzeptiven Bereich, wahrscheinlich sogar nur unter Berücksichtigung der räumlichen Lokationen der virtuellen Tasten. Offensichtlich kann auch das Auge die Evolution der Hand nachvollziehen, zu einem Instrument neuer menschlicher Fertigkeiten, die durch technologischen Fortschritt ins Leben gerufen wurden.

Danksagungen

Unser Dank gilt Mitarbeitern und Kollegen in Dresden, Bielefeld und Toronto, die an diesen Untersuchungen teilgenommmem haben, insbesondere Frank Binkowski, Rosalía Hillers, Constanze Liebers, Eyal Reingold und Helge Ritter. Wir bedanken uns auch bei Nancy und Dixon Cleveland (LC Technologies, Inc., Fairfax, VA) für die Information über Trainierbarkeit der menschlichen Okulomotorik.

Literaturverzeichnis

[1] Baluja S. & Pomerleau D.: Non-intrusive gaze-tracking using artificial neural networks. In: Neural information processing systems 6, New York, 1994: Morgan Kaufman Publishers.

[2] Blaxton, T.A., Bookheimer, S.Y., Zeffiro, T.A., Figlozzi, C.M., Gaillard, W.D. & Theodore, W.H.: Functional mapping of human memory using PET: Comparison of conceptual and perceptual tasks In: Canadian Journal of Experimental Psychology, 50 (1996), 42-56.

[3] Challis, B.H., Velichkovsky, B.M. & Craik, F.I.M. Levels-of-processing effects on a variety of memory tasks: New findings and theoretical implications. In: Consciousness & Cognition, 5 (1996), 63-89.

[4] Cleveland, D. & Cleveland, N.: Persönliche Mittelung, 1996 September.

[5] Deacon, T. W.: Prefrontal cortex and symbolic learning. In: B.M.Velichkovsky and D.M.Rumbaugh (Eds.), Communicating meaning. Mahwah, NJ, 1996: Lawrence Erlbaum Associates.

[6] Engell-Nielsen, T., Glenstrup, A.J. & Hansen, J.P.: Eye-gaze interaction: A new media – not just a fast mouse. In: International Journal of Human-Computer Interaction Studies, 1996 (in press).

[7] Fischer, B. & Weber, H.: Express saccades and visual attention. In: Behavioral and Brain Sciences, 16, (1993), 553-610.

[8] Hansen, J.P., Andersen, A.W. & Roed, P. Eye-gaze control of multimedia systems. In Y.Anzai, K.Ogawa and H.Mori (eds), Symbiosis of human and artifact. Proceedings of the 6th international conference on human computer interaction. Amsterdam, 1995: Elsevier Science Publisher.

[9] Helmholtz, H.v.: Handbuch der physiologischen Optik, 3.Auflage. Hamburg und Leipzig, 1896: L.Voss.

[10] Hering, E.: The theory of binocular vision. New York, 1977: Plenum Press [Erste deutsche Ausgabe, 1868]

[11] Hoffmann, J.E.: Visual attention and eye movements. In: H.Pashler (Ed.), Attention. London, 1996 in press: University College London Press.

[12] Jacob, R.J.K.: Eye tracking in advanced interface design. In: W.Barfield and T.Furness (Eds.), Advanced interface design and virtual environments. New York, 1995: Oxford University Press.

[13] Joch, A.: What pupils teach computers. In: Byte 21 (1996), 99-102.

[14] Nielsen, J.: Noncommand user interfaces. In: Communications of the ACM, 36(4), 83-99, 1993.

[15] Pomplun, M., Ritter, H. & Velichkovsky, B.M. Disambiguating complex visual information: Towards communication of personal views of a scene. In: Perception (UK) **25 (8)**, 1996.

[16] Pomplun, M., Velichkovsky, B.M. & Ritter, H.: An artificial neural network for high precision eye movement tracking. In B.Nebel & L.Drescher-Fischer (Eds.), Lectures notes in artificial intelligence. Berlin, 1994: Springer Verlag.

[17] Pomplun, M., Sichelschmidt, L., Wagner, K., Cleremont, C., Rickheit, G. & Ritter, H: Comparative visual search. In: Journal of Experimental Psychology: Human Perception & Performance, 1996 submitted.

[18] Raeithel, A. & Velichkovsky, B.M.: Joint attention and co-construction of tasks. B.Nardi (Ed.), Context and consciousness: Activity theory and human-computer interaction. Cambridge MA, 1995: MIT Press. 199-233.

[19] Ritter, H. Parametrized self-organizing maps. In S.Gielen and B.Kappen (Eds.), ICANN93-Proceedings, Berlin, 1993: Springer.

[20] Stampe, D.M. & Reingold, E. Eye movement as a response modality in psychological research. In Proceedings of the 7th European conference on eye movements, Durham, University of Durham, 31st of August-3d of September, 1994.

[21] Starker, I. & Bolt, R. A. A gaze-responsive self-disclosing display. In CHI 90 Proceedings. New York, 1990: ACM Press.

[22] Velichkovsky, B.M.: The levels endeavour in psychology and cognitive science. In: P.Bertelson, P.Eelen and G.d'Ydewalle (Eds.), International perspectives on psychological sciences: Leading themes.Howe UK, 1994: Lawrence Erlbaum Associates.

[23] Velichkovsky, B.M.: Communicating attention: Gaze position transfer in cooperative problem solving. In: Pragmatics and Cognition, 3 (1995), 199-222. 23.

[24] Velichkovsky, B.M., Challis, B.H. & Pomplun, M.: Arbeitsgedächtnis und Arbeit mit dem Gedächtnis: Visuell-räumliche und weitere Komponente der Verarbeitung. In: Zeitschrift für experimentelle Psychologie, 42 (1995), 672-701.

[25] Velichkovsky, B.M. & Hansen, J.P.: New technological windows into mind. In: CHI 96 Proceedings: Human factors in computing systems. New York, 1996: ACM Press. 496-504.

[26] Velichkovsky, B.M., Hillers, R. & Sprenger, A.: Enkodierung, Augenbewegungen und Gedächtnisleistung bei der Bearbeitung unbekannter Gesichter. 1996 in Vorbereitung.

[27] Velichkovsky, B.M., Pomplun, M. and Rieser. H.: Attention and communication: Eye-movement-based research paradigms. In: W.H.Zangemeister et al. (Eds.), Visual attention and cognition. Amsterdam, 1996: Elsevier Science Publisher. 125-154.

[28] Velichkovsky, B.M., Klemm, T., Dettmar, P. & Volke, H.-J.: Evozierte Kohärenz des EEG II: Kommunikation der Hirnareale und Verarbeitungstiefe. In: Zeitschrift für Elektroenzephalographie, Elektromyographie und verwandte Gebiete, 27 (1996), 111-118.

[29] Yarbus, A.L.: Eye movements and vision, New York, 1967: Plenum Press [Erste russische Ausgabe, 1965].

Adressen der Autoren

Prof. Dr. habil. Boris M.Velichkovsky
Dipl.-Psych. Andreas Sprenger
Technische Universität Dresden
Fachbereich Psychologie
AE Angewandte Kognitionsforschung
D-01062 Dresden
Email velich@psy1.psych.tu-dresden.de

Dipl.-Inf. Marc Pomplun
Universität Bielefeld
Technische Fakultät
AE Neuroinformatik
D-33615 Bielefeld
Email impomplu@techfak.uni-bielefeld.de

Voice Mail versus Type Mail

Hartmut Wandke, Anja Dresenkamp und Ute Janßen
Humboldt-Universität zu Berlin, Institut für Psychologie

Zusammenfassung

In einer experimentellen Untersuchung hatten zehn Paare von Versuchspersonen gemeinsam Aufgaben mit Hilfe eines elektronischen Organizers zu lösen. Die beiden Versuchspersonen waren räumlich getrennt. Eine Person verfügte über das Gerät, die andere über das Handbuch. Sie hatten die Möglichkeit über Email miteinander zu kommunizieren und konnten bei jeder Nachricht zwischen Voice Mail und Type Mail wählen oder beide Formen kombinieren. Überwiegend wurde Type Mail gewählt. Von den im Versuchsplan variierten und kontrollierten Bedingungen haben folgende einen Einfluß auf die Wahl der Mailform: Tastaturfertigkeit (langsame Vpn nutzen häufiger Voice Mail), Form der aktuell erhaltenen Nachricht (Beibehalten dieser Form), Inhalt der Nachricht (für Beziehungsinformation häufiger Voice Mail) und Länge der Nachricht (lange Nachrichten werden eher gesprochen). Die Bedingungen persönliche Bekanntschaft, Schwierigkeit der Aufgabe und Form der selbst zuletzt gesendeten Nachricht hatten keinen Einfluß auf die Wahl der Mailform. In der Interpretation wird darauf eingegangen, welche Effekte nur im Kontext der Versuchssituation gelten und welche verallgemeinert werden können.

1. Problemlage

In den letzten Jahren hat der Einsatz von Personalcomputern als Mittel der Telekommunikation [10], [1], [15] zwischen Personen eine bedeutende Ausweitung erfahren, insbesondere nachdem mit dem Internet in seiner heutigen Form eine universelle und globale Verbindungsaufnahme möglich war [6], [14]. Email wurde neben anderen noch schneller wachsenden Kommunikationsmitteln wie Mobiltelefonie und Fax zu einer wichtigen Form der Übermittlung von Nachrichten.

Die Benutzungsschnittstelle von Email-Systemen wurde u.a. Gegenstand software-ergonomischer Untersuchungen und Modellbildungen. [16], [22]. Dennoch weisen viele der gegenwärtig verbreiteten Mail-Programme immer noch einige software-ergonomische Mängel auf [11], [4]. Auch die in [20] berichteten Wissensdefizite von Email-Benutzern dürften z.T. auf software-ergonomische Mängel zurückzuführen sein.

Dies mag vielleicht eine Ursache dafür sein, daß die Anwendung und die Akzeptanz von Email hinter den Erwartungen und etwa im Vergleich zur Verbreitung von Fax oder Mobiltelefonie zurückgeblieben ist. So bringt z.B. [21] den Siegeszug von Faxgeräten mit der Unmöglichkeit zusammen, Grafiken mit einer Standard-Email zu versenden.

Durch eine weitere Verbesserung der Benutzungsschnittstellen wird eine größere Akzeptanz und eine stärkere Verbreitung von Email-Systemen erwartet. Insbesondere, wenn das leidige Tippen wegfallen würde (z.B. durch Spracheingabe - Voice Mail genannt), könnte es zu einer besseren Benutzbarkeit kommen (aus einer aktuellen Diskussion in der CWSW-SIG-Liste des Mailbase-Listserv).

Die Spracheingabe ist schon seit längerer Zeit [7] ein häufig gewünschtes Merkmal für eine ergonomische Benutzungsschnittstelle. Gedacht   war dabei weniger an   Email-Systeme, sondern eher an eine „hörende und verstehende Schreibmaschine". Wenn auch in den letzten Jahren erhebliche Fortschritte bei dieser Technologie erreicht wurden, so ist es dennoch verfrüht, von einem Durchbruch zu sprechen.

Anders ist die Situation jedoch bei Voice Mail Systemen: hier muß der gesprochene Text nicht erkannt und in eine schriftliche Form umgewandelt werden, sondern es reicht aus, ihn zu digitalisieren, zu übertragen, zu speichern und ihn schließlich wiederum akustisch auszugeben.

Obwohl dies seit einigen Jahren möglich ist, hat sich die Technik nicht sehr verbreitet. Ein möglicher Grund könnte sein: „Voice mail is effectively sent by the normal telephone networks, but the inclusion of a voice annotation in an email message will certainly become a more widespread possibility." [21, S. 373].

Ziel unserer Untersuchung war, herauszufinden, unter welchen Bedingungen Benutzer Type Mail oder Voice Mail wählen, wenn sie miteinander kommunizieren und innerhalb eines Email-Systems frei über das Medium der Kommunikation entscheiden können. Wir untersuchten natürlich nicht alle möglichen Einflußfaktoren, sondern wählten für das Experiment solche aus, die uns zum einen aufgrund von Alltagserfahrungen, Ergebnissen von Grundlagenexperimenten und psychologischen Theorien relevant erschienen, und zum anderen solche, die im Rahmen einer Pilostudie mit begrenztem Aufwand realisierbar waren.

Bevor die im Experiment variierten bzw. kontrollierten Bedingungen mit ihren hypothetisch erwarteten Bedingungen gegenüber gestellt werden, sollen einige allgemeine Vor- und Nachteile von Voice und Type Mail aufgelistet werden, die im Experiment nur z.T. berücksichtigt werden konnten.

Vorteile von Type Mail	Nachteile von Type Mail
Hoch geübte Schreiber können mit einer Tastatur schneller schreiben als sprechen.	Ungeübte Tastatur-Schreiber brauchen sehr viel Zeit zur Texteingabe.
Der Absender kann sicher sein, daß der Empfänger schriftliche Texte empfangen kann, wenn er über eine Email-Adresse verfügt.	Die Schriftsprache ist ausdrucksärmer als die gesprochene Sprache. Zeitliche Charakteristiken, Pausen, Stimmlagen, Intonation fehlen.
Der Text kann beim Absender sehr leicht kontrolliert und korrigiert werden.	Es können mehr Fehler gemacht werden (Tippfehler, Interpunktion, Groß- und Kleinschreibung, falsche Schreibweise von Fremdworten, die korrekt ausgesprochen werden).
Eine Nachricht kann direkt oder indirekt aus anderen Anwendungen (z.B. Textverarbeitungssystem) versandt werden.	Texte müssen oft auf die ASCII- Zeichen beschränkt werden. Umlaute, ß und andere Zeichen müssen umständlich transkribiert werden.
Der Text kann beim Empfänger sehr leicht weiterverarbeitet werden, z.B. in ein Textverarbeitungssystem integriert werden.	Absender mit motorischen Behinderungen können den Text nicht oder nur mit großen Aufwand eingeben.
Für den Empfänger ist Lesen einfacher als Hören.	Blinde Empfänger können den Text nicht lesen.
Der Text kann beim Absender und beim Empfänger ausgedruckt werden.	Die Texteingabe bindet die Hände und kann nicht parallel zu anderen manuellen Aktivitäten erfolgen.
Tabellen, Bilder und Grafiken können in Texte integriert werden.	

Tab. 1: Vor- und Nachteile von Type Mail

Andere Merkmale sind eher ambivalent und nur situationsabhängig als Vor- oder als Nachteil einzustufen: z.B. empfinden viele Benutzer, daß schriftliche Texte einen höheren Grad von

Verbindlichkeit haben und sind geneigt, Nachrichten eher als Dokumente wahrzunehmen, während verbal-akustische Nachrichten eher informellen Charakter tragen.

Die Vor- und Nachteile von Voice Mail Systemen stehen in der Regel in einem Kontrastverhältnis zu den Nach- und Vorteilen der Type Mail Systeme. Deshalb werden nur die bisher noch nicht erörterten Aspekte in der Tabelle 2 erfaßt.

Natürlich können auch Voice Mail und Type Mail kombiniert werden, um die verschiedenen Nachteile auszugleichen.

Vorteile von Voice Mail	Nachteile von Voice Mail
Die meisten Benutzer können wesentlich schneller sprecher als Zeichen mit einer Tastatur eingeben.	Der Absender weiß oft nicht, ob der Empfänger über Voice Mail verfügt.
Neben Sprache können andere akustische Daten übertragen werden.	Spezielle Hard- und Software bei Sender und Empfänger notwendig.
Bei bekannten Absendern kann der Empfänger sofort dessen Identität erkennen (erspart Verschlüsselungsprozeduren).	Sprachein- und ausgabe können störend wirken, wenn sich Sender oder Empfänger mit anderen Personen in einem Raum befinden.
Die Ähnlichkeit mit Telefon-Anrufbeantwortern erleichtert die Erstbenutzung.	Umgebungslärm wirkt sich störend beim Senden und Empfangen aus.
Spracheingabe erfordert sehr wenig Interaktionswissen.	

Tab. 2: Vor- und Nachteile von Voice Mail

Trotz der oben oft diskutierten Akzeptanzschwelle, haben Email-Systeme in speziellen Benutzergruppen großen Erfolg. Bei einer Umfrage, die 1995 an der Humboldt-Universität zu Berlin durchgeführt wurde, waren es bereits 70 %, die Email-Benutzer waren. Email war eine der häufigsten Funktionen (nach Textverarbeitung, Berechnungen und Grafikerstellung), für die Personalcomputer eingesetzt wurden [9].

Die Benutzung von Email-Systemen ist vor allem unter organisationspsychologischen Gesichtspunkten erforscht worden [20], Sproull und Kiesler, Bikson und Eveland, zitiert nach [8] und [17]. Spezielle Untersuchungen zu Voice Mail gibt es kaum. Zwei Forscher [18] untersuchten in einer Arbeit die Nutzung von und die Einstellung zu telefonbasierten Voice Mail Systemen in zwei Organisationen. Ein Ergebnis ihrer Untersuchungen ist, daß organisatorische Rahmenbedingungen (Innovationsbereitschaft) und Aufgabencharakteristika bestimmen, inwieweit Voice Mail tatsächlich genutzt wird. Je intensiver die Nutzung ist, desto besser fallen dann die subjektiven Bewertungen aus. In einer von uns durchgeführten Befragung zu Benutzererwartungen hinsichtlich der auf einem neuen Universitätscampus verfügbaren IuK-Technologien [9] nannten 75% der Befragten Email (Type Mail) als unverzichtbar (19% wünschenwert), während nur 1% Voice Mail als unverzichtbar bezeichnete (17% wünschenswert).

Die geringe Nachfrage hängt wahrscheinlich mit der geringen oder fehlenden Erfahrung mit dieser Mailform zusammen. Aus dem Alltagserleben weiß man, daß manche Personen eine Scheu davor haben, auf einen Anrufbeantworter zu sprechen. Eine mögliche Ursache besteht darin, daß in der natürlichen Umgebung des Menschen das Medium verbal-akustische Sprache mit einer synchronen Form der Kommunikation verbunden ist, während das Medium Schriftsprache für eine asynchrone Kommunikationsform eingesetzt wird. Eine weitere Ursache besteht wahrscheinlich darin, daß die einmal geäußerten Mitteilungen durch den Absender nicht mehr gelöscht oder geändert werden können. Da jede Nachricht neben einer Sachinformation auch Information über den Absender enthält (z.B. über sein Wisssen, seine

Sprachkompetenz, seine sprachliche Ausdrucksfähigkeit und Eloquenz), ist das (meist durch eine überraschende Aufforderung ausgelöste) Sprechen auf einen Anrufbeantworter z.T. von unangenehmen Emotionen begleitet. Bei Voice Mail dürfte dieser Effekt nicht auftreten, da ja Funktionen zum Löschen und Korrigieren *vor* dem Absenden zur Verfügung stehen. Ähnliches gilt für Benutzung eines Diktiergerätes.

Ein weiterer Hinweis auf die mögliche Verbesserung der Benutzbarkeit von Email-Systemen durch die Verwendung von Voice Mail läßt sich aus Untersuchungen zu (allerdings synchronen) Videokonferenzsystemen ableiten: Der Audio-Kanal ist bedeutend wichtiger als der Videokanal und wird auch subjektiv durch die Kommunikationspartner als bedeutsamer eingeschätzt [5], [13].

Eine andere Quelle, die zur Ableitung der Fragestellung und der Hypothesen herangezogen werden kann, sind Untersuchungen zu Multimedia-Systemen, bei denen die Sprach*ausgabe* im Mittelpunkt stand. Untersuchungen von [12] und [19] zeigen, daß die Informationsaufnahme beim Hören gesprochener Texte wesentlich langsamer erfolgt als beim Lesen der gleichen Texte. Die würde auf Seiten des Empfängers für Type Mail sprechen. Allerdings bestimmt in der Regel der Absender und nicht der Empfänger die Mailform.

2. Methodik und Hypothesen

2.1 Grundsituation

Die von uns untersuchte Situation entsprach der, die in der mittlerweile klassisch zu nennenden Arbeit von [3] entwickelt wurde: zwei Personen haben gemeinsam ein Problem zu lösen und können auf unterschiedliche Weise miteinander kommunizieren. (Übrigens hatte auch [3] herausgefunden, daß eine reine verbal-akustische Kommunikation fast so gut ist wie eine face-to-face Kommunikation.) An jedem Versuch nahmen zwei Personen teil. Die Versuchspersonen befanden sich in zwei verschiedenen Räumen. Eine Person erhielt ein kleines elektronisches Gerät, einen sogenannten Electronic Organizer (Casio, Digital Diary SF 5500B). Dieses Gerät weist eine Vielzahl von Funktionen auf, die in die Rubriken Terminplaner, Kalender, Uhr, Rechner, Telefonregister, Datenbank, Notizen eingeteilt sind. Insgesamt verfügt der Organizer über 35 Funktionen und ist mit Hilfe von 84 Tasten zu bedienen, die teilweise mehrfach belegt sind. Die zweite Person erhielt das zu dem Electronic Organizer gehörende Handbuch. Das Handbuch umfaßte 107 Seiten. Die Person mit dem Organizer erhielt nacheinander vier Aufgaben in aufsteigender Schwierigkeit, die sie mit dem Organizer lösen sollte. Um sie lösen zu können, wurde ihr empfohlen, über Email Kontakt mit der zweiten Versuchsperson aufzunehmen, die im Besitz des Handbuches war. In der Praxis finden sich solche Situation häufig als Hotline-Konsultation (dort allerdings als synchrone Kommunikation und meist über Telefon).

Jede Versuchsperson wurde durch eine gesonderte Versuchsleiterin betreut, die sie instruierte, während des Versuchs ihr Verhalten registrierte und das Abschlußinterview durchführte.

2.2 Stichprobe

An dem Versuch nahmen zwanzig Vpn in zehn Zweiergruppen teil. Es handelt sich um Studierende der Psychologie (mittleres Alter: 24,5 Jahre, 15 weiblich, 5 männlich). Durch Befragung und gegebenenfalls Selektion der Vpn wurde abgesichert, daß alle Vpn zuvor schon wenigstens einmal an einem Computersystemen gesessen hatten (meist nur

Textverarbeitung) und daß keine Versuchsperson zuvor das untersuchte Email-System benutzt hatte.

2.3 Hard- und Software zur Versuchsdurchführung

Die Versuche wurden an zwei Next-Workstations durchgeführt. Als Mailsystem dient Next-Mail. Für jeden Versuchsteilnehmer wurde eine gesonderte Mailbox angelegt, in der die eingegangenen Nachrichten gesammelt wurden. Type Mail Nachrichten wurden nach dem Versuch ausgedruckt und ausgewertet. Voice Mail Nachrichten wurden manuell transkribiert, um sie weiter auswerten zu können.

2.4 Unabhängige Variablen und Hypothesen

Es wurden zwei unabhängige Variable in der Versuchsplanung berücksichtigt, die erste im Sinne von getrennten Teilstichproben, die zweite im Sinne verbundener Stichproben.

Persönliche Bekanntschaft der Kommunikationspartner

Die Teilstichprobe wurde so ausgewählt, daß zehn der Versuchspersonen (fünf Paare) miteinander gut bekannt oder befreundet waren, während die anderen nicht oder nur flüchtig miteinander bekannt waren.

Hypothese 1: Persönliche Bekanntschaft führt zu einer häufigeren Verwendung von Voice Mail, da die Personen neben der Sachinformation auch Beziehungsinformation (z.B. Gefühle, ausgelöst durch die Kommunikation) übermitteln wollen und die Ausdrucksmöglichkeiten in der Voice Mail größer sind.

Schwierigkeit der Aufgabe

Die Aufgaben besaßen einen unterschiedlichen Schwierigkeitsgrad, der über die Anzahl der notwendigen Interaktionsschritte am Organizer operationalisiert wurde. Damit verbunden war eine größere Menge von Information, die die Person mit dem Handbuch an den Kommunikationspartner zu übermitteln hatte (mehr und/oder längere Nachrichten).

Hypothese 2: Bei schwierigen Aufgaben wird häufiger Type Mail benutzt, weil der Empfänger die umfangreichen Informationen besser aufnehmen kann, wenn sie in schriftlicher Form vorliegen [12], [19].

2.5 Kontrollierte Variablen und Hypothesen

Nach dem Versuch sollten die Ergebnisse in bestimmten Variablen des Versuchspersonen-verhaltens dazu benutzt werden, um die Stichprobe (der Personen und der ausgetauschten Nachrichten) neu zu unterteilen und Hypothesen hinsichtlich des Einflusses des tatsächlichen Verhaltens auf die Wahl des Kommunikationsmediums zu prüfen.

Tastaturfertigkeit

Alle Vpn hatten zu Beginn einen Standardtext, der gleichzeitig als Testnachricht diente, einzutippen. Dafür hatten sie vier Minuten Zeit. Der Median der Anzahl der korrekt getippten Zeichen in diesem Zeitraum diente als Kriterium für die Einteilung der Stichprobe in schnelle und langsame Schreiber.

Hypothese 3: Schnelle Schreiber bevorzugen Type Mail, langsame Voice Mail.

Länge der aktuellen Nachricht

Die Stichprobe aller ausgetauschten Nachrichten wurde nach dem Median aufgeteilt.

Hypothese 4a: Lange Nachrichten werden häufiger gesprochen, da Sprechen für den Absender einfacher ist.

Hypothese 4b: Lange Nachrichten werden häufiger getippt, da der Absender weiß, daß Lesen für den Empfänger einfacher ist.

Mailform der gerade erhaltenen Nachricht

Die Stichprobe aller Nachrichten wurde danach unterteilt, ob die zuvor erhaltene Nachricht schriftlich oder verbal-akustisch übermittelt wurde.

Hypothese 5: Benutzer beantworten eine Nachricht mit der Mailform, in der sie sie empfangen haben, da 1. ein Wechsel zusätzliche Operationen erfordert und 2. ein Beibehalten einer impliziten Abstimmung des Kommunikationsverhaltens entspricht.

Mailform der zuvor selbst gesendeten Nachricht

Die Stichprobe aller Nachrichten wurde danach unterteilt, ob die in einem Kommunikationszyklus zuvor selbst gesendete Nachricht schriftlich oder verbal-akustisch übermittelt wurde.

Hypothese 6: Benutzer bleiben häufiger bei einer einmal gewählten Mailform, da jeder Wechsel zusätzlichen Aufwand bedeutet (Vorteil der Gewohnheitsbildung).

Inhalt der ausgetauschten Information

Die Stichprobe aller ausgetauschten Nachrichten wurde von zwei Auswerterinnen unabhängig voneinander in verschiedene Kategorien (Frage, Bitte, Anweisung, Antwort, persönliche Anrede, persönliche Bemerkung) eingeordnet. Für die Hypothesenbildung fassen wir alle Kategorien mit ausschließlich Sachinformationen einerseits und Kategorien der Beziehungsebene, sowie Kategorien mit emotionalen Inhalten andererseits zusammen.

Hypothese 7: Für Nachrichten mit sachlichen Inhalten wird häufiger Type Mail verwendet, während bei Nachrichten mit Inhalten zur Beziehung und mit emotialen Anteilen häufiger mit Voice Mail (wegen der größeren Ausdrucksmöglichkeit) verwendet wird.

2.6 Versuchsablauf

Alle Versuchspersonen absolvierten den Versuch in folgendem Ablauf:

1. Allgemeine Instruktion

In der Instruktion wurde das Versuchsziel nicht genannt. Die Versuchspersonen wußten zu Beginn nicht, daß wir herausbekommen wollten, unter welchen Bedingungen welche Mailform verwendet wird.

2. Erläuterung der Benutzung des Email-Programms (Type und Voice Mail)

3. Versuchsperson A: Erläuterungen der Basisfunktionalität und der wichtigsten Interaktionstechniken des Organizers (Cursorsteuerung, Shift, Space, Escape, Modustasten)

Versuchsperson B: Erläuterung des Aufbaus des Handbuches, Aufforderung zum Blättern im Handbuch

4. Präsentation der ersten Aufgabe, während der Aufgabenbearbeitung protokollierte eine Versuchsleiterin die Interaktionsschritte der Versuchsperson A mit dem Electronic

Organizer.
5.-7. Aufgaben zwei bis vier
8. Getrenntes Abschlußinterview mit beiden Versuchspersonen, in dem u.a. auf die subjektive Bevorzugung oder Ablehnung der beiden Mailformen eingegangen wurde.

3. Ergebnisse

Bevor die Resultate hypothesenbezogen dargestellt werden, sollen einige generelle Aspekte der Versuchsergebnisse wiedergegeben werden. Insgesamt haben die Vpn 161 Nachrichten ausgetauscht, davon 128 in schriftlicher Form und 26 in verbal-akustischer. Nur sehr wenige Nachrichten (sieben Stück) waren ein Mix aus Voice Mail und Type Mail.
Die Variationen beim Gebrauch von Voice Mail und Type Mail waren sehr groß (siehe Abbildung 1). Es gab zwei Paare, die nur schriftlich miteinander kommunizierten und eines, das überwiegend verbal-akustische Nachrichten austauschte. Es gab kein Paar, das ausschließlich die Spracheingabe benutzte.

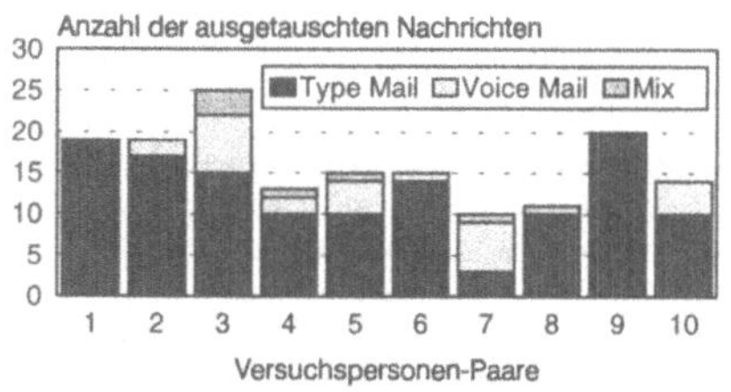

Abb. 1: Verteilung der Nachrichten nach dem Typ der benutzten Mail-Funktion. Überwiegend wird Type Mail benutzt. Nur ein Versuchspersonen-Paar (Nr. 7) benutzt Voice Mail häufiger als Type Mail.

Pro Versuchspersonen-Paar ergeben sich damit im Mittel 16,3 Nachrichten (Standardabweichung 4,6 Nachrichten). Beide Partner sendeten in etwa die gleiche Anzahl von Nachrichten, im Mittel waren es 8,7 bei den Versuchspersonen mit dem Organizer und 7,6 bei den Versuchspersonen mit dem Handbuch.
Eine Nachricht war im Mittel 23 Worte lang. Allerdings war es so, daß die Kommunikationsanteile ungleich verteilt waren. Die Versuchspersonen mit dem Handbuch sendeten im Mittel längere Texte als die Versuchspersonen mit dem Organizer (30,4 Worte zu 16,8 Worte, dieser Effekt ist statistisch signifikant, Chi^2-Test, 1%-Niveau). Das Ergebnis ist leicht erklärlich, da die Personen mit dem Handbuch ja die Auskunftgebenden waren.

Im folgenden werden die Ergebnisse hypothesenbezogen dargestellt.
Hypothese 1: Persönliche Bekanntschaft führt zu einer häufigeren Verwendung von Voice Mail.

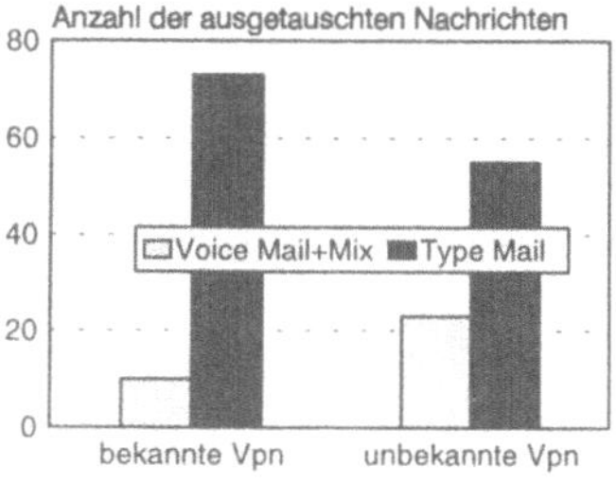

Abb. 2: Die Häufigkeitsdaten zeigen, daß die Hypothese nicht bestätigt werden kann, sondern daß scheinbar das genaue Gegenteil zutrifft: Voice Mail wird von den unbekannten Personen häufiger benutzt als von den bekannten.

Bei diesem überraschenden Effekt (signifikant auf dem 1%-Niveau, Chi^2-Test), dürfte es sich jedoch um einen statistischen Artefakt handeln: zum einen dominiert in beiden Gruppen ganz deutlich die Verwendung von Type Mail, zum anderen entsteht der Effekt durch die relativ große Anzahl von ausgetauschten Nachrichten. Wenn man die *Anzahl der Paare*, die überhaupt Voice Mail+Mix verwenden, betrachtet, so sind es fünf bei den unbekannten Vpn und drei bei den bekannten. Auf dieser Basis ist natürlich kein statistischer Vergleich möglich. Als Fazit bleibt festzuhalten, daß unabhängig von der persönlichen Beziehung der Personen Type Mail bevorzugt wird.

Hypothese 2: Bei schwierigen Aufgaben wird häufiger Type Mail benutzt, weil der Empfänger die umfangreichen Informationen besser aufnehmen kann, wenn sie in schriftlicher Form vorliegen.
Wir stellen in Abhängigkeit von den Aufgaben die Häufigkeiten von Voice Mail und Type Mail gegenüber.

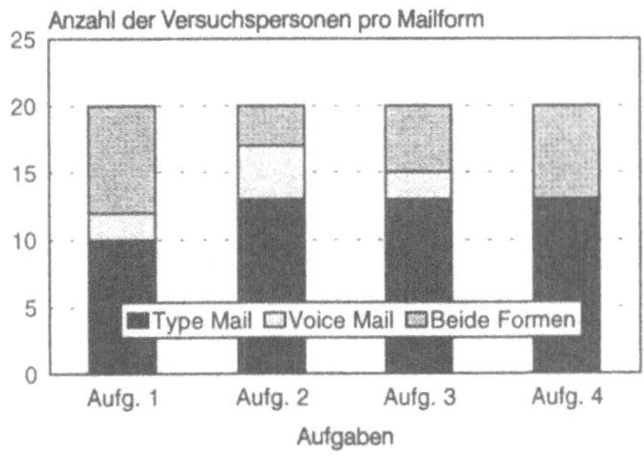

Abb. 3: Für alle Aufgaben wird mindestens von der Hälfte der Versuchspersonen allein Type Mail zur Kommunikation verwendet. Für die erste Aufgabe wählen 10 Versuchspersonen auch oder allein Voice Mail, während für die letzte Aufgabe von allen Vpn immer auch Type Mail verwendet wird. Statistisch gesehen, sind jedoch die Wahlhäufigkeiten für alle vier Aufgaben nicht signifikant voneinander unterschieden.

Die Hypothese zum Einfluß der Aufgabenschwierigkeit ist somit nicht bestätigt worden. Die Versuchspersonen wählen für alle Aufgaben größtenteils Type Mail. Einige Versuchspersonen verwenden unabhängig von den Aufgaben zusätzlich Voice Mail. Nur vier von zwanzig Versuchspersonen kommunizieren durchgängig während der Lösung einer Aufgabe mit Voice Mail.

Hypothese 3: Schnelle Schreiber bevorzugen Type Mail, langsame Voice Mail.
Der Median der Tastenanschläge, die innerhalb von vier Minuten ausgeführt wurden, lag bei 407. Die Tastaturfertigkeit der Versuchspersonen variierte nicht sehr stark. Die schnellste Versuchsperson schaffte 529 Anschläge, die langsamste 105. (Standardabweichung: 123,4)
Die Abb. 4 zeigt die Häufigkeitsverteilung von geschriebenen und gesprochenen Worten für schnelle (> 407 Anschläge) und langsame (< 407 Anschläge) Tipper.

Die Hypothese 3 kann bestätigt werden. Die Unterschiede bei der Wahl der Mailform können zum Teil auf die Tastaturfertigkeiten der kommunizierenden Versuchspersonen zurückgeführt werden. In der Abb. 4 werden die Häufigkeit der Mailformen und die Länge der ausgetauschten Nachrichten in der Anzahl der Worte über alle Nachrichten zusammengefaßt. Obwohl langsamere Vpn stärker zur Voice Mail tendieren, ist ihre dominierende Mailform immer noch die Type Mail.

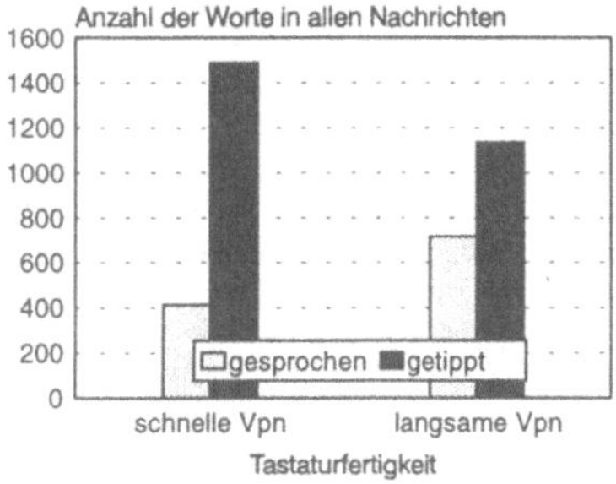

Abb. 4: Versuchspersonen mit geringeren Tastaturfertigkeiten (langsame Vpn) verwenden in stärkerem Maße Voice Mail als Versuchspersonen mit hohen Tastaturfertigkeiten (schnelle Vpn). Dieser Unterschied ist statistisch signifikant (Chi^2-Test 1%-Niveau).

Hypothese 4: Die Länge der Nachricht beeinflußt die Auswahl der Mailform.

4a: Lange Nachrichten werden häufiger gesprochen, da Sprechen für den Absender einfacher ist.

4b: Lange Nachrichten werden häufiger getippt, da der Absender weiß, daß Lesen für den Empfänger einfacher ist.

Zur Prüfung dieser Hypothesen wurden die Mittelwerte der Nachrichtenlängen für alle Voice- und alle Type Mail-Nachrichten bestimmt. Wie die Abbildung 5 zeigt, sind Voice Mail-Nachrichten deutlich länger.

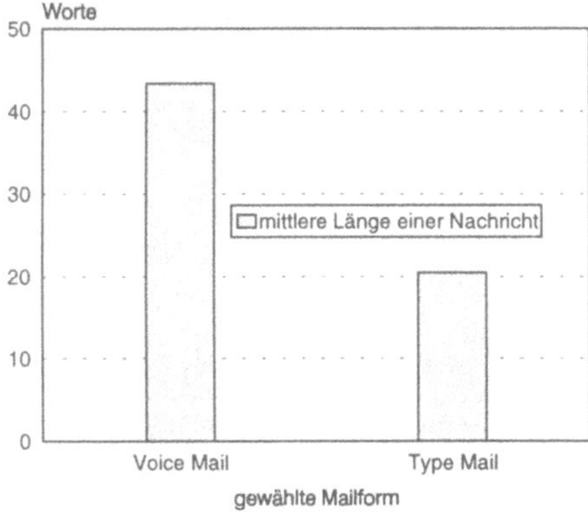

Abb. 5: Voice Mail Nachrichten sind mehr als doppelt so lang als Type Mail Nachrichten. Dieser Unterschied ist statistisch signifikant (Chi^2-Test 1%-Niveau).

Damit scheint 4a zuzutreffen. Wir können jedoch nicht unterscheiden, ob Benutzer Voice Mail gewählt haben, weil sie beabsichtigten, eine längerer Nachricht zu senden, oder ob die erfolgte Wahl dazu führte, daß einfach länger gesprochen wurde. Die Beobachtungen der Versuchsleiterinnen legen letzteres nahe, allerdings ohne es mit Daten belegen zu können.

Hypothese 5: Benutzer beantworten eine Nachricht mit der Mailform, in der sie sie empfangen haben.

Zu diesem Zwecke wurden die relativen Häufigkeiten für die vier möglichen Abfolgen Type Mail → Type Mail, Voice Mail → Voice Mail, Voice Mail → Type Mail und Type Mail → Voice Mail bestimmt. Diese relativen Häufigkeiten werden verglichen mit den theoretisch bestimmbaren relativen Abfolgehäufigkeiten, wobei angenommen wird, es handle sich um unabhängige Ereignisse. Die Abb. 6 zeigt die Ergebnisse zum Wechsel, und zum Beibehalten der Mailformen.

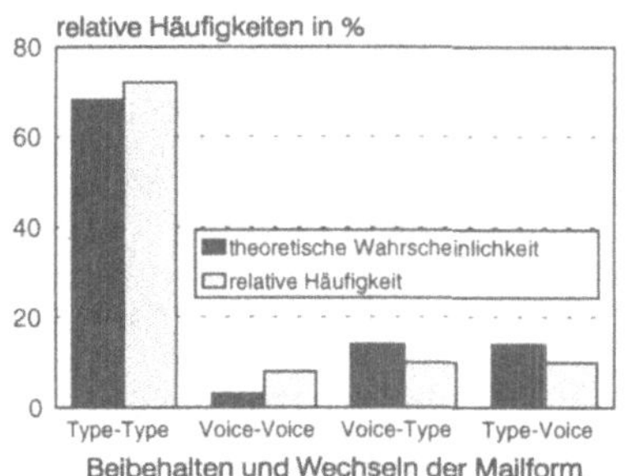

Abb. 6: Relative Häufigkeiten des Beibehaltens oder des Wechselns der Mailform in Abhängigkeit von der gerade erhaltenen Nachricht. Es zeigt sich, daß die empirisch ermittelten relativen Häufigkeiten beim Beibehalten größer sind als die theoretisch bestimmten, während die Wechselhäufigkeiten kleiner sind. Die empirisch ermittelten Häufigkeiten unterscheiden sich von den theoretisch angenommenen signifikant (Chi^2-Test, Irrtumswahrscheinlichkeit kleiner als 6%).

Es ist in der Tat so, daß es eine Tendenz zur Beibehaltung der Mailform gibt. Die Hypothese 5 kann somit als angenommen betrachtet werden, wenngleich die klassische 5%-Signifikanzgrenze knapp verfehlt wurde.

Hypothese 6: Benutzer bleiben häufiger bei einer einmal gewählten Mailform, da jeder Wechsel zusätzlichen Aufwand bedeutet.
Im Unterschied zu der vorgehenden Hypothese geht es nicht um ein Beibehalten der Mailform bezüglich der gerade empfangenen Nachricht, sondern bezüglich der letzten selbst gesendeten Nachricht.
Die Auswertungsprozedur lief ähnlich wie bei der Prüfung der Hypothese 5 ab. Die Abbildung 7 zeigt die Ergebnisse.

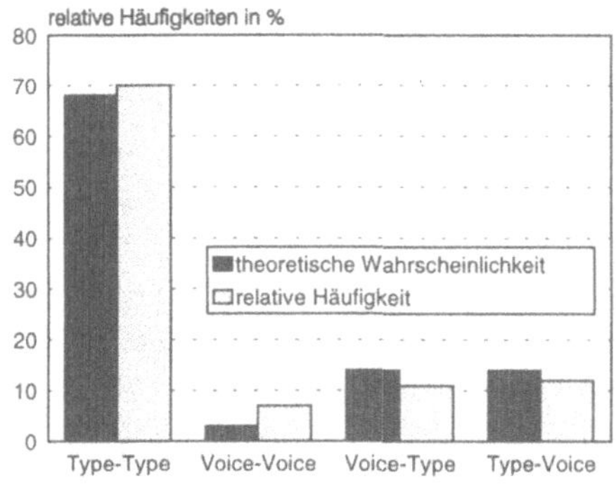

Abb. 7: Relative Häufigkeiten des Beibehaltens oder des Wechselns der Mailform in Abhängigkeit von der selbst zuletzt *gesendeten* Nachricht. Es zeigt sich ein ähnliches Bild wie bei der Abhängigkeit von der zuletzt *empfangenen* Nachricht, allerdings sind die (sehr geringen) Unterschiede zwischen den empirisch ermittelten Häufigkeiten und den theoretisch angenommenen nicht mehr signifikant (Chi^2-Test, Irrtumswahrscheinlichkeit größer als 30%).

Die Hypothese 6 muß somit abgelehnt werden. Das heißt, für die Wahl einer Mailform spielt es keine Rolle, ob die zuvor gesendete Mail in derselben oder einer anderen Form verfaßt wurde.

Hypothese 7: Für Nachrichten mit sachlichen Inhalten wird häufiger Type Mail verwendet, während für Nachrichten mit Inhalten zur Beziehung und mit emotialen Anteilen häufiger Voice Mail verwendet wird.

Jede Nachricht wurde von zwei Beurteilern unabhängig voneinander in die beiden Kategorien Sachebene und Beziehungsebene eingestuft. Die Einstufung stimmte in 94,6% aller Fälle überein. Die so klassifizierten Nachrichten wurden dahin gehend überprüft ob sie per Type Mail oder per Voice Mail gesendet wurden. Die Abb. 8 zeigt die Ergebnisse.

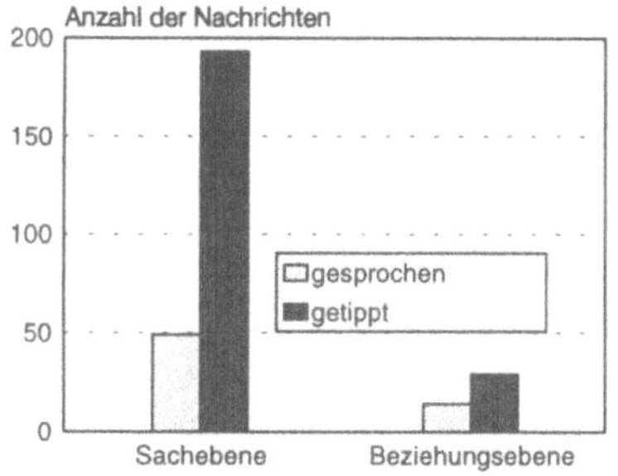

Abb. 8: Insgesamt wurden wenige Nachrichten (genau waren es 43) ausgetauscht, die auch Informationen zur Beziehung zwischen den beiden Versuchspersonen enthielten. Der relative Anteil von Voice Mail war mit 33% jedoch signifikant höher als der Voice Mail-Anteil auf der Sachebene (insgesamt 242 Nachrichten, davon 20% Voice Mail). Die Hypothese 7 kann damit angenommen werden (Chi^2-Test, 5%-Niveau).

Bei diesen Ergebnissen gilt, analog zu den vorausgegangenen Auswertungsschritten, daß Voice Mail relativ, aber nie absolut, bevorzugt wird, wenn die Nachrichten Informationen zur Beziehungsebene enthalten.

4. Schlußfolgerungen

Die Ergebnisse zeigen, daß sowohl Type Mail als auch Voice Mail von den Versuchspersonen relativ einfach benutzt werden können und daß unter den gegebenen Versuchsbedingungen eine eindeutige Bevorzugung von Type Mail auftritt. Wie die Ergebnisse des abschließenden Interviews zeigen, scheint die Bevorzugung von Type Mail eher eine generalisierbare Tendenz, denn ein versuchsspezifischer Effekt zu sein. Einige Versuchspersonen äußerten Unbehagen darüber „in eine Maschine hineinzusprechen" oder erklärten, daß sie diese Kommunikationsform als „unnatürlich" empfinden. Möglicherweise ist die Benutzung der verbal-akustischen Sprache, die ja in der interpersonalen *synchronen* Kommunikation entstanden ist, für die prizipiell *asynchron* arbeitenden Email-Programme nicht geeignet.
Hier wäre ein Vergleich zu der sich entwickelnden Internet-Telefonie bzw. zu Diktiersystemen interessant.
Die größere Ausdruckskraft der verbal-akustischen Sprache führt nicht zu einer Bevorzugung dieser Mailform. Dieser Effekt ist aber wahrscheinlich versuchsspezifisch. In der vorliegenden Situation waren hauptsächlich sachliche Informationen zu übertragen, die dann unmittelbar an Operationen mit dem Gerät umgesetzt wurden. Diskussionen, Verhandlungen, Überzeugung und Überredung des Kommunikationspartners, Umgang mit unsicherem Wissen, Streitgespräche, Kompromißsuche oder andere stärker emotional beeinflußte Kommunikationen fanden nicht statt. In diesen Fällen könnte es jedoch zu einer stärkeren Verwendung von Voice Mail kommen[1].

---

[1] Der Versuch, ein konflikthaltiges Element und damit negative Emotionen wie Enttäuschung oder Ärger in die Kommunikation einzuführen (die letzte Aufgabe war nicht lösbar und die Versuchsperson mit dem Electronic Organizer sollte auf vermeintlich falsche Instruktionen der Versuchsperson mit dem Handbuch hin die Kommunikation abbrechen), erwies sich als nicht praktikabel.

In diesen Zusammenhang gehört auch das Ergebnis, daß die persönliche Bekanntschaft keinen Einfluß auf die Wahl der Mailform hat. Es ist wahrscheinlich wichtig, welchen Inhalt eine Nachricht hat, und nicht so wichtig, ob man mit dem Empfänger persönlich gut bekannt ist. Dafür spricht auch das Ergebnis, daß der Inhalt (Sach- oder Beziehungsebene) durchaus die Wahl der Mailform beeinflußt.

Der Einfluß der Tastaturfertigkeit ist offensichtlich. Daß aber auch langsamere Schreiber überwiegend Type Mail verwendeten, ist dagegen überraschend. Hier spielen möglicherweise Stichprobenmerkmale eine Rolle. Alle Vpn (Studenten) konnten zumindestens recht und schlecht tippen. Die langsamen Schreiber benötigten im Mittel 0,80 Sekunden für einen Tastenanschlag. Das ist zwar sehr viel mehr als geübte Schreibkräfte benötigen, aber auch wesentlich weniger als die von [2] angegebenen 1,20 Sekunden pro Anschlag für Personen, die die QWERTY-Tastatur nicht kennen.
Möglicherweise ist Voice Mail eine gute Alternative für Personen, die überhaupt nicht mit einer Tastatur Texte schreiben. Die Ergebnisse legen zudem den Schluß nahe: wenn einmal Voice Mail gewählt wurde, dann wird auch diese Kommunikationsform intensiver genutzt, d.h. es wird länger gesprochen.
Die Ergebnisse belegen, daß Personen bei der elektronischen Kommunikation auf das Kommunikationsverhalten des Partners eingehen, so bei der Wahl der Mailform. Sie sind eher geneigt, in der Mailform zu antworten, in der sie eine Nachricht empfangen haben. Dieses Phänomen kann auch im Sinne einer generellen kognitiven Aufwandssenkung interpretiert werden: jeder Wechsel erfordert zusätzliche kognitive Operationen, die vermieden werden sollen. Interessant ist, daß das Beibehalten sich eher auf die letzte (empfangene) Nachricht bezieht und nicht auf die vorletzte (letzte gesendete) Nachricht. Dies spricht dafür, daß eine Anpassung an den Partner stärker ist als die persönliche Bevorzugung einer Mailform. Allerdings muß hier einschränkend festgehalten werden, daß aufgrund der Neuheit des Mediums Email und der geringen Dauer des Versuchs keine persönlichen Bevorzugungen gebildet werden konnten.
Ob die genannten Ergebnisse verallgemeinerbar sind, läßt sich nur in weiteren Untersuchungen feststellen. Insbesondere halten wir empirische Untersuchungen zur praktischen Benutzung von Voice und Type Mail für notwendig, um etwas zur Bevorzugung der unterschiedlichen Techniken sagen zu können. Aber auch weitere experimentelle Untersuchungen, bei denen z.B. unterschiedliche Gruppen *entweder nur* mit Type Mail *oder nur* mit Voice Mail kommunizieren, können weitere Aufklärung zu den benutzerbezogenen Vor- und Nachteilen erbringen.
Eines scheint jedoch schon jetzt festzustehen: allein die Verfügbarkeit von Voice Mail löst nicht die Probleme, die gegenwärtig bei der Benutzung von Email-Systemen konstatiert werden.

5. Literaturverzeichnis

[1] Bannon, L.J. (1986). Computer-Mediated Communication. in D. A. Norman and St.W. Draper (eds.) User centered system design. Hillsdale : Lawrence Erlbaum Ass. 433-542
[2] Card, S., T. Moran and A. Newell (1983). The Psychology of human-computer interaction. Hillsdale : Lawrence Erlbaum Ass.
[3] Chapanis, A. (1975). Interactive human communication. Scientific American, 323, 36-42

[4] Chingaira, N. (1995). Ergonomical comparison of Email-Systems. Semesterarbeit am Institut für Informatik der Humboldt-Universität zu Berlin, unveröffentlicht

[5] Gale, St. (1990). Human aspects of interactive multimedia communication. Interacting with Computers, 2, 153-172

[6] Gaffin, A. (1993). Big Dummies Guide to the Internet. Washington, D.C. : Electronic Frontier Foundation

[7] Gould, J. D., J. Conti and T. Hovanyecz (1983): Composing letters with a simulated listening typewriter. Communcations of the ACM. 26, 295- 308

[8] Hasebrook, J. (1995). Multimedia-Psychologie. Heidelberg : Spektrum Akademischer Verlag

[9] Hensel, J. (1995). Informations- und Kommunikationstechnik in Adlershof: Eine Befragung an den mathematisch-naturwissenschaftlichen Fakultäten der Humboldt-Universität zu Berlin. Semesterarbeit am Institut für Psychologie der Humboldt-Universität zu Berlin, unveröffentlicht

[10] Hiltz, S.R. and M. Turoff (1978): The network nation: human communication via computer. London : Addison-Wesley

[11] Hoang Thi, T. H. (1995). Schwierigkeiten beim Umgang mit verschiedenen Email-Systemen. Semesterarbeit am Institut für Informatik der Humboldt-Universität zu Berlin, unveröffentlicht

[12] Ito, N., S. Inoue, M. Okhura and W. Masada (1989). The effect of voice messages on the interactive computer system. in: Klix, F., N.A. Streitz, Y. Waern and H. Wandke (eds.): Man-Computer Interaction Research - MACINTER II, Amsterdam : North-Holland, 245-252

[13] Kawalek, J. (1994b). Der Einsatz von Videokonferenzen in Schulungen: Welche Konsequenzen hat der Einsatz unterschiedlicher Konferenzsysteme? Vortrag gehalten auf dem 39. Kongreß der Deutschen Gesellschaft für Psychologie. 25.-29. September 1994 in Hamburg.

[14] Kehoe, B.P. (1994). Zen und die Kunst des Internet. New York u.a. : Prentice Hall

[15] Maaß, S. (1991). Computerunterstützte Kommunikation und Kooperation. in: Oberquelle, H. (Hrsg.) Kooperative Arbeit und Computerunterstützung. Göttingen : Verlag für Angewandte Psychologie, 11-35

[16] Moran, T. P. (1981). The Command Language Grammar: a representation for the user interface of interactive computer systems. Int. J. Man-machine Studies, 15, 3-50

[17] Olaniran, B. A. (1996). A model of group satistaction in computer-mediated communication and face-to-face meetings. Behaviour and Information Technology, 15, 24-63

[18] Rice, R. E. and Tyler, J. (1995). Individual and organizational influences on voice mail use and evaluation. Behaviour and Information Technology, 14, 329-341

[19] Rinck, M. und U. Glowalla (im Druck). Die multimediale Darstellung quantitativer Daten. Zeitschr. f. Psychologie

[20] Scholl, W., J. Pelz, J. Rade (1996). Computer-vermittelte Kommunikation in der Wissenschaft. Münster : Waxmann-Verlag

[21] Shneiderman, B. (1992). Designing the user interface. Reading : Addison-Wesley

[22] Veer, G.C. van der, S. Guest, P. Haselanger, P. Innocent, E. McDaid, L. Oesterreicher, M. J. Tauber, U. Vos, Y. Waern (1990). Designing for the mental model: an interdisciplinary approach to the definition of a user interface for electronic mail systems. in: D. Ackermann amd M. J. Tauber (eds): Mental Models and Human-Computer-Interaction 1.: Amsterdam u.a. : North Holland

Adresse der Autoren

Prof. Dr. Hartmut Wandke • cand. psych. Anja Dresenkamp • cand. psych. Ute Janßen
Humboldt-Universität zu Berlin
Institut für Psychologie
Oranienburger Str. 18
10178 Berlin
E mail: hwandke@rz.hu-berlin.de

ViewNet - Konzeptionelle Gestaltung und Modellierung von Navigationsstrukturen

Jürgen Ziegler

Fraunhofer-Institut für Arbeitswirtschaft und Organisation

Zusammenfassung

Der Beitrag stellt einen Ansatz zur Modellierung von Navigationsstrukturen bei datenbankorientierten Anwendungen mit graphischen Benutzungsschnittstellen sowie hypertext-artigen Strukturen dar. Im Gegensatz zu existierenden Techniken, die sich eher auf die Spezifkation der Dynamik des Dialogablaufs konzentrieren, steht bei ViewNets die Darstellung der Semantik der einzelnen Knoten der Dialogstruktur im Vordergrund. Hierzu werden Sichtentypen eingeführt und durch graphische Symbole veranschaulicht. Bausteinkonzepte und komplexe Sichten bieten die Möglichkeit zur Darstellung auch umfangreicher Navigationsstrukturen in übersichtlicher Form.

1 Einleitung

Eines der Hauptprobleme bei der Entwicklung komplexer Anwendungssysteme mit graphischen Benutzungsschnittstellen liegt heute in zunehmendem Maß in der Auslegung der Dialogstruktur einer Anwendung „im Großen", d. h. der Gesamtarchitektur der Benutzungsschnittstelle. Während für die Verwendung und Gestaltung der Interaktionselemente graphischer Benutzungsschnittstellen Gestaltungsregeln z. B. in Form von Style Guides vorliegen, ist die Frage der Konzeption geeigneter Dialogstrukturen vom Entwickler für die jeweilige Anwendung meist spezifisch zu lösen und wird bestenfalls durch anwendungs- oder firmenspezifische Richtlinien unterstützt. Gerade die Auslegung der Dialogpfade durch die unterschiedlichen Bereiche des Systems ist aber für die Benutzbarkeit von zentraler Bedeutung, insbesondere wenn der Benutzer die grundlegende Handhabung graphischer Oberflächen bereits beherrscht.

Zur besseren Eingrenzung gegenüber den Dialogteilen, die zur Manipulation von Interaktionsobjekten (Listen, Felder, Buttons etc.) dienen, soll hier im folgenden der Begriff *Navigation* für die Bewegung des Benutzers durch die Sichten eines System verwendet werden. Unter einer *Sicht* soll eine Zusammenfassung von Elementen verstanden werden, die ein oder mehrere zugrundeliegende Anwendungsobjekte oder Aufgaben in zusammenhängender Weise z. B. in einem Fenster oder einer Maske für den Benutzer sichtbar macht.

Zur Modellierung von Navigationsabläufen wurde bislang eine Reihe von Techniken entwickelt, die im wesentlichen eine genaue Beschreibung des Systemverhaltens in Abhängigkeit von Systemzustand und Benutzereingaben zum Ziel haben. Diese Techniken basieren typischerweise auf Zustandsübergangsdiagrammen [1], [2] oder Petrinetzen [3]. Durch die detaillierte Spezifikation von auslösenden Ereignissen, Bedingungen etc. bergen sie jedoch die Gefahr in sich, den Blick auf die Gesamtstruktur für den Entwickler eher zu verstellen als offenzulegen. Im Bereich von Hypermedia-Systemen gibt es zwar verschiedene Ansätze zum strukturellen Entwurf der Navigation [4], [5], diese sind aber für datenbank-orientierte Anwendungen wenig geeignet, bei denen spezielle Abläufe aufgrund der Kardinalitäten von Objektklassen (viele Instanzen gleichen Typs) erforderlich werden.

Die angeführten Methoden sagen auf der anderen Seite wenig über Art und Eigenschaften der bei der Navigation zu durchlaufenden Sichten aus (in der Regel nur den Namen). Sie können daher als reich an Transitionsinformationen, aber arm an Zustandsinformationen bezeichnet werden. Diese Information ist aber für eine konsistente und verständliche Systemgestaltung wichtig, in den frühen Entwurfsphasen meist wichtiger als eine genaue Ablaufspezifikation. Es fehlt bislang eine Methode, die beim Navigationsentwurf beiden Aspekten Rechnung trägt und besonders die Bedeutung der Sichten, aus denen die Navigationsstruktur aufgebaut ist, deutlich macht. In diesem Beitrag wird die Methode **ViewNet** vorgestellt, die diese Anforderungen berücksichtigt und den Entwickler beim konzeptionellen Entwurf von Navigationsstrukturen unterstützt.

2 Prinzipien der Navigationsstrukturierung

Navigationsdialoge können nach unterschiedlichen Prinzipien strukturiert werden. Ausschlaggebend ist dabei der jeweils dominante Aspekt der Aufgabenstellung des Benutzers, der als Ausgangspunkt zur Auslegung des Dialogpfades für die Aufgabenerledigung dient [6]. Grundlegende Aufgabenaspekte sind das zu bearbeitende Objekt, der Zustand des Objektes oder die auszuführende Funktion. Hinzu kommen bei reinen Informationsbeschaffungsaufgaben Art und Beziehung der gewünschten Informationen. Entsprechend lassen sich folgende Navigationsprinzipien unterscheiden:

Funktionsorientierte Navigation: Ausgangspunkt für die Navigation bildet die Auswahl einer bestimmten Funktion (z. B. in einem konventionellen hierarchischen Menüsystem). Die eigentliche Datensicht wird erst nach einem oder mehreren Folgeschritten sichtbar bzw. manipulierbar.

Prozessorientierte Navigation kann als Erweitung des funktionsorientierten Prinzips gesehen werden und bietet Unterstützung für die Ausführung einer zu einem Arbeitsprozess gehörenden Menge von Aufgaben. Das System kann den Status des Prozesses kontrollieren und zustandsal..ngig den Zugang zu den benötigten Objekten und Funktionen ermöglichen. Während solche Systeme früher den Benutzer oft zwangen, eine fixe Schrittfolge einzuhalten, erlauben heutige graphische Oberflächen eine flexiblere Gestaltung, z. B. durch Visualisierung der gegenwärtig verfügbaren Aufgaben in einer Task-Leiste.

Objektorientierte Navigation verwendet die Objekte der Anwendung und deren semantische Beziehungen zur Auslegung von Dialogpfaden. Die Funktionen werden erst bei Sichtbarkeit des zu bearbeitenden Objekts verfügbar. Objektorientierte Navigation ist eines der Grundprinzipien graphischer Benutzungsschnittstellen dar und zeichnet sich besonders durch hohe Einheitlichkeit und Flexibilität aus.

Assoziationsorientierte Navigation soll die typische Navigationsform in Hypertext-/Hypermedia-Systemen bezeichnen. Hierbei stellen die Knoten der Navigationsstruktur beliebige einzelne Informationseinheiten dar (im Gegensatz zu festen Objekttypen mit beliebig vielen Instanzen bei objektorientierter Navigation). Die möglichen Übergänge werden durch die Assoziationen zwischen diesen Informationseinheiten bestimmt.

Diese unterschiedlichen Navigationsformen haben bei gegebenen Aufgabenstellungen unterschiedliche Profile hinsichtlich Benutzbarkeitskriterien wie Effizienz, Verständlichkeit oder Flexibilität. In realen Systemen treten daher oft Mischformen auf. Auf der Basis der aufgeführten Navigationsgrundformen sollen nun im folgenden unterschiedliche Typen von Sichten entwickelt werden, die die Knoten der Navigationstruktur darstellen.

3 Sichtentypen

Sichten bilden Anwendungsobjekte, Aufgaben oder allgemeine Informationseinheiten in einer zusammenhängenden Darstellung ab, wobei die Ausgangsobjekte ganz oder teilweise repräsentiert werden können. Mehrfache Sichten auf ein Objekt sind möglich. Sichten können sich hierarchisch aus Teilsichten zusammensetzen. Im folgenden werden die wesentlichen Typen von Sichten beschrieben und entsprechende graphische Symbole eingeführt, aus denen ein ViewNet-Diagramm aufgebaut wird.

Objektsichten stellen eine Instanz einer bestimmten Objektklasse (z. B. Kunde) in unterschiedlichen Formen dar (Abb. 1). Zu unterscheiden sind hier die Komplettsicht, meist in Icon-Form (Icon-Sicht), die Attributsicht, die alle oder einige der Attribute darstellt und die Graphiksicht, die zur Darstellung beliebiger graphischer Repräsentationsformen einer Instanz herangezogen wird (z.B. als Karte, Geschäftsgrafik etc.).

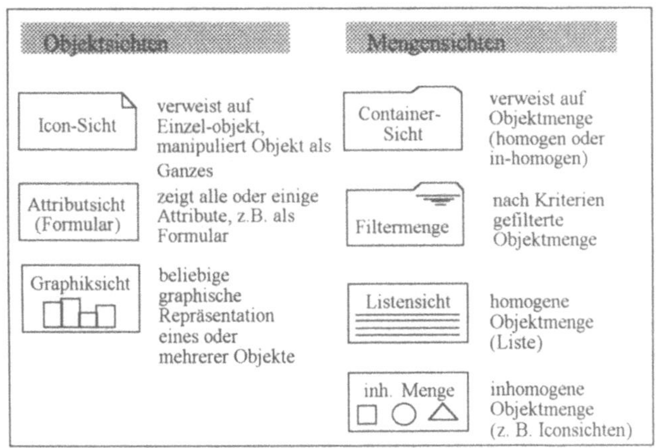

Abb.1: Objekt- und Mengensichten

Zur Strukturierung des objektorientierten Dialogs sind *Mengensichten* erforderlich, die entweder eine Gesamt- oder Teilmenge der Instanzen einer Objektklasse abbilden (homogene Mengen) oder Objekte unterschiedlichen Typs zusammenfassen (inhomogene Menge). Ein spezieller Fall einer Mengensicht sind gefilterte Mengen, durch die der Benutzer auf vor- oder eigendefinierte Untermengen der Datenobjekte zugreifen (z. B. alle fälligen Rechnungen). Ggf. kann der Benutzer solche Filterobjekte an der Benutzungsschnittstelle in flexibler Weise selbst anlegen und verwalten.

Funktionssichten zeigen eine oder mehrere Funktionen, ggf. mit Parametern, wie sie z. B. durch Menümasken oder modale Dialogboxen repräsentiert werden (Abb. 2). Objektattribute können in einer Funktionssicht mit angezeigt werden, sind aber nur in Verbindung mit der gewählten Funktion zu sehen (z. B. Suchattribute in einer Dialogbox 'Suchen'). *Aktorensichten* sind komplexe Funktionssichten, die den Benutzer durch einen zusammenhängenden Vorgang führen und in sehr unterschiedlicher Formen realisiert werden können(To-Do-Listen, Task-Leiste oder ein Assistenten-Fenster wie z.B. heute bei vielen Setup-Programmen verwirklicht).

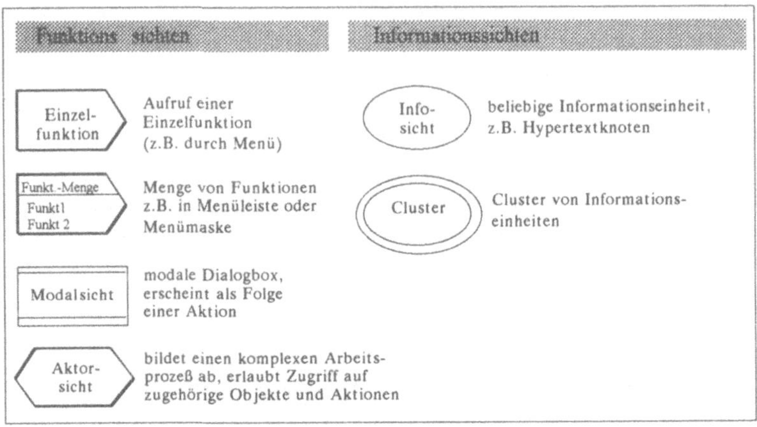

Abb.2: Dialog- und Informationssichten

Informationssichten werden in assoziationsorientierten Navigationen verwendet und können als Einzelknoten oder als Cluster dargestellt werden. Beispiele sind Inhalte eines Hilfesystems oder eines Hypermedia-Lernprogramms.

Abb. 3 zeigt *zusammengesetzte Sichten*. Typische, bei graphischen Schnittstellen häufig verwendete Konstrukte wie Master-Detail-Sichten oder Notebooks sind zwecks besserer Anschaulichkeit mit einem eigenen Symbol aufgeführt, das sich aus der konkreten visuellen Darstellung einer solchen Sicht ableitet. *Generische Bausteine* erlauben es, Teile einer Navigationsstruktur in einem Element zusammenzufassen und mit einem Parameter zu versehen. So kann z.B. ein aus mehreren Sichten bestehende Suchdialog für Instanzen einer Klasse mit dem Klassennamen parametrisiert werden. Gleichartige Suchdialoge für unterschiedliche Klassen (z.B. Auftrag, Kunde, Produkt) können dann durch ein einzelnes Bausteinsymbol mit Angabe der entsprechenden Klasse dargestellt werden. Hierdurch wird die Übersichtlichkeit verbessert und eine konsistente Navigationsentwicklung deutlich unterstützt.

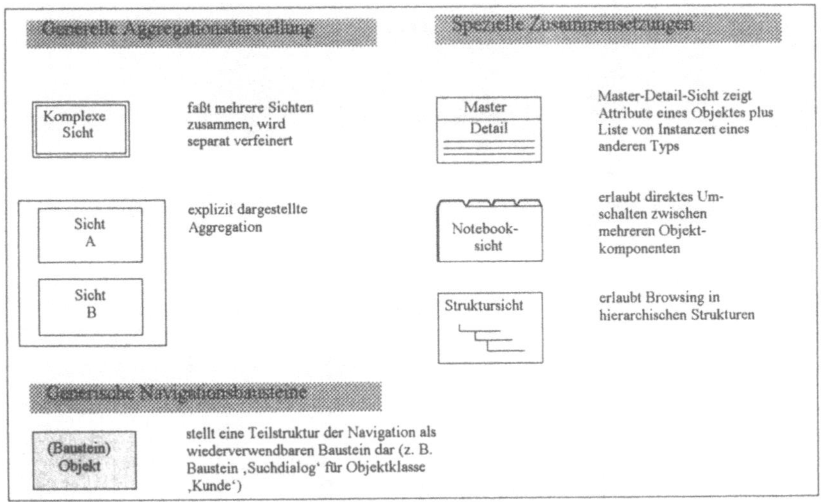

Abb. 3: Zusammengesetzte Sichten

4 Navigationsdarstellung mit ViewNets

Mit Hilfe der definierten Sichtentypen können Navigationsstrukturen in Diagrammform darge-stellt werden. Für eine übersichtliche Darstellung der Navigation werden ViewNets in ihrer einfachen Form als Erreichbarkeitsgraphen dargestellt. Hierbei werden Ereignisse, die die Na-vigationsschritte auslösen, sowie eventuelle Bedingungen und Aktionen nicht dargestellt. Er-weiterungen für eine detaillierte Spezifikation der Transitionen, etwa in Form von Dialognet-zen [3], lassen sich jedoch in die Darstellung ohne Schwierigkeiten einführen. Für den Zweck der konzeptionellen Gestaltung des Dialoges ist es aber meist als günstiger anzusehen, zu-nächst von den Details der Dialogführung zu abstrahieren.

Abb. 4 zeigt eine typische objektorientierte Dialogstruktur, wie sie z. B. beim Entwurf eines Auftragsverwaltungsystemes auftreten könnte. Die oberste Ebene der Navigation (z. B. ein Desktop) entspricht zwecks Vereinfachung der Darstellung der Zeichenfläche. Ausgehend von einer Icon-Darstellung von Objektmengen (Kunden) auf dem Desktop kann über entsprechen-de Listenauswahl ('Öffnen') oder Suchmechanismen ('Suchen') ein Zugriff auf die Attribut-sicht eines bestimmten Objektes erreicht wird, in der lokal entsprechende Operationen (z. B. als Buttons oder Menüeinträge) zur Verfügung stehen. Die Attributsicht für Kunde ist hier als Notebook-Sicht dargestellt, da angenommen werden soll, daß es sich um umfangreiche Daten handelt. Von den Details dieser zusammengesetzten Struktur wird in diesem Schritt noch ab-strahiert. Die Beziehungen zwischen Objektklassen (z. B. Kunde-Auftrag) werden durch Navi-gationspfade realisiert, wobei die Kardinalitäten der Beziehungen ggf. durch Mengensichten aufgelöst werden (hier z. B. durch die Liste 'Aufträge').

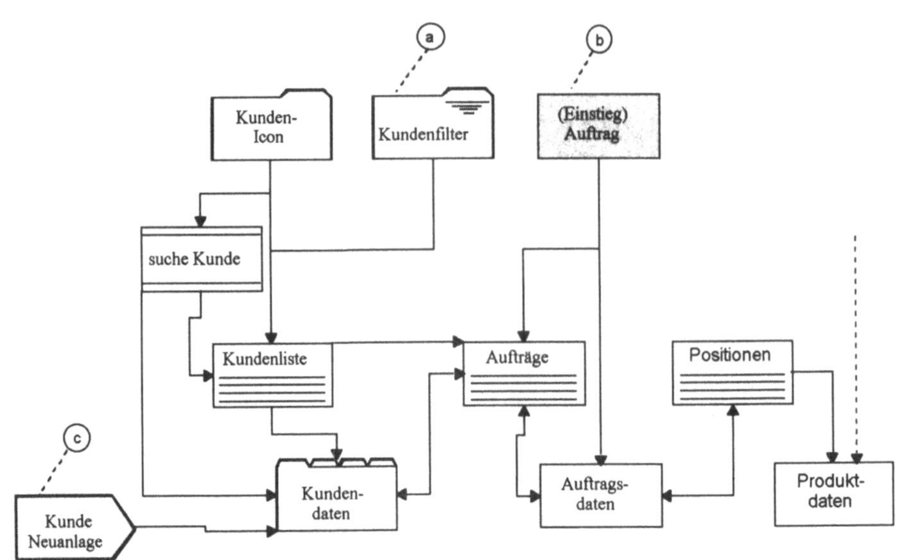

Abb. 4: Beispiel für die Darstellung einer Navigationsstruktur mit einem ViewNet

Das Element (a) zeigt zusätzlich die Möglichkeit, über gefilterte Objektmengen (z. B. Kunden Süddeutschland), die der Nutzer selbst definieren kann, wiederholt auf interessierende Objekte zuzugreifen. Die gesamte Substruktur für den Einstiegsdialog für Kunden (Icon-Suchdialog-Filter) kann zu einem generischen Baustein zusammengefaßt und in gleicher Weise für die Klasse 'Auftrag' zur Verfügung gestellt werden (b). Bausteindefinitionen erfolgen in separaten Diagrammen, so daß sie für unterschiedliche Dialoge wiederverwendet werden können. (c) zeigt einen zusätzlich zur objektorientierten Navigation eingeführten funktionalen Einstieg (z. B. über ein Icon oder einen Menüeintrag 'Kunde Neuanlage'). Funktionale Einstiege werden meist auf Effizienzgründen für häufig wiederkehrende Aufgaben eingeführt und können eine objektorientierte Grundstruktur überlagern.

Basierend auf einer anfänglichen ViewNet-Darstellung können die Navigationsabläufe im Bezug auf die Benutzeraufgaben weiter optimiert werden, wobei sich auch der Aufbau einzelner Sichten noch verändern kann. So wird man im Beispiel bezüglich der Darstellung von Auftragsdaten und Auftragspositionen bei üblichen Mengengerüsten zu dem Schluß kommen, daß eine Darstellung in einer Sicht (Master-Detail-Sicht) übersichtlicher ist und Dialogschritte sparen kann (Abb. 5). Die ViewNet-Darstellung stellt für solche Optimierungen eine anschauliche Grundlage dar, insbesondere auch für die Überführung von Objektbeziehungen mit Kardinalitäten 1:n und m:n in entsprechende Navigationsabläufe.

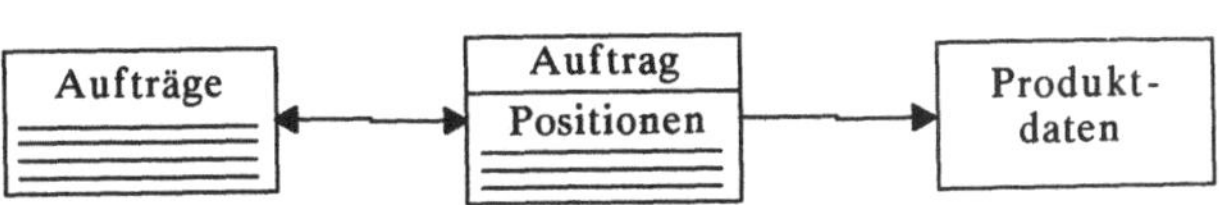

Abb. 5: Optimierung von Sichten und Abläufen für eine Teilstruktur des Beispiels in Abb. 4

ViewNets können nicht nur für Anwendungen mit Klasse-Instanz-Konstellationen, bei denen feste Sichtenstrukturen (Masken) mit wechselnden Inhalten gefüllt werden, sondern auch für Navigationen zwischen hypertext-organisierten Informationseinheiten oder Mischformen aus diesen Ansätzen verwendet werden. Abb. 6 zeigt einen möglichen Ausschnitt aus einem Internet-basierten Produktinformationssystem, bei dem Datenbank- und Hypertext-Komponenten gemeinsam auftreten.

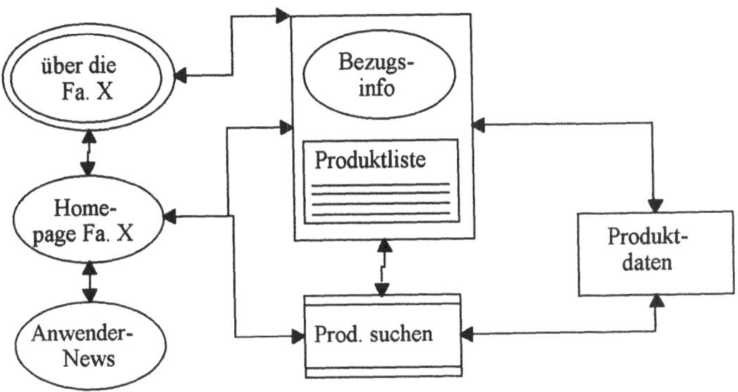

Abb. 6: Beispiel für eine gemischte Hypertext-/Datenbank-Navigation

Der Inhalt der einzelnen Sichten wird nicht in der graphischen Darstellung definiert, sondern z. B. in einem separaten Sichtendefinitionsschema aufgeführt. Dieses gibt die in einer Sicht dargestellten Attribut, ggf. die Darstellungsform sowie die für die Sicht verfügbaren Operationen an.

Diskussion

Die vorgestellte Modellierungstechnik ViewNet liefert im Gegensatz zu existierenden Dialogmodellierungstechniken eine reichhaltigere Beschreibung der Semantik der Sichten. Die Einführung unterschiedlicher Typen von Sichten unterstützt die Entwicklung der konzeptionellen Struktur einer Anwendung. Durch die Verwendung möglichst anschaulicher piktogrammhafter Symbole für die Darstellung der Sichten werden dem Entwickler (wie auch dem beteiligten Benutzer) visuelle Merkmale angeboten, die für ein schnelles Erfassen der angezeigten Struktur hilfreich sind. Der Entwickler kann in systematischer Weise Art und Funktion der verwendeten Sichten reflektieren und das der Navigationsstruktur zugrundeliegende Prinzip leichter erkennen.

Das Problem der Dialogstrukturierung kann damit leichter in einzelne Teilschritte zerlegt werden. So kann z. B. die Entwicklung der objektorientierten Navigationsgrundstruktur einer Anwendung (die bei graphischen Oberflächen meist sinnvoll ist) getrennt werden von der Erstellung zusätzlicher aufgaben- oder prozeßbezogener Dialogpfade. Die Ableitung des Navigationsmodells aus Objekt- und Aufgabenmodellen der Anforderungsanalyse kann in einer systematischen, regelgestützten Weise erfolgen [7], [8].

Der Übergang zu einer detaillierten Beschreibung des dynamischen Dialogverhaltens kann in die verwendete Diagrammtechnik ohne Schwierigkeiten integriert werden, z. B. mit den Konstrukten von (Petri-Netz-basierten) Dialognetzen. Der vorgestellte Ansatz geht allerdings von der Annahme aus, daß insbesondere in den frühen Gestaltungsphasen die konzeptionelle Struktur des Dialogs Vorrang vor der Dynamik des Dialogablaufs hat.

Die ViewNet-Technik bzw. Vorläuferversionen wurden in einer Reihe von Entwicklungsprojekten angewendet und qualitativ bewertet. Als besonders nützlich erwies sich die Darstellung dabei in Gruppendiskussionen, in denen gemeinsam die Struktur einer Anwendung erarbeitet wurde. Zukünftige Arbeiten werden sich auf eine entsprechende Werkzeugunterstützung, gerade auch für Gruppensituationen konzentrieren.

Literatur

[1] Denert, E. (1977): Specification and design of dialog system with state transition diagrams. In Morlet, E. & Ribbens, D. (Eds.), Proc. Int. Computing Symposium. Amsterdam: Noth-Holland, 417-427.

[2] Jacob, R.J.K. (1986): A specification language for direct manipulation user interfaces. ACM Transactions on Graphics, Vol. 5, 283-317.

[3] Janssen, Chr. (1993): Dialognetze zur Beschreibung von Dialogabläufen in graphisch-interaktiven Systemen. In K.-H. Rödiger (Hrsg.): Software-Ergonomie '93. Stuttgart: Teubner, 67-76.

[4] Nielsen, J. (1990): The art of navigating through hypertext. Communications of the ACM, March 1990, Vol. 33, No. 3, 296-310

[5] Berk, E. & Devlin, J. (1991): Hypertext/Hypermedia Handbook. New York: McGraw-Hill.

[6] Ziegler, J. & Janssen, Chr. (1995): Aufgabenbezogene Dialogstrukturen für Dialogsysteme. In: Böcker, H.-D. (Hrsg.): Software-Ergonomie '95 - Mensch-Computer-Interaktion - Anwendungs-bereiche lernen voneinander. Fachtagung Germ. Chapt. ACM , GI u. a. (20.-23.2.1995, Darmstadt), 395-406.

[7] Greutmann, Th. (1993): Datenmodellierung und aufgabengerechte Dialoge: ein Synchronisierungsproblem. In: Rödiger, K.-H. (Hrsg.): Software-Ergonomie '93 (Bremen 15.-17.3.1993), Stuttgart: Teubner.

[8] Janssen, C.; Weisbecker, A. & Ziegler, J. (1993): Generating user interfaces from data models and dialogue net specifications. In Proceedings of INTERCHI (Amsterdam, 24-29 April), New York: ACM, 418-423.

Adresse des Autors

Dr.-Ing. Jürgen Ziegler
Fraunhofer-Institut für Arbeitswirtschaft und Organisation
Nobelstraße 12, 70569 Stuttgart
Email:juergen.ziegler@iao.fhg.de

Das CAD-Referenzmodell auf dem Weg in die Praxis
Ein Beitrag zur Software-Ergonomie in der Konstruktion

Olaf Abeln

Forschungszentrum Informatik, Karlsruhe
http://www.igd.fhg.de/cadrm

Das vom Bundesministerium für Forschung und Technik im Rahmen des Programms „Arbeit und Technik" geförderte Projekt „CAD - Referenzmodell" greift die aktuellen Probleme der rechnerunterstützten Konstruktionsarbeit auf und erarbeitet Vorschläge in Richtung Organisationsmodell, Architektur und Einsatzmethoden, wie diese in Zukunft gestaltet werden kann. Erreicht werden soll die Neugestaltung einer zukunftsorientierten, flexiblen und computergestützten Produktentwicklung, die dem Anspruch einer menschengerechten Arbeits- und Technikgestaltung im Konstruktionsprozeß genügt.

Aus einer fundierten Analyse der wichtigsten Schwachstellen heutiger Konstruktionsarbeit wurde ein integriertes Organisations- und Technikkonzept entwickelt, aus dem Lösungsvorschläge für eine neue Systemarchitektur und neue Arbeitsweisen der computerunterstützten Konstruktionsarbeit abgeleitet wurden. Zur Zeit wird dieses Konzept in verschiedenen Unternehmen modellhaft umgesetzt, um zu zeigen, daß die zugrundeliegenden Lösungsvorschläge in besonderer Weise geeignet sind, die Anforderungen an eine innovative, flexible und vor allem menschengerechte Arbeitsgestaltung der Zukunft zu erfüllen

In dem Workshop sollen in verschiedenen Beiträgen Ergebnisse aus dem laufenden Projekt vorgestellt werden, wie die rechnerunterstützte Arbeit der Unternehmen im Engineering-Bereich verbessert werden kann. Praxisbeispiele und Demonstrationen sollen diese verdeutlichen und anschaulich machen, in Diskussionen sollen die Teilnehmer miteinbezogen werden.

- Die Funktionen und die Ziele des CAD-Referenzmodells *(Prof. O. Abeln)*

- Benutzerpartizipation bei der Anpassung von CAD-Systemen *(Dipl.-Ing. Thorsten Siodla)*

- CAD-Benutzungsoberflächen: Arbeitswissenschaftliche Anforderungen und Möglichkeiten der informationstechnischen Umsetzung *(Dr. Peter Martin, Dipl.-Inf. Andre Sork)*

- Telekooperationstechniken in CAD *(Dipl.-Ing. Ute Dietrich)*

- Von Produktmodellen zu Wissensmodellen bei der Lösungsfindung im Produktentwicklungsprozeß *(Dr. Stefan Rude)*

- Benutzungsgerechte Gestaltung und Handhabung von Produktmodellen in Produktentwicklungsprozessen *(Dipl.-Ing. Haygazun Hayka)*

Sinn und Unsinn bei der Umsetzung der EU- Richtlinie 90/270 für Bildschirmarbeitsplätze

Ulrich Klotz

IG Metall
Ulrich_Klotz@magicvillage.de

Marion Wittstock

Berlin
triwitt@berlin.snafu.de

Matthias Rauterberg

ETH Zürich
rauterberg@ifap.bepr.ethz.ch

Das zur Zeit vielleicht wichtigste Diskussionsthema im Bereich Software-Ergonomie ist die EU- Richtlinie 90/270 für Bildschirmarbeitsplätze.

Bei der Umsetzung und Anwendung gehen die Meinungen der Experten weit auseinander: Von der Low- Cost- Lösung der Arbeitgeber- und Herstellerseite bis hin zur arbeitsintensiven Meßmethodik der Wissenschaftler.

Der Workshop soll Vertreter der verschiedenen Lager an einen Tisch bringen, um die Diskussion mittels Öffentlichkeit zu versachlichen.

Ziel dieses Workshops ist es *nicht*, die Inhalte der EU- Richtlinie und ihre Einbettung in die anderen Regelwerke darzustellen, sondern eine Diskussionsplattform zu schaffen, auf der die *verschiedenen* Sichten deutlich werden:

- Wer überprüft die Zertifizierer?
- Wie können die Defizite im Bereich der praktischen Umsetzung behoben werden?
- Wie kann verhindert werden, daß dieBildschirmrichtlinie auf eine „Möbelrichtlinie" reduziert wird?

Im Rahmen des Workshops werden verschiedene eingeladene Fachexperten ihre Standpunkte zum Thema darstellen.
Neben einer kurzen Standortbestimmung aller Diskutanden soll der moderierten Diskussion möglichst viel Platz eingeräumt werden.
Die Workshop - Besucher werden an der Diskussion beteiligt.

Zielgruppen:
- Soft- und Hardwarehersteller,
- Prüfinstitutionen,
- Anwender (Arbeitgeber),
- Benutzer (Arbeitnehmer),
- Berater

Aufgabenbasierte Gestaltung der Benutzungs-schnittstellen: Vom Aufgaben- zum Dialogmodell

Birgit Bomsdorf, Gerd Szwillus

Universität - GH Paderborn
kne@hni.uni-paderborn.de
szwillus@uni-paderborn.de

Die Entwicklung von Benutzungsschnittstellen hat sich zunehmend von einem technischen Softwareproblem zu einer umfassenden Gestaltungsaufgabe gewandelt. Dabei überwiegt heutzutage die Meinung, daß ein wesentliches Kriterium bei der Gestaltung einer Benutzungsschnittstelle die Aufgaben sind, die der Benutzer an bzw. mit dieser Schnittstelle durchführen soll. Somit wird die Aufgabenmodellierung zentraler Bestandteil bei der Erstellung einer aufgaben- und benutzungsorientierten Mensch-Maschine-Schnittstelle. Auf dieser Idee basierende Ansätze verfolgen die Zielsetzung, den Übergang von einer Betrachtung der Aufgaben zur konstruktiven Gestaltung der Schnittstelle systematisiert zu behandeln.

Eine aufgaben- und benutzerorientierte Schnittstellengestaltung erfordert die gesamtheitliche Berücksichtigung verschiedener Methoden und Modelle. In diesem Workshop möchten wir uns auf die Fragestellung konzentrieren, wie die Ergebnisse der Aufgabenanalyse (insbesondere Aufgabenmodelle) in das Dialogdesign (bzw. in das Dialogmodell) miteingehen können. Dieses Vorgehen beinhaltet die Problematik einer möglichst kontinuierlichen Unterstützung verschiedener Schritte und die Integration unterschiedlicher Techniken:

- Die semi-formale Beschreibung der Aufgaben, die sich in ihrerer Formulierung an der Sichtweise und dem Vokabular des Benutzers orientiert, muß in eine präzise Spezifikation des Verhaltens der Benutzungsschnittstelle transformiert werden, deren Terminologie von der softwaretechnischen Realisierung geprägt ist.
- Zu den in der Aufgabenmodellierung auftretenden Aufgaben müssen geeignete Interaktionsobjekte und Interaktionssequenzen zugeordnet werden, mit denen der Benutzer die jeweilige Aufgabe durchführen kann.

Ziel des Workshops ist es, Arbeiten und Erfahrungen der Teilnehmer bei der Behandlung dieser Thematik aufzuarbeiten und miteinander in Bezug zu setzen. Wir sind dabei an allen Ansätzen - ob werkzeugunterstützt oder nicht - interessiert, die zum Ziel haben, das Aufgabenmodell systematisch in ein Dialogmodell zu überführen. Darüberhinaus möchten wir gemeinsam überlegen, um welche Aspekte die Aufgabenmodellierung erweitert werden müßte, um die Dialogspezifikation adäquat zu unterstützen bzw. vorzubereiten. Dabei könnte das Ergebnis ein Meinungsbild darüber sein,

- welche Aspekte des Dialogentwurfs auf naheliegende Weise aus dem Aufgabenmodell abgeleitet werden könnten,
- welche Aspekte des Dialogentwurfs sich einer solchen Ableitung aus systematischen Gründen entziehen und demgemäß grundsätzlich individuell entwickelt werden müssen,
- welche Konsequenzen man aus diesen Überlegungen für eine mögliche Werkzeugunterstützung ziehen muß,

insgesamt also eine Einschätzung darüber, wie groß sich die Kluft zwischen der Modellierung der Benutzeraufgaben und dem Entwurf des Dialogs darstellt.

User- Perceived Quality of Interactive Systems
Eine Bilanz nach 20 Jahren Software-Ergonomie in der BRD

Paul Hubert Vossen   Günther Gediga   Kai-Christoph Hamborg

Fraunhofer IAO, Stuttgart   Universität Osnabrück   Universität Osnabrück

Paulus.Vossen@iao.fhg.de

In dem Workshop werden der Werdegang von Ansätzen zur Normierung und Bewertung der Benutzerfreundlichkeit bzw. Benutzbarkeit von Software dargestellt und eine kritische Bestandsaufnahme bisheriger Arbeiten zu diesem Thema gemacht.
Es sollen aktuelle Ansätze in Hinblick auf ihre wissenschaftliche Fundierung und ihre Praktikabilität diskutiert werden.

Neben der Darstellung der Geschichte der Software- Ergonomie und einer Bestandsaufnahme bisheriger Arbeiten zu diesem Thema, soll auch über aktuelle Fragestellungen und Herausforderungen diskutiert werden, wie z.B.:

- Woher und wohin hat sich die „User- Perceived Quality" (UPQ) entwickelt und welche signifikanten Entwicklungen sind zu verzeichnen?

- Gibt es eine theoretische Einbindung der UPQ- Konzepte und in wieweit ist dies sinnvoll?

- Ist eine summative UPQ- Bewertung sinnvoll - wenn ja - für welchen Zweck kann und sollte man diese benutzen?

- Was können formative UPQ- Instrumente leisten? Sind aufwendige formative UPQ- Instrumente vom pragmatischen Gesichtspunkt her tragbar?

- Wie ist es um die Validierung (im weitesten Sinne) von UPQ- Instrumenten bestellt?

Derzeit ist es nicht befriedigend gelöst, wie man bei der Anwendung und Umsetzung der Norm im Spannungsfeld zwischen *empirischer Forschung* (methodologisch sauberer Ansatz) und *pragmatischer Beratung* (Kosten vs. Nutzen) einen „vernünftigen" Mittelweg wählen könnte.

Es wird eine Übersichtspräsentation zur historischen Entwicklung gegeben und es werden Ergebnisse einer Expertenbefragung zur Diskussion gestellt.

Zielgruppen:
Psychologen, Ergonomen, Informatiker, die sich (intensiv) mit der Bewertung und Überprüfung von Informationssystemen (MMK, HCI) befassen, insbesondere diejenigen, die sich zu diesem Zweck dem etablierten Repertoire der empirischen Forschung bedienen (möchten).

Usability elektronischer Dokumente für blinde Menschen

Wolfgang Wünschmann,Thomas Kahlisch
Technische Universität Dresden
wuenschmann@inf.tu-dresden.de, kahlisch@inf.tu-dresden.de

Der Workshop soll Gelegenheit bieten, aktuelle Trends der Erstellung und Präsentation elektronischer Dokumente vor dem Hintergrund ihrer Gebrauchstauglichkeit für blinde Menschen zu vermitteln und einer kritischen Wertung zu unterziehen.

Besonderer Schwerpunkt ist dabei das Puplizieren im WWW einschließlich der Vermittlung graphischer Bestandteile in elektronischen Dokumenten.

An Beispielen ausgewählter Lehrmaterialien soll die Erkenntnis vermittelt werden, daß bei der Gestaltung elektronischer Dokumente hohe Gebrauchswerteigenschaften erzielbar sind, die sowohl für sehende als auch für blinde Menschen gleichermaßen mit Vorteil nutzbar sind.

Es wird für erforderlich gehalten, dabei die software-ergonomischen Aspekte in den umfassenderen Rahmen der Gestaltung einer Studierumgebung einzuordnen.
Folgende Teilprobleme sollen ausführlicher diskutiert werden:

- Richtlinien zur Dokumentgestaltung
- Arbeit mit Tabellen
- Mathematische Notationen
- Tastbare Abbildungen
- Unterstützungssysteme mit auditiven Komponenten

Es ist vorgesehen, die aktuelle Situation der Nutzbarkeit von Tabellen innerhalb von HTML-Dokumenten anhand eines ausgewählten Lösungsansatzes praxisnah zu diskutieren.

Zusammenfassend soll eine Einschätzung der Workshopteilnehmer über die Anwendbarkeit der Norm ISO 9241/Teil 11 bezüglich der interaktiven Nutzung elektronischer Dokumente durch blinde Menschen erarbeitet werden.
Es wird angestrebt, dabei Aspekte der Gebrauchstauglichkeit elektronischer Dokumente unter sonstigen besonderen Bedingungen einzubeziehen.

Zielgruppen:
- Gestalter und Evaluatoren für Büro- und Lernumgebungen
- Gestalter von Benutzungsoberflächen für Bibliothekssysteme, Infotheken..
- Mitwirkende in Standardisierungsgremien mit Bezug zu elektronischen Dokumenten
- Software-Ergonomen mit Spezialisierung auf Informationsdienste in Datennetzen
- Interessenten für assistive Technologien
- Computerbenutzer mit besonderen Bedürfnissen

Modellbasiertes Prototyping zur Unterstützung einer aufgabenorientierten Schnittstellengestaltung

Birgit Bomsdorf

Universität - GH Paderborn
kne@hni.uni-paderborn.de

Gerd Szwillus

Universität - GH Paderborn
szwillus@uni-paderborn.de

Bei der Gestaltung von Benutzungsschnittstellen wächst die Bedeutung von aufgabenbasierten Ansätzen. Hierbei werden typischerweise zwei Aufgabenmodelle erstellt: Eines für die Beschreibung der Aufgaben *vor* Einführung des Systems und eines für die, im Hinblick auf das Zielsystem, *zukünftigen* Benutzeraufgaben. Trotz der unbestrittenen grundsätzlichen Bedeutung von Aufgabenmodellen werden sie derzeit in der Praxis nur in geringem Umfang eingesetzt. Dies ist auf eine unzureichende Integration mit bisherigen Methoden und Techniken der Softwaretechnik, auf mangelnde Ausdruckskraft für realistisch dimensionierte Projekte, als auch auf das Fehlen einer adäquaten Werkzeugunterstützung zurückzuführen. Auch ist eine direkte Weiterverwendung der erstellten Modelle im folgenden Entwicklungsprozeß im Sinne einer teilweisen Übernahme der erarbeiteten Informationen in Dokumente späterer Entwicklungsphasen nicht möglich.

Die hier vorgestellte statische Aufgabenmodellierung erlaubt die Beschreibung verschiedener, wesentlicher Beziehungen zwischen Aufgaben, wie etwa die hierarchische Strukturierung in Teilaufgaben, quantitative und qualitative temporale Relationen sowie Vor- und Nachbedingungen. Zusätzlich zur Beschreibung der Aufgabenstruktur wird ein Objektmodell entwickelt, welches die mit den Aufgaben verbundenen Objekte repräsentiert. Demgemäß bilden Querbezüge zwischen Aufgaben- und Objektmodell in unserem Ansatz einen expliziten Modellierungsgegenstand, wodurch eine parallele Entwicklung und frühzeitige Integration beider Modelle unterstützt wird. Zusätzlich können hier immer wiederkehrende Aufgaben allgemein formuliert und in verschiedenen Aufgabenkontexten situationsabhängig eingebunden werden.

Es gilt generell als schwierig, dynamische Systeme mittels statischer Mittel, wie Diagramme oder Text, adäquat zu repräsentieren und daraus ein korrektes und umfassendes Verständnis des Verhaltens unter Berücksichtigung aller Zusammenhänge zu gewinnen. Zur Unterstützung des Verständnisses der globalen Dynamik wird in unserem Ansatz aus der statischen Beschreibung eines Aufgabenmodells ein korrespondierendes, verhaltensäquivalentes dynamisches System generiert. Die verschiedenen Aspekte des Modells werden hierdurch zeitpunktgerecht zusammengefaßt und einander gegenüber gestellt.

In dem auf diesem Konzept basierenden Prototyp werden jeweils nach der simulierten Durchführung einer Aufgabe die aktuelle Situation und die möglichen bzw. notwendigen Folgeaufgaben ersichtlich. Insgesamt wird hiermit dem Entwickler ein Prototyping-Verfahren an die Hand gegeben, daß ein Testen der Aufgabenstruktur im Rahmen verschiedener Aufgabenszenarien - noch bevor die Entwicklung der Bedienoberfläche überhaupt begonnen hat - erlaubt. Ausgehend von einer abstrakten Darstellung der Aufgaben können mit fortschreitendem Entwicklungsprozeß sukzessive Oberflächenobjekte bzw. graphische Darstellungen eingebunden werden, wodurch das Testen früher Layoutentscheidungen für die Oberfläche im Kontext der Aufgaben ermöglicht wird.

HEPHAISTOS
Multimodale und multimediale Steuerung von Heimgeräten für ältere und behinderte Menschen

Michael Burmester, Franz Koller und Joachim Machate

Fraunhofer-Institut für Arbeitswirtschaft und Organisation

Email: Franz.Koller@iao.fhg.de
Email: Joachim.Machate@iao.fhg.de

Das Projekt HEPHAISTOS (Home Environment Private Help AssISTant fOr elderly and diSabled) wird im Rahmen des europäischen TIDE Programms (Technology Initiative for Disabled and Elderly) gefördert.

Ziel des HEPHAISTOS Projektes ist es, eine Fernbedienung zur Steuerung elektronischer Heimgeräte zu entwickeln, die speziell für die Bedürfnisse älterer und behinderter Menschen angepaßt ist. Die Fernbedienung besteht aus einem hochauflösenden und berührungsempfindlichen Grafikdisplay. Ferner können alle Funktionen der Heimgeräte mit Sprachkommandos angesprochen werden.

Die multimodale Benutzungsschnittstelle wurde so gestaltet, daß ältere Menschen und spezielle Gruppen behinderter Menschen (z.B. Multiple Sklerose, Querschnittslähmung) optimal mit dieser Benutzungsschnittstelle interagieren können. Die Anpassung der Benutzungsschnittstelle an die Bedürfnisse von älteren und behinderten Menschen wurde durch intensive Anforderungsanalysen in der Startphase des Projektes und durch einen iterativen Entwicklungsansatz mit Benutzerbeteiligung abgesichert. In dem Projekt wurden vier Entwicklungszyklen aus Design, Prototyping und Evaluation mit den Endbenutzern durchgeführt. Ergebnis des Projektes ist ein Demonstrator des HEPHAISTOS Systems, bestehend aus der Fernbedienung mit multimodaler Benutzungsschnittstelle und 5 angeschlossenen Heimgerätetypen (Fernseher, Videorecorder, Waschmaschine, Herd und eine Lampe).

SHIVA
Structured Human Interface VAlidation

Michael Burmester, Franz Koller und Jürgen Ziegler

Fraunhofer Institut für Arbeitswirtschaft und Organisation

Email: Franz.Koller@iao.fhg.de
Email: Juergen.Ziegler@iao.fhg.de

SHIVA ist ein rechnergestütztes Software-Evaluationsverfahren für Software-Ergonomie-Experten. Es beinhaltet die Analyse statischer Elemente anhand einer View-By-View-Vorgehensweise und Ansatzpunkte für Walkthroughkonzepte und Konsistenztests, um die dynamischen Strukturen eines Computerprogrammes zu evaluaieren. View-By-View konzentriert sich dabei auf den Aufbau einer Bildschirmseite und die Anordnung der darin enthaltenen Elemente. Analysegegenstände sind Menüs, Buttons, Beschriftungen etc. Dagegen wird mit dem Walkthroughansatz nach Schwachstellen bei der Bearbeitung von Aufgaben, wie zum Beispiel dem Erstellen einer Rechnung, gesucht. Der Konsistenztest untersucht auf beiden Ebenen, ob gleichartige Elemente (View by View) oder Abläufe (Walkthrough) auch gleich gestaltet sind.

Die Datenerfassung erfolgt in einer Datenbank, die neben Texten auch Bilder der einzelnen Bildschirmansichten (Views) aufnimmt und in Kopien dieser Originalaufnahmen grafische Veränderungsmöglichkeiten anbietet, um für das Software-Design Verbesserungsvorschläge zu erstellen. Die dynamischen Komponenten einer Software können in SHIVA in einem Navigationsschema, der mit der Datenbankverbunden ist, erfaßt werden.

SHIVA unterstützt unterschiedliche Evaluationstiefen. Die tiefste Analyseebene stellt dabei das strukturierte Abarbeiten aller Normvorschriften der ISO 9241 Teil 10 bis 17 dar. Auf einer mittleren Ebene werden Prüfthemen benutzt und entdeckte Unstimmigkeiten in freier Formulierung protokolliert. Die Expertenebene stellt schließlich nur eine Datenabfrage in der ISO-Normendatenbank zur Verfügung. Alle drei Vorgehensweisen lassen sich auch kombinieren. Am Ende erstellt Shiva in Word einen aus den bearbeiteten und protokollierten Prüfthemen und -fragen automatisch einen Evaluationsbericht.

MOON: Ein interaktives, natürlichsprachliches Programmiersystem für Software-Anwender und Entwickler

Markus Dahm

Rheinisch-Westfälische Technische Hochschule Aachen
www.lfm.rwth-aachen.de

Eine Erfahrung aus *interdisziplinären Projekten* zeigt, daß die Verständigung zwischen den technisch orientierten und den sozialwissenschaftlichen Mitarbeitern bei der *Umsetzung von Gestaltungsideen* in eine Programm-Konzeption und Implementation schwierig ist. Das ist zum großen Teil auf die Werkzeuge der Umsetzung (Sprache, Editor, Compiler, Debugger, etc.) zurückzuführen, die erst nach langer Einarbeitung beherrscht werden und somit nur den Computer-Experten zugänglich sind. Um dem abzuhelfen, wurde das interaktive, objektorientierte Programmiersystem MOON entwickelt, welches erlaubt *in Sätzen natürlicher Sprache zu programmieren*, z.B.:

> copy the medical record of patient Current from folder Old to New
> zeige den patientenkalender von Patient auf dem monitor Sonie

Dadurch wird ein *leichterer Zugang und höhere Transparenz* der Software-Entwicklung ermöglicht. Arbeitswissenschaftler und spätere Anwender der Software können somit leichter und früher aktiv an der Gestaltung der Arbeitsplätze mitwirken und so deren *Akzeptanz und Gebrauchstauglichkeit erhöhen*. Die Entwicklung der Software wird ebenfalls durch besser lesbaren und damit wiederverwendbaren Code unterstützt. MOON integriert damit textorientierte und fehlertolerante Mensch-Computer-Interaktion und Werkzeuge zur Software-Entwicklung in ein und dem selben Programmiersystem.

Ein wesentlicher Vorteil dieser Art der Programmierung ist, daß die *Parameter im Sinnzusammenhang* eines "natürlichsprachlich" formulierten Satzes erscheinen und so in ihrer Bedeutung direkt verstanden werden können, was die Erfüllung der immer wieder auftauchenden und berechtigten Forderung nach "sprechenden", bzw. beschreibenden Namen für Funktionen und Prozeduren deutlich erleichtert. Um dem Ideal von *leicht lesbarem, selbstdokumentierendem Code* möglichst nahe zu kommen, werden kaum syntaktische Restriktionen für die Formulierung von eigenen Sätzen auferlegt. Insbesondere ist die *Sprache umschaltbar*, z.B. von Englisch nach Deutsch.

Eine tragende Rolle für die Gebrauchstauglichkeit spielt die *mehrstufige Fehlertoleranz* und die *konstruktive Rückkopplung* des Interpreters, der im Fehlerfall versucht, über einen Dialog mit dem Benutzer nicht verstandene Eingaben zu klären.

MOON bietet als *vollständig objektorientiertes System* Polymorphismus und late binding durch message passing, einfache Vererbung sowie für maximale Sicherheit Datenkapselung, strong typing und automatische garbage collection. Es verfügt über eine *interaktive, graphische Benutzungsschnittstelle*, in der Objekte und Methoden (Funktionen) definiert oder modifiziert werden. Diese können dann sofort aufgerufen oder in der laufenden Applikation angewendet werden, ohne einen externen Compiler/Linker aufrufen zu müssen.

MOON läuft auf den *Betriebssystemen* Windows 95, Windows NT, Solaris, Linux und Macintosh OS. Ab Frühjahr 1997 wird ein in Netscape Navigator integriertes interaktives *Tutorial* verfügbar sein

Die Anwendung eines rekursiven Informationskonzeptes beim Redesign existierender Benutzungsoberflächen

Gunter Dubrau

Fakultät Informatik, Technische Universität Dresden.

dubrau@inf.tu-dresden.de

Bei der Gestaltung von Benutzungsoberflächen von Lehr-/ Lernsystemen lassen sich drei klassische Herangehensweisen unterscheiden. Eine experten-, eine mensch- und eine technik-orientierte. Dem soll eine informations-orientierte hinzugefügt werden. Damit soll gezeigt werden, daß die Informatik, als die Wissenschaft von der systematischen und maschinellen Informationsverarbeitung, in der Lage ist, in die Gestaltung von Benutzungsoberflächen von Lehr-/ Lernsystemen einen informationsspezifischen Anteil einzubringen.

Im Kontext der immer weiter andauernden Diskussion um einen einheitlichen Informations-begriff wurde aus einem rekursiven Ansatz heraus eine Redesign-Systematik für exisitierende Benutzungsoberflächen in Lehr-/Lernsystemen abgeleitet. Hauptanliegen der auf dieser Basis durchgeführten Redesign-Arbeiten ist die Verbesserung der Anwender-Akzeptanz unter Wahrung des Charakters des jeweiligen Lehr-/ Lernsystems. Im ersten Schritt dieser Redesign-Systematik geht es darum, die informationellen Träger der untersten Ebene im konkreten Anwendungsfall zu definieren (holistischer Ansatz). Dann werden die Möglich-keiten betrachtet, mit den vorhandenen informationellen Trägern neue, auf der Benutzungs-oberfläche noch nicht dargestellte Informationen zu erzeugen. Dabei wird durch eine effektivere Auslastung der bereits vorhandenen informationellen Träger eine Verbreiterung der genutzten Informationskanäle erreicht.

Die bereits angesprochene Redesign-Systematik wurde auf ihre Anwendbarkeit und ihre Effekte für die Benutzerakzeptanz mit Hilfe von Redesigns existierender Benutzungsober-flächen in Lehr-/ Lernsystemen untersucht und ist ebenfalls innerhalb eines Prototyping-Ansatzes anwendbar. Dabei wurden unterschiedlichste Systeme betrachtet, um die Unab-hängigkeit der Theorie von der Realisierungs-Plattform zu demonstrieren. Konkret wurden Internet-, ToolBook- und Hilfeanwendungen für MS Windows überarbeitet. Durchgeführte Akzeptanz-Untersuchungen ergaben fast durchgängig positive Tendenzen. Die auf der Tagung präsentierten Werte erheben aber auf Grund des geringen Stichprobenumfanges keinen Anspruch auf Allgemeingültigkeit. Vielmehr sind sie als Anreiz für weitere, vor allem interdisziplinäre Forschungen gedacht.

TWINS: Training Wheels für Windows-Applikationen

Peter W. Fach
Produktplanung / Software-Egonomie
RWG GmbH, Stuttgart
Email: fach@rwg.c-mail.com

Maria Bannert
Zentrum für empirische pädagogische
Forschung
Universität Koblenz-Landau
Email: bannert@ps2.zepf.uni-landau.de

Ein entscheidender Nachteil von gängigen computerunterstützten Lernprogrammen zur Einarbeitung in eine komplexe Computeranwendung besteht in didaktischer Hinsicht darin, daß den Lernenden überhaupt kein oder nur wenig Handlungsspielraum zugestanden wird. Häufig wird die Handhabung der zu erlernenden Funktionen lediglich simuliert und darüber hinaus werden vom Lernprogramm nur richtige Benutzereingaben zugelassen. Wird hingegen - um dieses Manko auszugleichen - in dem Lernprogramm ein Zugang zur Originalsoftware hergestellt, wo der Lerner sofort die erlernten Funktionen ausprobieren und das reale System weitgehend eigenständig explorieren kann, ist wiederum keine vollständige Kontrolle über den aktuellen Systemzustand möglich. Zwar kann dabei der Lerner durch verschiedene Funktionsbereiche „geführt", zur Ausführung bestimmter Kommandos aufgefordert, über potentielle Fehlerzustände informiert und letztlich auch über deren mögliche Beseitigung aufgeklärt werden. Jedoch bleibt der Umgang mit fehlerhaften Benutzereingaben bzw. deren systemseitigen Konsequenzen problematisch. *Training Wheels* [1] stellen einen interessanten Ansatz der Gestaltung von computerunterstützten Lernumgebungen zur Einarbeitung in komplexe EDV-Programme dar, welche genau diese Schwachstelle umgehen.

Die Posterpräsentation soll nicht nur auf dieses software-ergonomische Gestaltungskonzept von Benutzeroberflächen aufmerksam machen, sondern darüber hinaus Anregung bieten, diese besondere Form der Adaptierbarbarkeit von computerunterstützten Lernumgebungen zu diskutieren, die speziell zur Einarbeitung in komplexe Computer-Applikationen entworfen wurde. Zu diesem Zweck stellen wir das von uns entwickelte Programm TWINS (Training Wheels für Windows-Applikationen) vor, das eine Training Wheels-Schnittstelle zur Einarbeitung in das Textverarbeitungsprogramm Word für Windows 6.0 bereitstellt. An dieser konkreten Ausarbeitung werden die Hintergründe und die verschiedenen Gestaltungsaspekte von Training Wheels aufgezeigt sowie wesentliche Ergebnisse ihres Einsatzes [2] im Rahmen eines Word für Windows-Einführungskurses präsentiert.

Software-Ergonomie
in mittelständischen Softwarehäusern -
eine Befragung auf der CeBIT' 96

Jens Hüttner

Institut für Psychologie, Humbolt-Universität zu Berlin

Die Überführung von Software-Ergonomie Wissen aus der Welt der Wissenschaft in die Praxis wird immer wieder beklagt. Die Verbesserung der ergonomischen Qualität von Produkten auf dem Softwaremarkt wird immer wieder an den Produkten großer Unternehmen gemessen. Die Fortschritte in den letzten Jahren sind auch nicht zu übersehen. Über die Qualität der Produkte vieler kleiner und mittelständischer Unternehmen ist wenig bekannt.

Auf der CeBIT' 96 wurden im „Software-Zentrum Mittelstand" (Halle 5) mit 49 von 250 Ausstellern Interviews durchgeführt. Die Hersteller beantworteten Fragen zu ihren Produkten, zu firmeninternen Standards, zum Kontakt Entwicker-Kunde und zur Benutzungsfreundlichkeit ihrer Entwicklungen. Dabei konnte ein breites Spektrum an Lösungen für verschiedene Branchen einbezogen werden (z.B. 25 Produkte für den Dienstleistungsbereich, 18 für die Produktion). Die Unternehmen entwickeln vorrangig Produkte für DOS und Windows-Systeme, 67% der Produkte sind eigenständige Entwicklungen. Rund ein Drittel der Unternehmen verkauft zur Software spezielle Hardware. Die Größe der Unternehmen variiert stark, 55% haben bis zu 10 Angestellte, 4 Software-Häuser haben mehr als 100 Angestellte. Fast alle Befragten (90%) waren an der Entwicklung der ausgestellten Software beteiligt.

Fast 90% der Befragten liefern mit der Software auch Handbücher aus, ein Viertel von diesen hat einen Umfang von mehr als 600 Seiten. 88% bieten spezielle Schulungen für ihre Produkte an, mit einer Dauer zwischen einem Tag und einem Monat. Drei Viertel der Befragten konnten auf ein Hilfesystem für die Software verwiesen. Usability-tests werden in 38 Unternehmen durchgeführt, allerdings waren einem Viertel der Befragten die Ergebnisse der Tests nicht bekannt (auf der Messe!). Alle Unternehmen gaben an, Kontakt zu ihren Kunden zu haben, 39% haben diesen Kontakt „sehr selten".

Der Aufwand für die Entwicklung der Benutungsoberfläche wird wie folgt eingeschätzt:

Aufwand	0-20	20-40 %	40-60 %	60-80 %	80-100 %
Prozent der Befragten	31	35	26	8	-

Mit der Befragung konnte festgestellt werden, daß das Themenfeld Benutzerfreundlichkeit in den Software-Häusern durchaus bekannt ist. Die Unternehmen sind bestrebt, auf ergonomische Aspekte ihrer Software zu achten. Die wird von den Entwicklern allerdings so „nebenbei" miterledigt. Wie weit das positive Bild der Selbstauskünfte auch objektiven Kriterien standhalten würde, muß offen bleiben.

Das Verbandsinformationssystem ELVIRA

Jürgen Krause*, Thomas Mandl und Maximilian Stempfhuber
* Fachbereich Informatik, Universität Koblenz-Landau
und Informationszentrum Sozialwissenschaften, Bonn
ma@bonn.iz-soz.de

Im Projekt ELVIRA wurde ein online-Verbandsinformationssystem für die Firmen des ZVEI (Zentralverband der Elektroindustrie entwickelt, mit dem Zeitreihen recherchiert werden können. An dem Projekt waren die Wirtschaftsforschungsinstitute DIW und IFO, der ZVEI und das Informationszentrum Sozialwissenschaften (IZ) beteiligt. Das IZ erstellte die Benutzungsoberfläche und die Kommunikationsroutinen auf der Clientseite. Schwerpunkt des Projektes war die Erleichterung des Zugangs zu elektronischen Medien für die Unternehmen der Elektroindustrie zum einen durch die hohe Qualität und Relevanz der Informationen des Verbandes und zum anderen durch die Entwicklung einer benutzerfreundlichen Oberfläche. Dazu wurden die Benutzerbedürfnisse in Interviews mit potentiellen Anwenden untersucht. Das System wurde in der Pilotphase bei 15 Pilotanwendern eingesetzt und getestet.

Beim Design der Oberfläche gelang es, die spezifische Situation der Fragestellungen auf einen Ausschnitt der Booleschen Logik einzuschränken und für diesen einen optimierten Zugang zu schaffen. Als Basis diente das WOB-Modell (auf der Werkzeugmetapher basierende strikt objektorientierte grafisch-direktmanipulative Benutzungsoberflächen), ein aufeinander abgestimmtes Bündel softwareergonomischer Vorschläge, die in ihrer Gesamtheit zu effzienten Oberflächen führen sollen. Im Falle von ELVIRA wurde besonders dynamische Anpassung eingesetzt, um die Zusammenhänge zwischen den Daten für den Benutzer sichtbar zu machen. ELVIRA unterstützt weiterhin die im WOB-Modell enthaltene kognitive Strategie des iterativen Retrieval durch entsprechende Gestaltung des Ergebnisobjekts.

InterBook-S - Ein multimediales und interaktives Lernprogramm für das Feldbussystem INTERBUS-S

Reinhard Langmann

FH Düsseldorf
E-Mail: R.Langmann@t-online.de
http://www.fh-duesseldorf.de/WWW/DOCS/FB/ETECH/DOCS/langmann/prospekt.html

Das Lernsystem *InterBook-S* bietet als multimediales und interaktives Buch anhand von 65 Displayseiten einen ca. 6 stündigen Grundkurs zum Feldbussystem INTERBUS-S (IBS). Der IBS ist ein in der Fertigungsautomatisierung (Automobilindustrie) weit verbreiteter Feldbus, mit dem sehr effizient moderne Automatisierungsstrukturen aufgebaut werden können.

Das Lernsystem erläutert in drei elektronischen Büchern die Grundlagen und den Nachrichtenaustausch im IBS sowie den Hardwareaufbau der Baugruppen einschließlich ihrer Installation. Ein viertes Buch beinhaltet einen Feldbus-Monitor, mit dem graphisch und durch Hilfen unterstützt ein IBS-System mit max. 40 Modulen aus dem Lernsystem heraus komfortabel bedient bzw. getestet werden kann.

InterBook-S beinhaltet 13 Videosequenzen mit insgesamt 7 min Spieldauer. Hier findet der Lernende z.B. Erläuterungen zur IBS-Hardware, Hinweise zur Installation sowie Informationen zu konstruktiven Besonderheiten der IBS-Baugruppen.

Das Lernen mit *InterBook-S* besitzt eine Reihe von Vorteilen, wie z.B. eine selbstbestimmte und systematische Einarbeitung in das Feldbussystem, eine explorative und anschauliche Vermittlung von Lerninhalten sowie Praxisnähe durch interaktive Eingriffsmöglichkeiten.

Neue Formen der computergestützten Arbeit - neue Gesundheitsrisiken? Eine Längsschnittsstudie mit Operatoren-Managern in der Telekommunikation

Anna B. Leonova

Moskauer Lomonossov-Universität
Lehrstuhl für Arbeits- und Ingenieurpsychologie
Email aleon@chair.cogsci.msu.su

Mit dem fortschreitenden Vorgang des technologischen Wandels verändern sich die traditionellen Formen der Mensch-Maschine- und Mensch-Computer-Interaktion. Anstelle der haupsächlich sensomotorischen und elementaren kognitiven Belastungen der industriellen Ära, kommen neue Formen des Stresses, die mit den höheren kognitiven und metakognitiven Leistungen verbunden sind. Es gibt einige Andeutungen, daß damit auch neuere gesundheitliche Risiken, z.B. im kardiovaskulären Bereich, verbunden sind (Nickerson, 1994; Leonova, 1997 in press). Diese eventuellen Risiken wurden bis jetzt aber kaum erforscht, da entsprechende Untersuchungen extrem komplexe Untersuchungspläne erfordern. Wir berichten über eine der ersten solchen Untersuchungen. Eine Gruppe von hochqualifizierten Operatoren eines Netzes der Satellitenkommunikation wurde dabei systematisch über 7 Jahre hinweg untersucht. In der Mitte dieser Periode erfolgte eine komplette Umstellung der alten Formen der Arbeit auf computergestützte Überwachungs- und Entscheidungstätigkeit, so daß die Operatoren den Status von Operatoren-Managern bekamen. Über die ganze Periode der Untersuchung verfolgten wir im Rahmen des von uns entwickelten „widespread-netting" Designs das subjektive Befinden, objektive Leistungen, elementare psychophysiologische Parameter und den gesundheitlichen Zustand der Operatoren (siehe dazu Leonova, 1996).

Die Automatisierung verbesserte zusehends die objektiven Leistungsparameter der Tätigkeit und reduzierte den Ausprägungsgrad des Ermüdung-Monotonie Syndroms. Gleichzeitig wurde aber eine wesentliche Zunahme der kognitiven und emotionalen Spannung gefunden. Ein interessantes Ergebnis dieser Untersuchung besteht in einer gewissen Abspaltung der subjektiven Einschätzung des eigenen Zustandes von psychophysiologischen und gesundheitlichen Parametern. Während sich das subjektive Befinden bei der Einführung neuer Technologien oft verbessert, beanspruchen die damit verbundenen Formen der kognitiven und emotionalen Belastung die Operatoren und können durchaus ihre tatsächliche gesundheitliche Situation verschlechtern.

Literaturverzeichnis

[1] Leonova, A.B.: Occupational stress, personnel adaptation, and health. In Ch.D.Spielberger and I.G.Sarason (Eds.), Stress and Emotion. Vol.16., Washington, DC: Taylor & Francis, 1996, 109-126.

[2] Leonova, A.B.: Basic issues in occupational stress research. In J.Adair and F.I.M.Craik (Eds.), State-of-the-art lectures of the XXVIth International Congress of Psychology. Mahwah, NJ: Lawrence Erlbaum Assiociates, 1997 (in press).

[3] Nickerson, R.: Looking ahead: Human factors challenges in a changing world. Hillsdale, NJ: Lawrence Erlbaum Associates, 1994.

Das BEBA - Verfahren: Ermittlung psychischer Belastungsrisiken bei der Büroarbeit

Christine Maßloch, Perry Jordan, Andreas Pohlandt
Institut für Psychologie, Technische Universität Dresden
pj1@rcs.urz.tu-dresden.de

Die neuen Gesetze zum Arbeits- und Gesundheitsschutz, wie das Arbeitsschutzgesetz (*ArbSchG*) und die Bildschirmarbeitsverordnung (*BildscharbV*) stellen die Unternehmer vor neuartige Anforderungen, die sich aus der zukünftigen Verpflichtung des Arbeitgebers zur Durchführung von Arbeitsplatzanalysen und zur Gestaltung der Bildschirmarbeitsplätze gemäß bestimmter Mindestanforderungen ergeben. Für die Analyse psychischer Belastungsrisiken wurde deshalb im Rahmen des BMBF-Verbundprojekts **SANUS** ("Sicherheit und Gesundheitsschutz bei der Arbeit an Bildschirmgeräten auf der Basis internationaler Normen und Standards") ein Screening-Verfahren geschaffen, mit dem die Mitarbeiter anhand bestimmter Merkmale der Arbeitsinhalte und Arbeitsbedingungen wesentliche arbeitsbedingte Quellen von Gesundheitsbeeinträch-tigungen ermitteln können. Es werden Zusammenhänge zu wirtschaftlichen und sozialen Problemen im Unternehmen verdeutlicht und Vorschläge zur Verbesserung der Gestaltungsgüte der Tätigkeiten und Arbeitsabläufe unterbreitet.

Das BEBA-Verfahren besteht aus drei Verfahrensteilen:

Teil A: Erfassung wirtschaftlicher und sozialer Probleme durch den Vorgesetzten
Teil B: Information der Mitarbeiter und Erhebung von Belastungsrisiken durch die Mitarbeiter
Teil C: Auswertung der erhobenen Daten und Ableitung von Gestaltungserfordernissen

Folgende Tätigkeitsmerkmale werden in Form einer einfachen Beurteilungsskala abgeprüft:

- Aufgabenganzheitlichkeit
- Aufgabenvielfalt
- Entscheidungsmöglichkeiten
- Kommunikationserfordernisse
- Qualifikationsnutzung- und -entwicklung
- körperliche Abwechslung
- Beteiligung an der Arbeitsgestaltung
- soziale Unterstützung
- zeitlicher Spielraum
- Transparenz der Arbeitsabläufe
- Arbeitsvolumen
- Aufgabenschwierigkeit
- Rückmeldungen
- Informationsverfügbarkeit
- Störungsfreiheit
- Möglichkeit zu Kurzpausen
- Aufgabenangemessenheit der Technik
- Benutzerfreundlichkeit der Technik
- Zuverlässigkeit der Technik

Das BEBA-Verfahren wurde anhand von Arbeitsuntersuchungen in 40 Unternehmen bei 250 Tätigkeiten validiert. Es bestehen signifikante Zusammenhänge zwischen kritischen Ausprägungen der Tätigkeitsmerkmale und Fehlbeanspruchungen (psychische Ermüdung, Monotonie, psychische Sättigung und Streß; vgl. DIN ISO 10075) bzw. körperlichen Beschwerden. Darüber können Zusammenhänge zwischen kritischen Ausprägungen der Tätigkeitsmerkmale einer Organisationseinheit und kritischen Einschätzungen des Erfüllungsgrades wirtschaftlicher und sozialer Ziele in dieser Organisationseinheit belegt werden.

PEGASUS
„Partizipative Evaluation und Gestaltung aufgabenorientierter Software unter Berücksichtigung europäischer Standards"

Jochen Prümper
Fachhochschule für Technik
und Wirtschaft Berlin

Uta Sailer
Ludwig-Maximilians-
Universität München

Marc Hassenzahl
Technische Hochschule
Darmstadt

Gegenstand des PEGASUS Projektes ist eine Datenbank zur Verwaltung und Bearbeitung von Forschungsanträgen einer Institution der öffentlichen Verwaltung. Ziel des Gestaltungsprozesses ist eine nach ISO 9241 normkonforme Software. Die Ergonomieevaluierung der Software erfolgt in drei Schritten: (1) Bewertung des IST-Zustands und Sammlung von Hinweisen auf mögliche Problembereiche anhand einer Gebrauchstauglichkeitsstudie, (2) iterative Verbesserung des Systems anhand software-ergonomischer Kriterien unter Benutzerbeteiligung und (3) erneute Bewertung des Systems zu einem zweiten Zeitpunkt und damit Überprüfung der umgesetzten Gestaltungsvorschläge bzw. der gesamten Qualitätssicherungsmaßnahmen.

Kriterien zur Bewertung wurden direkt aus den ISO-Normen 9241/10 und 9241/11 in Form von Performanzdaten abgeleitet. Um umfassende Hinweise auf Problembereiche der Software zu erhalten, wurden zwei Ansatzpunkte gewählt: neben einer Gebrauchstauglichkeitsstudie kamen verschiedene Verfahren zur Erfassung der Nutzungskontexte „Benutzer" und „Aufgabe" sowie zur Gebrauchstauglichkeit der Software zum Einsatz. Im Rahmen der Gebrauchstauglichkeitsstudie bearbeiteten 25 (zukünftige) Benutzer konkrete Arbeitsaufgaben mit dem zu evaluierenden System. Der Prozeß der Aufgabenbearbeitung wurde anhand verschiedener Protokollinstrumente aufgezeichnet und im weiteren analysiert.

Die Ergebnisse aller Erhebungen bzw. Analysen erfüllten einen doppelten Zweck: Zum einen wurden sie in die Ergonomieworkshops eingespeist. So bildeten sie die Grundlage zur Identifikation von Problembereichen, zur Priorisierung dieser Probleme nach ihrer Bedeutung und damit zur Erstellung einer Bearbeitungsreihenfolge, und zur Erarbeitung von Lösungsvorschlägen. Die Teilnehmer der Ergonomieworkshops setzten sich aus Benutzern, Entwicklern und Projektverantwortlichen zusammen, die in zwei Gruppen Lösungsvorschläge zum jeweiligen Themenbereich erarbeiteten und diese anschließend im Plenum vorstellten. Diese Lösungen wurden unter Anleitung der Moderatoren zu einem gemeinsamen, detaillierten Konzept integriert und als „Arbeitsanweisung" an die Entwickler weitergegeben. Ausgehend von einem Teil der Ergebnisse wurden zunächst einzelne Verbesserungsvorschläge erarbeitet, die dann zu einem übergreifenden Konzept integriert wurden. Die erfolgten Umsetzungen wurden im jeweils darauffolgenden Workshop begutachtet und ggf. weiter verbessert.

Zum anderen erlauben die Ergebnisse der ersten Gebrauchstauglichkeitsstudie durch einen Vergleich mit der zweiten Messung zu einem späteren Zeitpunkt die Dokumentation von Veränderungen sowie eine abschließende Evaluierung der umgesetzten Gestaltungsvorschläge. Alle Daten werden in bezug auf den ersten und den zweiten Zeitpunkt der Gebrauchstauglichkeitsstudie miteinander verglichen, um zu überprüfen, ob sich die Verbesserung der Software in der Performanz der Teilnehmer niederschlägt.

Inhalt des Posters ist neben einer Vorstellung der Vorgehensweise und der eingesetzten Instrumente die Präsentation einiger beispielhafter Ergebnisse.

Psychische Belastungen bei der Bildschirmarbeit in Abhängigkeit vom Softwareeinsatz

Ursula Schmitz und Marc Rose

Bergische Universität-Gesamthochschule Wuppertal

Email: schmitz@sanus.erziwi.uni-wuppertal.de

Im Rahmen des SANUS-Projektes zur Umsetzung der Bildschirmrichtlinie 90/270/EWG wurde das SynBA-Verfahren zur Analyse der psychischen Belastung an 176 Bildschirmarbeitsplätzen eines führenden Elektronikkonzerns eingesetzt. Darüberhinaus wurden u.a. die quantitative und qualitative Softwareausstattung der einzelnen Arbeitsplätze erfaßt. Die Studie zeigt, inwieweit psychische Belastungen auf Gestaltungsmängel an der Mensch-Maschine-Schnittstelle zurückgeführt werden können und wie sich die Art und Anzahl der eingesetzten Software auf das Belastungserleben der Stelleninhaber auswirkt. Aus den Erkenntnissen läßt sich zeigen, unter welchen Bedingungen die Berücksichtigung der Dialogprinzipien der ISO 9241-10 zur Optimierung der psychischen Belastung an der Mensch-Maschine-Schnittstelle führt.

Negotiating Quality Models during Software Development

Marcin Sikorski

Technical University of Gdansk, Faculty of Management and Ergonomics, Gdansk Poland
msik@pg.gda.pl

During software development it is usually problematic to indentify which features of software products are really important for users for their subjective „feeling" that they operate high-quality product. Product usability is too general term to be easily used but it is easier to measure impact of various components of software product on user's opinion about the product.

Current research project conducted in Ergonomics Dept. at the Technical University of Gdansk is aimed to specify:

- differences in understanding the term „quality" for users and softwaredevelopers,
- identifying relationships between the „quality of construction" and „quality of use" for particular software products,
- developing methods how software quality models of designers and of users

can be negotiated and merged during specification requirements phase and later during product development and testing.

This research project includes surveying the users and software developers, as well as usability tests of selected software products for management support. The results of the tests are expected to show relationships with the technology used for product development as well as with degree to which users' opinions were included during the product development. The degree of user participation is to be measured and relationships with product quality to be identified. The idea of negotiating quality models of users and of designers in a participative way is aimed to contribute to software quality assurance programs in sofware houses.

Semantische Inhaltsanalyse
Software-ergonomischer Evaluierungsverfahren

Christian Stary, Alexandra Totter

Institut für Wirtschaftsinformatik, Communications Engineering, Universität Linz

Christian.Stary@ce.uni-linz.ac.at

Im Rahmen der Verfahrensbildung basierend auf den verschiedenen Software-ergonomischen Merkmalen von Benutzungsschnittstellen werden Statements bzw. Fragen (Testitems) definiert, welche einer ersten qualitativen Analyse, nämlich einer semantischen Inhaltsanalyse, hinsichtlich testtheoretischer Güte nicht standhalten. Die Verfahren weisen geringe Validität, Reliabilität und Objektivität auf. Eine Methode, die zur Identifikation dieser Mängel verwendet werden kann, die semantische Inhaltsanalyse, wurde zur Untersuchung der Operationalisierung von Aufgabenangemessenheit herangezogen. Dieses Merkmal wurde ausgewählt, da es (wie auch andere, z.B. Adaptivität) aufgrund seiner Komplexität nicht durch unmittelbar operational faßbaren Eigenschaften meßbar gemacht werden kann.

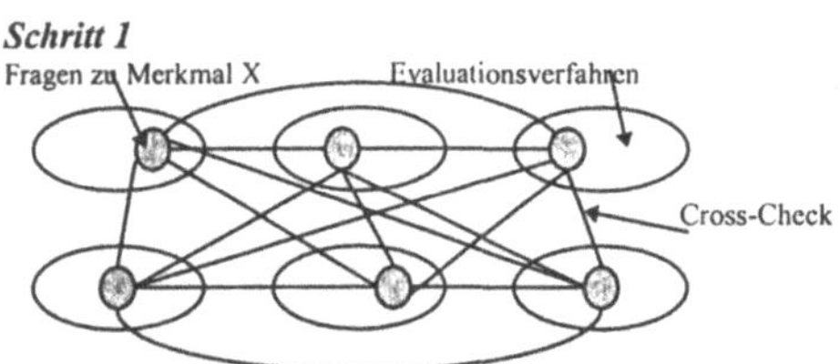

Analyse semantischer Übereinstimmung der Fragen zu einem Merkmal zwischen den Verfahren

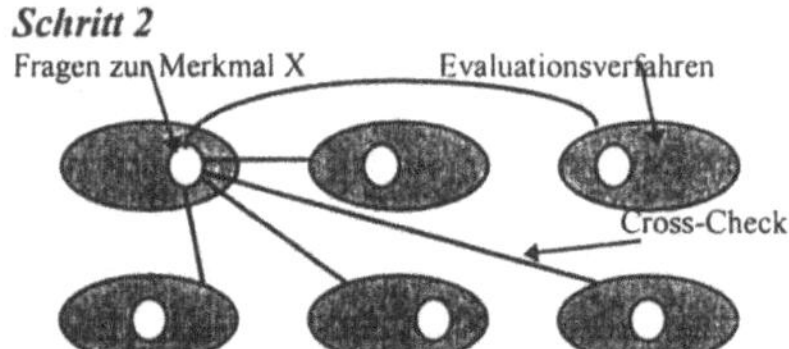

Analyse semantischer Übereinstimmung der Fragen zu einem Merkmal & unter anderen Merkmalen gestellten Fragen über alle Verfahren hinweg

In einem ersten Schritt werden Fragen zur Aufgabenangemessenheit eines Verfahrens mit den Fragen zur Aufgabenangemessenheit eines zweiten bzw. dritten etc. Verfahrens hinsichtlich semantischer Übereinstimmung verglichen (Schritt 1). Dieser Schritt wird so lange durchgeführt, bis jedes Verfahren mit jedem verglichen wurde. Um die Objektivität zu gewährleisten, wird dieser Schritt von mindestens zwei unabhängigen Beurteilern durchgeführt. Im zweiten Schritt werden die Fragen zur Aufgabenangemessenheit eines Verfahrens mit allen, unter anderen Merkmalen gestellten Fragen des zweiten bzw. dritten etc. Verfahrens hinsichtlich semantischer Übereinstimmung verglichen. Dieser Prozeß endet, wenn alle Fragen, die unter dem Merkmal Aufgabenangemessenheit gestellt werden, mit allen, unter anderen Merkmalen gestellten Fragen hinsichtlich semantischer Übereinstimmung überprüft worden sind. Mit Hilfe dieser Methode konnte gezeigt werden, daß der Mangel an umfassenden Grundlagen, im Sinne von theoretischen Konstrukten, zu einer enormen Begriffsverwirrung und in weiterer Folge zu einem Mangel an Validität, Reliabilität und Objektivität der Evaluierungsverfahren führt. Die semantische Inhaltsanalyse trägt damit zu einer allgemeinen Diskussion über die Forderung nach exakter Bedeutungsanalyse und Operationalisierung bei.

Bauknecht/Zehnder

Grundlagen für den Informatikeinsatz

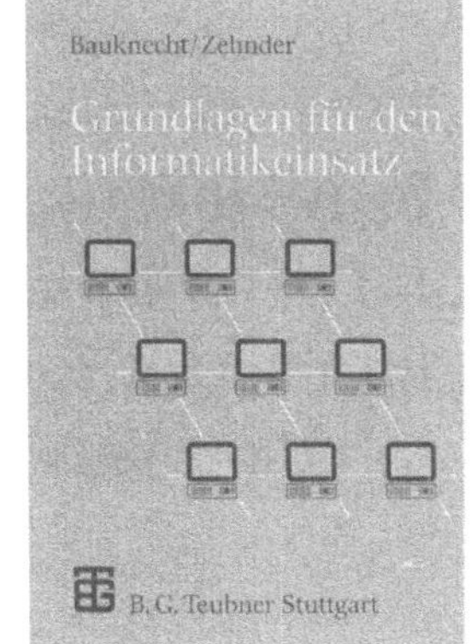

Computeranwendungen gehören heute zum Alltag. Viele Informatikanwender begnügen sich dabei mit oberflächlichen Kenntnissen bestimmter Anwenderprogramme und mit Ratschlägen wohlwollender Informatikberater. Andere wollen aber die Welt der Informatik besser verstehen und Zusammenhänge sehen, ohne sich in technischen Einzelheiten dieses noch immer stürmisch wachsenden Fachgebiets zu verlieren. An diese wendet sich dieses Buch. Studierende aller Fachgebiete, namentlich in Natur-, Ingenieur- und Wirtschaftswissenschaften, aber auch Lehrkräfte verschiedenster Stufen erhalten hier jene Grundlagen, auf denen der moderne Informatikeinsatz in Industrie, Dienstleistungen und Verwaltung aufbaut.

Der Text vermittelt einen Überblick über das gesamte Gebiet der Informatik (Geräte, Programme, Daten, Netze, Methoden), soweit dies für das bessere Verständnis des praktischen Informatikeinsatzes hilfreich ist. Viele, oft verblüffend einfache Beispiele (vom Telefonbuch bis zum Bancomat) schaffen den Bezug zur Praxis; über 130 Figuren unterstützen die Erläuterungen. Auch neueste Entwicklungen (z.B. Internet - World Wide Web) werden vorgestellt, immer aber ausgerichtet auf Grundsätzliches und in einfacher, verständlicher Sprache.

Aus dem Inhalt

Erste Übersicht und Grundlagen – Arbeiten mit Informatikanwendungen: Mensch und Maschine, Computerdialog am Beispiel Textverarbeitung, Dateneingabe und Datenpräsentation – Realisierung von Informatikanwendungen: Informatikprojekte, Datenbanken, Informatikführung – Computerprogramme: Programmiersprachen, Methoden und Werkzeuge, Software-Qualität – Computersysteme: Digitaltechnik, Computerarchitektur, Datenspeicher, Datenein- und -ausgabegeräte – Telekommunikation: Verbindungen, Datennetze, Internet – Datensicherheit und Datenschutz – Maßeinheiten

Von Prof. Dr. sc. techn.
Kurt Bauknecht
Universität Zürich
und Prof. Dr. sc. math.
Carl August Zehnder
ETH Zürich

5., völlig neubearbeitete und erweiterte Auflage. 1997. 334 Seiten. 16,2 x 22,9 cm. Kart. DM 49,80 ÖS 364,–/ SFr 45,– ISBN 3-519-42450-9

(Leitfäden der Informatik)

Preisänderungen vorbehalten.

B. G. Teubner Stuttgart · Leipzig

Oberweis

Modellierung und Ausführung von Workflows mit Petri-Netzen

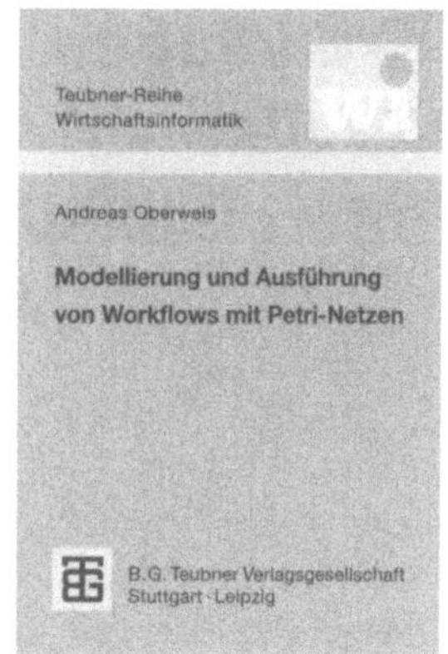

Dieses Buch beschreibt einen auf Petri-Netzen basierenden integrierten Ansatz zur Modellierung von betrieblichen Abläufen und Objekten. Es wird eine evolutionäre Vorgehensweise zur Ablaufbeschreibung vorgestellt, welche von einer anwendungsnahen Notation zu einer präzisen und für die Ausführung mit Workflow-Managementsystemen geeigneten Notation führt.

Basierend auf den Sprachkonzepten für die Ablaufbeschreibung wird die Architektur eines Workflow-Managementsystems konzipiert.

Die Besonderheit dieses Systems besteht darin, daß Petri-Netze unmittelbar als Grundlage für die Ablaufkontrolle und -steuerung eingesetzt werden. Die beschriebenen Workflow-Engine kann auch als frei konfigurierbare Ablaufsteuerung von Standard-Softwaresystemen eingesetzt werden. Die Abläufe sind hier nicht »fest verdrahtet« in der Software enthalten, sondern als flexible Beschreibungen in Form eines Ablaufschemas gegeben, das im Rahmen der Ablaufmodellierung erstellt und optimiert worden ist.

Von Prof. Dr.
Andreas Oberweis
Universität Frankfurt/Main

1996. 302 Seiten mit
105 Bildern.
16,2 x 22,9 cm.
Kart. DM 54,–
ÖS 394,– / SFr 49,–
ISBN 3-8154-2600-6

(Teubner-Reihe
Wirtschaftsinformatik)

Preisänderungen vorbehalten.

B.G.Teubner Stuttgart · Leipzig

Band 34: **Friedrich/Rödiger, Computergestützte Gruppenarbeit (CSCW)**
Fachtagung vom 30. 9. bis 2. 10. 1991 in Bremen. 314 Seiten, DM 69,–/ÖS 504,–/SFr. 62,–

Band 35: **Hoffmann, Eiffel**
Fachtagung am 25./26. 5. 1992 in Darmstadt. 112 Seiten, DM 42,–/ÖS 307,–/SFr. 38,–

Band 36: **Schweiggert, Wirtschaftlichkeit von Software-Entwicklung und -Einsatz**
Fachtagung am 21./22. 9. 1992 in Ulm. 272 Seiten, DM 62,–/ÖS 453,–/SFr. 56,–

Band 37: **Ludewig/Schneider, Software Engineering im Unterricht der Hochschulen SEUH '92**
Workshop am 27./28. 2. 1992 in Stuttgart. 132 Seiten, DM 46,–/ÖS 336,–/SFr. 41,–

Band 38: **Raasch/Bassler, Software Engineering im Unterricht der Hochschulen SEUH '93**
Workshop am 25./26. 2. 1993 in Hamburg. 190 Seiten, DM 49,–/ÖS 358,–/SFr. 44,–

Band 39: **Rödiger, Software-Ergonomie '93**
Fachtagung vom 15. bis 17. 3. 1993 in Bremen. 330 Seiten, DM 78,–/ÖS 569,–/SFr. 70,–

Band 40: **Coy/Gorny/Kopp/Skarpelis, Menschengerechte Software als Wettbewerbsfaktor**
Arbeitstagung am 27./28. 1. 1993 in Bonn. 647 Seiten, DM 138,–/ÖS 1007,–/SFr. 124,–

Band 41: **Züllighoven/Altmann/Doberkat, Requirements Engineering '93: Prototyping**
Fachtagung vom 25. bis 27. 4. 1993 in Bonn. 383 Seiten, DM 88,–/ÖS 642,–/SFr. 79,–

Band 43: **Hußmann/Paech, Software Engineering im Unterricht der Hochschulen SEUH '94**
Workshop am 24./25. 2. 1994 in München. 178 Seiten, DM 54,–/ÖS 394,–/SFr. 49,–

Band 44: **Spillner/Breymann, Software Engineering im Unterricht der Hochschulen SEUH '95**
Workshop am 23./24. 2. 1995 in Bremen. 141 Seiten, DM 48,–/ÖS 350,–/SFr. 43,–

Band 45: **Böcker, Software-Ergonomie '95**
Fachtagung vom 20. bis 23. 2. 1995 in Darmstadt. 424 Seiten, DM 98,–/ÖS 715,–/SFr. 88,–

Band 46: **Daldrup, Menschengerechte Softwaregestaltung**
252 Seiten, DM 54,–/ÖS 394,–/SFr.49,–

Band 47: **Schweiggert/Stickel, Informationstechnik und Organisation**
Fachtagung am 28./29. 9.1995 in Ulm. 276 Seiten, DM 64,–/ÖS 467,–/SFr. 58,–

Band 48: **Forbrig/Riedewald, Software Engineering im Unterricht der Hochschulen SEUH '97**
Workshop am 27./28. 2.1997 in Rostock. 124 Seiten, DM 54,–/ÖS 394,–/SFr. 49,–

Band 49: **Liskowsky / Velichkovsky / Wünschmann, Software Ergonomie '97**
Fachtagung vom 3. bis 6. 3. 1997 in Dresden. 370 Seiten, DM 108.–/ÖS 788,–/SFr. 97,–

Preisänderungen vorbehalten

B. G. Teubner Stuttgart